Improving nitrogen use efficiency in crop production

It is widely recognised that agriculture is a significant contributor to global warming and climate change. Agriculture needs to reduce its environmental impact and adapt to current climate change whilst still feeding a growing population, i.e. become more 'climate-smart'. Burleigh Dodds Science Publishing is playing its part in achieving this by bringing together key research on making the production of the world's most important crops and livestock products more sustainable.

Based on extensive research, our publications specifically target the challenge of climate-smart agriculture. In this way we are using 'smart publishing' to help achieve climate-smart agriculture.

Burleigh Dodds Science Publishing is an independent and innovative publisher delivering high quality customer-focused agricultural science content in both print and online formats for the academic and research communities. Our aim is to build a foundation of knowledge on which researchers can build to meet the challenge of climate-smart agriculture.

For more information about Burleigh Dodds Science Publishing simply call us on +44 (0) 1223 839365, email info@bdspublishing.com or alternatively please visit our website at www.bdspublishing.com.

Related titles:

Achieving sustainable crop nutrition sustainability
Print (ISBN 978-1-78676-308-2); Online (ISBN 978-1-78676-310-5, 978-1-78676-311-2)

Advances in crop modelling for a sustainable agriculture
Print (ISBN 978-1-78676-240-5); Online (ISBN 978-1-78676-242-9, 978-1-78676-243-6)

Advances in sensor technology for sustainable crop production
Print (ISBN 978-1-78676-977-0); Online (ISBN 978-1-78676-979-4, 978-1-78676-231-3)

Assessing the environmental impact of agriculture
Print (ISBN 978-1-78676-228-3); Online (ISBN 978-1-78676-230-6, 978-1-80146-100-9)

Improving soil health
Print (ISBN 978-1-78676-670-0); Online (ISBN 978-1-78676-672-4, 978-1-78676-673-1)

Understanding and improving crop root function
Print (ISBN 978-1-78676-360-0); Online (ISBN 978-1-78676-362-4, 978-1-78676-363-1)

Chapters are available individually from our online bookshop:
https://shop.bdspublishing.com

climate•SMART•publishing
IN AGRICULTURAL SCIENCE

BURLEIGH DODDS SERIES IN AGRICULTURAL SCIENCE
NUMBER 150

Improving nitrogen use efficiency in crop production

Edited by Adjunct Professor Jagdish Kumar Ladha,
University of California-Davis, USA

burleigh dodds
SCIENCE PUBLISHING

Published by Burleigh Dodds Science Publishing Limited
82 High Street, Sawston, Cambridge CB22 3HJ, UK
www.bdspublishing.com

Burleigh Dodds Science Publishing, 1518 Walnut Street, Suite 900, Philadelphia, PA 19102-3406, USA

First published 2024 by Burleigh Dodds Science Publishing Limited

Library of Congress Control Number: 2023952377

British Library Cataloguing in Publication Data
A catalogue record for this book is available from the British Library

ISBN 978-1-80146-470-3 (print)
ISBN 978-1-80146-472-7 (PDF)
ISBN 978-1-80146-471-0 (ePub)
ISSN 2059-6936 (print)
ISSN 2059-6944 (online)

DOI: 10.19103/AS.2024.0135

Typeset by Deanta Global Publishing Services, Dublin, Ireland

Contents

Part 3 Organic sources of nitrogen

Series list

Title | Series number

Dedication

This book is dedicated in loving memory of Dr Pallavolu Maheswara Reddy, whose passing on October 25, 2023, deeply grieved the global biological nitrogen fixation community. Dr Reddy, affiliated with The Energy and Resources Institute, New Delhi, India, was a luminary in the field, leaving an indelible mark through his numerous contributions.

Born in Bangalore, India in 1947, Dr. Reddy earned his Ph.D. in microbial ecology from Banaras University, Varanasi in 1978. His doctoral research, which delved into the biochemical control of heterocyst formation in blue-green algae, marked the beginning of a distinguished scientific journey. His groundbreaking investigations encompassed the genetic potential of rice in forming nitrogen-fixing symbiosis with rhizobia, pioneering studies that illuminated rice's ability to perceive and transduce rhizobial Nod signals, akin to legume nodulation. Additionally, his discovery of rice's homologues of legume nodulation genes revolutionized our understanding of this crucial biological process. Dr. Reddy's pioneering research continues to inform and inspire scientific research worldwide, particularly in enhancing rice's symbiotic relationship with rhizobia.

Beyond his scientific ability, Dr. Reddy was renowned for his kindness, strength and unwavering spirit. His exceptional professional achievements were paralleled by his ability to forge deep connections with colleagues and students alike, leaving an enduring impact on all who crossed his path. His legacy extends through the ongoing work of his graduate students and colleagues, who continue to champion biological nitrogen fixation for sustainable agricultural development.

In honoring Dr Reddy's memory, we pay tribute to a brilliant scientist and an extraordinary human being whose legacy will continue to inspire generations to come.

Introduction

In the last 60 years fertiliser use in agriculture has increased by 900%. However, it's been reported that up to 80% of these fertilisers are not utilised by crops but are lost to the environment as nitrous oxide, ammonia and nitrate. Improving nitrogen use efficiency is recognised as one possible solution to reducing the sector's environmental impact and optimising its productivity and sustainability in the face of increasing pressure to feed a growing population.

This volume reviews recent advances in understanding nitrogen cycling in soil, best practices to assess crop nitrogen status and different ways of optimising nitrogen use efficiency. The chapters are split into three parts: Part 1 chapters explore topics such as nitrogen cycling in soil, nitrogen use efficiency and crop plant growth, breeding and other molecular interventions for improving crop nitrogen use efficiency, and improving nitrogen use efficiency in major field crops. Chapters in Part 2 discuss developments in proximal sensors to detect crop nitrogen, synchronising nitrogen fertiliser application with crop nitrogen needs, and enhanced efficiency nitrogen fertilisers, as well as deep banding and sub-surface nitrogen fertiliser application, modelling and decision support systems for optimising nitrogen application, and economic perspectives on nitrogen cropping systems. Finally, Part 3 chapters review optimising livestock manure as a source of nitrogen, improving soil management to optimise nitrogen use, optimising service crops as a source of nitrogen, and the role of rotations and break crops in optimising nitrogen.

Part 1 Understanding nitrogen cycling in crop production

The first chapter of the book focuses on the advances in understanding nitrogen (N) cycling in soil. Chapter 1 begins by describing the knowledge surrounding nitrogen cycling processes and bringing into focus the main issues researchers and farmers face when attempting to optimise productivity while minimising adverse nitrogen losses. This covers: nitrogen fixation, mineralisation, plant nitrogen uptake, volatilisation, nitrification, nitrate leaching, and denitrification. Next, three case studies focusing on the use of cover crop mixtures to mitigate nitrate leaching are explored. These are: the response of soil N-cycling genes to nitrification and urease inhibitors, field-scale nitrous oxide fluxes and microbial processes under two management systems, and finally, nitrous oxide flux during spring thaw.

Chapter 2 examines the role of ammonium transport proteins in improving nitrogen use efficiency in crop production. The chapter begins with a description of setting up the nodule nitrogen-fixing system which is followed by sections covering transport activities of the symbiosome membrane, nitrogen fixation and delivery in nodules, and the requirement for ammonium transporters. Next, the chapter explores ammonium transport in legumes, ammonium transporter functionality from *Arabidopsis thaliana*, and altered ammonium transporter activity in plants. A section on the ammonium major facilitator is also included. Finally, the chapter explores the physiological role of the ammonium major facilitator in root nodules.

The following chapter concerns molecular interventions for improving crop nitrogen use efficiency, specifically trends, opportunities and challenges in rice. Chapter 3 begins by detailing a selection of nitrogen use efficiency indices, which is followed by a description of nitrogen management practices for improved nitrogen use efficiency. Next, biological intervention for nitrogen use efficiency improvement is explored, covering: nitrogen use efficiency phenotypes, quantitative trail loci mapping, whole genome re-sequencing and genome-wide association study for nitrogen use efficiency candidates. Genetic manipulation of genes involved in nitrogen use efficiency is also reviewed. Biological processes regulated by nitrogen use efficiency genes are also covered alongside genetic and genomic strategies for manipulation of nitrogen use efficiency genes, and molecular markers for nitrogen use efficiency genes in rice.

The final chapter of Part 1 discusses improving the effective use of nitrogen on major field crops across the globe. Chapter 4 begins with an outline of the traditional nitrogen framework for targeting optimal nitrogen supply before detailing a more functional nitrogen indicator for evaluating different genotype-environment-management scenarios. A section regarding the interactions of plant nitrogen status with other nutrients and water is also included, as well as a section on the potential use of the nitrogen nutrition index across genotype-environment-management scenarios. This is followed by a retrospective analysis of true gains of nitrogen over time for major field crop and an exploration of a path forward for enhancing nitrogen uptake and improve yields. Finally, the chapter examines using the nitrogen nutrition index as a holistic approach to defining crop nitrogen status.

Part 2 Monitoring and optimising nitrogen use

Part 2 opens with a chapter about developments in proximal sensors to detect crop nitrogen status. Chapter 5 begins with an exploration of the principles of proximal sensing, and proximal sensor types and applications for crop nitrogen status assessment. The sensors examined are the leaf colour chart, chlorophyll meters, fluorescence meters, and reflectance sensors. A section on the integration of crop data to develop algorithms to predict nitrogen requirements is also included.

The subject of Chapter 6 is synchronising nitrogen fertiliser application to crop nitrogen needs. It begins with an explanation of the asynchrony between crop nitrogen demand, soil nitrogen supply and fertiliser nitrogen application. The chapter then moves on to discuss improving fertiliser nitrogen use efficiency through synchronising fertiliser nitrogen application with soil nitrogen and crop nitrogen demand. After this, a series of innovative nitrogen management strategies are explored, including: in-season monitoring of crop nitrogen and/or biomass status, site-specific nutrient management using split applications, management zones, and enhanced efficiency fertilisers.

Chapter 7 highlights developments in the use of enhanced efficiency nitrogen fertilisers. It starts with a discussion of enhanced efficiency fertilisers, covering urease inhibitors, nitrification inhibitors and controlled-release fertilisers. Next the chapter examines the performance of these enhanced efficiency fertilisers using insights from meta-analyses, whilst also taking into account the need to design new enhanced efficiency fertilisers. Finally, the societal costs and benefits of enhanced efficiency fertiliser adoption are considered.

The next chapter covers banding nitrogen fertilisers and the implications for enhanced efficiency fertiliser technology. Chapter 8 begins by describing the nitrogen dynamics within bands of urea, going over the initial reactions on urea fertiliser in soil, the impact of banding on urea-nitrogen reactions, factors that influence the 'band effect', and the implications for nitrogen use efficiency and nitrogen losses. This is followed by a section on the nitrogen dynamics and implications for controlled-release fertilisers, reviewing synthetic polymer and biodegradable coatings. The chapter then moves on to discuss chemically-stabilised fertilisers, in particular synthetic polymer coatings and nitrification inhibitors. Finally, a case study of plant uptake and soil nitrogen dynamics from banded enhanced efficiency fertilisers is examined.

The subject of Chapter 9 is addressing the complexities of nitrogen management in maize production using dynamic models. The chapter begins by outlining the complexity of nitrogen management, covering 4R decision support, nitrogen rate and precision, computing optimum rates, and factors that affect optimum nitrogen rate. Next, the chapter brings focus to improved nitrogen rate prediction technologies, such as adaptive approaches and dynamic simulations models that incorporate the complexity of production environments. Finally, an Adapt-N case study is explored.

The final chapter of Part 2 draws attention to the economics of nitrogen in farming systems and beyond. Chapter 10 opens with an investigation of the economics of nitrogen as an input to production and nitrogen and economic risk at the farm level. The economics of nitrogen fixation by legumes is then examined, followed by the existence of flat payoff functions, which often allow flexibility in decisions about nitrogen fertiliser rates. After this, the chapter offers explanations for the over-application of nitrogen fertilisers by some farmers, the market for nitrogen fertiliser, and, finally, the economics of nitrogen pollution.

Part 3 Organic sources of nitrogen

Part 3 begins with a chapter on optimising livestock manure as a source of nitrogen and other nutrients. Chapter 11 provides a comprehensive overview of manure management. The basics of manure composition, collection, storage, and application are summarised with an emphasis on both scientific and practical considerations. Following this, long-standing challenges in manure management and the history of responses are reviewed, as well as two case studies which examine managing high carbon horse manure and reducing ammonia volatilisation.

Chapter 12 focuses on characterising soil nitrogen availability to improve nitrogen fertiliser recommendations. This chapter reviews a series of studies aiming to understand how inherent soil nitrogen supply might supplement the need for nitrogen fertiliser inputs in corn and forage management systems. The chapter also discusses research on the correlation and calibration of soil-test biological activity with soil nitrogen availability. Laboratory incubations, greenhouse growth trials and yield response trials in the field are also outlined, as well as a new approach for adjusting nitrogen fertiliser recommendations in the light of soil-available nitrogen.

Chapter 13 concerns service crops (which commonly include cover crops) as a source of nitrogen in temperate Europe. The chapter includes a discussion on service crops grown in the period between two main crops and those intercropped with the main crop. It also describes the effects of service crops on yields, covering undersown service crops and the effects on the main crop, pre-emptive competition effects on the following main crop, and as living mulches. Next, the residual effects of service crops are examined, followed by a section on nitrate leaching.

The final chapter of the book reviews how crop rotations affect nitrogen flows in organic farming. Chapter 14 begins by outlining the relationship between crop rotations and nitrogen flows, then introduces a comprehensive case study assessing 91 temperate field studies published between 1990 and 2020 that included at least one organic production system. The case study collates nitrogen input, output, and balance for the whole crop rotation and uses this data to analyse the nitrogen use efficiency of organic farming under different management regimes. Finally, the chapter offers ways of improving the use of rotations to optimise nitrogen flows and nitrogen use efficiency.

Part 1

Understanding nitrogen cycling in crop production

Chapter 1

Advances in understanding nitrogen cycling in soil

Samantha Earl-Goulet, Claudia Wagner-Riddle and Laura Van Eerd, University of Guelph, Canada; and Kate Congreves, University of Saskatchewan, Canada

1 Introduction

Scientific advances in understanding nitrogen cycling within agricultural soils have centered around the overriding challenge of optimizing crop production while minimizing environmental consequences. Despite dinitrogen being a dominant gas in the atmosphere, accounting for 78% of dry air by volume, fixed organic nitrogen and ammonium are often the limiting factors for primary productivity (Tate, 2020). The evolution of microbial metabolic pathways, including fixation, mineralization, and nitrification, supports the movement of nitrogen through the atmosphere, hydrosphere, and lithosphere alongside primary production (i.e. biosphere) (Canfield et al., 2010). However, increasing demand for crop production since the nineteenth century has resulted in anthropogenic alterations to the nitrogen cycle, such as the application of anthropogenic fertilizers, which have increased the supply of reactive nitrogen (Galloway et al., 2013). Despite this, crops' nitrogen use efficiency (NUE) has remained relatively constant at 50% (Fageria and Baligar, 2005). Consequently, nitrogen loss via ammonia volatilization, nitrate leaching, and nitrous oxide emissions have increased, presenting environmental concerns (Fowler et al., 2013).

Re-synchronizing the nitrogen supply with crop nitrogen requirements in agricultural systems will support increased productivity, reduced economic losses, and environmental remediation. To accomplish this, researchers and farmers require a detailed understanding of nitrogen fixation, mineralization, plant nitrogen uptake, volatilization, nitrification, and denitrification, to predict the natural supply and losses of nitrogen to the system. This chapter outlines

http://dx.doi.org/10.19103/AS.2024.0135.03

current knowledge, recent advances in these research areas, and the main issues researchers and farmers are presented with when attempting to optimize productivity while minimizing environmental impacts.

1.1 Nitrogen fixation: moving nitrogen (N_2) from the atmosphere to the lithosphere

The fixation of nitrogen is the transformation of dinitrogen (N_2) to ammonium (NH_4^+), which can be carried out by the Haber–Bosch process or through biological nitrogen fixation (BNF) by free-living, associative, or symbiotic diazotrophs.

The Haber–Bosch process replicates the natural biological nitrogen fixation reaction using high temperature and pressure systems alongside iron oxide (Fe_3O_4) as a catalyst to replicate the biological process (McLaren and Cameron, 1996).

$$3H_2 + N_2 \rightleftharpoons 2NH_3$$

The increased use of synthetic nitrogen fertilizers has directly altered the rate at which nitrogen enters the soil system. By increasing the nitrogen input, crop production has significantly risen, with synthetic fertilizer use supporting 48% of the global food supply in 2008 (Erisman et al., 2008). However, the low NUE and fertilizer recovery rate of key crops mean that approximately 57% of applied nitrogen is lost via nitrate leaching and gaseous emissions (Udvardi et al., 2021, Zhang et al., 2015). Therefore, recent literature has suggested that utilizing BNF as a natural source of nitrogen in the soil can reduce the requirement for fertilizer inputs while maintaining productivity.

Several bacterial groups contain the nitrogenase enzyme, which catalyzes the energy-expensive reduction of N_2 to NH_3.

$$N_2 + 8H^+ + 8e^- \rightarrow 2NH_3 + H_2$$

Free-living diazotrophs, including *Azotobacter, Bacillus,* and *Clostridium* species, source carbon through the oxidation of organic molecules released by other organisms and plants or through chemolithotrophic capabilities, which allow for the utilization of inorganic molecules. However, due to the scarcity of suitable carbon sources, these species' contribution to global agricultural nitrogen fixation remains relatively low (Wagner, 2011; Norman and Friesen, 2017). Similar limitations occur with associative BNF, observed through species of *Azospirillum* which form associations with several grasses, including rice, wheat, and corn (Vlassak and Reynders, 1979).

Symbiotic nitrogen fixation is the direct exchange of carbon and nitrogen between a bacteria species and a host plant. A series of complex signal transfers ensure that the two organisms are compatible, a biological mechanism to control the energy-expensive nodulation process further described in Ferguson et al. (2010) and Sprent et al. (2017). Following root nodulation, the bacteria infect the host plant root system, from which nitrogen fixation and the passive diffusion of essential nutrients can occur (Udvardi and Day, 1997). The dominant symbiotic relationship within global agricultural systems is between legume plants and *Rhizobium* bacteria (Ray and Daniel, 2004; Lugtenberg, 2016).

Global estimates of nitrogen input via BNF are challenging to obtain and often have significant uncertainty. Recent literature indicates that the likely range is between 52 Tg and 130 Tg N per year, with the majority of legume BNF occurring in tropical or subtropical regions (Davies-Barnard and Friedlingstein, 2020; Herridge et al., 2008). The input through agriculture is reported at 50–70 Tg N per year (Zeng et al., 2022).

Utilizing BNF within a cropping system can be achieved through several methods, including crop rotations, simultaneous cropping, legumes as a cover or catch crop, and green manures (Reinprecht et al., 2020; Kebede, 2021). The current challenge, however, is understanding how the rate of BNF is impacted by soil and environmental conditions and ensuring that the production and environmental benefits outweigh the increased costs associated with the above practices.

High concentrations of readily available nitrogen suppress the rate of BNF, acting as a biological control to limit the occurrence of the energy-expensive process (Zheng et al., 2019). A multi-year study observed that the application of 100 kg N ha^{-1} reduced the percentage of nitrogen derived from BNF from common bean (*Phaseolus vulgaris* L.) by 47% in the first year and 55% in the second and third years (Reinprecht et al., 2020). Additional studies have observed similar reductions alongside decreased nodule numbers and weight across several legume species and fertilizer rates (Argaw and Muleta, 2018; Chekanai et al., 2018; Pereira and Bliss, 1989). Reducing fertilizer application rate to optimize BNF has been shown to have either a negligible or positive impact on crop yield production (Mesfin et al., 2020; Peoples et al., 1995). However, post-harvest increases in nitrogen losses have been observed following the decomposition of legume residue due to its high nitrogen content (Jensen and Hauggaard-Nielsen, 2003; Tamagno et al., 2018). Considerable research has also focused on the response of BNF bacteria to soil water content, nutrient levels, pH, texture, salinity, temperature, and soil organic matter (SOM) content (Rice et al., 1977; Khosro et al., 2012; Vitousek et al., 2013). Understanding BNF's spatial and temporal variations are critical to achieving accurate fertilizer budgets.

Further advancements are required to assess the economic and environmental benefits of improved BNF utilization on an annual farm scale, accounting for the increased labor, machine use, and potential post-harvest nitrogen loss. Current research also focuses on understanding mycorrhizae's role in the symbiotic relationship, nodulation genetic regulations, and the potential to promote BNF within non-legume species (Rosenblueth et al., 2018; Khosro et al., 2012).

1.2 Mineralization: changing nitrogen forms in the lithosphere

Nitrogen mineralization and immobilization processes maintain the interchange between organic and inorganic forms of nitrogen.

$$Organic\,Nitrogen \rightleftharpoons NH_4^+ + NO_3^-$$

Soil nitrogen is dominantly found in the organic form, as SOM, following animal manure or plant residue (legume and non-legume) addition. However, as plants can only utilize a small portion of low-molecular-weight organic compounds, converting organic N to inorganic N through mineralization is critical for ecosystem productivity (Li et al., 2019). Therefore, accurately predicting the mineralization rate and plant nitrogen requirements across various environmental conditions will provide a significant stepping stone for re-synchronizing nitrogen supply to plant demand, reducing fertilizer requirements, and ultimately improving NUE.

Heterotrophic microorganisms carry out mineralization using enzymes to break down SOM, assimilate organic nitrogen compounds to produce microbial tissue, and utilize carbon as an energy source. If the nitrogen content of SOM exceeds microbial assimilation needs, excess nitrogen is released as ammonia (NH_3) via ammonification and rapidly converted from NH_4^+ to nitrate (NO_3^-) via nitrification (Janssen, 1996; Ma et al., 1999; Zhang et al., 2019). Immobilization occurs when the nitrogen content of SOM does not meet microbial assimilation requirements. Consequently, inorganic nitrogen in the soil is absorbed by microbes and utilized to synthesize proteins and other nitrogen-containing compounds, decreasing the plant available soil nitrogen content. This continuous transfer of nitrogen through mineralization and immobilization is often referred to as mineralization-immobilization turnover, with net mineralization occurring when the rate of gross mineralization exceeds that of gross immobilization (Benbi and Richter, 2002).

Over time, the view of nitrogen mineralization has shifted from an extracellular process controlling nitrogen cycling to an intracellular process driven by microbial nitrogen demand and soil nitrogen supply. Therefore, the microbial NUE has become a critical tool for estimating the total nitrogen availability for plants and potential ecosystem losses (Mooshammer et al., 2014). Under nitrogen-limiting conditions, most acquired organic nitrogen is utilized for microbial growth and biomass, leading to a high NUE and immobilization to meet assimilation demands. Meanwhile, a low microbial NUE has been observed under excess nitrogen, or carbon limiting, conditions resulting in high mineralization rates. Ensuring the regulation of microbial NUE is critical to controlling mineralization and immobilization rates, system productivity, and environmental losses of inorganic nitrogen (Li et al., 2019; Mooshammer et al., 2014; Zhang et al., 2019). The challenge being tackled by researchers and crop producers is how to manage microbial mineralization to optimize plant nitrogen supply under a variety of field conditions.

The mineralization rate is a central supply of inorganic nitrogen to the soil and, therefore, a critical factor in determining fertilizer requirements. Understanding how soil, environmental, and management factors impact the natural supply of inorganic nitrogen is essential to reassessing fertilizer requirements for cropping systems at a localized scale.

The composition of organic residue within a system impacts the direction of the mineralization-immobilization reaction. A carbon-to-nitrogen (C:N) ratio of 24:1 for plant residue is considered optimum for stimulating net mineralization, reducing fertilizer requirements, and maintaining a soil-protection residue cover (Ma et al., 1999; Cabrera et al., 2005). Low C:N ratios promote net mineralization; however, the increased decomposition rates will lessen the soil-protection period. Meanwhile, crops with high C:N ratios will stimulate net immobilization and potentially require the addition of nitrogen fertilizers to maintain production. Additional factors regarding plant residue that impact the mineralization rate are pH, salinity, heavy metal concentration, and different compounds present in organic residues, such as proteins, lignin or cellulose-like compounds, and soluble carbohydrates. The impact of these factors is discussed in the literature and often results in the rate of mineralization varying between residues despite the same C:N ratio (Khan and Scullion, 2002; Rowell et al., 2001).

Variations in mineralization rates are also attributed to soil characteristics such as clay content impacting the rate of organic nitrogen adsorption, aeration in sandier soils, variations in microbial populations, and salinity (Cabrera et al., 2005). Several studies have assessed the relationship between net mineralization rates and soil water content or temperature. Under favorable conditions, net mineralization increases with the soil temperature reaching an optimum rate at 25–35°C, depending on soil nitrogen availability and microbial community dynamics (Chapin et al., 2002; Guntiñas et al., 2012). The relationship between soil water holding capacity (WHC) and mineralization is linear, increasing to an optimum rate at 60% WHC with thresholds below 20% and above 80% WHC (Li et al., 2014). The interaction between soil moisture, residue composition, mineralization, and immobilization is complex and strongly affected by management practices (Benbi and Richter, 2002; Janssen, 1996). These interactions must be studied further and appropriately modeled for accurate mineralization and immobilization rate predictions.

The influence of synthetic fertilizer use on the nitrogen mineralization rate is debated in the literature. Stoichiometric decomposition theory predicts an immediate positive increase in mineralization following fertilizer application owing to increased microbial biomass and activity (Harris, 2003; Chen et al., 2014). In addition, increased rhizodeposition and the positive priming of microbes following fertilizer application have also been linked to increased net primary productivity (NPP) (Cheng, 2009). However, literature has indicated that this increase in NPP can result in reduced soil moisture and temperature, decreasing SOM mineralization. Similarly, the theory of microbial nitrogen mining suggests that fertilizer use will decrease the decomposition of energy-poor SOM substrates, thereby decreasing total mineralization (Moorhead and Sinsabaugh, 2006; Craine et al., 2007). Mahal et al. (2019) studied the effect of fertilizer use on SOM mineralization in two long-term experiments, observing that synthetic fertilizers caused a reduction in total SOM mineralization. However, it suggested that using fertilizers that do not contain NH_4^+ had a less significant impact. The prioritization of mineralization as a source of inorganic nitrogen, and reduction in fertilizer use, can result in reduced soil nitrogen and carbon (SOC) concentrations. While fertilization can provide an opportunity to increase SOC, Mahal et al. (2019) noted that this was due to reduced mineralization.

The variation in methods used to estimate fertilizer NUE (FNUE), and their accuracy, has been attributed to the conflicting results regarding the response in mineralization rate to fertilizer application. Studies either use an indirect *N difference* method (Varvel and Peterson, 1990), in which differences in nitrogen uptake between fertilized and unfertilized crops are recorded, or a direct *^{15}N tracer* method (Hauck and Bremner, 1976), recording plant uptake of isotopically labeled fertilizer. However, under and over-estimations are associated with each measurement technique; therefore, future research is required to achieve accurate mineralization measurements.

1.3 Plant nitrogen uptake: moving nitrogen (NH_4^+, NO_3^-) from the lithosphere to the biosphere

Once mineralization has occurred, and nitrogen is available in the soil, it can be removed via plant uptake, volatilization, leaching, or denitrification. The latter three processes are linked to environmental consequences and economic losses; therefore, the issue presented is how to maximize plant uptake of nitrogen. Confronting this issue requires a detailed understanding of how field conditions impact nitrogen uptake processes.

In nitrogen-limiting conditions, soil temperature indirectly affects plant uptake of nitrogen via alterations in plant roots' growth and physiological characteristics and directly affects the biogeochemical processes regulating nitrogen and soil water availability (Bassirirad, 2000; Laine et al., 1993). Several studies have observed increases in soil temperature correlating with improved nutrient uptake and changes in NH_4^+ root transport properties (Bassirirad et al., 1993; Cleve et al., 1990). However, the mechanisms responsible for these variations have not been identified. Increased root respiration under higher temperatures has been suggested as a controlling mechanism, owing to its role in mediating ion movement across the root. However, it cannot be solely responsible, as increased temperatures alter several additional root functions (Glass, 2003). Additional factors such as salinity, water content, and pH have also been shown to alter plant nitrogen uptake (Pessarakli, 2019).

Understanding environmental and soil conditions' impact on plant nitrogen uptake is critical for modeling nutrient processes. However, these factors have large spatial and temporal variability, which creates difficulty in modeling these processes in time for fertilizer application. Therefore the utilization of enhanced efficiency fertilizers (EEF), such as nitrification and urease inhibitors (NUI), and software for precision agriculture provides an opportunity to re-synchronize nitrogen supply and demand in a more straightforward format. In addition, these practices aim to reduce the percentage of fertilizer nitrogen lost following an application by slowing the release of NH_4^+ and NO_3^- into the soil system following application and improving fertilizer requirement calculations.

Detailed digital soil mapping of farm properties provides information on changes in soil properties, mineralization rates, and, therefore, fertilizer requirements at a smaller scale (Sela et al., 2016; Melkonian et al., 2008). This ensures that fertilizer is applied at a rate suitable for current growing conditions rather than the ideal ones. However, the significant knowledge gap regarding how various soil conditions impact the mineralization rate and plant nutrient uptake provides limitations regarding this software. Additional information about digital soil mapping and precision agriculture, alongside enhanced efficacy fertilizers, and their role in improving FNUE is discussed in the following chapters.

1.4 Volatilization: moving nitrogen (NH_3) from the lithosphere to the atmosphere

The volatilization of ammonia (NH_3) into the atmosphere occurs following nitrogen fertilizer application to agricultural soils. Urea applied to the soil is rapidly hydrolyzed to ammonium carbonate ($(NH_4)_2CO_3$) by the urease enzyme; owing to the unstable nature of $(NH_4)_2CO_3$, it is decomposed to form NH_3 gas, which is emitted into the atmosphere (Sherlock, 1984).

$$CO(NH_2)_2 + 2H_2O \underset{Urease}{\rightarrow} (NH_4)_2 CO_3 + H^+ \rightarrow 2NH_4^+ + HCO_3^-$$

$$NH_4^+ \rightarrow NH_{3(g)} + H^+$$

The emission of NH_3 is dominantly an economic loss for cropping systems, as the environmental consequences of air pollution and ground level smog are considered less significant compared to nitrate leaching or nitrous oxide emissions (Sharpe and Harper, 1995). Therefore, managing systems to reduce the ammonia volatilization rate provides an economic benefit by reducing the net nitrogen loss from the system. However, this subsequently increases the concentration of NO_3^-, requiring careful management of nitrification and denitrification processes in parallel.

Volatilization is a dominantly physical process, with the rate, source, and placement of nitrogen impacting the rate of nitrogen conversion and soil pH and buffering capacity controlling the rate of NH_3 loss. Reducing fertilizer application rates to meet plant demand and reduce excess NH_4^+ provides the largest reduction in NH_3 volatilization (Wan et al., 2021). Therefore, understanding and measuring nitrogen supply alongside plant requirements is critical to mitigating volatilization and increasing NUE. In addition, EEF using urease inhibitors has also shown significant reductions in volatilization rates and improved NUE owing to the slowing of the release of NH_4^+ and synchronizing the release with plant demand (Christianson et al., 1993). These concepts are discussed in further chapters

While not the only soil characteristic influencing NH_3 emissions, soil pH is essential in controlling the equilibrium between NH_4^+ and NH_3 in solution, with higher soil pH corresponding to increased NH_3 concentrations and, therefore, volatilization. Several studies have assessed the change in NH_3 concentration with pH, showing the percentage of un-ionized NH_3 at pH 6, 7, 8, and 9 being approximately 0.1%, 1%, 10%, and 50%, respectively (Freney et al., 1981). Within cropping soils, the soil pH is typically lower than that required for high concentrations of NH_3 emissions. However, temporary increases in soil pH following the application of urea-based fertilizers stimulate the volatilization process, accounting for 19–23% of agricultural NH_3 emissions (Rochette et al., 2009; Skorupka and Nosalewicz, 2021). This increase in pH around fertilizer granules can be resisted through the soil-buffering capacity and cation exchange capacity (CEC). Soils with high clay and organic matter content can buffer pH; a high CEC can reduce volatilization by lowering the concentration of NH_4^+ in soil solution (Keller and Mengel, 1986).

The current knowledge surrounding volatilization is extensive because the process is dominantly physical and primarily associated with fertilizer applications. Reducing the volatilization rate requires a detailed understanding of crop fertilizer requirements and using EEF. The direct implication of reducing volatilization and the NUE of cropping systems is favorable; however, increased soil NH_4^+ concentration indirectly increases the risk of nitrate leaching and nitrous oxide emissions. Therefore, managing nitrification and denitrification processes is critical to advancing our knowledge about nitrogen cycling and environmental losses, alongside increasing the NUE of agricultural systems.

1.5 Nitrification: changing nitrogen forms in the lithosphere

Nitrification, the oxidation of NH_4^+ to NO_3^-, is a central component of the nitrogen cycle as it links plant nitrogen uptake and nitrogen losses from the system. Nitrate is more mobile in soil solution than NH_4^+, resulting in improved nitrogen transport to root surfaces. However, this also means NO_3^- can be readily leached from the soil system or utilized by microorganisms to produce and emit N_2O. Therefore, appropriately managing nitrification within a system is critical to maximizing productivity while minimizing adverse environmental consequences. A detailed understanding of the nitrification pathway, including the response of microbe abundance, community structure, and relative contribution of nitrification in different field conditions, is critical to confronting this challenge.

The large body of literature surrounding nitrification discusses the two-step process driven by ammonia-oxidizing bacteria (AOB) and archaea (AOA) alongside nitrite-oxidizing bacteria (NOB). However, several additional nitrification pathways have been identified within the literature, including heterotrophic nitrification and comammox. Heterotrophic nitrification is performed by a wide range of eukaryotes and bacteria but does not contribute significantly to the soil NO_3^- pool in cropping systems as the process is not required for microbe energy conservation (De Boer and Kowalchuk, 2001). Single microorganisms of the *Nitrospira* genus carry the required enzymes to complete both processes of the nitrification reaction, resulting in the development of the comammox nitrification pathway (Hu and He, 2017; Daims et al., 2015; Van Kessel et al., 2015). Under low nitrogen conditions, comammox bacteria have a competitive advantage due to lower energy requirements and N_2O production; however, no literature has identified the pathways as a significant NO_3^- contributor in agricultural systems (Costa et al., 2006). The continual advancement in knowledge of the nitrification process highlights the evolution of individual pathways to ensure efficiency in all ecosystems. While also providing expanding research gaps surrounding how to manage nitrification to re-synchronize plant nitrogen supply and uptake.

The microbial pathways involved in the two-step nitrification process, divided by primary and secondary nitrification, are well discussed in the literature. Ammonia is oxidized to hydroxylamine (NH_2OH), catalyzed by the ammonia monooxygenase (AMO) enzyme, during primary nitrification (Arp et al., 2002). Within agricultural systems, the AOB populations, namely

Nitrosopira and *Nitrosomonas* genera (Hayatsu et al., 2021), are dominant nitrifiers as synthetic fertilizers suppress the AOA population (Di and Cameron, 2016). Hydroxylamine is further reduced to nitrite (NO_2^-) by the hydroxylamine oxidoreductase enzyme (HAO), with nitroxyl produced as an intermediate product (Caranto and Lancaster, 2017). Nitrate-oxidizing bacteria carry out secondary nitrification in which NO_2^- is converted to NO_3^- by the nitrite oxidoreductase enzyme (NXR). The NOB population is a phylogenetically diverse function group in which *Nitrobacter* and *Nitrospira* are typically in greater abundance within agricultural soils (Han et al., 2018). The role of secondary nitrification and NOB populations within the nitrogen cycle is critical for providing plant-available NO_3^- and regulating the soil NO_2^- concentrations, the latter of which is toxic to aquatic life and inhibits bacterial growth (Daims et al., 2016). Nitrous oxide is also a by-product of nitrification through abiotic reactions involving NH_2OH, NO_2^-, and NO (Charpentier et al., 2007; Kozlowski et al., 2014).

$$NH_3 + 3O_2 \underset{AMO}{\rightarrow} NH_2OH \underset{HAO}{\overset{N_2O \atop \uparrow}{\rightarrow}} [NOH] \rightarrow 2NO_2^- + O_2 \underset{NOR}{\overset{N_2O \atop \uparrow}{\rightarrow}} 2NO_3^-$$

As nitrification is a microbial process, its rate is strongly impacted by several soil and environmental conditions, with soil aeration being a significant controlling factor. The relative activity of nitrifying microbes increases between water-filled pore space (WFPS) of 10% and 60%, with the optimum water content varying with soil texture, 55% WFPS in fine-textured soils, and approximately 40% WFPS in coarse-textured soils (Robertson and Groffman, 2015; Norton and Ouyang, 2019). Soil aeration also impacts each microbial community's relative contribution toward total nitrification rates, with AOA populations having the highest affinity for oxygen, followed by AOB and NOB (Wen et al., 2020). Understanding the role of soil aeration and water content in determining the nitrification rate is critical for improving NUE and reducing the risk of fertilizer loss via leaching or denitrification.

The nitrification process can occur across various soil pH values due to the ecological and physiological variation between different nitrifying microorganisms. Nevertheless, it is important to note that soil pH has a significant influence on the extent to which the populations of AOB, AOA, or comammox bacteria contribute to the overall process of nitrification. The ionization of NH_3 to NH_4^+ and each species' affinity for NH_4^+ dictate the activity of each nitrifying population; therefore, soil pH commonly has a similar impact on nitrification rates and nitrifier niche separation as synthetic fertilizer use. In low nitrogen or pH environments, AOA and comammox populations are the dominant nitrifiers, often resulting in a lower total nitrification rate (Hu et al.,

2013; Prosser and Nicol, 2012). Therefore, in environmental systems with no anthropogenic nitrogen additions, AOA populations have a high contribution to total nitrification as the supply of nitrogen is provided slowly through mineralization (Ouyang et al., 2017; Hayatsu et al., 2021). The increased use of synthetic fertilizer has shifted AOB populations into the dominating nitrifiers and increased total nitrification rates, resulting in NO_3^- supply surpassing plant requirements (Di et al., 2016).

Owing to the variety of nitrification pathways and microbial species involved, significant research gaps remain about how the total nitrification rates vary under different field conditions and the impact of management practices on the bacterial community structure. A discussion surrounding EEF, specially NUI additives, is supplied in later chapters, providing insight into the suitability of these management practices to improve the NUE of cropping systems. These products work by either slowing the release of nitrogen into the soil system following application or temporally inhibiting nitrifying microbes, respectively (Akiyama et al., 2010; Cameron and Di, 2002). Literature has identified potential decreases in environmental losses of nitrogen, alongside yield increases, as a result of appropriate utilization of these products (Halvorson et al., 2014; Linquist et al., 2013). However, evidence also shows that these practices' suitability and success are strongly dictated by soil temperature, water content, texture, fertilization rates, and the type of EEF or NI used (Gilsanz et al., 2016; Misselbrook et al., 2014; Shepherd et al., 2014; Woodward et al., 2016; Tindaon et al., 2013).

1.6 Nitrate leaching: moving nitrogen (NO_3^-) from the lithosphere to the hydrosphere

Nitrate leaching is the downward movement of NO_3^- anions through the soil profile following significant water input events such as rainfall, snow melt, or irrigation. Negatively charged surfaces of clay minerals and SOM repel the water-soluble anion allowing for its rapid displacement and leaching into groundwater systems following large water fluxes (Padilla et al., 2018). This has a significant environmental concern due to the eutrophication of surface water systems. Eutrophication is the increase in one or more growth factors required for photosynthesis, consequently increasing algae and macrophyte production, resulting in depleted dissolved oxygen levels and the deterioration of water quality (Schindler, 2006). Increased NO_3^- concentrations also alter the ecological balance of water systems, having toxic impacts on fish, invertebrates, ruminants, and infants (Kirchmann et al., 2002; Beauchemin et al., 2020).

In addition to water movement through the soil profile, the dominant conditions dictating leaching rates are the supply of NO_3^- and its rate of utilization by plants. Yang et al. (2013) observed that residual soil nitrogen doubled, and leaching losses increased by 42% nationally in Canada between 1981 and 2006 due to large increases in fertilizer use without significant increases in crop nitrogen uptake. Eggleston et al. (2006) recorded the estimated fraction of nitrogen lost via leaching as 0.3 kg per kg N applied. The literature identifying the relationship between nitrogen fertilizer use and leaching is extensive (Sacco et al., 2003; Perego et al., 2012; Mantovi et al., 2006); identifying methods to improve the NUE and recovery of nitrogen fertilizers is critical to reducing the rate of nitrate leaching.

Late autumn and winter are critical for nitrate leaching in humid, temperate climates. During this time, the rate of precipitation typically exceeds evapotranspiration resulting in saturated soils and downward water movement. At the same time, reduced temperatures and post-harvest fallow reduce crop growth, causing the rate of mineralization and nitrification to exceed crop nitrogen uptake. Therefore, large concentrations of excess NO_3^- are readily displaced following the heavy rainfalls associated with these seasons (Kirchmann et al., 2002). Various studies have identified that periods of abnormally high rainfalls, which coincide with summer fallow, also result in increased nitrate leaching (Wang et al., 2014, 2013). Therefore, when implementing management practices to alter the nitrification rate and supply of NO_3^- to the soil system, such as EEFs, the seasonal variation in nitrate leaching must be understood. Crop rooting depth influences the length of time in which NO_3^- can be utilized by plants and, therefore, not leached. Cropping systems with a greater density of low rooting depth crops typically experience higher leaching losses as NO_3^- is moved below the root system faster than in a field with deeper rooting systems (Wang and Li, 2019). Leaching losses are also lower in fine-textured soils than in coarse-textured soils, as slower drainage provides a greater opportunity for plant uptake or nitrogen loss via denitrification (Padilla et al., 2018).

Improving the NUE of cropping systems will decrease the nitrate leaching rate; however, variations in soil hydrological, topographic, and management properties between farms and fields mean that nitrate leaching mitigation must be developed and implemented on a site-specific basis to account for spatial variation in the nitrogen cycle (Kirchmann et al., 2002). Current research focuses on mitigating and reducing leaching losses within agricultural systems, including improved nitrogen recovery, crop diversification through catch crops, and nitrification inhibitors. Each of these practices aims to synchronize the supply of NO_3^- with plant uptake, reducing excess anions in the soil solution.

1.7 Denitrification: moving nitrogen (N_2O, N_2) from the lithosphere to the atmosphere

Denitrification is the reduction of NO_3^- or NO_2^- to N_2O or N_2, which occurs as either a biological or chemical process. Within agricultural soils, biological or heterotrophic denitrification is the dominant pathway, involving the reduction of NO_3^- under anaerobic soil conditions. The low oxygen availability and redox potential of anaerobic soils result in specific heterotrophic bacteria and fungi utilizing NO_3^- as an alternate electron acceptor, allowing respiration and microbial growth to continue (Clough, 1994; McLaren and Cameron, 1996).

$$2NO_3^- \underset{nar}{\rightarrow} 2NO_2^- \underset{nir}{\rightarrow} 2NO_{(g)} \underset{nor}{\rightarrow} N_2O_{(g)} \underset{nos}{\rightarrow} N_{2(g)}$$

A complex multisite metalloenzyme catalyzes each step within the sequential denitrification reaction scheme. The nitrate reductase (*nar*), nitrite reductase (*nir*), nitric oxide reductase (*nor*), and nitrous oxide reductase (*nos*) enzymes each have different eco-physical characteristics allowing each one to catalyze a specific stage of the reaction (Robertson and Groffman, 2015; Samad et al., 2016). These variations also cause a lag time between the production and consumption of intermediate substrates. The occurrence and duration of this lag time are significant contributors to incomplete denitrification, the emission of N_2O before its reduction (Bothe et al., 2006; Haynes, 1986).

A secondary denitrification pathway called nitrifier denitrification has been identified as having a critical role within the nitrogen cycle, ensuring that the production of N_2 continues under various environmental conditions. Nitrifier denitrification is carried out by specific nitrifying bacteria containing the denitrification reductase enzymes, allowing NO_2^- reduction directly following the oxidation of NH_4^+ (Kool et al., 2011). Identifying the nitrifier denitrification pathway provides a critical understanding of how each phase of the nitrogen cycle is connected and how the cycle has evolved to fill each ecological niche.

Research focused on determining the relative proportion of N_2 and N_2O emitted from denitrification is limited owing to the high atmospheric N_2 concentration introducing uncertainties into measurement procedures. However, the current literature indicated that up to 90% of the global denitrification flux from agricultural soils results in the emission of N_2 (Weitzman et al., 2021; Scheer et al., 2020). While the emission of N_2O from incomplete denitrification accounts for a small percentage of the total denitrification flux, its properties as a potent greenhouse gas mean the environmental concerns of N_2O emissions are disproportionate to the emission rate. Nitrous oxide has a global warming potential 298 times that of carbon dioxide (CO_2) over 100 years (Liu et al., 2017). The increased rate of N_2O emission into the atmosphere has also been identified as the most significant threat to the ozone layer in the twenty-first century (Myers et al., 2017; Pachauri et al., 2014; Ravishankara et al., 2009). Therefore, management practices must aim to balance NO_3^- supply with plant demand, for example, the application of nitrification inhibitors, or alter the N_2O:N_2 ratio by increasing the rate of complete denitrification.

The latter of these methods is complex due to high variability in the field conditions controlling microbial denitrification process. This variability across a spatial and temporal scale can result in temporary increases in N_2O emission, commonly described as 'hot spots' and 'hot moments', respectively (Wagner-Riddle et al., 2020). Practices which temporarily alter the chemical and physical properties of soils, such as irrigation, fertilizer application, and cultivation, can result in changes in N_2O emissions (Chahal et al., 2021) by creating or removing hot spots and hot moments. Understanding these impacts are critical to improving on-farm management practices alongside increasing the accuracy of management and environmental-based models. Monitoring the occurrence of high denitrification and N_2O emissions provides important guidelines for assessing the most suitable times and rates of fertilizer application to increase crop NUE.

Spatial changes in aeration status, microbial population, and mineralizable carbon and nitrogen content influence the occurrence of hot spots, with soil aeration being the primary factor dictating denitrification rates. As NO_3^- is a less efficient electron acceptor than oxygen, the biosynthesis of each denitrifying enzyme is suppressed in well-aerated systems. The sensitivity of each enzyme to oxygen increases along the reaction scheme, with *nor* and *nos* enzymes inhibited at lower oxygen contents compared to *nar* and *nir* enzymes (Haynes, 1986). This biological control ensures that nitrification and aerobic respiration, the more energy-efficient pathways, occur in optimum conditions. The trigger point for heterotrophic denitrification is 60% WFPS, with the rate significantly increasing beyond a WFPS of 80%. Nitrifier denitrification occurs between WFPS of 50% and 70%, ensuring constant nitrogen cycling as the dominance of nitrification and denitrification switch (Kool et al., 2011). Between a WFPS of 60% and 90%, incomplete denitrification occurs in soils owing to the high sensitivity of *nos* to oxygen. For the *nos* enzyme to be activated, WFPS must be higher than 90% and held for an extended period, ensuring that activating the enzyme will result in a net energy gain (Freney and Simpson, 1983). As most agriculturally managed soil supports short periods of waterlogging via drainage systems, the $N_2O:N_2$ ratio favors N_2O production. The waterlogged soil conditions associated with denitrification typically occur following a significant rainfall or irrigation event. However, denitrification can occur within aerobic soils due to anaerobic microsites (Groffman et al., 2009).

Variations in microbial community sizes, mineralizable carbon, and NO_3^- concentration across a spatial scale also influence the spatial variability in the denitrification rate. Management practices, including fertilizer applications, NI use, cultivation, and crop diversification, have been shown to influence each of these factors, as discussed in several case studies. Typically, the denitrification rate increases with carbon and NO_3^- concentrations; however, high concentrations of NO_3^- in agriculturally managed soil have an inhibitory effect on the *nos* enzyme, resulting in incomplete denitrification (Freney and Simpson, 1983). Therefore, implementing practices to reduce the release of NO_3^- into the soil system improves NUE and increases the $N_2O:N_2$ ratio.

Hot moments for N_2O emissions occur due to sudden changes in the rate of denitrification through either management or weather events. The use of nitrogen fertilizer is a dominant source of N_2O hot moments, with large surface fluxes being recorded several days following application. Nitrous oxide emissions from agricultural soils contribute approximately 50% to the total global anthropogenic flux, with nitrogen fertilizers and animal manures being the dominant N_2O sources (Shcherbak et al., 2014; Bouwman et al., 2002). Weather-driven hot moments result from freeze-thaw (FT) and wet-dry (WD) cycles, in which N_2O emissions may increase up to 1000-fold and 4500-fold, respectively, compared to pre-event rates (Kim et al., 2012). Large N_2O pulses following WD cycles are attributed to the onset of anaerobic conditions combined with accumulated substrates over prolonged dry periods, creating conditions conductive to denitrification. Therefore, estimating and mitigating DW-related N_2O emissions require detailed knowledge of how soil moisture, temperature, and substrate concentration are impacted under different field and management conditions.

In cold climates where soils experience a prolonged freezing period during the winter, FT cycles are a significant source of N_2O emissions. Two dominant mechanisms contribute to FT cycle-related N_2O emissions, the physical release of previously trapped N_2O produced in the unfrozen soil subsurface and enhanced denitrification due to the release of substrates with aggregate breakdown or microbial lysis and increased soil water content following snow melt (Risk et al., 2013; Teepe et al., 2001; King et al., 2021). These FT cycles represent 17–28% of global agricultural N_2O emissions; however, significant knowledge gaps exist surrounding how these FT emissions mechanisms may vary over a spatial and temporal context (Wagner-Riddle et al., 2017).

The implementation of management practices, such as crop diversification that alter the concentration of available nitrogen in the soil system before winter freezing and global climate change, which is predicted to increase winter temperatures and reduce snowpack depth, are both expected to alter the pattern and rate of FT-induced N_2O emissions (Brin et al., 2018; Kaye and Quemada, 2017; Li et al., 2016; Ruan et al., 2017). Therefore, advancing knowledge within this area of the nitrogen cycle is critical to improving N_2O emission models, providing a tool to enhance the accuracy of greenhouse gas budgets and nitrogen loss pathways.

2 Case studies

The body of literature surrounding the nitrogen cycling process is extensive and ongoing. Continual advancements in the knowledge of each process from the microbial to production scale are being made, alongside developments in practices to improve NUE and mitigate environmental losses of nitrogen. Here, we discuss case studies focusing on the different knowledge gaps examined previously.

2.1 The use of cover crop mixtures to mitigate nitrate leaching

The environmental and production concerns associated with high concentrations of soil NO_3^- coinciding with drainage events are discussed earlier. Previous literature has focused on potential growing season (GS) reductions in NO_3^- leaching with the use of a leguminous/non-leguminous cover crop mixture, disregarding the substantial amount of leaching that occurs during the non-growing season (NGS) in cold and humid environments (Abdalla et al., 2019). Developing strategies to reduce surplus water and NO_3^- within the soil system during the NGS is a significant advancement in increasing the NUE of cropping systems.

Lapierre et al. (2022) observed that utilizing a leguminous/non-leguminous cover crop mixture had no annual effect on soil drainage, unlike previous studies; however, substantial reductions in NO_3^- concentrations were observed across the whole profile when cover crop establishment was successful. An 80% reduction in NO_3^- concentration at 90 cm, the depth used for leaching calculations, was observed across soil types between after-harvest and pre-planting of annual crops with the inclusion of a cover crop mixture compared to no cover crops. This reduction in post-harvest soil NO_3^- was attributed to enhanced root uptake of NO_3^- and the cover crop mixture's high C:N ratio. Additional benefits of cover crop utilization were increased surface soil temperature during the winter due to the insulation effect of crop residue. Providing an insulating layer to the soil surface was also attributed to potential reductions in mineralization rates and microbial cytoplasmatic release owing to the regulation of soil freezing and thawing events.

Reductions in soil NO_3^- concentrations with cover crop use translated to an average of 67% decrease in NO_3^- leaching (across two soil types). A cover crop mixture provided a two-fold benefit as the non-leguminous crops provided a larger leaching reduction benefit than legumes alone, owing to the lack of biological nitrogen fixation. In contrast, the leguminous crop can prevent a potential reduction in succeeding cash crop yield due to the natural supply of nitrogen via BNF. Across the annual experiment, winter and spring-thaw measurements accounted for 54–90% of the total leaching losses, highlighting the critical nature of accurately measuring NGS nitrogen losses. Ultimately, using cover crops in cold and humid environments significantly reduces NO_3^- leaching, thereby improving water quality and NUE of systems. Further advancements in this area of research should focus on enhancing cover crop establishment through cold short-season cultivars and improved agronomic practices.

2.1.1 Soil N-cycling genes response to nitrification and urease inhibitors

Understanding how NUI impact nitrogen losses provides information surrounding the potential to increase NUE or reduce environmental outputs through management practices. Therefore, conducting research assessing soil microbial communities and N-cycling groups' responses to these practices is essential to understanding their short- and long-term efficacy. Tosi et al. (2020) conducted a single-year study to measure the short-term changes in N-cycling genes and transcripts from fertilizer corn with and without the addition of common urease and nitrification inhibitors, and compare the differences in microbial dynamics to N_2O emissions.

The application of urease and nitrification inhibitors aimed to delay the NH_4^+ transformation process, inhibiting urea hydrolysis and NH_4^+ oxidation, respectively. The overall decrease in field-scale N_2O emissions from NUI-treated plots was 68%, with the most significant reduction in emission duration and magnitude observed between 5 days and 8 days after fertilization. Low abundance and transcription of *nir:nos* genes were observed during moments of peak N_2O emissions. However, this may have resulted from unmeasured N_2 emissions or a lag time between N_2O surface flux and transcript detection. The application of the NUI also resulted in a 45–57% reduction in clade II *nos*Z transcripts, recorded 9 days after application. The delay in microbial response to NUI may have resulted from the variation in measurement methods, with N_2O measurements reflecting microbial activity at the field scale while microbial measurements were only taken between crop rows and side-dressed fertilizer lines.

Comparing N-cycling gene activity to N_2O emissions can be challenging owing to alternative, non-target N-transformation pathways, primer specificity and coverage limitations, and large temporal variability in microbial activity and emissions. However, these studies are essential to understanding the link between soil microbial communities and ecosystem functions.

2.1.2 Field-scale N_2O fluxes and microbial processes under two management systems

Crop residue provides a critical opportunity to manage N-cycling processes and N_2O emissions from cropping systems. Removing residue from systems has been identified as a potential direct and indirect source of N_2O. Meanwhile, incorporating residue into the soil or leaving residue as a ground cover post-harvest may provide the system with an organic nitrogen source, reducing fertilizer requirements. The study by Congreves et al. (2017) aimed to understand how residue management impacts direct N_2O emission from cropping systems during the spring thaw and how results varied across different tillage systems.

Irrespective of residue management and tillage systems, overwinter, and spring-thaw N_2O emissions were large contributors to the total annual emissions (28–77%). Micrometeorological data showed that no-tillage practices resulted in lower N_2O emissions over 5 years, with differences in N_2O production mechanisms occurring between seasons. During winter and spring thaw, increased *de novo* denitrification was responsible for N_2O production and surface flux; meanwhile, enhanced nitrification-denitrification following fertilizer application in spring was observed as the major N_2O source. Consequently, showing that nitrifier communities are more affected by the cold temperatures and that N_2O response to tillage and residue practices may result from seasonal domination of microbial pathways. Removing residue increased N_2O emissions in both tillage systems; however, to a greater extent in conventional tillage systems, attributed to the difference in soil temperature and moisture dynamics.

An additional study by Németh et al. (2014), conducted at the same research site, focused on assessing nitrifiers'' and denitrified communities'' abundance and gene expression during the FT cycle. Reductions in *amo*A gene expression in plots with crop residue removed were attributed to lower dissolved organic carbon (DOC) and inorganic nitrogen concentrations in the soil. No difference in *nir* gene copies was observed, but *nos*Z communities were lower in the residue-removed plots. The dominance of *de novo* denitrification as an N_2O production mechanism at spring thaw was confirmed through a temporary increase in the denitrifier population containing *nos*Z. Meanwhile, the rise in *nos*Z denitrifiers in plots where residue was not removed indicates that the lower N_2O emissions may be a result of increased reduction of N_2O to N_2

Overall, overwinter and spring thaw N_2O emissions were greater in conventional tillage systems where crop residue was removed following harvest. A shift in the primary N_2O production pathway may be why these differences were only observed during the NGS, highlighting the requirements for year-round measurements. Keeping crop residue on the soil surface reduced N_2O emissions by providing an insulation layer to the soil overwinter and altering microbial community abundance and gene expression. Further studies assessing the relationship between crop residue management, nitrifier and denitrifier microbal activity, and N_2O emissions are required across various soil types, and management practices.

2.1.3 Nitrous oxide flux during spring thaw

Studies aimed at increasing the understanding of N_2O freeze-thaw cycle mechanisms are critical to improving process-based modeling of national and global N_2O emissions. The physical release of trapped N_2O gas upon thawing and *de novo* denitrification following soil thawing have been proposed mechanisms; however, they are very different processes. Linking N_2O surface fluxes during spring thaw to each mechanism provides an opportunity to advance knowledge of N_2O flux events, annual emissions, and temporal variation.

The physical release process involves microbial processes; however, the processes occur over the entire time of freezing, with N_2O accumulation occurring at depth and in unfrozen water films. Meanwhile, enhanced denitrification is observed through increased biological activity in the top layer of soil following thaw. Wagner-Riddle et al. (2008) applied a ^{15}N tracer to different depths in the soil column, and soil gas samples at depth were collected alongside surface flux data. The ^{15}N tracer allowed N_2O fluxes to be traced to an origin depth while also identifying the dominant mechanisms responsible for N_2O production.

Peak N_2O fluxes occurred following the thawing of surface soil layers, despite the soil remaining frozen at depth. This increase in N_2O emission was attributed to production at the surface soil layer, with fluxes being 1.5–5 times greater than when the tracer had been applied to the deeper soil layer. The $^{15}N_2O$ flux data indicate that the high $^{15}N_2O$ concentrations at depth made a limited contribution to the surface flux. While conditions were conducive for denitrification at both the soil surface and at depth, the results of this study identified that enhanced denitrification in the soil surface layer during thaw is the dominant mechanism resulting in peak surface N_2O fluxes. Meanwhile, N_2O trapped at depth in the profile was likely further reduced to N_2 as it diffused from the profile or could have potentially been lost through leaching while contributing minimally to the surface flux.

This study saw results similar to the literature, in which no-tillage significantly reduced N_2O emissions during the thaw by decreasing soil freezing and improving soil insulation. However, there is a knowledge gap surrounding the quantification of each N_2O production mechanism to the total surface flux and how this is altered under different management practices. For example, utilizing cover crops to absorb excess NO_3^- post-harvest may reduce the contribution of *de novo* denitrification to the N_2O surface flux event. In addition, research assessing the spatial and temporal variations in N_2O production mechanisms during spring thaw is limited. Advancement in this knowledge area is required to improve N_2O modeling and FTC emission factors.

3 Conclusion and future research trends

Optimizing productivity while reducing cropping systems' environmental footprint proves to be a complex challenge. Manipulations made to one part of the nitrogen cycle, such as increased fertilizer application or reduced volatilization rates, have a flow-on effect that must be understood and considered with all management changes to a system.

The rate, source, placement, and timing of nitrogen application to the soil system, alongside soil water content, are the dominant factors influencing the microbial, chemical, and physical processes that make up the agricultural nitrogen cycle. Understanding how these properties alter nitrogen fixation, mineralization, nitrification, and denitrification reactions creates a fundamental opportunity to identify critical knowledge gaps and potentially re-synchronize the nitrogen supply with crop demand.

Management practices such as urease and nitrification inhibitors, enhanced efficiency fertilizers, and precision agricultural tools provide an opportunity to improve the NUE of cropping systems while reducing nitrate leaching and nitrous oxide losses. However, further understanding and modeling of spatial and temporal changes in nitrogen cycling processes are required to appropriately allocate management practices at a field- or farm-based scale. Similarly, the complex interactions between soil and environmental properties that result in significant variations in nitrate and nitrous oxide production must be studied to assess the potential synergistic effects of how multiple management practices enhance NUE.

4 Where to look for further information

- A comprehensive introduction to nitrogen cycling processes is 'Mineral nitrogen in the plant-soil system', by Haynes, R. J. (1986), Academic Press, 482 pp.
- An overview of the nitrogen cycle and potential future concerns is *Nitrogen in the Environment*, 2nd edition, by Hatfield, J. L. and Follett, R. F. (2008), Academic Press, 702 pp.
- A study comparing fertilizer management practices' impact on NUE is 'Comparing soluble to controlled-release nitrogen fertilizers: storage

cabbage yield, profit margins, and N use efficiency', in *Canadian Journal of Plant Science* 98, 815–829, by Van Eerd, L. L., Turnbull, J. J. D., Bakker, C. J., Vyn, R. J., McKeown, A. W. and Westerveld, S. M. (2017).

- A review of nitrogen loss processes is 'Nitrogen losses from the soil/plant system: a review', in *Annals of Applied Biology* 162, 145–173 by Cameron, K. C., Di, H. J. and Moir, J. L. (2013).

5 References

Abdalla, M., Hastings, A., Cheng, K., Yue, Q., Chadwick, D., Espenberg, M., Truu, J., Rees, R. M. and Smith, P. 2019. A critical review of the impacts of cover crops on nitrogen leaching, net greenhouse gas balance and crop productivity. *Global Change Biology* 25(8), 2530–2543.

Akiyama, H., Yan, X. and Yagi, K. 2010. Evaluation of effectiveness of enhanced-efficiency fertilizers as mitigation options for N_2O and NO emissions from agricultural soils: meta-analysis. *Global Change Biology* 16(6), 1837–1846.

Argaw, A. and Muleta, D. 2018. Effect of genotypes-Rhizobium-environment interaction on nodulation and productivity of common bean (*Phaseolus vulgaris* L.) in eastern Ethiopia. *Environmental Systems Research* 6, 1–16.

Arp, D. J., Sayavedra-Soto, L. A. and Hommes, N. G. 2002. Molecular biology and biochemistry of ammonia oxidation by *Nitrosomonas europaea*. *Archives of Microbiology* 178(4), 250–255.

Bassirirad, H. 2000. Kinetics of nutrient uptake by roots: responses to global change. *New Phytologist* 147(1), 155–169.

Bassirirad, H., Caldwell, M. M. and Bilbrough, C. 1993. Effects of soil temperature and nitrogen status on kinetics of $^{15}NO_3^-$ uptake by roots of field-grown Agropyron desertorum (Fisch. ex Link) Schult. *New Phytologist* 123(3), 485–489.

Beauchemin, K. A., Ungerfeld, E. M., Eckard, R. J. and Wang, M. 2020. Review: fifty years of research on rumen methanogenesis: lessons learned and future challenges for mitigation. *Animal* 14(S1), s2–s16.

Benbi, D. K. and Richter, J. 2002. A critical review of some approaches to modelling nitrogen mineralization. *Biology and Fertility of Soils* 35(3), 168–183.

Bothe, H., Ferguson, S., Newton, W. E. and Newton, W. E. 2006. *Biology of the Nitrogen Cycle*. Elsevier Science.

Bouwman, A. F., Boumans, L. J. M. and Batjes, N. H. 2002. Emissions of N_2O and NO from fertilized fields: summary of available measurement data. *Global Biogeochemical Cycles* 16(4), 6–1.

Brin, L. D., Goyer, C., Zebarth, B. J., Burton, D. L. and Chantigny, M. H. 2018. Changes in snow cover alter nitrogen cycling and gaseous emissions in agricultural soils. *Agriculture, Ecosystems and Environment* 258, 91–103.

Cabrera, M. L., Kissel, D. E. and Vigil, M. F. 2005. Nitrogen mineralization from organic residues: research opportunities. *Journal of Environmental Quality* 34(1), 75–79.

Canfield, D. E., Glazer, A. N. and Falkowski, P. G. 2010. The evolution and future of Earth's nitrogen cycle. *Science* 330(6001), 192–196.

Caranto, J. D. and Lancaster, K. M. 2017. Nitric oxide is an obligate bacterial nitrification intermediate produced by hydroxylamine oxidoreductase. *Proceedings of the National Academy of Sciences of the United States of America* 114(31), 8217-8222.

Chahal, I., Baral, K. R., Congreves, K. A., Van Eerd, L. L. and Wagner-Riddle, C. 2021. Opportunities to reduce nitrous oxide emissions from horticultural production systems in Canada. *Canadian Journal of Plant Science* 101(6), 999-1013.

Chapin, F. S., Matson, P. A., Mooney, H. A. and Vitousek, P. M. 2002. *Principles of Terrestrial Ecosystem Ecology*. Springer.

Charpentier, J., Farias, L., Yoshida, N., Boontanon, N. and Raimbault, P. 2007. Nitrous oxide distribution and its origin in the central and eastern South Pacific Subtropical gyre. *Biogeosciences* 4(5), 729-741.

Chekanai, V., Chikowo, R. and Vanlauwe, B. 2018. Response of common bean (*Phaseolus vulgaris* L.) to nitrogen, phosphorus and rhizobia inoculation across variable soils in Zimbabwe. *Agriculture, Ecosystems and Environment* 266, 167-173.

Chen, R., Senbayram, M., Blagodatsky, S., Myachina, O., Dittert, K., Lin, X., Blagodatskaya, E. and Kuzyakov, Y. 2014. Soil C and N availability determine the priming effect: microbial N mining and stoichiometric decomposition theories. *Global Change Biology* 20(7), 2356-2367.

Cheng, W. 2009. Rhizosphere priming effect: its functional relationships with microbial turnover, evapotranspiration, and C-N budgets. *Soil Biology and Biochemistry* 41(9), 1795-1801.

Christianson, C. B., Baethgen, W. E., Carmona, G. and Howard, R. G. 1993. Microsite reactions of urea-nBTPT fertilizer on the soil surface. *Soil Biology and Biochemistry* 25(8), 1107-1117.

Cleve, K. V., Oechel, W. C. and Hom, J. L. 1990. Response of black spruce (*Picea mariana*) ecosystems to soil temperature modification in interior Alaska. *Canadian Journal of Forest Research* 20(9), 1530-1535.

Clough, T. J. 1994. Fate of urine nitrogen applied to peat and mineral soils from grazed pastures. (Doctoral Theses). Doctor of Philosophy in Soil Science Doctoral Lincoln University.

Congreves, K. A., Brown, S. E., Németh, D. D., Dunfield, K. E. and Wagner-Riddle, C. 2017. Differences in field-scale N_2O flux linked to crop residue removal under two tillage systems in cold climates. *GCB Bioenergy* 9(4), 666-680.

Costa, E., Pérez, J. and Kreft, J. U. 2006. Why is metabolic labour divided in nitrification? *Trends in Microbiology* 14(5), 213-219.

Craine, J. M., Morrow, C. and Fierer, N. 2007. Microbial nitrogen limitation increases decomposition. *Ecology* 88(8), 2105-2113.

Daims, H., Lebedeva, E. V., Pjevac, P., Han, P., Herbold, C., Albertsen, M., Jehmlich, N., Palatinszky, M., Vierheilig, J., Bulaev, A., Kirkegaard, R. H., von Bergen, M., Rattei, T., Bendinger, B., Nielsen, P. H. and Wagner, M. 2015. Complete nitrification by Nitrospira bacteria. *Nature* 528(7583), 504-509.

Daims, H., Lücker, S. and Wagner, M. 2016. A new perspective on microbes formerly known as nitrite-oxidizing bacteria. *Trends in Microbiology* 24(9), 699-712.

Davies-Barnard, T. and Friedlingstein, P. 2020. The global distribution of biological nitrogen fixation in terrestrial natural ecosystems. *Global Biogeochemical Cycles* 34(3), e2019GB006387.

De Boer, W. and Kowalchuk, G. A. 2001. Nitrification in acid soils: micro-organisms and mechanisms. *Soil Biology and Biochemistry* 33(7-8), 853-866.

Di, H. J. and Cameron, K. C. 2016. Inhibition of nitrification to mitigate nitrate leaching and nitrous oxide emissions in grazed grassland: a review. *Journal of Soils and Sediments* 16(5), 1401–1420.

Di, H. J., Cameron, K. C., Podolyan, A., Edwards, G. R., De Klein, C. A. M., Dynes, R. and Woods, R. 2016. The potential of using alternative pastures, forage crops and gibberellic acid to mitigate nitrous oxide emissions. *Journal of Soils and Sediments* 16(9), 2252–2262.

Cameron, K. C. and Di, H. J. 2002. The use of a nitrification inhibitor, dicyandiamide (DCD), to decrease nitrate leaching and nitrous oxide emissions in a simulated grazed and irrigated grassland. *Soil Use and Management* 18(4), 395–403.

Eggleston, H., Buendia, L., Miwa, K., Ngara, T. and Tanabe, K. 2006. *2006 IPCC Guidelines for National Greenhouse Gas Inventories*. Japan.

Erisman, J. W., Sutton, M. A., Galloway, J., Klimont, Z. and Winiwarter, W. 2008. How a century of ammonia synthesis changed the world. *Nature Geoscience* 1(10), 636–639.

Fageria, N. K. and Baligar, V. C. 2005. Enhancing nitrogen use efficiency in crop plants. *Advances in Agronomy* 88, 97–185.

Ferguson, B. J., Indrasumunar, A., Hayashi, S., Lin, M. H., Lin, Y. H., Reid, D. E. and Gresshoff, P. M. 2010. Molecular analysis of legume nodule development and autoregulation. *Journal of Integrative Plant Biology* 52(1), 61–76.

Fowler, D., Coyle, M., Skiba, U., Sutton, M. A., Cape, J. N., Reis, S., Sheppard, L. J., Jenkins, A., Grizzetti, B. and Galloway, J. N. 2013. The global nitrogen cycle in the twenty-first century. *Philosophical Transactions of the Royal Society of London Series B* 368, 20130164.

Freney, J., Simpson, J. and Denmead, O. 1981. Ammonia volatilization. *Ecological Bulletins*, 291–302.

Freney, J. R. and Simpson, J. R. (Eds.) 1983. *Gaseous Loss of Nitrogen from Plant-Soil Systems*. Martinus Nijhoff & Dr W. Junk Publishers.

Galloway, J. N., Leach, A. M., Bleeker, A. and Erisman, J. W. 2013. A chronology of human understanding of the nitrogen cycle. *Philosophical Transactions of the Royal Society of London. Series B, Biological Sciences* 368(1621), 20130120.

Gilsanz, C., Báez, D., Misselbrook, T. H., Dhanoa, M. S. and Cárdenas, L. M. 2016. Development of emission factors and efficiency of two nitrification inhibitors, DCD and DMPP. *Agriculture, Ecosystems and Environment* 216, 1–8.

Glass, A. D. M. 2003. Nitrogen use efficiency of crop plants: physiological constraints upon nitrogen absorption. *Critical Reviews in Plant Sciences* 22(5), 453–470.

Groffman, P. M., Butterbach-Bahl, K., Fulweiler, R. W., Gold, A. J., Morse, J. L., Stander, E. K., Tague, C., Tonitto, C. and Vidon, P. 2009. Challenges to incorporating spatially and temporally explicit phenomena (hotspots and hot moments) in denitrification models. *Biogeochemistry* 93(1–2), 49–77.

Guntiñas, M. E., Leirós, M. C., Trasar-Cepeda, C. and Gil-Sotres, F. 2012. Effects of moisture and temperature on net soil nitrogen mineralization: a laboratory study. *European Journal of Soil Biology* 48, 73–80.

Halvorson, A. D., Snyder, C. S., Blaylock, A. D. and Del Grosso, S. J. 2014. Enhanced-efficiency nitrogen fertilizers: potential role in nitrous oxide emission mitigation. *Agronomy Journal* 106(2), 715–722.

Han, S., Zeng, L., Luo, X., Xiong, X., Wen, S., Wang, B., Chen, W. and Huang, Q. 2018. Shifts in Nitrobacter-and Nitrospira-like nitrite-oxidizing bacterial communities under long-term fertilization practices. *Soil Biology and Biochemistry* 124, 118–125.

Harris, G. 2003. *Ecological Stoichiometry: Biology of Elements from Molecules to the Biosphere*. Princeton University Press.

Hauck, R. D. and Bremner, J. M. 1976. Use of tracers for soil and fertilizer nitrogen research. *Advances in Agronomy* 28, 219-266.

Hayatsu, M., Katsuyama, C. and Tago, K. 2021. Overview of recent researches on nitrifying microorganisms in soil. *Soil Science and Plant Nutrition* 67(6), 619-632.

Haynes, R. J. 1986. *Mineral Nitrogen in the Plant-Soil System*. Academic Press.

Herridge, D. F., Peoples, M. B. and Boddey, R. M. 2008. Global inputs of biological nitrogen fixation in agricultural systems. *Plant and Soil* 311(1-2), 1-18.

Hu, H.-W. and He, J.-Z. 2017. Comammox–a newly discovered nitrification process in the terrestrial nitrogen cycle. *Journal of Soils and Sediments* 17(12), 2709-2717.

Hu, H.-W., Zhang, L.-M., Dai, Y., Di, H.-J. and He, J.-Z. 2013. PH-dependent distribution of soil ammonia oxidizers across a large geographical scale as revealed by high-throughput Pyrosequencing. *Journal of Soils and Sediments* 13(8), 1439-1449.

Janssen, B. 1996. Nitrogen mineralization in relation to C:N ratio and decomposability of organic materials. *Progress in Nitrogen Cycling Studies*. Springer.

Jensen, E. S. and Hauggaard-Nielsen, H. 2003. How can increased use of biological N_2 fixation in agriculture benefit the environment? *Plant and Soil* 252(1), 177-186.

Kaye, J. P. and Quemada, M. 2017. Using cover crops to mitigate and adapt to climate change. A review. *Agronomy for Sustainable Development* 37(1), 4.

Kebede, E. 2021. Contribution, utilization, and improvement of legumes-driven biological nitrogen fixation in agricultural systems. *Frontiers in Sustainable Food Systems* 5, 1-16.

Keller, G. D. and Mengel, D. B. 1986. Ammonia volatilization from nitrogen fertilizers surface applied to no-till corn. *Soil Science Society of America Journal* 50(4), 1060-1063.

Khan, M. and Scullion, J. 2002. Effects of metal (Cd, Cu, Ni, Pb or Zn) enrichment of sewage-sludge on soil micro-organisms and their activities. *Applied Soil Ecology* 20(2), 145-155.

Khosro, M., Yousef, S., Gholamreza, H., Shiva, K. and Mohammad, M. 2012. Effective factors on biological nitrogen fixation. *African Journal of Agricultural Research* 7, 1782-1788.

Kim, D.-G., Vargas, R., Bond-Lamberty, B. and Turetsky, M. R. 2012. Effects of soil rewetting and thawing on soil gas fluxes: a review of current literature and suggestions for future research. *Biogeosciences* 9(7), 2459-2483.

King, A. E., Rezanezhad, F. and Wagner-Riddle, C. 2021. Evidence for microbial rather than aggregate origin of substrates fueling freeze-thaw induced N2O emissions. *Soil Biology and Biochemistry* 160, 108352.

Kirchmann, H., Johnston, A. E. and Bergström, L. F. 2002. Possibilities for reducing nitrate leaching from agricultural land. *AMBIO: A Journal of the Human Environment* 31(5), 404-408.

Kool, D. M., Dolfing, J., Wrage, N. and Van Groenigen, J. W. 2011. Nitrifier denitrification as a distinct and significant source of nitrous oxide from soil. *Soil Biology and Biochemistry* 43(1), 174-178.

Kozlowski, J. A., Price, J. and Stein, L. Y. 2014. Revision of N_2O-producing pathways in the ammonia-oxidizing bacterium Nitrosomonas europaea ATCC 19718. *Applied and Environmental Microbiology* 80(16), 4930-4935.

Laine, P., Ourry, A., Macduff, J., Boucaud, J. and Salette, J. 1993. Kinetic parameters of nitrate uptake by different catch crop species: effects of low temperatures or previous nitrate starvation. *Physiologia Plantarum* 88, 85-92.

Lapierre, J., Machado, P. V. F., Debruyn, Z., Brown, S. E., Jordan, S., Berg, A., Biswas, A., Henry, H. A. L. and Wagner-Riddle, C. 2022. Cover crop mixtures: A powerful strategy to reduce post-harvest surplus of soil nitrate and leaching. *Agriculture, Ecosystems and Environment* 325, 107750.

Li, W., Wu, J., Bai, E., Guan, D., Wang, A., Yuan, F., Wang, S. and Jin, C. 2016. Response of terrestrial nitrogen dynamics to snow cover change: a meta-analysis of experimental manipulation. *Soil Biology and Biochemistry* 100, 51-58.

Li, Y., Liu, Y., Wang, Y., Niu, L., Xu, X. and Tian, Y. 2014. Interactive effects of soil temperature and moisture on soil N mineralization in a Stipa krylovii grassland in Inner Mongolia, China. *Journal of Arid Land* 6(5), 571-580.

Li, Z., Tian, D., Wang, B., Wang, J., Wang, S., Chen, H. Y. H., Xu, X., Wang, C., He, N. and Niu, S. 2019. Microbes drive global soil nitrogen mineralization and availability. *Global Change Biology* 25(3), 1078-1088.

Linquist, B. A., Liu, L., Van Kessel, C. and Van Groenigen, K. J. 2013. Enhanced efficiency nitrogen fertilizers for rice systems: meta-analysis of yield and nitrogen uptake. *Field Crops Research* 154, 246-254.

Liu, T., Liang, Y. and Chu, G. 2017. Nitrapyrin addition mitigates nitrous oxide emissions and raises nitrogen use efficiency in plastic-film-mulched drip-fertigated cotton field. *PLoS ONE* 12(5), e0176305.

Lugtenberg, B. 2016. *Principles of Plant-Microbe Interactions*. Springer.

Ma, B. L., Dwyer, L. M. and Gregorich, E. G. 1999. Soil nitrogen amendment effects on seasonal nitrogen mineralization and nitrogen cycling in maize production. *Agronomy Journal* 91(6), 1003-1009.

Mahal, N. K., Osterholz, W. R., Miguez, F. E., Poffenbarger, H. J., Sawyer, J. E., Olk, D. C., Archontoulis, S. V. and Castellano, M. J. 2019. Nitrogen fertilizer suppresses mineralization of soil organic matter in maize agroecosystems. *Frontiers in Ecology and Evolution* 7, 59.

Mantovi, P., Fumagalli, L., Beretta, G. P. and Guermandi, M. 2006. Nitrate leaching through the unsaturated zone following pig slurry applications. *Journal of Hydrology* 316(1-4), 195-212.

Mclaren, R. and Cameron, K. 1996. *Soil Science: Sustainable Production and Environmental Protection*. Oxford University Press.

Melkonian, J., Van Es, H., Degaetano, A., Joseph, L. and ADAPT, N. 2008. Adaptive nitrogen management for maize using high-resolution climate data and model simulations. In: *Proceedings of the 9th International Conference on Precision Agriculture*. Denver, CO: International Society of Precision Agriculture, 20-23.

Mesfin, S., Gebresamuel, G., Haile, M., Zenebe, A. and Desta, G. 2020. Mineral fertilizer demand for optimum biological nitrogen fixation and yield potentials of legumes in Northern Ethiopia. *Sustainability* 12(16), 6449.

Misselbrook, T. H., Cardenas, L. M., Camp, V., Thorman, R. E., Williams, J. R., Rollett, A. J. and Chambers, B. J. 2014. An assessment of nitrification inhibitors to reduce nitrous oxide emissions from UK agriculture. *Environmental Research Letters* 9(11), 1-11.

Moorhead, D. L. and Sinsabaugh, R. L. 2006. A theoretical model of litter decay and microbial interaction. *Ecological Monographs* 76(2), 151-174.

Mooshammer, M., Wanek, W., Hämmerle, I., Fuchslueger, L., Hofhansl, F., Knoltsch, A., Schnecker, J., Takriti, M., Watzka, M. and Wild, B. 2014. Adjustment of microbial nitrogen use efficiency to carbon: nitrogen imbalances regulates soil nitrogen cycling. *Nature Communications* 5, 1-7.

Myers, S. S., Smith, M. R., Guth, S., Golden, C. D., Vaitla, B., Mueller, N. D., Dangour, A. D. and Huybers, P. 2017. Climate change and global food systems: potential impacts on food security and undernutrition. *Annual Review of Public Health* 38, 259-277.

Németh, D. D., Wagner-Riddle, C. and Dunfield, K. E. 2014. Abundance and gene expression in nitrifier and denitrifier communities associated with a field scale spring thaw N_2O flux event. *Soil Biology and Biochemistry* 73, 1-9.

Norman, J. S. and Friesen, M. L. 2017. Complex N acquisition by soil diazotrophs: how the ability to release exoenzymes affects N fixation by terrestrial free-living diazotrophs. *The ISME Journal* 11(2), 315-326.

Norton, J. and Ouyang, Y. 2019. Controls and adaptive management of nitrification in agricultural soils. *Frontiers in Microbiology* 10, 1931.

Ouyang, Y., Norton, J. M. and Stark, J. M. 2017. Ammonium availability and temperature control contributions of ammonia oxidizing bacteria and archaea to nitrification in an agricultural soil. *Soil Biology and Biochemistry* 113, 161-172.

Pachauri, R. K., Allen, M. R., Barros, V. R., Broome, J., Cramer, W., Christ, R., Church, J. A., Clarke, L., Dahe, Q. and Dasgupta, P. 2014. *2014: Synthesis Report. Contribution of Working Groups I, II and III to the Fifth Assessment Report of the Intergovernmental Panel on Climate Change*. Geneva, Switzerland: IPCC.

Padilla, F. M., Gallardo, M. and Manzano-Agugliaro, F. 2018. Global trends in nitrate leaching research in the 1960-2017 period. *Science of the Total Environment* 643, 400-413.

Peoples, M. B., Herridge, D. F. and Ladha, J. K. 1995. *Biological Nitrogen Fixation: An Efficient Source of Nitrogen for Sustainable Agricultural Production? Management of Biological Nitrogen Fixation for the Development of More Productive and Sustainable Agricultural Systems*. Springer.

Perego, A., Basile, A., Bonfante, A., De Mascellis, R., Terribile, F., Brenna, S. and Acutis, M. 2012. Nitrate leaching under maize cropping systems in Po valley (Italy). *Agriculture, Ecosystems and Environment* 147, 57-65.

Pereira, P. A. A. and Bliss, F. A. 1989. Selection of common bean (*Phaseolus vulgaris* L.) for N_2 fixation at different levels of available phosphorus under field and environmentally-controlled conditions. *Plant and Soil* 115(1), 75-82.

Pessarakli, M. 2019. *Handbook of Plant and Crop Stress*. CRC Press.

Prosser, J. I. and Nicol, G. W. 2012. Archaeal and bacterial ammonia-oxidisers in soil: the quest for niche specialisation and differentiation. *Trends in Microbiology* 20(11), 523-531.

Ravishankara, A. R., Daniel, J. S. and Portmann, R. W. 2009. Nitrous oxide (N_2O): the Dominant Ozone-Depleting Substance Emitted in the 21st Century. *Science* 326(5949), 123-125.

Ray, D. and Daniel, K. 2004. Genetic regulation of biological nitrogen fixation. *Nature Reviews in Microbiology* 2, 621-631.

Reinprecht, Y., Schram, L., Marsolais, F., Smith, T. H., Hill, B. and Pauls, K. P. 2020. Effects of nitrogen application on nitrogen fixation in common bean production. *Frontiers in Plant Science* 11, 1172.

Rice, W. A., Penney, D. C. and Nyborg, M. 1977. Effects of soil acidity on rhizobia numbers, nodulation and nitrogen fixationby alfalfa and red clover. *Canadian Journal of Soil Science* 57(2), 197-203.

Risk, N., Snider, D. and Wagner-Riddle, C. 2013. Mechanisms leading to enhanced soil nitrous oxide fluxes induced by freeze-thaw cycles. *Canadian Journal of Soil Science* 93(4), 401-414.

Robertson, G. P. and Groffman, P. M. 2015. Nitrogen transformations. In: Paul, E. A. (Ed.). *Soil Microbiology, Ecology and Biochemistry* (4th edn.). Academic Press.

Rochette, P., Macdonald, J. D., Angers, D. A., Chantigny, M. H., Gasser, M. O. and Bertrand, N. 2009. Banding of urea increased ammonia volatilization in a dry acidic soil. *Journal of Environmental Quality* 38(4), 1383–1390.

Rosenblueth, M., Ormeño-Orrillo, E., López-López, A., Rogel, M. A., Reyes-Hernández, B. J., Martínez-Romero, J. C., Reddy, P. M. and Martínez-Romero, E. 2018. Nitrogen fixation in cereals. *Frontiers in Microbiology* 9, 1794.

Rowell, D. M., Prescott, C. E. and Preston, C. M. 2001. Decomposition and nitrogen mineralization from biosolids and other organic materials: relationship with initial chemistry. *Journal of Environmental Quality* 30(4), 1401–1410.

Ruan, L., Robertson, G. P., Robertson, G. P. and Robertson, G. P. 2017. Reduced snow cover increases wintertime nitrous oxide (N2O) emissions from an agricultural soil in the upper U.S. Midwest. *Ecosystems* 20(5), 917–927.

Sacco, D., Bassanino, M. and Grignani, C. 2003. Developing a regional agronomic information system for estimating nutrient balances at a larger scale. *European Journal of Agronomy* 20(1–2), 199–210.

Samad, M. S., Bakken, L. R., Nadeem, S., Clough, T. J., De Klein, C. A., Richards, K. G., Lanigan, G. J. and Morales, S. E. 2016. High-resolution denitrification kinetics in pasture soils link N_2O emissions to pH, and denitrification to C mineralization. *PLoS ONE* 11(3), e0151713.

Scheer, C., Fuchs, K., Pelster, D. E. and Butterbach-Bahl, K. 2020. Estimating global terrestrial denitrification from measured N2O:(N2O+ N2) product ratios. *Current Opinion in Environmental Sustainability* 47, 72–80.

Schindler, D. W. 2006. Recent advances in the understanding and management of eutrophication. *Limnology and Oceanography* 51(1part2), 356–363.

Sela, S., Van Es, H. M., Moebius-Clune, B. N., Marjerison, R., Melkonian, J., Moebius-Clune, D., Schindelbeck, R. and Gomes, S. 2016. Adapt-N outperforms grower-selected nitrogen rates in Northeast and Midwestern United States strip trials. *Agronomy Journal* 108(4), 1726–1734.

Sharpe, R. R. and Harper, L. 1995. Soil, plant and atmospheric conditions as they relate to ammonia volatilization. In: *Nitrogen Economy in Tropical Soils*. Springer.

Shcherbak, I., Millar, N. and Robertson, G. P. 2014. Global metaanalysis of the nonlinear response of soil nitrous oxide (N_2O) emissions to fertilizer nitrogen. *Proceedings of the National Academy of Sciences of the United States of America* 111(25), 9199–9204.

Shepherd, M., Welten, B., Wyatt, J. and Balvert, S. 2014. Precipitation but not soil texture alters effectiveness of dicyandiamide to decrease nitrate leaching from dairy cow urine. *Soil Use and Management* 30(3), 361–371.

Sherlock, R. R. 1984. Dynamics of ammonia volatilization and nitrous oxide production from urine patches in grazed pastures. (Doctoral Theses). Lincoln College, University of Canterbury.

Skorupka, M. and Nosalewicz, A. 2021. Ammonia volatilization from fertilizer urea–a new challenge for agriculture and industry in view of growing global demand for food and energy crops. *Agriculture* 11(9), 822.

Sprent, J. I., Ardley, J. and James, E. K. 2017. Biogeography of nodulated legumes and their nitrogen-fixing symbionts. *New Phytologist* 215(1), 40–56.

Tamagno, S., Sadras, V. O., Haegele, J. W., Armstrong, P. R. and Ciampitti, I. A. 2018. Interplay between nitrogen fertilizer and biological nitrogen fixation in soybean: implications on seed yield and biomass allocation. *Scientific Reports* 8(1), 17502.

Tate, R. L. 2020. *Soil Microbiology*. John Wiley & Sons.

Teepe, R., Brumme, R. and Beese, F. 2001. Nitrous oxide emissions from soil during freezing and thawing periods. *Soil Biology and Biochemistry* 33(9), 1269–1275.

Tindaon, F., Benckiser, G. and Ottow, C. G. 2013. Side effects of nitrification inhibitors on non target microbial processes in soils. *Journal of Tropical Soils* 16(1), 7–16.

Tosi, M., Brown, S., Machado, P. V. F., Wagner-Riddle, C. and Dunfield, K. 2020. Short-term response of soil N-cycling genes and transcripts to fertilization with nitrification and urease inhibitors, and relationship with field-scale N_2O emissions. *Soil Biology and Biochemistry* 142, 107703.

Udvardi, M., Below, F. E., Castellano, M. J., Eagle, A. J., Giller, K. E., Ladha, J. K., Liu, X., Maaz, T. M., Nova-Franco, B., Raghuram, N., Robertson, G. P., Roy, S., Saha, M., Schmidt, S., Tegeder, M., York, L. M. and Peters, J. W. 2021. A research road map for responsible use of agricultural nitrogen. *Frontiers in Sustainable Food Systems* 5, 660155.

Udvardi, M. K. and Day, D. A. 1997. Metabolite transport across symbiotic membranes of legume nodules. *Annual Review of Plant Physiology and Plant Molecular Biology* 48, 493–523.

Van Kessel, M. A., Speth, D. R., Albertsen, M., Nielsen, P. H., Op Den Camp, H. J., Kartal, B., Jetten, M. S. and Lücker, S. 2015. Complete nitrification by a single microorganism. *Nature* 528(7583), 555–559.

Varvel, G. E. and Peterson, T. A. 1990. Nitrogen fertilizer recovery by corn in monoculture and rotation systems. *Agronomy Journal* 82(5), 935–938.

Vitousek, P. M., Menge, D. N., Reed, S. C. and Cleveland, C. C. 2013. Biological nitrogen fixation: rates, patterns and ecological controls in terrestrial ecosystems. *Philosophical Transactions of the Royal Society of London. Series B, Biological Sciences* 368(1621), 20130119.

Vlassak, K. and Reynders, L. 1979. Factors effecting biological dinitrogen fixation by associative symbiosis in temperate regions. *Isotopes and Radiation in Research on Soil-Plant Relationships* 11, 137–146.

Wagner-Riddle, C., Baggs, E. M., Clough, T. J., Fuchs, K. and Petersen, S. O. 2020. Mitigation of nitrous oxide emissions in the context of nitrogen loss reduction from agroecosystems: managing hot spots and hot moments. *Current Opinion in Environmental Sustainability* 47, 46–53.

Wagner-Riddle, C., Congreves, K. A., Abalos, D., Berg, A. A., Brown, S. E., Ambadan, J. T., Gao, X. and Tenuta, M. 2017. Globally important nitrous oxide emissions from croplands induced by freeze–thaw cycles. *Nature Geoscience* 10(4), 279–283.

Wagner-Riddle, C., Hu, Q. C., Van Bochove, E. and Jayasundara, S. 2008. Linking nitrous oxide flux during spring thaw to nitrate denitrification in the soil profile. *Soil Science Society of America Journal* 72(4), 908–916.

Wagner, S. 2011. Biological nitrogen fixation. *Nature Education Knowledge* 3, 15.

Wan, X., Wu, W. and Liao, Y. 2021. Mitigating ammonia volatilization and increasing nitrogen use efficiency through appropriate nitrogen management under supplemental irrigation and rain-fed condition in winter wheat. *Agricultural Water Management* 255,10750–10761.

Wang, G., Chen, X., Cui, Z., Yue, S. and Zhang, F. 2014. Estimated reactive nitrogen losses for intensive maize production in China. *Agriculture, Ecosystems and Environment* 197, 293–300.

Wang, Z., Li, Q., Li, M. and Li, F. 2013. Effects of nitrogen application rate on winter wheat yield and soil nitrate nitrogen during summer fallow season on dryland. *Acta Pedologica Sinica* 50, 956–965.

Wang, Z.-H. and Li, S.-X. 2019. Nitrate N loss by leaching and surface runoff in agricultural land: a global issue (a review). *Advances in Agronomy* 156, 159–217.

Weitzman, J. N., Groffman, P. M., Adler, P. R., Dell, C. J., Johnson, F. E., Lerch, R. N. and Strickland, T. C. 2021. Drivers of hot spots and hot moments of denitrification in agricultural systems. *Journal of Geophysical Research: Biogeosciences* 126(7), 1-18.

Wen, J., Lechevallier, M. W. and Tao, W. 2020. Nitrification kinetics and microbial communities of activated sludge as a full-scale membrane bioreactor plant transitioned to low dissolved oxygen operation. *Journal of Cleaner Production* 252, 119872.

Woodward, E. E., Hladik, M. L. and Kolpin, D. W. 2016. Nitrapyrin in Streams: the first study documenting off-field transport of a nitrogen stabilizer compound. *Environmental Science and Technology Letters* 3(11), 387–392.

Yang, J. Y., Drury, C. F., De Jong, R., Huffman, E. C., Yang, X. M. and Reid, K. 2013. Sensitivity analysis for nitrogen inputs, nitrogen outputs, and changes in biofuel crop acreages for predicting residual soil nitrogen and nitrate leaching in Canadian agricultural soils. *Ecological Modelling* 267, 26–38.

Zeng, Q., Ding, X., Wang, J., Han, X., Iqbal, H. M. N. and Bilal, M. 2022. Insight into soil nitrogen and phosphorus availability and agricultural sustainability by plant growth-promoting rhizobacteria. *Environmental Science and Pollution Research International* 29(30), 45089–45106.

Zhang, S., Zheng, Q., Noll, L., Hu, Y. and Wanek, W. 2019. Environmental effects on soil microbial nitrogen use efficiency are controlled by allocation of organic nitrogen to microbial growth and regulate gross N mineralization. *Soil Biology and Biochemistry* 135, 304–315.

Zhang, X., Davidson, E. A., Mauzerall, D. L., Searchinger, T. D., Dumas, P. and Shen, Y. 2015. Managing nitrogen for sustainable development. *Nature* 528(7580), 51–59.

Zheng, M., Zhou, Z., Luo, Y., Zhao, P. and Mo, J. 2019. Global pattern and controls of biological nitrogen fixation under nutrient enrichment: a meta-analysis. *Global Change Biology* 25(9), 3018–3030.

Chapter 2

The role of ammonium transport proteins in improving nitrogen use efficiency in crop production

Muhammad K. Uddin, Francine Perrine-Walker and Brent N. Kaiser[1]*, University of Sydney, Australia*

1 Introduction

Plants depend on nitrogen (N) for growth, development and reproduction (Tegeder and Masclaux-Daubresse 2018). In an agricultural context, soil N availability becomes a vital factor in any successful crop production cycle (Barker and Bryson 2016). N deficiencies quickly restrict plant growth and reduce potential crop yields (Lemaire and Gastal 2019). Significant quantities of synthetic N fertilisers are applied annually (120 million tonnes per annum; http://www.fao.org/faostat/) to achieve economic yields. The substantial use of N fertilisers does have consequences on agricultural systems and the planet, including a significant cost burden to growers and a considerable liability on

[1] Correspondence: brent.kaiser@sydney.edu.au.

http://dx.doi.org/10.19103/AS.2024.0135.05

the environment through N-related pollution (Lal and Stewart 2018; Lee et al., 2019). A balance in crop production and agronomic practices that minimises N fertiliser requirements while retaining yield and quality remains a global priority (McLellan et al., 2018).

Biological nitrogen fixation (BNF) is an alternative N source to commercial synthetic N fertilisers (Herridge et al., 2008; Vitousek et al., 2013). BNF involves the energy-intensive reduction of atmospheric nitrogen (N_2) to ammonia (NH_3) through a bacterial enzyme called nitrogenase (Seefeldt et al., 2018). Most members of the Papilionaceae family (i.e. legumes) can form a symbiotic partnership with rhizobia bacteria found in the soil (Babalola et al., 2017). When in symbiosis with legumes, rhizobia express the enzyme nitrogenase that can reduce atmospheric N_2 to NH_3, a form of N readily assimilated by the plant. In return, the plant provides carbon that is respired by the symbiotic bacteria to produce energy in the form of adenosine triphosphate (ATP) used for nitrogenase activity (Foyer et al., 2019). Through this activity, legume N_2 fixation provides a source of reduced N that will allow the plant to grow and produce seed (Tegeder and Masclaux-Daubresse 2018). It has been estimated that N_2-fixing legumes contribute about 200 million tons of reactive N per annum (Hunt 1997; Unkovich et al., 2008). This nitrogen advantage is a significant benefit to the overall farming system and a direct path towards improving farm-based agricultural sustainability. The role of legumes in sustainable agriculture is well documented (Herridge et al., 2008; Biswas and Gresshoff 2014; Atienza and Rubiales 2017; Peoples 2017; Stagnari et al., 2017; Meena and Lal 2018) where they play a key part in the overall cropping system when used as a rotation crop or when intercropped with non-legume plants (Blair et al., 2016). Crop rotations are useful to break disease cycles, manage weeds and in the case of legumes, provide soil N inputs from N_2 fixation that enhances growth and yield of follow on non-legume crops (Foyer et al., 2016). In some cases, cereals display greater N use efficiencies after a legume crop than when supplied annual N fertilisers (Lassaletta et al., 2014). N management by legumes involves conflicting processes, a pathway which caters to the symbiotic pressure to develop a N_2-fixing symbiosis (indirect N acquisition through nodules) and a direct N acquisition pathway, which involves N acquisition strategies involving root N uptake. The plant balances the two, relative to need and the underlying cost (energy) of the N acquisition processes.

Legume N_2 fixation involves a symbiosis with N_2-fixing rhizobia across a wide range of different species. In general, rhizobia enter legume roots through a complex signal-exchange process, resulting in the formation of a specialised structure called the root nodule. Within the nodule, replicated and modified bacteria (bacteroids) are housed inside cortical cells (Geddes and Oresnik 2016) from where biological nitrogen fixation takes place. The nodule structure provides a compatible environment for the rhizobia with a low oxygen environment maintained by bacterial respiration and through the delivery of oxygen to bacteroids by the oxygen-binding plant protein, leghemoglobin (Singh and Varma 2017). This low oxygen environment is required by the oxygen-labile protein nitrogenase to fix atmospheric N_2 (Oldroyd et al., 2011; Geddes and Oresnik 2016). N_2 fixation is an energy-intensive process that requires 16 ATP, $8H^+$ and $8e^-$ to produce $2NH_3$ from one molecule of N_2. Plants supply reduced carbon and essential nutrients to rhizobia for growth and the ability to reduce N_2 (Udvardi and Day 1997; Hoffman et al., 2014). The plant symbiont is thus considered the principal regulator of the symbiosis (Ferguson et al., 2019). A number of studies have suggested the control of nutrient supply (i.e. branched-chain amino acids, the production of homocitrate (FEN1) and cysteine-rich peptides) as strong evidence of established plant-level controls and potential bacterial dependency on the symbiosis (Hakoyama et al., 2009; Prell et al., 2009; Van de Velde et al., 2010).

2 The beginning: setting up the nodule nitrogen-fixing system

Rhizobial entry into legume roots is initiated with the secretion of flavonoids from plant roots into the rhizosphere (Liu et al., 2019). Flavonoids are low-molecular-weight secondary metabolites synthesised in legume roots that can act as an attractant to select bacteria (Liu and Murray 2016). Plants which grow under low N conditions tend to secrete flavonoids to attract N_2-fixing diazotrophs to legume roots, this includes *Rhizobium* sp. (Mierziak et al., 2014). In legumes, flavonoid release is used to attract complementary rhizobia (Wang et al., 2012; Coronado et al., 1995; Liu et al., 2019), which, when successful, initiates the transcriptional activation of bacterial *NodD* genes that result in the production and secretion of lipo-chitooligosaccharides (LCOs), commonly named Nod factor (NF) (Oldroyd 2013). The release of NF and recognition by nearby roots initiates a signalling pathway that enables rhizobia infection (Mabood et al., 2014). The signal transduction pathways leading to nodule initiation have been mostly investigated using the model species, *Lotus japonicus* and *Medicago truncatula* (Oldroyd and Downie 2008; Oldroyd 2013; Ferguson et al., 2010; Madsen et al., 2010; Biswas and Gresshoff 2014; Downie 2014; Baral et al., 2016). In brief, nodulation signalling begins with compatible bacterial NF perceived by the *LysM* (lysin motif) domain of *NFR1/NFP* and *NFR5/LYK3* (NF receptor kinase) on the plasma membrane (PM) of root hairs (Limpens et al., 2003; Madsen et al., 2003; Radutoiu et al., 2003; Smit et al., 2007). A leucine-rich receptor (LRR), called *SYMRK/DMI2* (Ané et al., 2004; Stracke et al., 2002) acts as a co-receptor downstream of the *LysM* domain. *SYMRK/DMI2* kinase interacts with 3-hydroxy-3-methylglutaryl-CoA-reductase (*HMG-CoA-R*), which is responsible for mevalonate biosynthesis (Kevei et al., 2007; Venkateshwaran et al., 2015). Venkateshwaran et al. (2015) suggested that mevalonate, or a related metabolite, acts as a secondary signal to guide NF perception to the nucleus. This secondary signal assumed to be responsible for the opening of the potassium channel *CASTOR/POLLUX/DMI1* (Ané et al., 2004; Charpentier et al., 2008; Imaizumi-Anraku et al., 2005) and nuclear envelope localised calcium channels leading to nuclear Ca^{2+} oscillation (Oldroyd and Long 2003). Recently three cyclic nucleotide-gated channels have been identified which forms a complex with K^+ channels for symbiotic Ca^{2+} oscillation in *M. truncatula* (Charpentier et al., 2016). *MCA8*, an ATPase pump (Capoen et al., 2011) transports Ca^{2+} back into the lumen creating a repeated opening and closure of Ca^{2+} and K^+ channels, a phenomenon known as Ca^{2+} spiking. Three nucleoporins, *NENA*, *NUP133* and *NUP85* are also required for Ca^{2+} spiking (Groth et al., 2010; Kanamori et al., 2006; Saito et al., 2007). Ca^{2+} spiking activates a calcium and calmodulin-dependent protein kinase (*CCaMK/DMI3*) (Horváth et al., 2011; Lévy et al., 2004) that phosphorylates the transcription factor, *CYCLOPS/IPD3* (Messinese et al., 2007; Ovchinnikova et al., 2011; Yano et al., 2008). *CYCLOPS/IPD3* is thought to bind with the GRAS domains, NSP1 and NSP2 (Nodulation signal pathway) (Kaló et al., 2005; Smit et al., 2005) to activate the transcription factors, *NIN* (nodule inception protein) (Marsh et al., 2007) and *ERN1* (ethylene response for nodulin1) (Cerri et al., 2012; Middleton et al., 2007). *NIN* activates the nodulation gene (*ENOD*) in the nucleus, leading to bacterial infection in the epidermis (Catoira et al., 2000; Oldroyd and Downie 2008; Oldroyd and Long 2003). The activation of the *CCaMK-CYCLOPS* signal transduction pathway involves cytokinins perceived by the *LHK1/CRE1* receptor (Eckardt 2006; Tirichine et al., 2007). The cytokinin signalling pathway involves the response regulator (*RR1*, *RR4*) and transcription factors (*NSP1*, *NSP2* and *NIN*) which eventually suppress polar auxin transport to initiate the nodule primordia (Gonzalez-Rizzo et al., 2006).

Nod factor perception and signal transduction events initiate entry of rhizobia which are either entrapped in the pocket of a curling root hair tip or penetrate directly through permeable cracks within the endodermal cell wall (Perret et al., 2000; Murray 2011). Localised NF secretion by attached bacteria polarises the root hair cell, resulting in root hair swelling and or infection pocket formation (Laplaze et al., 2015). Root hair infection varies with plant species (Laplaze et al., 2015; Turgeon and Bauer 1985) but in general, an invagination of cell PM proceeds resulting in the formation of an infection thread (Fournier et al., 2015; Gage 2004; Liu et al., 2019; Murray 2011). The infection thread is a tube-like structure full of multiplying bacteria that grow towards the de-differentiating cortex eventually ramifying cortical cells (Liu et al., 2019; Oldroyd et al., 2011). During the growth and branching of the infection thread, bacteria remain isolated from the host cell. After reaching the developing nodule primordia, infection thread releases the bacteria inside an endocytotic vesicle 'droplet' into the cortical cell cytosol (Verma and Hong 1996; Crespi and Frugier 2008; Ferguson et al., 2010; Laplaze et al., 2015). Once in the cytosol, the vesicles are commonly known as symbiosomes consisting of the symbiosome membrane (inside out plasma membrane), bacteroid and intervening peribacteroid space (Whitehead and Day 1997; Emerich and Krishnan 2014; Noor et al., 2015; Coba de la Peña et al., 2018). Symbiosomes differentiate with changes in the SM involving a further exocytosis process that directs new protein cargoes to the developing symbiosomes (Ivanov et al., 2012; Gavrin et al., 2014). In an infected nodule cortical cell, up to 25 000 symbiosomes may be present which may contain up to 1–20 mature bacteroids per symbiosome depending on the plant species (Bergersen 1982; Emerich and Krishnan 2014). After infection, cells containing symbiosomes are referred to as infected cells. Depending on the legume symbiosis, bacteria stop dividing or continue to multiply (e.g. soybean) before finally differentiating into a N_2-fixing bacteroid (Perret et al., 2000; Downie 2014; Alaswad et al., 2019). Infected cells are normally larger than non-infected cells and connected via plasmodesmata pores. The infected region is surrounded by a layer of endodermal cortical cells which ends up with a layer of vascular cells that provides metabolites to the nodule and a route for exchange with the root and shoot (Downie 2014).

3 Transport activities of the symbiosome membrane

Bacteroids depend entirely on the plant for nutrients and particularly reduced carbon to generate ATP for their respirational needs (Day et al., 1995; Clarke et al., 2014, 2015). C_4-dicarboxylates are the predominant product supplied by the plant to bacteroids (Udvardi and Day 1997). The SM is the effective gatekeeper regulating all nutrient exchange between bacteroids and the plant and vice versa, including N release to support plant growth. Several transport systems on the SM have been characterised biochemically, but only a few transporters characterised genetically (Clarke et al., 2014; Noor et al., 2015). Biochemically characterised transporters include a C_4-dicarboxylate transporter (Udvardi et al., 1988; Udvardi and Day 1997); P-type H^+-ATPase (Blumwald et al., 1985; Udvardi and Day 1989; Fedorova et al., 1999); Ca^{2+}-ATPase (Andreev et al., 1999; Krylova et al., 2013); Fe^{2+} transporter (Moreau et al., 1998; Moreau et al., 1995); Fe^{3+} transporter (LeVier et al., 1996); aquaporin (*Nod26*) (Dean et al., 1999; Rivers et al., 1997; Niemietz and Tyerman 2000) and a voltage-gated cation channel for ammonium (NH_4^+), potassium and sodium (Mouritzen and Rosendahl 1997; Roberts and Tyerman 2002; Tyerman et al., 1995). SM transporters characterised at the molecular level, include *GmZIP1*, a Zn^{2+} transporter (Moreau et al., 2002); Fe^{2+} transporters *GmDMT1* and *GmVTL1* (Kaiser et al., 2003; Brear et al., 2020); *LjSST1*, an SO_4^- transporter

(Krusell et al., 2005); *N70*, a putative NO_3^- transporter (Vincill et al., 2005) and a branched-chain amino acid transporter (Prell et al., 2009). Other nutrients like phosphorus, manganese, cobalt, copper, nickel, molybdenum and selenium are also essential for bacteroids but have not been identified or characterised on the PBM (Clarke et al., 2014; Noor et al., 2015; Udvardi and Poole 2013).

4 Nitrogen fixation and nitrogen delivery in nodules

Most of the NH_3 generated by nitrogenase is transferred to the PBS by diffusion following concentration gradients established by an expected NH_3 concentration in bacteroids of greater than 12 mM and a pHΔ between the bacteroid (alkaline) and the peribacteroid space (acidic) (Streeter 1989; Day et al., 2001). It is assumed the elevated ammonium concentrations in bacteroids represses the high-affinity ammonium transporter AMTB (influx and efflux) activity preventing recycling of fixed N to support bacteroid metabolism (Taté et al., 1998; Atkins 1991). Efflux of NH_3 from the bacteroid thus remains unidirectional (Udvardi and Day 1997). NH_3 released into the PBS is assumed to be rapidly protonated to NH_4^+. The PBS is acidic relative to the cell cytosol and bacteroids due to the presence of a P-type H^+-ATPase located on the SM delivering H^+ into the PBS (Udvardi and Day 1997). PBS NH_4^+ creates a NH_3 concentration gradient that promotes further efflux of the uncharged NH_3 ion from bacteroids into the PBS (Niemietz and Tyerman 2000). Accumulated NH_4^+ is then expected to translocate into the infected cell cytosol across the SM following its own concentration gradient (Niemietz and Tyerman 2000). This process may involve a biochemically characterised voltage-gated cation channel which resides on the SM (Tyerman et al., 1995; Mouritzen and Rosendahl 1997; Roberts and Tyerman 2002). Other routes may involve the aquaporin protein nodulin 26 (NOD26) located on the SM which can act as an NH_3 transport channel when expressed in proteoliposomes (Hwang et al., 2010). This would be in line with TIP aquaporins from Arabidopsis that facilitate NH_3 into vacuoles (Loqué et al., 2005). Interestingly, the SM localised NOD26 is associated with a cytosolic glutamine synthetase (GS), attached to its C-terminus (Niemietz and Tyerman 2000; Hwang et al., 2010; Masalkar et al., 2010). NH_4^+ is thought to be converted to glutamine (Gln) by GS immediately after entering the cytosol due to the high activities of measured GS activities from the cytosolic fraction of nodules (Hoelzle et al., 1992; Roche et al., 1993). In determinate-type nodules (e.g. soybean, lotus), Gln is transported to plastids and mitochondria for *de novo* purine biosynthesis (Smith and Atkins 2002). Purines are later reduced to xanthine, which diffuses to nearby uninfected cells as uric acid. In the peroxisome of uninfected cells, uric acid is converted to allantoin (Van den Bosch and Newcomb 1986; Datta et al., 1991). Allantoin is then converted to allantoic acid in the endoplasmic reticulum (ER) (Werner et al., 2008). Allantoin and allantoic acid collectively are called ureides which diffuse to the vascular system and are exported to the root and aerial parts of the plant (Brown et al., 1995). In indeterminate nodules (e.g. *Medicago*) Gln is converted to asparagine (Asn) and both Gln and Asn are assimilated in the vascular system as amides (Atkins and Smith 2007).

5 The requirement for ammonium transporters

Ammonium (NH_4^+) and nitrate (NO_3^-) are the two preferred sources of inorganic N for plant growth (Bloom 2015; Xuan et al., 2017). Many plants have shown a preference for NH_4^+ over NO_3^- (~20-fold), when both molecules are supplied at the same time and concentration (Kronzucker et al., 1997; Gazzarrini et al., 1999). Nitrogen selectivity is linked to the reduced energy requirements for the uptake and assimilation of NH_4^+ relative to the more oxidised NO_3^- form (Bloom et al., 1992). Using split-root assays, it has been shown that N acquisition pathways (NO_3^- uptake, NH_4^+ uptake and N_2 fixation) are controlled by the plant's N status through a systemic signalling mechanism(s) regardless of N source (Gansel et al., 2001; Ruffel et al., 2008; Jeudy et al., 2010). The systemic signals regulating N uptake have been suggested to involve more than 3000 genes (including transcription factors, transporters, channels and assimilatory enzymes) (Ruffel et al., 2008; Jeudy et al., 2010).

NH_4^+ transport by plants involves the two available forms of ammonium (NH_3 and NH_4^+). NH_4^+ is the predominant ion at physiological pH due to its pH-dependent equilibrium (pKa = 9.25) (Bloom 2015). NH_3 concentrations increase at alkaline pH and at elevated NH_4^+ concentrations in pH buffered soils or in pH neutral cell compartments such as the cytosol (von Wirén et al., 2000; Schjoerring et al., 2002). At high concentrations, NH_4^+ can be toxic to plants, causing a reduction in root and shoot growth and induced chlorosis of the leaves (Britto and Kronzucker 2002; Marschner 1995). The mechanism of the toxicity phenotype is still being defined but has been aligned with the rapid influx and storage of ammonium/ammonia into the cell and the disruption in transport of important cations including Ca^{2+}, Mg^{2+} and K^+ through both cellular alkalisation with the uptake of NH_3 and by localised extra cellular acidification with the uptake of NH_4^+ (Britto and Kronzucker 2002). To protect from NH_4^+ toxicities, the cell actively regulates NH_4^+ concentrations through changes in NH_4^+ uptake, storage in vacuoles, efflux from the cell and assimilatory processes to convert NH_4^+ into amino acids (Wang et al., 1993; Kronzucker et al., 1996).

NH_4^+ transport in plants generally involves two physiological processes: (i) a low-affinity transport system (LATS), operating at high concentrations (>0.5 mM) and (ii) a high-affinity transport system (HATS) active at low NH_4^+ concentrations (<0.5 mM) (Wang et al., 1993; Sohlenkamp et al., 2000; Ludewig et al., 2007). LATS activity is often non-saturable, resulting in a linear uptake pattern as external concentrations increase (Wang et al., 1993). On the other hand, HATS is a saturable system that operates at low external concentrations (Straub et al., 2014). Molecular processes controlling NH_4^+ transport have been identified in plants including the AMT1, AMT2 and AMF transporters while select aquaporins (TIPS) targeting the tonoplast can transport NH_3. The members of the *AMT1* and *AMT2* family of ammonium transporters (AMT) are considered the primary mechanisms responsible for the transport of NH_4^+ across the PM. AMT transporters are generally proteins of 45–50 kDa, but size may vary due to an extended C-terminus (Straub 2016). AMT1 transporters were first identified in a functional complementation study using an NH_4^+ transport defective yeast mutant (26972c) (Marini et al., 1994; Ninnemann et al., 1994). Subsequently, multiple forms of AMT1 proteins have been identified, tested in yeast or *Xenopus laevis* oocytes for functional properties as well as functionally expressed *in planta*. Based on these activities, AMTs are now considered the primary contributor to plant NH_4^+ HATS activity across a range of plant species including legumes (Sohlenkamp et al., 2000; Becker et al., 2002; Sonoda et al., 2003; D'Apuzzo et al., 2004; Loqué et al., 2006). In contrast, AMF transport proteins remain poorly understood. Identification of *ScAMF1* and *GmAMF3* and their functional characterisation as low-affinity NH_4^+ transport properties in both yeast and *X. laevis* oocytes has opened up the idea that secondary NH_4^+ transport proteins exist, while sequence information would suggest *AMF* genes are ubiquitous in all plants and often represented by multigene families particularly in legume orientated clades (Chiasson et al., 2014).

6 Ammonium transport in legumes

In legume nodules, NH_3 passively diffuses from the bacteroids to the PBS following concentration gradients and differences in pHΔ between the SM separated compartments. Once in the acidic PBS, NH_3 is rapidly protonated to NH_4^+. Transfer of NH_4^+ across the SM to the cell cytosol will follow established concentration gradients, where NH_4^+ concentrations are considered low in the cytosol as a result of active GS activity (Streeter 1989; Shelp and Atkins 1984; Masalkar et al., 2010). NH_3 may also diffuse to the plant cytosol if PBS NH_4^+ concentrations are high and NH_3 in the cytosol remains low (Day et al., 2001). Several attempts have been made to define the NH_4^+ and NH_3 transport pathways across the SM. Using the patch-clamp technique, a cation channel permeable to NH_4^+ and other monovalent cations that was regulated by Ca^{2+} and Mg^{2+} was identified in soybean, pea and *Lotus* PBM (Tyerman et al., 1995; Mouritzen and Rosendahl 1997; Roberts and Tyerman 2002). Unfortunately, the molecular identity of this transport activity has yet to be defined. One would expect that AMT and or AMF-type transport proteins may be responsible for

NH_4^+ transport across the SM or work in association with other transporters known to transport NH_3 (such as aquaporins). Both the *AMT* and *AMF* gene families are conserved amongst the sequenced legumes but no AMT or AMF

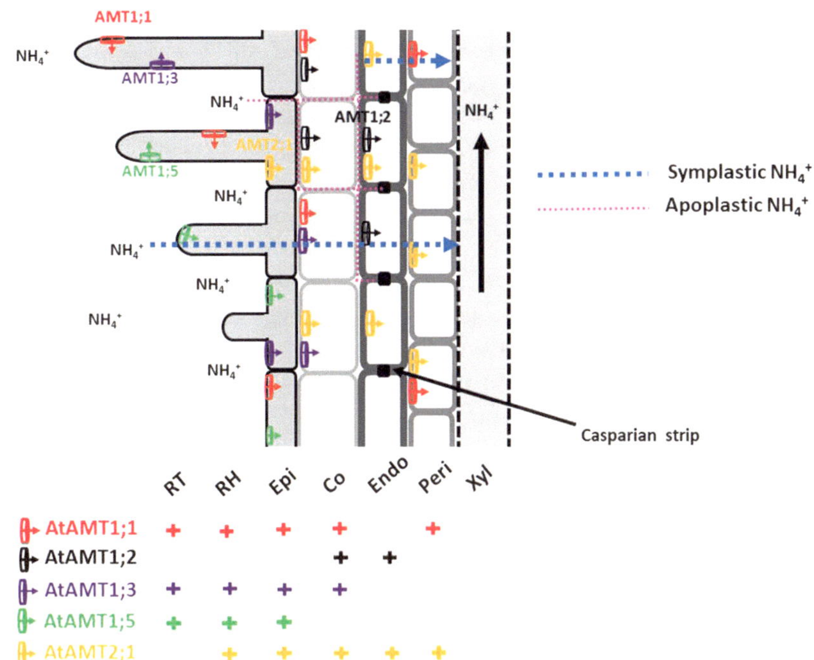

Figure 1 Model summarising the functions of AMT1-type transporters in high or low-affinity ammonium uptake in *Arabidopsis* roots. Schematic representation of the contribution to ammonium uptake and spatial expression in root tissues of AtAMT1;1 (in red ,), AtAMT1;2 (in black ,), AtAMT1;3 (in purple ,), AtAMT1;5 (in green ,), and AtAMT2;1 (in yellow ,) under nitrogen deficiency or under the influence on the form of N supply. Ammonium can enter the symplastic route (dashed blue line with arrowhead) for radial transport toward the root stele via AMTs which are localised at the plasma membrane of various cell types in the root. Ammonium can also bypass outer root cells via the apoplastic transport route (pink dashed line) and subsequently enter the root symplast by AMT-mediated transport across the plasma membrane of endodermal (in the root hair zone) and cortical (in more basal root zones) cells and pericycle cells. In the symplast, ammonium can either be assimilated into amino acids or pass the Casparian strip by diffusing through an endodermal cell as a solute in water and be subsequently loaded into the xylem by another transporter (e.g. AtAMT2;1) for long-distance transport to the shoot loaded into the xylem according to Giehl et al. (2017) and Palmgren (2018). (**+**) denotes the presence of AMTs in specific Arabidopsis root cell types from localisation studies by Loqué et al. (2005) and Mayer and Ludewig (2006) for AtAMT1;1, Duan et al. (2018), Neuhäuser et al. (2007) and Yuan et al. (2007) for AtAMT1;2, Duan et al. (2018), Lima et al. (2010) and Loqué et al. (2007) for AtAMT1;3, Yuan et al. (2007) for AtAMT1;5, and Neuhäuser et al. (2009) in lateral roots and Giehl et al. (2017) for AtAMT2;1 respectively. The schematic representation was adapted from Yuan et al. (2007). RT, root tip; RH, root hair; rhizo, rhizodermis; co, cortex; endo, endodermis; peric, pericycle; xyl, xylem.

protein has been identified on the SM to date using proteomic-based peptide analysis (Clarke et al., 2014).

Like other AMTs, legume variants are integral transmembrane proteins of ~50–55 kDa in size with 9–12 transmembrane domains and an extracellular N-terminus and a cytoplasmic C-terminus. An evolutionary analysis of legume AMT proteins suggests two separate ancestors for AMT1 and AMT2 and a separation of activity depending on the symbiotic partner where mycorrhizal induced AMTs and rhizobial induced AMTs clearly separate, possibly to accommodate new roles in the selected interacting tissues with the symbionts (see Fig. 1). The amino acid sequence analysis of characterised legume AMT proteins revealed that the AMT1 subfamily members retained 70–91% sequence homology, while AMT2 subfamily members have broader sequence identities, between 58% and 91%. Sequence homology suggests the two AMT subfamilies remain distantly related (20–22% sequence identity).

AMT NH_4^+ transport proteins have been characterised in common bean, *Lotus*, *Medicago* and soybean. *LjAMT1*;1 was the first functionally characterised legume AMT1-type transporter, expressed in leaves, roots and nodules (Salvemini et al., 2001). Two other AMT1-type transporters, *LjAMT1*;2 and *LjAMT1*;3 and one AMT2-type transporters, *LjAMT2*;1, have also been characterised (Simon-Rosin et al., 2003; D'Apuzzo et al., 2004). *LjAMT1;1* expression was found to be the highest in leaves relative to *LjAMT1;2* and *LjAMT1;3*. *LjAMT1;3* was found expressed highly in flowers (D'Apuzzo et al., 2004). When plants were exposed to elevated CO_2, in an attempt to repress photorespiration, only *LjAMT1;1* transcript levels decreased (D'Apuzzo et al., 2004). Promoter-GUS studies in nodules showed that activation of the *LjAMT1;1* promoter occurred in vascular tissues and the central infected cortical cells. *LjAMT1;2* promoter activity was limited to the outer cortical cells and central tissue periphery while *LjAMT1;3* was expressed only in the stele of the root (D'Apuzzo et al., 2004).

In the context of NH_4^+ transport in legume nodules, the varied expression pattern of *LjAMT1's* suggests different roles across the organ including the potential for retrieval of lost NH_4^+ from infected nodule cells or in the initial uptake of NH_4^+ directly lost to the nodulesphere outside the nodule. A GFP fusion tag of LjAMT1;1 revealed a PM localisation in root hairs indicating a potential role in NH_4^+ uptake and or N signalling in roots (Rogato et al., 2008). Loss of *LjAMT1;1* expression enhanced nodule numbers and N_2 fixation rates (Rogato et al., 2008). *LjAMT1;3* expression was found to be dependent on the concentration of external NH_4^+, suggesting a role as a putative intracellular NH_4^+ sensor to optimise root development (Rogato et al., 2010a,b). *LjAMT2;1* was found to be expressed in both shoot and root tissues (Simon-Rosin et al., 2003). In nodules, *in situ RNA* experiments revealed *LjAMT2;1* was expressed broadly across all nodule cell types, while a LjAMT2;1-GFP fusion protein showed localisation to the PM and the nuclear envelope (Simon-Rosin et al., 2003). A mycorrhizal-induced *LjAMT2;2* expressed in roots suggest a role in NH_3/NH_4^+ transport across the arbuscule apoplastic interfacial compartment potentially for release of NH_4^+ into the plant cytoplasm (Guether et al., 2009). We cannot rule out the possibility of similar functional transporter(s) in the rhizobia induced nodule.

In *Medicago*, eight putative *AMT*'s have been identified with the majority expressed in shoot tissues except for *MtAMT2;3*, which is expressed only in the root. Genes identified as *MtAMT1;3*, *MtAMT2;2* and *MtAMT2;4* were not expressed in either root or shoot tissues (Straub et al., 2014). Using yeast complementation assays of an NH_4^+ transport-deficient strain, both *Mt*AMT1;1 and *Mt*AMT2;1 were found to be NH_4^+ transporters with K_M values of ~89 μM and 123 μM, respectively. Using *X. laevis* oocytes, expression of *MtAMT1;1* and *MtAMT2;1* also resulted in the accumulation of $^{15}NH_4^+$ (Straub et al., 2014). Using native promoters, MtAMT1;1::GFP and MtAMT2;1::GFP were found targeted to the PM of epidermal and cortical root cells. The characterised NH_4^+ transporter from *Medicago* were cloned from root cDNA but was not found localised in nodules suggesting a different role in NH_4^+ management. *MtAMT2;3* has been found localised in cortical and endodermal cells of non-colonised *Medicago* roots, but upon mycorrhizal colonisation, promoter activity changed where localisation was restricted only to cells containing arbuscules (Breuillin-Sessoms et al., 2015). In common bean, a high-affinity pH-dependent H^+/NH_4^+ symport transporter (PvAMT1;1) has been characterised as a nodule expressed gene (through cDNA cloning) but lacks any detailed cellular or intercellular identity (Ortiz-Ramirez et al., 2011). Mycorrhiza-induced GmAMT4;1 has been found localised in the peri-arbuscular membrane and expressed in the arbuscular cortical cell but lacks any evidence of an association with root nodules (Kobae et al., 2010). In a recent bioinformatics study, the *Vigna subterranean* NH_4^+ transporter, *VsAMT1* with 89% homology to the soybean *AMT1* was identified (Adetunji et al., 2015). Although there is a growing number of identified *AMT* genes in legumes, more work is required to relate gene expression to function. Future studies need to define the cellular and subcellular localisation of other AMT1s and AMT2s so that individual roles in NH_4^+ transport across the nodule and root system can be identified.

7 Understanding ammonium transporter functionality from a model plant: *Arabidopsis thaliana*

Our basic understanding of legume AMTs is still developing. To get a good interpretation of the diversity of roles AMT genes may have on NH_4^+ transport in legume cells, an examination of AMT activity in Arabidopsis provides a good platform to interpret potential activities that may occur across both root and nodules of legumes. The best-understood AMT proteins are those described in Arabidopsis, consisting of six homologs subdivided into two sub-families, AMT1 and AMT2. AMT1 consists of AtAMT1;1-5 while AtAMT2;1 is the sole member of the AMT2 sub-family. The *AMT*'s have a diverse expression pattern across both tissues and cell types in Arabidopsis. *AtAMT1;1* was the first plant NH_4^+ transporter (AMT1) cloned from an Arabidopsis cDNA library through functional complementation of a mutant yeast strain (26972c) defective in NH_4^+ uptake (Ninnemann et al., 1994). It showed a high affinity for NH_4^+ and was shown to be the major contributor of Arabidopsis root NH_4^+ acquisition (Kaiser et al., 2002). Functional analysis of AtAMT1;1 revealed it is responsible for root NH_4^+ absorption, long-distance transport to the shoot and potentially the recapture of NH_4^+ in the shoot during photorespiration (Mayer and Ludewig 2006). *AtAMT1;1* is localised to the PM of epidermal cells and root hairs and transports NH_4^+ via a symplastic pathway (see Fig. 2; Kaiser et al., 2002; Loqué et al., 2006; Mayer and Ludewig 2006; Yuan et al., 2007b). *AtAMT1;1* is also reported to be localised in pericycle, leaf epidermal and mesophyll cells (Glass et al., 2002; Mayer and Ludewig 2006). *AtAMT1;1* expression is particularly responsive to N starvation and diurnal light cycles (Engineer and Kranz 2007; Gazzarrini et al., 1999; Shelden et al., 2001). Transcript levels increase in roots with N starvation, while high expression is observed in shoots when grown

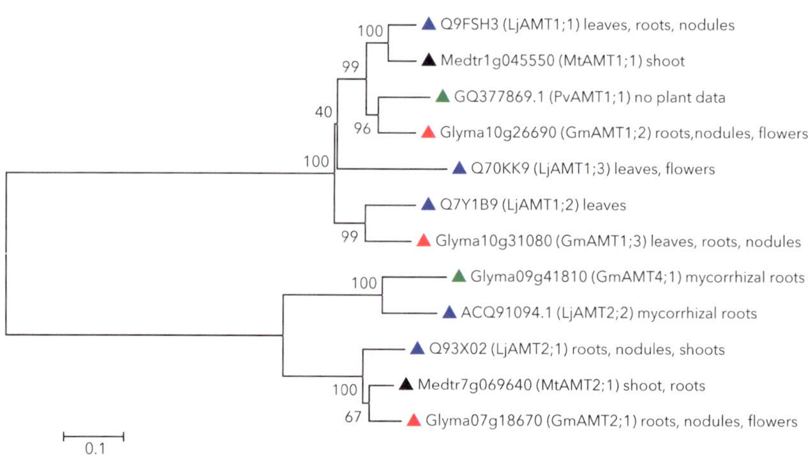

Figure 2 Evolutionary analysis of characterised legume AMTs. Analysis of the amino acid sequence of high-affinity AMT proteins from *Lotus* (blue triangle), *Medicago* (black triangle), common bean (green triangle) and soybean (red triangle). Evolutionary analyses were conducted in MEGA7 (Kumar et al., 2016), distances were computed using the Poisson correction method (Zuckerkandl and Pauling 1965), and the tree was generated using the neighbour-joining method (Saitou and Nei 1987). The tree is drawn to scale, and preferential expression tissues are listed next to gene names.

under sufficient N (Engineer and Kranz 2007; Gazzarrini et al., 1999). *AtAMT1;1* transcript levels are expressed highly during the light period, but levels are low compared to *AtAMT1;3* (Gazzarrini et al., 1999), while another study revealed that *AtAMT1;1* mRNA levels are tightly regulated by glutamine (Gln) (Rawat et al., 1999). It was also demonstrated that upregulation and downregulation of NH_4^+ influx ($^{13}NH_4^+$) was correlated with *AtAMT1;1* mRNA status in roots. Post-translational regulation of AtAMT1;1 has been shown to involve a phosphorylation event in the C-terminus hinge at residue T460 (Lanquar et al., 2009). Phosphorylation inactivates the protein by disrupting the trimeric AtAMT1;1 complex (Loqué et al., 2007; Neuhäuser et al., 2007), consequently reducing NH_4^+ transport. The kinase responsible for phosphorylation of AtAMT1;1 is linked to an accumulation of NH_4^+ in the cytosol. CIPK23, a calcineurin B-like interacting protein kinase 23 linked to the phosphorylation AKT1 and AtNPF6;3 also phosphorylates the C-terminus of AtAMT1;1 and AtAMT1;2 (Straub et al., 2017). The allosteric regulation of AMT1;1 activity may also be controlled by variations in homo- and heterotrimers amongst AMT1 proteins, where AMT1;3 can trimerise with AtAMT1;1 and influence NH_4^+ transport in yeast and in planta (Yuan et al., 2013). In addition, Lanquar et al. (2009) suggested that AtAMT1;1 could serve as a transceptor with a dual function in transport and signalling.

AtAMT1;2 is localised to the PM of root endodermal and cortical cells although it contains a putative plastid transit peptide (Shelden et al., 2001; Neuhäuser et al., 2007; Yuan et al., 2007a). The localisation of the AtAMT1;2:GFP protein changed depending on the root zones (Fig. 2). In the root hair differentiation zone, longitudinal and radial sections showed the AtAMT1;2:GFP protein to be confined mainly to endodermal cells and then extended to cortical cells higher up in the root hair zone (Yuan et al., 2007a). In the zone of emerging lateral roots, GFP-dependent fluorescence was exclusively restricted to the cortical cells (Yuan et al., 2007a). A similar pattern of localisation was observed in the main root by Duan et al. (2018) and the same authors demonstrated that apoplastic transport controlled by AtAMT1;2 at the endodermis prevailed at high external ammonium while the epidermal AtAMT1;3 mediated symplastic ammonium transport across the root tissue when external ammonium was low. It is shown that mutations in a putative phosphorylation site of the AtAMT1;2 carboxyl tail significantly reduce its activity and that of other NH_4^+ transporters by inactivating them to protect the cell from excess influx of cytotoxic NH_4^+ (Neuhäuser et al., 2007). Under N starvation, *AtAMT1;2* is expressed in plant roots (Gazzarrini et al., 1999; Shelden et al., 2001). Expression of *AtAMT1;2* in an NH_4^+ transport yeast mutant showed a dual affinity transport activity for MA uptake with K_M values of 36 μM and 3.0 mM, respectively, for high- and low-affinity transport (Shelden et al., 2001). In *X. laevis* oocytes, AtAMT1;2

behaves as a monophasic transporter with a K_M value of 140 µM and 1.9 mM, respectively, for NH_4^+ and MA (Neuhäuser et al., 2007). AtAMT1;2 may also act as H^+/NH_3 co-transporter when non-electrogenic transport was detected in mutants with amino acid disruptions amongst the subunit contact sites of the AtAMT1;2 trimer (Neuhäuser and Ludewig 2014).

AtAMT1;3 is exclusively root-based and involved in third-order root proliferation (Lima et al., 2010) (Fig. 2). In AMT1;3 promoter:AMT1;3 gene:GFP plants, GFP-fluorescence signals were observed in the lateral root cap and epidermal cells of first-order lateral root tips and the cortex and rhizodermis within the differentiation zone of second-order lateral roots with third-order lateral root primordia (Lima et al., 2010). Previous work by Loqué et al. (2006), using transgenic lines carrying AMT-promoter–GFP construct, i.e. AMT1;3pro:GFP, found GFP-fluorescence in the cortical cells and an enhanced fluorescence signal in the rhizodermis, including root hairs and in the root tip under nitrogen-deficiency conditions. It was characterised by functional complementation of a yeast strain defective in three endogenous NH_4^+ transporters (Gazzarrini et al., 1999). ^{14}C-MA uptake by AtAMT1;3 shows saturable concentration-dependent kinetics with a K_M value of 11 µM. AtAMT1;3 is localised in the PM of epidermal cells and root hairs which participate in the uptake of NH_4^+ from the soil solution (Yuan et al., 2007a). *AtAMT1;3* transcript levels increase with NH_4^+ uptake under N starvation conditions. *AtAMT1;3* transcript levels increase at the end of light period indicating a link between carbon provision and N assimilation (Gazzarrini et al., 1999; Lejay et al., 2003). Loss of *AtAMT1;3* reduced the high-affinity NH_4^+ uptake approximately 30% under N-deficient conditions (Loqué et al., 2006), a similar impact to net NH_4^+ uptake with the loss of *AtAMT1;1* (Kaiser et al., 2002). Wang et al. (2013) utilised a single particle imaging approach to describe how AMT1;3 plant cells could avoid accumulation of toxic levels of NH_4^+ by monitoring dynamic changes in EGFP-tagged AMT1;3 on the plasma membrane of Arabidopsis epidermal root cells at various ammonium levels. Moving fluorescent spots of AMT1;3-EGFP dynamically appeared and disappeared on the plasma membrane under N-deprived and N-sufficient conditions and when under external high-ammonium stress, the moving fluorescent spots amassed into clusters, which were subsequently internalised into the cytoplasm (Wang et al., 2013). Thus, by the clustering and endocytosis of AMT1;3, NH_4^+ transport/accumulation to toxic levels could be regulated. Recent work investigating how AtAMT1;3 could function structurally as a transporter was explored by creating point mutations located within conserved regions of the binding and substrate permeation tunnel of AtAMT1;3 and by expressing these mutants in *Xenopus* oocytes (Hao et al., 2020). In oocyte transport studies, the P181A mutation in AtAMT1;3 approximately doubled the amplitude of both NH_4^+ and MeA^+ induced currents, whereas the F178I and Q365K mutations reduced these currents by about 50%

suggesting that P181 and F178 amino acid residues contributed to the tuning of the substrate binding as well as to permeation properties, and that Q365 modulated substrate permeation (Hao et al., 2020).

AtAMT1;4 is expressed exclusively in pollen grains and in the pollen tube after germination (Yuan et al., 2009). It has been suggested that AtAMT1;4 is localised in the PM of pollen and regulates NH_4^+ uptake and/or retrieval. Overexpression and heterologous expression in plants and yeast revealed that *AtAMT1;4* is a functional high-affinity NH_4^+ transporter with a K_M of 17 μM. Recently, an intracellular localisation study showed that *AtAMT1;4* mRNA inside sperm cells of Arabidopsis pollen grains using atomic force microscopy (AFM) (Kim et al., 2015). The authors demonstrated localisation of *AtAMT1;4* mRNA in the nucleus of the sperm cell rather than the cytosol. It would be interesting to explore what role *AtAMT1;4* orthologues (Mazurkiewicz, 2008) have in legumes and their possible involvement in NH_4^+ uptake or retrieval in the above-ground parts of the plant.

AtAMT1;5 is a root-based high-affinity AMT1-type NH_4^+ transporter in Arabidopsis (Yuan et al., 2007a). AMT1;5 complemented a yeast strain defective in NH_4^+ uptake. *AtAMT1;5* is transcriptionally upregulated in N-deficient conditions but less effective in complementing AMT mutants in Arabidopsis relative to *AtAMT1;3*. *AtAMT1;5* is expressed in epidermal cells and in root hairs (Yuan et al., 2007a).

AtAMT2;1 represents the lone AMT2-type transporter in Arabidopsis. Phylogenetic analysis revealed *AtAMT2;1* only having 23–25% sequence identity with *AMT1*-type transporters while showing high sequence homology with yeast NH_4^+ transporters, MEP and prokaryotic homologs, AMTB. It was characterised as functionally similar to the AMT1-type transporters by its ability to complement NH_4^+ transport in yeast, although it could not transport detectable levels of ^{14}C-MA (Sohlenkamp et al., 2000). *AtAMT2;1* expressed in both shoot and root tissues but de-repressed in roots when N-starved similar to *AtAMT1;1* (Sohlenkamp et al., 2000). It is a high-affinity NH_4^+ transporter with K_M of 21 μM (Sohlenkamp et al., 2002). When expressed in yeast and *X. laevis* oocytes, electrophysiology analysis suggests NH_3 is transported across the membrane after NH_4^+ is recruited (Neuhäuser et al., 2009). Using the GUS reporter, *AMT2* promoter activity was found localised to the vascular tissues of roots, leaves and flowers (Sohlenkamp et al., 2002). Work by Giehl et al. (2017) highlighted the discrepancies in the AtAMT2 localisation in roots due to different AtAMT2 constructs in transgenic lines and growth media containing different forms and concentrations of N in previous studies. So in their study, the use of 1883 bp of the 50-upstream sequence of AMT2;1 fused with GFP mapped AMT2;1 promoter activity in the mature zone of roots and its absence from root tips regardless of the N treatment (Giehl et al., 2017). Furthermore, they demonstrated a change in the proAMT2;1-GFP expression in transgenic plants depending on the source of N. Under N deficiency, though AMT2;1-driven GFP expression was detected in the other cell types, it was most pronounced in cortical cells. When nitrate was the sole N source, AMT2;1 promoter activity shifted slightly towards the endodermis, becoming almost undetectable in epidermal cells and when the N supply was only ammonium, AMT2;1 expression became more confined to endodermal and especially to pericycle cells (Giehl et al., 2017). Combined with ^{15}N-labeled ammonium transport studies in *amt2;1* mutant lines and *amt1;1 amt1;2 amt1;3 amt2;1* quadruple mutant (*qko*), Giehl et al (2017) suggested that AtAMT2;1 functioned mainly in root-to-shoot translocation of ammonium in response to the plant nutritional status and to local ammonium gradients in a cell-type-specific manner (see Fig. 2).

8 Altered ammonium transporter transporter activity in plants

Gene silencing is an excellent way to help understand gene function and has been used broadly in plant science (Gilchrist and Haughn 2010; Guo et al., 2016; Zhang et al., 2019). In Arabidopsis, it has been demonstrated that the loss of *AtAMT1;1* activity by T-DNA mutagenesis, reduced high-affinity NH_4^+ uptake into roots by 30% (Kaiser et al., 2002). This result was further supported by another study that demonstrated loss of either *AtAMT1;1* or *AtAMT1;3* reduced the high-affinity NH_4^+ uptake by 30% while loss of both *AtAMT1;1* and *AtAMT1;3* reduced the high-affinity uptake up to 70% compared to wild-type plant (Loqué et al., 2006). A later study demonstrated that T-DNA-tagged *Atamt1;2* reduced high-affinity NH_4^+ uptake by 18–26% and that a quadruple mutant (*qko*) consisting of *amt1;1 amt1;2*, *amt1;3* and *amt2;1* showed a severe growth depression under sufficient NH_4^+ and reduced uptake (5–10%) of NH_4^+ compared to wild-type plants (Yuan et al., 2007a). In contrast, RNAi-induced silencing of *AtAMT2* resulted in no obvious phenotypic change in growth or NH_4^+ uptake (Sohlenkamp et al., 2002). Legume has similar *AMT* genes compared to Arabidopsis, some of which may play a similar role to genes in Arabidopsis while others may be novel. In the model legume *L. japonicus*, RNAi-mediated gene silencing of *LjAMT1;1* resulted in increased nodule numbers on inoculated roots but a decrease in nodule and overall nitrogenase activity (Rogato et al., 2008). Other AMT activities in the Lotus have yet to be defined, while silencing of soybean *AMTs* in transgenic root showed controlling of nodulation (Uddin, unpublished data). This difference could indicate a difference in AMT activities between *LjAMT1;1* and the soybean orthologues. Apart from the previously characterised nodulation signalling genes controlling infection and nodule development (e.g. *GmSymRK*, *GmCND* and *GmNORK*) (Libault et al., 2009; Wang et al., 2016; Wang et al., 2017), no

AMT-type transporter has ever reported influencing nodule development. The author suggested *GmAMT* may have a role in nodule development through a root-specific activity possibly involving NH_4^+ uptake-mediated apoplastic pH control and auxin redistribution for secondary root initiation as recently observed in *Arabidopsis thaliana AMT* mutants (Meier et al., 2020; Motte and Beeckman 2020). In an earlier study, the same group demonstrated that loss of *AtAMT1;3* function significantly reduced the lateral root branching compared to wild type while a quadruple mutant (*amt1;1 amt1;2 amt1;3* and *amt2;1*) lost lateral root development completely, but a reconstitution of *AMT1;3* functions in the mutant could restore the lateral root development (Lima et al., 2010). Loss of select *AMT*'s in Arabidopsis influences auxin redistribution altering where secondary root primordia develop in the presence of NH_4^+. This could indicate legume *AMT*'s may also have a role in nodule development possibly in an NH_4^+ dependent manner. Further research is required to link the impact of NH_4^+ in the rhizosphere and the location where nodules can develop along the root.

9 Ammonium major facilitator

Glycine max symbiotic ammonium transporter1 (GmSAT1) was initially characterised as a functional NH_4^+ transport protein localised in symbiosome membrane (Kaiser et al., 1998). A later study confirmed GmSAT1 was not an NH_4^+ transporter and suggested GmSAT1 as a potential transcription factor (Marini et al., 2000) which was established in the subsequent study and *GmSAT1* renamed as *G. max* basic helix-loop-helix membrane1 (*GmbHLHm1*) (Chiasson et al., 2014). The characterisation of GmbHLHm1 pioneered a new class of NH_4^+ transport protein, ammonium major facilitator1 (AMF1) which shared chromosomal synteny with *Gm*bHLHm1. AMF1 was first identified in yeast as *Saccharomyces cerevisiae* ammonium major facilitator1 (*ScAMF1*) through the interaction with the heterologously expressed GmbHLHm1 in yeast (Chiasson et al., 2014). ScAMF1 facilitated the transport of high concentrations of the NH_4^+-analogue, MA and when expressed in *X. laevis* oocytes displayed NH_4^+ induced currents and the uptake of ^{15}N-labelled NH_4^+ (Chiasson et al., 2014). A soybean homologue of *ScAMF1*, *GmAMF3*, was identified by sequence analysis. Functional studies in yeast and *Xenopus* oocytes confirmed that *Gm*AMF3 is also an NH_4^+ transporter localised to parenchyma and vascular cells of roots and nodules suggested a potential role of exporting fixed NH_4^+ from the infected region of the nodule (Chiasson et al., 2014). What role AMF proteins have in NH_4^+ transport in plants and more specifically legume nodules is yet to be answered in detail, but it was demonstrated that the loss of GmbHLHm1 has a detrimental effect on the legume-rhizobia symbiosis as it impaired nodule fitness and growth (Chiasson et al., 2014). Like AMT1 and AMT2, AMF1 homologs require more detailed investigation to link gene activity to function and their individual and net role in NH_4^+ transport. A recent study (Carter and Tegeder 2016), showed overexpression of the ureide export protein UPS1 in cortical and endodermis cells can facilitate the transport of ureides from soybean nodules to the root. This change enhanced N_2 fixation and seed N deposition and interestingly both *GmAMT2;1* and *GmAMF3* transcript levels were strongly upregulated in the nodules, suggesting a role of the two NH_4^+ transporters in supporting enhanced N_2 fixation and N export.

10 The physiological role of ammonium major facilitator in root nodules

In contrast to *AMT* gene expression patterns, we know very little about the transcriptional phenotype of *AMF*-type NH_4^+ transporter genes. In an earlier study, it was demonstrated that silencing of the *GmAMF3* linked symbiosome membrane (SM) bound transcription factor (TF), *GmSAT1* (*GmbHLHm1*) impaired nodule development and reduced availability of N_2-fixation products

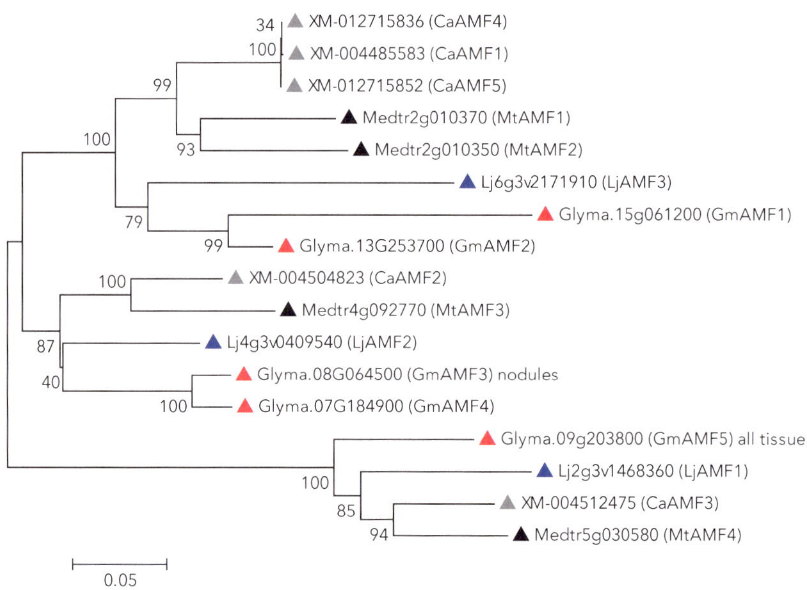

Figure 3 Evolutionary analysis of characterised and putative AMF proteins. Phylogenetic tree of characterised and putative AMF amino acid sequences from *Lotus* (blue triangle), *Medicago* (black triangle), chickpea (grey triangle) and soybean (red triangle). Evolutionary analyses were conducted in MEGA7 (Kumar et al., 2016), distances were computed using the Poisson correction method (Zuckerkandl and Pauling 1965), and the tree was generated using the neighbour-joining method (Saitou and Nei 1987). The tree is drawn to scale, and preferential expression tissues are listed next to gene names.

to the plant, but what role AMF proteins have in NH_4^+ transport in plants and more specifically legume nodules is yet to be answered in detail (Chiasson et al., 2014). In soybean nodules, it has been found that *GmAMF3* expression was high in the nodule relative to other organs tested (shoots, roots and flowers) (Chiasson et al., 2014). Its role in plants is still undefined, but it may be involved in scavenging and exporting fixed N from the infected region of the nodule or could also play a role in photo-respiratory N management of soybean leaves.

Evolutionary analysis of characterised and putative AMF proteins of different legumes (see Fig. 3) showed AMF proteins also subdivided into two main clusters. For example, among five soybean AMF paralogue, AMF5 form a cluster with LjAMF1, CaAMF3 and MtAMF4, while GmAMF1-4 form a separate cluster with AMFs of other legumes. These evolutionary data support the previous studies, which suggested like AMTs, AMFs also subdivided into two subfamilies (AMF1 and AMF2) and might play a different role in NH_4^+ regulation in legume nodules.

11 Conclusion and future direction

High-affinity AMT and low-affinity AMF transmembrane protein family members have been linked to NH_4^+ transport systems in legume root nodules. Though AMT-type NH_4^+ transporters have been characterised across many plant species, their study is minimal when it comes to legume root nodules (see Fig. 4). Although there is a growing number of identified *AMT* and *AMF* genes in legume genomic sequence annotations, the lack of functional identity and cellular resolution within nodule cell types and roots limits a full picture of their role in nodule and root N metabolism. Our current understanding of high-affinity AMTs has been useful but not enough to interpret every NH_4^+ transport activity convincingly in legumes and in other plants. For example, what is the degree of selectivity to NH4$^+$ (K_M and V_{MAX}) or do they transport other cations in a similar fashion to non-selective channels (NSCs). More importantly, do AMF1 proteins work as uniporters or is the transport coupled to the movement of other ions either in symport or antiport? It will be interesting to resolve the role of AMT and AMF activities relative to nodule N_2 fixation and the redistribution of N and their potential role in NH_4^+ transport across symbiotic membranes. This will confirm how AMT and AMF activities influence the movement of NH_4^+ within or potentially out of the nodule and whether other transporters need to be identified or allocated with the role. As many AMTs are well studied, it would be interesting to see further analysis of AMF activity in Arabidopsis to allow us to quantify the contribution of the AMF family to net NH_4^+ uptake, organelle partitioning and redistribution of NH_4^+ in non-symbiotic tissues (i.e. non-mycorrhizal and non-legume) and quantify their overall impact on plant growth and development, which will be a prelude to

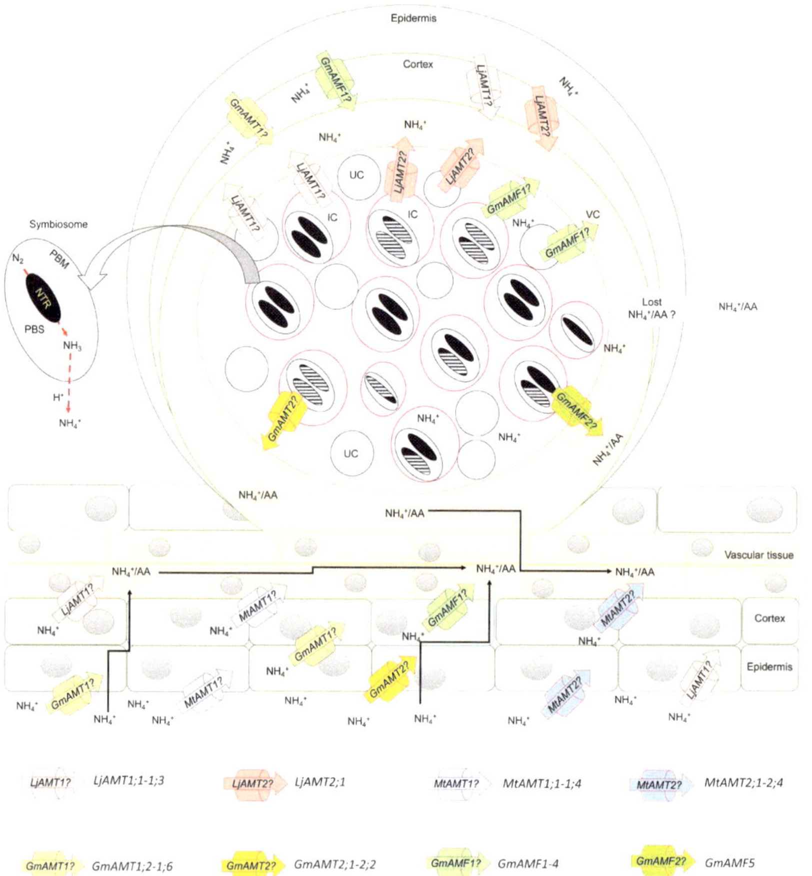

Figure 4 Schematic of legume root nodule with proposed *AMT* and *AMF activities*. Schematic illustrating studied *AMT*- and *AMF*-type transporters in legume root and nodule based on published and unpublished studies. AA: amino acid, NTR: nitrogenase, PBM: peribacteroid membrane, PBS: peribacteroid space, IC: infected cell, UC: uninfected cell and VC: vascular tissue. We used question mark (?) as some data based on quantitative analysis and missing experimental data.

future investigations in agriculturally important crops including other legumes (chickpea, lentil, faba) and cereals (wheat, rice and maize).

Phylogenetic tree of characterised and putative AMF amino acid sequences from lotus (blue triangle), *Medicago* (black triangle), chickpea (grey triangle) and soybean (red triangle). Evolutionary analyses were conducted in MEGA7 (Kumar et al., 2016), distances were computed using the Poisson correction method (Zuckerkandl and Pauling 1965), and the tree was generated using the neighbour-joining method (Saitou and Nei 1987).

The tree is drawn to scale, and preferential expression tissues are listed next to gene names.

12 References

Adetunji, A. T., Lewu, F. B. and Mundembe, R. (2015). *Vigna subterranea* ammonium transporter gene (VsAMT1): some bioinformatics insights. *Biotechnol. Rep.* 8 (Suppl. C), 88–93.

Alaswad, A. A., Oehrle, N. W. and Krishnan, H. B. (2019). Classical Soybean (*Glycine max* (L.) Merr) symbionts, *Sinorhizobium fredii* USDA191 and *Bradyrhizobium diazoefficiens* USDA110, reveal contrasting symbiotic phenotype on pigeon pea (*Cajanus cajan* (L.) Millsp). *Int. J. Mol. Sci.* 20(5), 1091. doi: 10.3390/ijms20051091.

Andreev, I. M., Dubrovo, P. N., Krylova, V. V. and Izmailov, S. F. (1999). Functional identification of ATP-driven Ca^{2+} pump in the peribacteroid membrane of broad bean root nodules. *FEBS Lett.* 447(1), 49–52.

Ané, J. M., Kiss, G. B., Riely, B. K., Penmetsa, R. V., Oldroyd, G. E., Ayax, C., Lévy, J., Debellé, F., Baek, J. M., Kalo, P., Rosenberg, C., Roe, B. A., Long, S. R., Dénarié, J. and Cook, D. R. (2004). *Medicago truncatula* DMI1 required for bacterial and fungal symbioses in legumes. *Science* 303(5662), 1364–1367.

Atienza, S. G. and Rubiales, D. (2017). Legumes in sustainable agriculture. *Crop Pasture Sci.* 68(11), i–ii.

Atkins, C. (1991). Ammonia assimilation and export of nitrogen from the legume nodule. Studies in plant science 1. In: Dilworth, M. J. and Glenn, A. R. (Eds). *Biology and Biochemistry of Nitrogen Fixation*, Elsevier, Amsterdam, pp 293–319.

Atkins, C. A. and Smith, P. M. C. (2007). Translocation in legumes: assimilates, nutrients, and signaling molecules. *Plant Physiol.* 144(2), 550–561.

Babalola, O. O., Olanrewaju, O. S., Dias, T., Ajilogba, C. F., Kutu, F. R. and Cruz, C. (2017). Biological nitrogen fixation: the role of underutilized leguminous plants. In: Panpatte, D. G., Jhala, Y. K., Vyas, R. V., Shelat, H. N. (Eds). *Microorganisms for Green Revolution*, Springer, Singapore, pp 431–443.

Baral, B., Teixeira da Silva, J. A. and Izaguirre-Mayoral, M. L. (2016). Early signaling, synthesis, transport and metabolism of ureides. *J. Plant Physiol.* 193, 97–109.

Barker, A. V. and Bryson, G. M. (2016). Nitrogen. In: Barker, A. V. and Pilbeam, D. J. (Eds). *Handbook of Plant Nutrition*, CRC Press, Boca Raton, FL, pp 37–66.

Becker, D., Stanke, R., Fendrik, I., Frommer, W. B., Vanderleyden, J., Kaiser, W. M. and Hedrich, R. (2002). Expression of the NH_4^+-transporter gene LeAMT1; 2 is induced in tomato roots upon association with N_2-fixing bacteria *Planta* 215(3), 424–429.

Bergersen, F. J. (1982). *Root Nodules of Legumes: Structure and Functions*, Research Studies Press, Letchworth, Hert.

Biswas, B. and Gresshoff, P. M. (2014). The role of symbiotic nitrogen fixation in sustainable production of biofuels. *Int. J. Mol. Sci.* 15(5), 7380–7397.

Blair, M. W., Wu, X., Bhandari, D., Zhang, X. and Hao, J. (2016). Role of legumes for and as horticultural crops in sustainable agriculture. In: Nandwani, D. (Ed.). *Organic Farming for Sustainable Agriculture*, Springer, Cham, pp 185–211.

Bloom, A. J. (2015). The increasing importance of distinguishing among plant nitrogen sources. *Curr. Opin. Plant Biol.* 25, 10–16.

Bloom, A. J., Sukrapanna, S. S. and Warner, R. L. (1992). Root respiration associated with ammonium and nitrate absorption and assimilation by barley. *Plant Physiol.* 99(4), 1294–1301.

Blumwald, E., Fortin, M. G., Rea, P. A., Verma, D. P. S. and Poole, R. J. (1985). Presence of host-plasma membrane type H^+-ATPase in the membrane envelope enclosing the bacteroids in soybean root nodules. *Plant Physiol.* 78(4), 665–672.

Brear, E. M., Bedon, F., Gavrin, A., Kryvoruchko, I. S., Torres-Jerez, I., Udvardi, M. K., Day, D. A. and Smith, P. M. C. (2020). GmVTL1a is an iron transporter on the symbiosome membrane of soybean with an important role in nitrogen fixation. *New Phytol.* 228(2), 667–681.

Breuillin-Sessoms, F., Floss, D. S., Gomez, S. K., Pumplin, N., Ding, Y., Levesque-Tremblay, V., Noar, R. D., Daniels, D. A., Bravo, A., Eaglesham, J. B., Benedito, V. A., Udvardi, M. K. and Harrison, M. J. (2015). Suppression of arbuscule degeneration in *Medicago truncatula* phosphate transporter4 mutants is dependent on the ammonium transporter 2 family protein AMT2;3. *Plant Cell* 27(4), 1352–1366.

Britto, D. T. and Kronzucker, H. J. (2002). NH_4^+ toxicity in higher plants: a critical review. *J. Plant Physiol.* 159(6), 567–584.

Brown, S. M., Oparka, K. J., Sprent, J. I. and Walsh, K. B. (1995). Symplastic transport in soybean root nodules. *Soil Biol. Biochem.* 27(4–5), 387–399.

Capoen, W., Sun, J., Wysham, D., Otegui, M. S., Venkateshwaran, M., Hirsch, S., Miwa, H., Downie, J. A., Morris, R. J., Ané, J. M. and Oldroyd, G. E. (2011). Nuclear membranes control symbiotic calcium signaling of legumes. *Proc. Natl. Acad. Sci. U. S. A.* 108(34), 14348–14353.

Carter, A. M. and Tegeder, M. (2016). Increasing nitrogen fixation and seed development in soybean requires complex adjustments of nodule nitrogen metabolism and partitioning processes. *Curr. Biol.* 26(15), 2044–2051.

Catoira, R., Galera, C., de Billy, F., Penmetsa, R. V., Journet, E. P., Maillet, F., Rosenberg, C., Cook, D., Gough, C. and Dénarié, J. (2000). Four genes of *Medicago truncatula* controlling components of a Nod factor transduction pathway. *Plant Cell* 12(9), 1647–1666.

Cerri, M. R., Frances, L., Laloum, T., Auriac, M. C., Niebel, A., Oldroyd, G. E., Barker, D. G., Fournier, J. and de Carvalho-Niebel, F. (2012). *Medicago truncatula* ERN transcription factors: regulatory interplay with NSP1/NSP2 GRAS factors and expression dynamics throughout rhizobial infection. *Plant Physiol.* 160(4), 2155–2172.

Charpentier, M., Bredemeier, R., Wanner, G., Takeda, N., Schleiff, E. and Parniske, M. (2008). *Lotus japonicus* CASTOR and POLLUX are ion channels essential for perinuclear calcium spiking in legume root endosymbiosis. *Plant Cell* 20(12), 3467–3479.

Charpentier, M., Sun, J., Martins, T. V., Radhakrishnan, G. V., Findlay, K., Soumpourou, E., Thouin, J., Véry, A. A., Sanders, D., Morris, R. J. and Oldroyd, G. E. D. (2016). Nuclear-localized cyclic nucleotide-gated channels mediate symbiotic calcium oscillations. *Science* 352(6289), 1102–1105.

Chiasson, D. M., Loughlin, P. C., Mazurkiewicz, D., Mohammadidehcheshmeh, M., Fedorova, E. E., Okamoto, M., McLean, E., Glass, A. D. M., Smith, S. E., Bisseling, T., Tyerman, S. D., Day, D. A. and Kaiser, B. N. (2014). Soybean *SAT1* (*Symbiotic Ammonium Transporter 1*) encodes a bHLH transcription factor involved in nodule growth and NH_4^+ transport. *Proc. Natl. Acad. Sci. U. S. A.* 111(13), 4814–4819.

Clarke, V. C., Loughlin, P. C., Day, D. A. and Smith, P. M. C. (2014). Transport processes of the legume symbiosome membrane. *Front. Plant Sci.* 5. doi: 10.3389/fpls.2014.00699.

Clarke, V. C., Loughlin, P. C., Gavrin, A., Chen, C., Brear, E. M., Day, D. A. and Smith, P. M. (2015). Proteomic analysis of the soybean symbiosome identifies new symbiotic proteins. *Mol. Cell. Proteomics* 14(5), 1301–1322.

Coba de la Peña, T., Fedorova, E., Pueyo, J. J. and Lucas, M. M. (2018). The symbiosome: legume and rhizobia co-evolution toward a nitrogen-fixing organelle? *Front. Plant Sci.* 8, 2229.

Coronado, C., Zuanazzi, J. S., Sallaud, C., Quirion, J. C., Esnault, R., Husson, H. P., Kondorosi, A. and Ratet, P. (1995). Alfalfa root flavonoid production is nitrogen regulated. *Plant Physiol.* 108(2), 533–542.

Crespi, M. and Frugier, F. (2008). De novo organ formation from differentiated cells: root nodule organogenesis. *Sci. Signal.* 1(49), re11–re11. doi: 10.1126/scisignal.149re11.

D'Apuzzo, E., Rogato, A., Simon-Rosin, U., El Alaoui, H., Barbulova, A., Betti, M., Dimou, M., Katinakis, P., Marquez, A., Marini, A. M., Udvardi, M. K. and Chiurazzi, M. (2004). Characterization of three functional high-affinity ammonium transporters in *Lotus japonicus* with differential transcriptional regulation and spatial expression. *Plant Physiol.* 134(4), 1763–1774.

Datta, D. B., Triplett, E. W. and Newcomb, E. H. (1991). Localization of xanthine dehydrogenase in cowpea root nodules: implications for the interaction between cellular compartments during ureide biogenesis. *Proc. Natl. Acad. Sci. U. S. A.* 88(11), 4700–4702.

Day, D. A., Kaiser, B. N., Thomson, R., Udvardi, M. K., Moreau, S. and Puppo, A. (2001). Nutrient transport across symbiotic membranes from legume nodules. *Funct. Plant Biol.* 28(7), 669–676.

Day, D. A., Whitehead, L., Hendriks, J. H. M. and Tyerman, S. D. (1995). Nitrogen and carbon exchange across symbiotic membranes from soybean nodules. In: Tikhonovich, I. A. Provorov, N. A., Romanov, V. I., Newton, W. E. (Eds). *Nitrogen Fixation: Fundamentals and Applications.* Proceedings of the 10th International Congress on Nitrogen Fixation, St. Petersburg, Russia, May 28, Springer Netherlands, Dordrecht, pp 557–564.

Dean, R. M., Rivers, R. L., Zeidel, M. L. and Roberts, D. M. (1999). Purification and functional reconstitution of soybean nodulin 26. An aquaporin with water and glycerol transport properties. *Biochemist* 38(1), 347–353.

Downie, J. A. (2014). Legume nodulation. *Curr. Biol.* 24(5), R184–R190.

Duan, F., Giehl, R. F. H., Geldner, N., Salt, D. E. and von Wirén, N. (2018). Rootzone-specific localization of AMTs determines ammonium transport pathways and nitrogen allocation to shoots. *PLoS Biol.* 16(10), e2006024.

Eckardt, N. A. (2006). Medicago truncatula CRE1 cytokinin receptor regulates nodulation and lateral root development. *Plant Cell* 18(10), 2419.

Emerich, D. W. and Krishnan, H. B. (2014). Symbiosomes: temporary moonlighting organelles. *Biochem. J.* 460(1), 1–11.

Engineer, C. B. and Kranz, R. G. (2007). Reciprocal leaf and root expression of AtAmt1. 1 and root architectural changes in response to nitrogen starvation. *Plant Physiol.* 143(1), 236–250.

Fedorova, E., Thomson, R., Whitehead, L. F., Maudoux, O., Udvardi, M. K. and Day, D. A. (1999). Localization of H^+-ATPases in soybean root nodules. *Planta* 209(1), 25–32.

Ferguson, B. J., Indrasumunar, A., Hayashi, S., Lin, M. H., Lin, Y. H., Reid, D. E. and Gresshoff, P. M. (2010). Molecular analysis of legume nodule development and autoregulation. *J. Integr. Plant Biol.* 52(1), 61-76.

Ferguson, B. J., Mens, C., Hastwell, A. H., Zhang, M., Su, H., Jones, C. H., Chu, X. and Gresshoff, P. M. (2019). Legume nodulation: the host controls the party, *Plant Cell Environ.* 42(1), 41-51.

Fournier, J., Teillet, A., Chabaud, M., Ivanov, S., Genre, A., Limpens, E., de Carvalho-Niebel, F. and Barker, D. G. (2015). Remodeling of the infection chamber before infection thread formation reveals a two-step mechanism for rhizobial entry into the host legume root hair. *Plant Physiol.* 167(4), 1233-1242.

Foyer, C. H., Lam, H. M., Nguyen, H. T., Siddique, K. H., Varshney, R. K., Colmer, T. D., Cowling, W., Bramley, H., Mori, T. A., Hodgson, J. M., Cooper, J. W., Miller, A. J., Kunert, K., Vorster, J., Cullis, C., Ozga, J. A., Wahlqvist, M. L., Liang, Y., Shou, H., Shi, K., Yu, J., Fodor, N., Kaiser, B. N., Wong, F. L., Valliyodan, B. and Considine, M. J. (2016). Neglecting legumes has compromised human health and sustainable food production. *Nat. Plants* 2, 16112.

Foyer, C. H., Nguyen, H. and Lam, H. M. (2019). Legumes–the art and science of environmentally sustainable agriculture. *Plant Cell Environ.* 42(1), 1-5.

Gage, D. J. (2004). Infection and invasion of roots by symbiotic, nitrogen-fixing rhizobia during nodulation of temperate legumes. *Microbiol. Mol. Biol. Rev.* 68(2), 280-300.

Gansel, X., Muños, S., Tillard, P. and Gojon, A. (2001). Differential regulation of the NO_3^- and NH_4^+ transporter genes *AtNrt2.1* and *AtAmt1.1* in *Arabidopsis*: relation with long-distance and local controls by N status of the plant *Plant J.* 26(2), 143-155.

Gavrin, A., Kaiser, B. N., Geiger, D., Tyerman, S. D., Wen, Z., Bisseling, T. and Fedorova, E. E. (2014). Adjustment of host cells for accommodation of symbiotic bacteria: vacuole defunctionalization, HOPS suppression, and TIP1g retargeting in *Medicago*. *Plant Cell* 26(9), 3809-3822.

Gazzarrini, S., Lejay, L., Gojon, A., Ninnemann, O., Frommer, W. B. and von Wirén, N. (1999). Three functional transporters for constitutive, diurnally regulated, and starvation-induced uptake of ammonium into *Arabidopsis* roots. *Plant Cell* 11(5), 937-948.

Geddes, B. A. and Oresnik, I. J. (2016). The mechanism of symbiotic nitrogen fixation. In: Hurst, C. J. (Ed.). *The Mechanistic Benefits of Microbial Symbionts*, Springer, Cham, pp 69-97.

Giehl, R. F. H., Laginha, A. M., Duan, F., Rentsch, D., Yuan, L. and von Wirén, N. (2017). A critical role of AMT2;1 in root-to-shoot translocation of ammonium in *Arabidopsis*. *Mol. Plant* 10, 1449-1460.

Gilchrist, E. and Haughn, G. (2010). Reverse genetics techniques: engineering loss and gain of gene function in plants. *Brief. Funct. Genom.* 9(2), 103-110.

Glass, A. D. M., Britto, D. T., Kaiser, B. N., Kinghorn, J. R., Kronzucker, H. J., Kumar, A., Okamoto, M., Rawat, S., Siddiqi, M. Y., Unkles, S. E. and Vidmar, J. J. (2002). The regulation of nitrate and ammonium transport systems in plants. *J. Exp. Bot.* 53(370), 855-864.

Gonzalez-Rizzo, S., Crespi, M. and Frugier, F. (2006). The *Medicago truncatula* CRE1 cytokinin receptor regulates lateral root development and early symbiotic interaction with *Sinorhizobium meliloti*. *Plant Cell* 18(10), 2680-2693.

Groth, M., Takeda, N., Perry, J., Uchida, H., Dräxl, S., Brachmann, A., Sato, S., Tabata, S., Kawaguchi, M., Wang, T. L. and Parniske, M. (2010). *NENA*, a *Lotus japonicus* homolog of *sec13*, is required for rhizodermal infection by arbuscular mycorrhiza fungi and rhizobia but dispensable for cortical endosymbiotic development. *Plant Cell* 22(7), 2509–2526.

Guether, M., Neuhäuser, B., Balestrini, R., Dynowski, M., Ludewig, U. and Bonfante, P. (2009). A Mycorrhizal-specific ammonium transporter from *Lotus japonicus* acquires nitrogen released by arbuscular mycorrhizal fungi. *Plant Physiol.* 150(1), 73–83.

Guo, Q., Liu, Q., Smith, N. A., Liang, G. and Wang, M. B. (2016). RNA silencing in plants: mechanisms, technologies and applications in horticultural crops. *Curr. Genomics* 17(6), 476–489.

Hakoyama, T., Niimi, K., Watanabe, H., Tabata, R., Matsubara, J., Sato, S., Nakamura, Y., Tabata, S., Jichun, L., Matsumoto, T., Tatsumi, K., Nomura, M., Tajima, S., Ishizaka, M., Yano, K., Imaizumi-Anraku, H., Kawaguchi, M., Kouchi, H. and Suganuma, N. (2009). Host plant genome overcomes the lack of a bacterial gene for symbiotic nitrogen fixation. *Nature* 462(7272), 514–517.

Herridge, D. F., Peoples, M. B. and Boddey, R. M. (2008). Global inputs of biological nitrogen fixation in agricultural systems. *Plant Soil* 311(1–2), 1–18.

Hoelzle, I., Finer, J. J., McMullen, M. D. and Streeter, J. G. (1992). Induction of glutamine synthetase activity in nonnodulated roots of *Glycine max*, *Phaseolus vulgaris*, and *Pisum sativum*. *Plant Physiol.* 100(1), 525–528.

Hoffman, B. M., Lukoyanov, D., Yang, Z. Y., Dean, D. R. and Seefeldt, L. C. (2014). Mechanism of nitrogen fixation by nitrogenase: the next stage. *Chem. Rev.* 114(8), 4041–4062.

Horváth, B., Yeun, L. H., Domonkos, A., Halász, G., Gobbato, E., Ayaydin, F., Miró, K., Hirsch, S., Sun, J., Tadege, M., Ratet, P., Mysore, K. S., Ané, J. M., Oldroyd, G. E. and Kaló, P. (2011). *Medicago truncatula IPD3* is a member of the common symbiotic signaling pathway required for rhizobial and mycorrhizal symbioses. *Mol. Plant Microbe Interact.* 24(11), 1345–1358.

Hunt, S. (1997). Measurement of nitrogenase activity in N_2-fixing nodules of soybean. In: Glase, J. C. (Ed.). *Tested Studies for Laboratory Teaching Vol. 18*. Proceedings of the 18th Workshop/Conference of the Association for Biology Laboratory Education (ABLE), Boston University, pp 125–141.

Hwang, J. H., Ellingson, S. R. and Roberts, D. M. (2010). Ammonia permeability of the soybean nodulin 26 channel. *FEBS Lett.* 584(20), 4339–4343.

Imaizumi-Anraku, H., Takeda, N., Charpentier, M., Perry, J., Miwa, H., Umehara, Y., Kouchi, H., Murakami, Y., Mulder, L., Vickers, K., Pike, J., Downie, J. A., Wang, T., Sato, S., Asamizu, E., Tabata, S., Yoshikawa, M., Murooka, Y., Wu, G. J., Kawaguchi, M., Kawasaki, S., Parniske, M. and Hayashi, M. (2005). Plastid proteins crucial for symbiotic fungal and bacterial entry into plant roots. *Nature* 433(7025), 527–531.

Ivanov, S., Fedorova, E. E., Limpens, E., De Mita, S., Genre, A., Bonfante, P. and Bisseling, T. (2012). *Rhizobium*-legume symbiosis shares an exocytotic pathway required for arbuscule formation. *Proc. Natl. Acad. Sci. U. S. A.* 109(21), 8316–8321.

Jeudy, C., Ruffel, S., Freixes, S., Tillard, P., Santoni, A. L., Morel, S., Journet, E. P., Duc, G., Gojon, A., Lepetit, M. and Salon, C. (2010). Adaptation of *Medicago truncatula* to nitrogen limitation is modulated via local and systemic nodule developmental responses. *New Phytol.* 185(3), 817–828.

Kaiser, B. N., Finnegan, P. M., Tyerman, S. D., Whitehead, L. F., Bergersen, F. J., Day, D. A. and Udvardi, M. K. (1998). Characterization of an ammonium transport protein from the peribacteroid membrane of soybean nodules. *Science* 281(5380), 1202–1206.

Kaiser, B. N., Moreau, S., Castelli, J., Thomson, R., Lambert, A., Bogliolo, S., Puppo, A. and Day, D. A. (2003). The soybean NRAMP homologue, GmDMT1, is a symbiotic divalent metal transporter capable of ferrous iron transport. *Plant J.* 35(3), 295–304.

Kaiser, B. N., Rawat, S. R., Siddiqi, M. Y., Masle, J. and Glass, A. D. (2002). Functional analysis of an Arabidopsis T-DNA "knockout" of the high-affinity $NH_4^{(+)}$ transporter AtAMT1;1. *Plant Physiol.* 130(3), 1263–1275.

Kaló, P., Gleason, C., Edwards, A., Marsh, J., Mitra, R. M., Hirsch, S., Jakab, J., Sims, S., Long, S. R., Rogers, J., Kiss, G. B., Downie, J. A. and Oldroyd, G. E. (2005). Nodulation signaling in legumes requires NSP2, a member of the GRAS family of transcriptional regulators. *Science* 308(5729), 1786–1789.

Kanamori, N., Madsen, L. H., Radutoiu, S., Frantescu, M., Quistgaard, E. M., Miwa, H., Downie, J. A., James, E. K., Felle, H. H., Haaning, L. L., Jensen, T. H., Sato, S., Nakamura, Y., Tabata, S., Sandal, N. and Stougaard, J. (2006). A nucleoporin is required for induction of Ca^{2+} spiking in legume nodule development and essential for rhizobial and fungal symbiosis. *Proc. Natl. Acad. Sci. U. S. A.* 103(2), 359–364.

Kevei, Z., Lougnon, G., Mergaert, P., Horváth, G. V., Kereszt, A., Jayaraman, D., Zaman, N., Marcel, F., Regulski, K., Kiss, G. B., Kondorosi, A., Endre, G., Kondorosi, E. and Ané, J. M. (2007). 3-hydroxy-3-methylglutaryl coenzyme A reductase1 interacts with NORK and is crucial for nodulation in *Medicago truncatula*. *Plant Cell* 19(12), 3974–3989.

Kim, J. S., Park, Y. S., Nam, H. G. and Park, J. W. (2015). Imaging a specific mRNA in pollen with atomic force microscopy. *RSC Adv.* 5(24), 18858–18865.

Kobae, Y., Tamura, Y., Takai, S., Banba, M. and Hata, S. (2010). Localized expression of arbuscular mycorrhiza-inducible ammonium transporters in soybean. *Plant Cell Physiol.* 51(9), 1411–1415.

Kronzucker, H. J., Siddiqi, M. Y. and Glass, A. D. (1996). Kinetics of NH_4^+ influx in spruce. *Plant Physiol.* 110(3), 773–779.

Kronzucker, H. J., Siddiqi, M. Y. and Glass, A. D. M. (1997). Conifer root discrimination against soil nitrate and the ecology of forest succession. *Nature* 385(6611), 59–61.

Krusell, L., Krause, K., Ott, T., Desbrosses, G., Krämer, U., Sato, S., Nakamura, Y., Tabata, S., James, E. K., Sandal, N., Stougaard, J., Kawaguchi, M., Miyamoto, A., Suganuma, N. and Udvardi, M. K. (2005). The sulfate transporter SST1 is crucial for symbiotic nitrogen fixation in *Lotus japonicus* root nodules. *Plant Cell* 17(5), 1625–1636.

Krylova, V., Andreev, I. M., Zartdinova, R. and Izmailov, S. F. (2013). Biochemical characteristics of the Ca^{2+} pumping ATPase in the peribacteroid membrane from broad bean root nodules. *Protoplasma* 250(2), 531–538.

Kumar, S., Stecher, G. andTamura, K. (2016). MEGA7: molecular evolutionary genetics analysis version 7.0 for bigger datasets. *Mol. Biol. Evol.* 33(7), 1870–1874.

Lal, R. and Stewart, B. A. (2018). *Soil Nitrogen Uses and Environmental Impacts* (1st edn.), CRC Press, Boca Raton, FL.

Lanquar, V., Loqué, D., Hörmann, F., Yuan, L., Bohner, A., Engelsberger, W. R., Lalonde, S., Schulze, W. X., von Wirén, N. and Frommer, W. B. (2009). Feedback inhibition of ammonium uptake by a phospho-dependent allosteric mechanism in *Arabidopsis*. *Plant Cell*, WB 21(11), 3610–3622.

Laplaze, L., Lucas, M. and Champion, A. (2015). Rhizobial root hair infection requires auxin signaling. *Trends Plant Sci.* 20(6), 332–334.

Lassaletta, L., Billen, G., Grizzetti, B., Anglade, J. and Garnier, J. (2014). 50 year trends in nitrogen use efficiency of world cropping systems: the relationship between yield and nitrogen input to cropland. *Environ. Res. Lett.* 9(10), 105011.

Lee, M., Shevliakova, E., Stock, C. A., Malyshev, S. and Milly, P. C. D. (2019). Prominence of the tropics in the recent rise of global nitrogen pollution. *Nat. Commun.* 10(1), 1437.

Lejay, L., Gansel, X., Cerezo, M., Tillard, P., Müller, C., Krapp, A., von Wirén, N., Daniel-Vedele, F. and Gojon, A. (2003). Regulation of root ion transporters by photosynthesis: functional importance and relation with hexokinase. *Plant Cell* 15(9), 2218–2232.

Lemaire, G. and Gastal, F. (2019). Crop responses to nitrogen. In: Savin, R. and Slafer, G. A. (Eds). *Crop Science. Encyclopedia of Sustainability Science and Technology Series*, Springer, New York, pp 159–184.

LeVier, K., Day, D. A. and Guerinot, M. L. (1996). Iron uptake by symbiosomes from soybean root nodules. *Plant Physiol.* 111(3), 893-900.

Lévy, J., Bres, C., Geurts, R., Chalhoub, B., Kulikova, O., Duc, G., Journet, E. P., Ané, J. M., Lauber, E., Bisseling, T., Dénarié, J., Rosenberg, C. and Debellé, F. (2004). A putative Ca^{2+} and calmodulin-dependent protein kinase required for bacterial and fungal symbioses. *Science* 303(5662), 1361–1364.

Libault, M., Joshi, T., Takahashi, K., Hurley-Sommer, A., Puricelli, K., Blake, S., Finger, R. E., Taylor, C. G., Xu, D., Nguyen, H. T. and Stacey, G. (2009). Large-scale analysis of putative soybean regulatory gene expression identifies a Myb gene involved in soybean nodule development. *Plant Physiol.* 151(3), 1207–1220.

Lima, J. E., Kojima, S., Takahashi, H. and von Wirén, N. (2010). Ammonium triggers lateral root branching in *Arabidopsis* in an ammonium TRANSPORTER1; 3-dependent manner. *Plant Cell* 22(11), 3621–3633.

Limpens, E., Franken, C., Smit, P., Willemse, J., Bisseling, T. and Geurts, R. (2003). LysM domain receptor kinases regulating rhizobial Nod factor-induced infection. *Science* 302(5645), 630–633.

Liu, C. W., Breakspear, A., Stacey, N., Findlay, K., Nakashima, J., Ramakrishnan, K., Liu, M., Xie, F., Endre, G., de Carvalho-Niebel, F., Oldroyd, G. E. D., Udvardi, M. K., Fournier, J. and Murray, J. D. (2019). A protein complex required for polar growth of rhizobial infection threads. *Nat. Commun.* 10(1), 2848.

Liu, C. W. and Murray, J. D. (2016). The role of flavonoids in nodulation host-range specificity: an update. *Plants* 5(3), 33.

Loqué, D., Lalonde, S., Looger, L. L., von Wiren, N. and Frommer, W. B. (2007). A cytosolic trans-activation domain essential for ammonium uptake. *Nature* 446(7132), 195–198.

Loqué, D., Ludewig, U., Yuan, L.and Wiren, N. (2005). Tonoplast intrinsic proteins AtTIP;1 and AtTIP2;3 facilitate NH_3 transport into the vacuole. *Plant Physiol.* 137(2), 671–680.

Loqué, D., Yuan, L., Kojima, S., Gojon, A., Wirth, J., Gazzarrini, S., Ishiyama, K., Takahashi, H. and von Wiren, N. (2006). Additive contribution of AMT1;1 and AMT1;3 to high-affinity ammonium uptake across the plasma membrane of nitrogen-deficient *Arabidopsis* roots. *Plant J.* 48(4), 522–534.

Ludewig, U., Neuhäuser, B. and Dynowski, M. (2007). Molecular mechanisms of ammonium transport and accumulation in plants. *FEBS Lett.* 581(12), 2301–2308.

Mabood, F., Zhou, X. and Smith, D. L. (2014). Microbial signaling and plant growth promotion. *Can. J. Plant Sci.* 94(6), 1051–1063.

Marini, A. M., Springael, J. Y., Frommer, W. B. and André, B. (2000). Cross-talk between ammonium transporters in yeast and interference by the soybean SAT1 protein. *Mol. Microbiol.* 35(2), 378–385.

Marschner, H. (1995). *Mineral Nutrition of Higher Plants*, Academic Press, London.

Marsh, J. F., Rakocevic, A., Mitra, R. M., Brocard, L., Sun, J., Eschstruth, A., Long, S. R., Schultze, M., Ratet, P. and Oldroyd, G. E. (2007). *Medicago truncatula* NIN is essential for rhizobial-independent nodule organogenesis induced by autoactive calcium/calmodulin-dependent protein kinase. *Plant Physiol.* 144(1), 324–335.

Masalkar, P., Wallace, I. S., Hwang, J. H. and Roberts, D. M. (2010). Interaction of cytosolic glutamine synthetase of soybean root nodules with the C-terminal domain of the symbiosome membrane nodulin 26 aquaglyceroporin. *J. Biol. Chem.* 285(31), 23880–23888.

Mayer, M. and Ludewig, U. (2006). Role of AMT1; 1 in NH_4^+ acquisition in *Arabidopsis thaliana*. *Plant Biol.* 8(4), 522–528.

Mazurkiewicz, D. (2003). Characterisation of a novel family of eukaryotic ammonium transport proteins, PhD thesis, University of Adelaide, Adelaide.

McLellan, E. L., Cassman, K. G., Eagle, A. J., Woodbury, P. B., Sela, S., Tonitto, C., Marjerison, R. D. and van Es, H. M. (2018). The nitrogen balancing act: tracking the environmental performance of food production. *BioScience* 68(3), 194–203.

Meena, R. S. and Lal, R. (2018). Legumes and sustainable use of soils. In: Meena, R. S., Das, A., Yadav, G. S. and Lal, R. (Eds). *Legumes for Soil Health and Sustainable Management*, Springer, Singapore, pp 1–31.

Meier, M., Liu, Y., Lay-Pruitt, K. S., Takahashi, H. and von Wirén, N. (2020). Auxin-mediated root branching is determined by the form of available nitrogen. *Nat. Plants* 6(9), 1136–1145.

Messinese, E., Mun, J.-H., Yeun, L. H., Jayaraman, D., Rougé, P., Barre, A., Lougnon, G., Schornack, S., Bono, J.-J. and Cook, D. R. (2007). A novel nuclear protein interacts with the symbiotic DMI3 calcium- and calmodulin-dependent protein kinase of *Medicago truncatula*. *Mol. Plant Microbe Interact.* 20(8), 912–921.

Middleton, P. H., Jakab, J., Penmetsa, R. V., Starker, C. G., Doll, J., Kaló, P., Prabhu, R., Marsh, J. F., Mitra, R. M., Kereszt, A., Dudas, B., VandenBosch, K., Long, S. R., Cook, D. R., Kiss, G. B. and Oldroyd, G. E. (2007). An ERF transcription factor in *Medicago truncatula* that is essential for Nod factor signal transduction. *Plant Cell* 19(4), 1221–1234.

Mierziak, J., Kostyn, K. and Kulma, A. (2014). Flavonoids as important molecules of plant interactions with the environment. *Molecules* 19(10), 16240–16265.

Moreau, S., Day, D. A. and Puppo, A. (1998). Ferrous iron is transported across the peribacteroid membrane of soybean nodules. *Planta* 207(1), 83–87.

Moreau, S., Meyer, J. M. and Puppo, A. (1995). Uptake of iron by symbiosomes and bacteroids from soybean nodules, *FEBS Lett.* 361(2–3), 225–228.

Moreau, S., Thomson, R. M., Kaiser, B. N., Trevaskis, B., Guerinot, M. L., Udvardi, M. K., Puppo, A. and Day, D. A. (2002). GmZIP1 encodes a symbiosis-specific zinc transporter in soybean. *J. Biol. Chem.* 277(7), 4738–4746.

Motte, H. and Beeckman, T. (2020). A pHantastic ammonium response. *Nat. Plants* 6(9), 1080–1081.

Mouritzen, P. and Rosendahl, L. (1997). Identification of a transport mechanism for NH_4^+ in the symbiosome membrane of pea root nodules. *Plant Physiol.* 115(2), 519–526.

Murray, J. D. (2011). Invasion by invitation: rhizobial infection in legumes. *Mol. Plant Microbe Interact.* 24(6), 631–639.

Neuhäuser, B., Dynowski, M. and Ludewig, U. (2009). Channel-like NH_3 flux by ammonium transporter *At*AMT2. *FEBS Lett.* 583(17), 2833–2838.

Neuhäuser, B., Dynowski, M., Mayer, M. and Ludewig, U. (2007). Regulation of NH_4^+ transport by essential cross talk between AMT monomers through the carboxyl tails. *Plant Physiol.* 143(4), 1651–1659.

Neuhäuser, B. and Ludewig, U. (2014). Uncoupling of ionic currents from substrate transport in the plant ammonium transporter *At*AMT1; 2. *J. Biol. Chem.* 289(17), 11650–11655.

Niemietz, C. M. and Tyerman, S. D. (2000). Channel-mediated permeation of ammonia gas through the peribacteroid membrane of soybean nodules. *FEBS Lett.* 465(2–3), 110–114.

Ninnemann, O., Jauniaux, J. and Frommer, W. (1994). Identification of a high affinity NH_4^+ transporter from plants. *EMBO J.* 13(15), 3464–3471.

Noor, S. N. M., Day, D. A. and Smith, P. M. (2015). The symbiosome membrane. In: de Bruijn, F. J. (Ed.). *Biological Nitrogen Fixation* (vol. 2), John Wiley & Sons, Inc., New York, pp 683–694.

Oldroyd, G. E. and Downie, J. A. (2008). Coordinating nodule morphogenesis with rhizobial infection in legumes. *Annu. Rev. Plant Biol.* 59, 519–546.

Oldroyd, G. E. and Long, S. R. (2003). Identification and characterization of nodulation-signaling pathway 2, a gene of *Medicago truncatula* involved in Nod factor signaling. *Plant Physiol.* 131(3), 1027–1032.

Oldroyd, G. E., Murray, J. D., Poole, P. S. and Downie, J. A. (2011). The rules of engagement in the legume-rhizobial symbiosis. *Annu. Rev. Genet.* 45, 119–144.

Oldroyd, G. E. D. (2013). Speak, friend, and enter: signalling systems that promote beneficial symbiotic associations in plants. *Nat. Rev. Microbiol.* 11(4), 252–263.

Ortiz-Ramirez, C., Mora, S. I., Trejo, J. and Pantoja, O. (2011). *Pv*AMT1; 1, a highly selective ammonium transporter that functions as H^+/NH_4^+ symporter. *J. Biol. Chem.* 286(36), 31113–31122.

Ovchinnikova, E., Journet, E. P., Chabaud, M., Cosson, V., Ratet, P., Duc, G., Fedorova, E., Liu, W., den Camp, R. O., Zhukov, V., Tikhonovich, I., Borisov, A., Bisseling, T. and Limpens, E. (2011). IPD3 controls the formation of nitrogen-fixing symbiosomes in pea and *Medicago* spp. *Mol. Plant Microbe Interact.* 24(11), 1333–1344.

Palmgren, M. (2018). Plant epithelia: what is the role of the mortar in the wall? *PLoS Biol.* 16(12), e3000073. doi: 10.1371/journal.pbio.3000073.

Peoples, M. B. (2017). Biological nitrogen fixation: contributions to agriculture. In Ed.: Lal, R. *Encyclopedia of Soil Science*, CRC Press, pp 223-226.

Perret, X., Staehelin, C. and Broughton, W. J. (2000). Molecular basis of symbiotic promiscuity. *Microbiol. Mol. Biol. Rev.* 64(1), 180–201.

Prell, J., White, J. P., Bourdes, A., Bunnewell, S., Bongaerts, R. J. and Poole, P. S. (2009). Legumes regulate *Rhizobium* bacteroid development and persistence by the supply of branched-chain amino acids. *Proc. Natl. Acad. Sci. U. S. A.* 106(30), 12477–12482.

Radutoiu, S., Madsen, L. H., Madsen, E. B., Felle, H. H., Umehara, Y., Gronlund, M., Sato, S., Nakamura, Y., Tabata, S., Sandal, N. and Stougaard, J. (2003). Plant recognition of symbiotic bacteria requires two LysM receptor-like kinases. *Nature* 425(6958), 585–592.

Rawat, S. R., Silim, S. N., Kronzucker, H. J., Siddiqi, M. Y. and Glass, A. D. (1999). *At*AMT1 gene expression and NH_4^+ uptake in roots of *Arabidopsis thaliana*: evidence for regulation by root glutamine levels. *Plant J.* 19(2), 143-152.

Rivers, R. L., Dean, R. M., Chandy, G., Hall, J. E., Roberts, D. M. and Zeidel, M. L. (1997). Functional analysis of nodulin 26, an aquaporin in soybean root nodule symbiosomes. *J. Biol. Chem.* 272(26), 16256-16261.

Roberts, D. M. and Tyerman, S. D. (2002). Voltage-dependent cation channels permeable to NH_4^+, K^+, and Ca^{2+} in the symbiosome membrane of the model legume *Lotus japonicus*. *Plant Physiol.* 128(2), 370-378.

Roche, D., Temple, S. J. and Sengupta-Gopalan, C. (1993). Two classes of differentially regulated glutamine synthetase genes are expressed in the soybean nodule: a nodule-specific class and a constitutively expressed class. *Plant Mol. Biol.* 22(6), 971-983.

Rogato, A., D'Apuzzo, E., Barbulova, A., Omrane, S., Stedel, C., Simon-Rosin, U., Katinakis, P., Flemetakis, M., Udvardi, M. and Chiurazzi, M. (2008). Tissue-specific down-regulation of *Lj*AMT1;1 compromises nodule function and enhances nodulation in *Lotus japonicus*. *Plant Mol. Biol.* 68(6), 585-595.

Rogato, A., D'Apuzzo, E., Barbulova, A., Omrane, S., Parlati, A., Carfagna, S., Costa, A., Schiavo, F. L., Esposito, S. and Chiurazzi, M. (2010a). Characterization of a developmental root response caused by external ammonium supply in *Lotus japonicus*. *Plant Physiol.* 154(2), 784-795.

Rogato, A., D'Apuzzo, E. and Chiurazzi, M. (2010b). The multiple plant response to high ammonium conditions: the *Lotus japonicus* AMT1; 3 protein acts as a putative transceptor. *Plant Signal. Behav.* 5(12), 1594-1596.

Ruffel, S., Freixes, S., Balzergue, S., Tillard, P., Jeudy, C., Martin-Magniette, M. L., van der Merwe, M. J., Kakar, K., Gouzy, J., Fernie, A. R., Udvardi, M., Salon, C., Gojon, A., Lepetit, M. (2008). Systemic signaling of the plant nitrogen status triggers specific transcriptome responses depending on the nitrogen source in *Medicago truncatula*. *Plant Physiol.* 146(4), 2020-2035.

Saito, K., Yoshikawa, M., Yano, K., Miwa, H., Uchida, H., Asamizu, E., Sato, S., Tabata, S., Imaizumi-Anraku, H., Umehara, Y., Kouchi, H., Murooka, Y., Szczyglowski, K., Downie, J. A., Parniske, M., Hayashi, M. and Kawaguchi, M. (2007). NUCLEOPORIN85 is required for calcium spiking, fungal and bacterial symbioses, and seed production in *Lotus japonicus*. *Plant Cell* 19(2), 610-624.

Saitou, N. and Nei, M. (1987). The neighbor-joining method: a new method for reconstructing phylogenetic trees. *Mol. Biol. Evol.* 4(4), 406-425.

Salvemini, F., Marini, A.-M., Riccio, A., Patriarca, E. J. and Chiurazzi, M. (2001). Functional characterization of an ammonium transporter gene from *Lotus japonicus*. *Gene* 270(1-2), 237-243.

Schjoerring, J. K., Husted, S., Mäck, G. and Mattsson, M. (2002). The regulation of ammonium translocation in plants. *J. Exp. Bot.* 53(370), 883-890.

Seefeldt, L. C., Hoffman, B. M., Peters, J. W., Raugei, S., Beratan, D. N., Antony, E. and Dean, D. R. (2018). Energy transduction in nitrogenase. *Acc. Chem. Res.* 51(9), 2179-2186.

Shelden, M. C., Dong, B., de Bruxelles, G. L., Trevaskis, B., Whelan, J., Ryan, P. R., Howitt, S. M. and Udvardi, M. K. (2001). *Arabidopsis* ammonium transporters, *At*AMT1;1 and *At*AMT1;2, have different biochemical properties and functional roles. *Plant Soil* 231(1), 151-160.

Shelp, B. J. and Atkins, C. A. (1984). Subcellular location of enzymes of ammonia assimilation and asparagine synthesis in root nodules of *Lupinus albus* L.. *Plant Sci. Lett.* 36(3), 225-230.

Simon-Rosin, U., Wood, C. and Udvardi, M. K. (2003). Molecular and cellular characterisation of LjAMT2; 1, an ammonium transporter from the model legume *Lotus japonicus*. *Plant Mol. Biol.* 51(1), 99-108.

Singh, S. and Varma, A. (2017). Structure, function, and estimation of leghemoglobin. In: Hansen, A. P., Choudhary, D. K., Agrawal, P. K. and Varma, A. (Eds). *Rhizobium Biology and Biotechnology*, Springer, Cham, pp 309-330.

Smit, P., Limpens, E., Geurts, R., Fedorova, E., Dolgikh, E., Gough, C. and Bisseling, T. (2007). *Medicago* LYK3, an entry receptor in rhizobial nodulation factor signaling. *Plant Physiol.* 145(1), 183-191.

Smit, P., Raedts, J., Portyanko, V., Debelle, F., Gough, C., Bisseling, T. and Geurts, R. (2005). NSP1 of the GRAS protein family is essential for rhizobial Nod factor-induced transcription. *Science* 308(5729), 1789-1791.

Smith, P. M. C. and Atkins, C. A. (2002). Purine biosynthesis. Big in cell division, even bigger in nitrogen assimilation. *Plant Physiol.* 128(3), 793-802.

Sohlenkamp, C., Shelden, M., Howitt, S. and Udvardi, M. (2000). Characterization of *Arabidopsis* *At*AMT2, a novel ammonium transporter in plants. *FEBS Lett.* 467(2-3), 273-278.

Sohlenkamp, C., Wood, C. C., Roeb, G. W. and Udvardi, M. K. (2002). Characterization of *Arabidopsis* *At*AMT2, a high-affinity ammonium transporter of the plasma membrane. *Plant Physiol.* 130(4), 1788-1796.

Sonoda, Y., Ikeda, A., Saiki, S., von Wirén, N., Yamaya, T. and Yamaguchi, J. (2003). Distinct expression and function of three ammonium transporter genes (*Os*AMT1; 1-1; 3) in rice. *Plant Cell Physiol.* 44(7), 726-734.

Stagnari, F., Maggio, A., Galieni, A. and Pisante, M. (2017). Multiple benefits of legumes for agriculture sustainability: an overview. *Chem. Biol. Technol. Agric.* 4(1), 2.

Stracke, S., Kistner, C., Yoshida, S., Mulder, L., Sato, S., Kaneko, T., Tabata, S., Sandal, N., Stougaard, J., Szczyglowski, K. and Parniske, M. (2002). A plant receptor-like kinase required for both bacterial and fungal symbiosis. *Nature* 417(6892), 959-962.

Straub, D., Ludewig, U. and Neuhäuser, B. (2014). A nitrogen-dependent switch in the high affinity ammonium transport in *Medicago truncatula*. *Plant Mol. Biol.* 86(4-5), 485-494.

Straub, T. (2016). Plant ammonium transporter (AMT) integration in regulatory networks. PhD Thesis, University of Hohenheim.

Straub, T., Ludewig, U. and Neuhäuser, B. (2017). The kinase CIPK23 inhibits ammonium transport in *Arabidopsis thaliana*. *Plant Cell* 29(2), 409-422.

Streeter, J. G. (1989). Estimation of ammonium concentration in the cytosol of soybean nodules. *Plant Physiol.* 90(3), 779-782.

Taté, R., Riccio, A., Merrick, M. and Patriarca, E. J. (1998). The *Rhizobium etli amtB* gene coding for an NH_4^+ transporter is down-regulated early during bacteroid differentiation *Mol. Plant Microbe Interact.* 11(3), 188-198.

Tegeder, M. and Masclaux-Daubresse, C. (2018). Source and sink mechanisms of nitrogen transport and use. *New Phytol.* 217(1), 35-53.

Tirichine, L., Sandal, N., Madsen, L. H., Radutoiu, S., Albrektsen, A. S., Sato, S., Asamizu, E., Tabata, S. and Stougaard, J. (2007). A gain-of-function mutation in a cytokinin receptor triggers spontaneous root nodule organogenesis. *Science* 315(5808), 104-107.

Turgeon, B. G. and Bauer, W. D. (1985). Ultrastructure of infection-thread development during the infection of soybean by *Rhizobium japonicum*. *Planta* 163(3), 328-349.

Tyerman, S. D., Whitehead, L. F. and Day, D. A. (1995). A channel-like transporter for NH_4^+ on the symbiotic interface of N_2-fixing plants. *Nature* 378(6557), 629-632.

Udvardi, M. and Poole, P. S. (2013). Transport and metabolism in legume-rhizobia symbioses. *Annu. Rev. Plant Biol.* 64, 781-805.

Udvardi, M. K. and Day, D. A. (1989). Electrogenic ATPase activity on the peribacteroid membrane of soybean (*Glycine max* L.) root nodules *Plant Physiol.* 90(3), 982-987.

Udvardi, M. K. and Day, D. A. (1997). Metabolite transport across symbiotic membranes of legume nodules. *Annu. Rev. Plant Physiol. Plant Mol. Biol.* 48(1), 493-523.

Udvardi, M. K., Price, G. D., Gresshoff, P. M. and Day, D. A. (1988). A dicarboxylate transporter on the peribacteroid membrane of soybean nodules. *FEBS Lett.* 231(1), 36-40.

Unkovich, M., Herridge, D., Peoples, M., Cadisch, G., Boddey, B., Giller, K., Alves, B. and Chalk, P. (2008). *Measuring Plant-Associated Nitrogen Fixation in Agricultural Systems*, Australian Centre for International Agricultural Research (ACIAR), Australia.

Van de Velde, W., Zehirov, G., Szatmari, A., Debreczeny, M., Ishihara, H., Kevei, Z., Farkas, A., Mikulass, K., Nagy, A., Tiricz, H., Satiat-Jeunemaître, B., Alunni, B., Bourge, M., Kucho, K.-i., Abe, M., Kereszt, A., Maroti, G., Uchiumi, T., Kondorosi, E. and Mergaert, P. (2010). Plant peptides govern terminal differentiation of bacteria in symbiosis. *Science* 327(5969), 1122-1126.

Van den Bosch, K. A. and Newcomb, E. H. (1986). Immunogold localization of nodule-specific uricase in developing soybean root nodules. *Planta* 167(4), 425-436.

Venkateshwaran, M., Jayaraman, D., Chabaud, M., Genre, A., Balloon, A. J., Maeda, J., Forshey, K., den Os, D., Kwiecien, N. W., Coon, J. J., Barker, D. G. and Ané, J. M. (2015). A role for the mevalonate pathway in early plant symbiotic signaling. *Proc. Natl. Acad. Sci. U. S. A.* 112(31), 9781-9786.

Verma, D. P. S. and Hong, Z. (1996). Biogenesis of the peribacteriod membrane in root nodules. *Trends Microbiol.* 4(9), 364-368.

Vincill, E. D., Szczyglowski, K. and Roberts, D. M. (2005). GmN70 and LjN70. Anion transporters of the symbiosome membrane of nodules with a transport preference for nitrate. *Plant Physiol.* 137(4), 1435-1444.

Vitousek, P. M., Menge, D. N., Reed, S. C. and Cleveland, C. C. (2013). Biological nitrogen fixation: rates, patterns and ecological controls in terrestrial ecosystems. *Philos. Trans. R. Soc. B* 368(1621), 20130119.

von Wirén, N., Gazzarrini, S., Gojon, A. and Frommer, W. B. (2000). The molecular physiology of ammonium uptake and retrieval. *Curr. Opin. Plant Biol.* 3(3), 254-261.

Wang, D., Yang, S., Tang, F. and Zhu, H. (2012). Symbiosis specificity in the legume - rhizobial mutualism. *Cell. Microbiol.* 14(3), 334-342.

Wang, L., Deng, L. and Jiao, Y. (2017). *GmFW1* expression decreased in *GmSymRK* knockdown transgenic soybean roots. *Acta Agric. Scand. B* 67(1), 28-36.

Wang, L., Wang, L., Tan, Q., Fan, Q., Zhu, H., Hong, Z., Zhang, Z. and Duanmu, D. (2016). Efficient inactivation of symbiotic nitrogen fixation related genes in *Lotus japonicus* using CRISPR-Cas9. *Front. Plant Sci.* 7, 1333. doi: 10.3389/fpls.2016.01333.

Wang, M. Y., Siddiqi, M. Y., Ruth, T. J. and Glass, A. D. (1993). Ammonium uptake by rice roots (II. Kinetics of $^{13}NH_4^+$ influx across the plasmalemma). *Plant Physiol.* 103(4), 1259-1267.

Werner, A. K., Sparkes, I. A., Romeis, T. and Witte, C. P. (2008). Identification, biochemical characterization, and subcellular localization of allantoate amidohydrolases from *Arabidopsis* and soybean. *Plant Physiol.* 146(2), 418-430.

Whitehead, L. F. and Day, D. A. (1997). The peribacteroid membrane. *Physiol. Plant* 100(1), 30-44.

Xuan, W., Beeckman, T. and Xu, G. (2017). Plant nitrogen nutrition: sensing and signaling. *Curr. Opin. Plant Biol.* 39, 57–65.

Yano, K., Yoshida, S., Müller, J., Singh, S., Banba, M., Vickers, K., Markmann, K., White, C., Schuller, B., Sato, S., Asamizu, E., Tabata, S., Murooka, Y., Perry, J., Wang, T. L., Kawaguchi, M., Imaizumi-Anraku, H., Hayashi, M. and Parniske, M. (2008). *CYCLOPS*, a mediator of symbiotic intracellular accommodation. *Proc. Natl. Acad. Sci. U. S. A.* 105(51), 20540–20545.

Yuan, L., Graff, L., Loqué, D., Kojima, S., Tsuchiya, Y. N., Takahashi, H. and von Wirén, N. (2009). AtAMT1; 4, a pollen-specific high-affinity ammonium transporter of the plasma membrane in *Arabidopsis*. *Plant Cell Physiol.* 50(1), 13–25.

Yuan, L., Gu, R., Xuan, Y., Smith-Valle, E., Loqué, D., Frommer, W. B. and von Wirén, N. (2013). Allosteric regulation of transport activity by heterotrimerization of *Arabidopsis* ammonium transporter complexes *in vivo*. *Plant Cell* 25(3), 974–984.

Yuan, L., Loqué, D., Kojima, S., Rauch, S., Ishiyama, K., Inoue, E., Takahashi, H. and von Wirén, N. (2007a). The organization of high-affinity ammonium uptake in *Arabidopsis* roots depends on the spatial arrangement and biochemical properties of AMT1-type transporters. *Plant Cell* 19(8), 2636–2652.

Yuan, L., Loqué, D., Ye, F., Frommer, W. B. and von Wirén, N. (2007b). Nitrogen-dependent posttranscriptional regulation of the ammonium transporter AtAMT1;1. *Plant Physiol.* 143(2), 732–744.

Zhang, W., Wang, S., Yu, F., Tang, J., Yu, L., Wang, H. and Li, J. (2019). Genome-wide identification and expression profiling of sugar transporter protein (STP) family genes in cabbage (*Brassica oleracea* var. *capitata* L.) reveals their involvement in clubroot disease responses. *Genes* 10(1), 71.

Zuckerkandl, E. and Pauling, L. (1965). Evolutionary divergence and convergence in proteins. In: Bryson, V. and Vogel, H. J. (Eds). *Evolving Genes and Proteins*, Academic Press, USA. pp 97–166.

Chapter 3

Molecular interventions for improving crop nitrogen use efficiency: trends, opportunities and challenges in rice

Dinesh Kumar Jaiswal and Nandula Raghuram[1], Centre for Sustainable Nitrogen and Nutrient Management, University School of Biotechnology, Guru Gobind Singh Indraprastha University, India

1 Introduction

Reactive nitrogen (N) is an indispensable macronutrient nutrient, which is part of important biomolecules that constitute both structural and non-structural parts of the cells. Soil contains a mixture of N forms, in which the most abundant inorganic N forms are nitrate and ammonium, whereas amino acids and peptides are the organic N forms. In Asia, the most widely used N fertilizer is urea, which can be utilized either directly by plants or in the split N forms (ammonium or nitrate) produced by bacterial nitrification in the soil. Reactive N regulates many physiological, biochemical, and molecular processes; therefore crop yield depends on N fertilizer input and the biological N use efficiency (NUE) of plants. These two parameters significantly contribute to the effective utilization of reactive N (Ladha et al., 1998; Raghuram and Sharma, 2019; Lee, 2021; Madan et al., 2022). Low crop NUE results in inefficient utilization of N fertilizers, and the unused N fertilizers cause air and water environmental pollution and climate change, apart from economic wastage

[1] Corresponding author: Prof. Nandula Raghuram.

http://dx.doi.org/10.19103/AS.2024.0135.06

(Sutton et al., 2013, 2021; Raghuram et al., 2021). The recognition that reactive N pollution adversely affects biodiversity, human health and climate change (Sutton et al., 2013, 2019) led governments to adopt resolutions on sustainable N management (UNEA4.14, 2019; Raghuram et al., 2021; Sutton et al., 2021). Recently, articles by Ladha et al. (2020) and Winiwarter et al. (2022) have comprehensively discussed on reactive N and UN sustainable development goals (SDGs). We need to improve crop NUE by manipulating the biological system and/or reduce the consumption of N fertilizers as a part of our wider plans for sustainable N management, in order to achieve all the relevant SDGs (Ladha et al., 2020).

Rice is considered a model crop plant and has many advantages over wheat and maize as the rice genome is smaller and was sequenced much earlier, offering many genetic and genomic resources as compared to other crops. Further, the improved rice transformation efficiency as compared to wheat and maize expedites the molecular studies in rice. Rice has diverse genetic resources, including locally adapted cultivars, wild relatives and landraces among the cereals. According to the Global Crop Diversity Trust (https://www.croptrust.org/), established in 2004 by the Food and Agriculture Organization, over 500 000 accessions of cultivated rice and wild relatives are known in the world, and the International Rice Research Institute holds over 130 000 rice accessions (Jamora and Ramaiah, 2022). Further, rice is the only crop in which genomic diversity is well characterized on over 3000 genomes (Li et al., 2014).

An increase in the application rate of N fertilizers improves crop yield up to a certain extent, but after that, the addition of extra N fertilizers does not increase the crop yield. Over half of the global N fertilizers are consumed by just three crops - rice, wheat and maize with 16% for rice alone (Ladha et al., 2016), though in India it is as high as 37% (Abrol et al., 2017). Yield-centric crop improvement programmes under high input conditions and the lack of low input programmes meant that NUE declined disproportionately even as yields increased slightly (Sutton et al., 2013; Abrol et al., 2017; Raghuram et al., 2022a; Kabange et al., 2022). Many rice varieties such as green revolution variety and super hybrid rice have been developed, but they are responsive to high N fertilizer input (Wang et al., 2021a; Wang and Peng, 2017). Therefore, it is very important to reduce N fertilizer usage while maintaining a similar or high yield, which can be achieved by improving the NUE of rice. Nitrogen processes such as N acquisition from the soil, translocation, assimilation and remobilization are affected by environmental conditions, which make NUE a highly complex trait (Ladha et al., 1998). To dissect the molecular aspects of NUE, many N-responsive or NUE genes associated with above-mentioned processes have been identified in rice (Sharma et al., 2022, 2023; Madan et al., 2022; Kumari et al., 2021; Sharma et al., 2020; Mandal et al., 2022; Pathak et al.,

2020; Sinha et al., 2020; Kumari and Raghuram, 2020; Raghuram and Sharma, 2019; Mandal et al., 2018).

Identification of NUE donor genotypes and associated information are necessary to expedite the rice breeding programmes. For this purpose, we must develop the molecular markers associated with NUE candidate genes for introgression into different rice varieties. In this context, we must also identify the allelic diversity responsible for NUE within rice subspecies germplasms. Quantitative trail loci (QTL) mapping and genome-wide association study (GWAS) have helped to identify the gene(s) and genomic regions responsible for NUE. In this chapter, we have covered these aspects for a better understanding of NUE in rice, with the ultimate goal of achieving rice genotypes with high yield at low N input to save both economic and environmental costs.

2 Definition and selection of nitrogen use efficiency indices

Ratio of input and output has been defined in different ways to calculate the agronomical performance of a crop under different N conditions (Ladha et al., 2005). NUE is the most important biological indicator to reduce N fertilizer input with similar or higher yield. NUE is a complex trait as various phenotypic, physiological, biochemical and molecular changes occur in plants to regulate reactive N homeostasis in plants. It is well known that even cultivars of the same species have significant difference in NUE, which reflects the complex nature of NUE (Ladha et al., 1998). Due to such complexity, various definitions have been proposed to define the NUE and the findings were summarized in different publications (Ladha et al., 2005; Pathak et al., 2008; Raghuram and Sharma, 2019; Briat et al., 2020; Lemaire and Ciampitti, 2020; Congreves et al., 2021; Ciampitti et al., 2022; Ciampitti and Lemaire, 2022; Madan et al., 2022). Congreves et al. (2021) have provided a comprehensive list of known NUE indices including their strength and weakness based on fertilizer, plants, soil and isotopes. To rank the N-efficient rice genotypes, a field study was conducted using medium- and long-duration rice and various agronomically important parameters such as physiological NUE, agronomic NUE and N productivity index (NPI) among others were measured. Analyses revealed that NPI was one of the most important parameters that can be used to rank the N-efficient rice genotypes (Singh et al., 1998). Recently, redefinition of NUE has been proposed (Ciampitti et al., 2022) in the context of N uptake efficiency and N conversion efficiency. Application of N fertilizer (Nf) maximizes the grain yield (Ymax), which is related to environmental N losses. To achieve Ymax, we must regulate Nf and contribution by soil N (Ns) in such a way that minimum N losses occur to the environment. They also proposed that previously known Nitrogen Nutrition Index (NNI), which is used to measure the plant and crop N status

Table 1 List of common NUE indices used in rice

Important NUE related indices	Abbreviation (unit)	Formula	Outcome
Nitrogen use efficiency	NUE (kg yield kg^{-1} N)	$= \frac{\Delta Y (or\, W)}{\Delta (Nf + Ns)}$	Maximizing grain yield driven by N fertilizer and soil N, while minimizing environmental N losses
Nitrogen nutrition index	NNI	$= \frac{PNC}{Nc}$	This reflects plant N status as the Nc changes throughout plant's growth. NNI < 1 (deficient plant N status), NNI = 1 (optimal plant N status), NNI > 1 (excess N status)
Nitrogen use efficiency (crop)	NUE_{crop} (kg yield kg^{-1} N)	$= \frac{Yield\, Nf}{Fertlizer\, N}$	Yield per unit of applied N fertilizer.
Partial-factor productivity	PFP (grain yield/ total N)	$= \frac{Yield\, f}{Fertlizer\, N}$	The fraction of N fertilizer utilized and allocated to yield N.
Agronomic efficiency	AE (kg grain kg^{-1} N)	$= \frac{Yield\, f - Yield\, 0}{Fertlizer\, N}$	The contribution of N fertilizer in yield.
Fertilizer-N recovery efficiency	REfertN (kg N in grain kg^{-1} N applied)	$= \frac{Plant\, Nf - Plant\, N0}{Fertlizer\, N} \times 100$	The percent of N fertilizer used by plants.
Physiological efficiency	PE (kg grain yield increase kg^{-1} N taken up)	$= \frac{Yieldf - Yield0}{Plant\, Nf - Plant\, N0}$	N fertilizer contribution from the plant tissues towards the yield.
N uptake efficiency	NUpE (kg N kg^{-1} N)	$= \frac{Plant\, N}{Fertilizer\, N + soil\, N} \times 100$	The percent of soil N utilized by the plant.
N utilization efficiency	NUtE (kg grain kg^{-1} N)	$= \frac{Yield}{Plant\, N}$	N fertilizer contribution from the plant tissues towards the yield irrespective of background N.

(Continued)

Table 1 (*Continued*)

Important NUE related indices	Abbreviation (unit)	Formula	Outcome
Nitrogen use efficiency (yield)	NUE_{yield} (kg grain kg^{-1} N)	$NUpE \times NUtE$	Basic NUE definition, which predicts the contribution of applied N to the yield N.
N harvest index	NHI (%)	$= \frac{Yield N}{Plant N} \times 100$	The fraction of plant tissue N present in yield (Percentage form).
Recovery efficiency of N fertilizer	^{15}NRE (%)	$= \frac{TNdff\ in Plant\ or\ Soil}{Fertilizer\ N applied} \times 100$	The percent recovery or utilization of N fertilizer.

Adapted from: Ladha et al., 2005; Pathak et al., 2011; Raghuram and Sharma, 2019; Congreves et al., 2021; Madan et al., 2022; Ciampitti et al., 2022.
f, applied fertilizer; 0, not applied fertilizer condition; Plant Nf, amount of N in a fertilized plant; Plantf, indicate the weight of a fertilized plant; TNdff, total N derived from fertilizer. Nf, application of N fertilizers; Ns, N contribution by soil; Δ*Y*, yield difference (*Y*max − *Y*0); *Y*max, maximizing grain yield; *Y*0, grain yield contributed by Ns; *W*max, maximizing biomass; PNC, plant nitrogen concentration; Nc, minimum PNC that can produce maximum aboveground biomass.

(Lemaire et al., 2008). The NNI index is the ratio between plant N concentration and critical N concentration, which reflects the minimum N uptake required for maximum aboveground biomass. The NNI has been used for estimation of N requirement, N partition, photosynthesis capacity and NUE (Hu et al., 2014; Zhao, 2014; Briat et al., 2020). Commonly NUE indices used for rice experiments are provided in Table 1, and specific NUE indices may be chosen based on the purpose of the experiments. An important consideration in this regard is the most abundant N form(s) available to the plant and the role of soil/root microbiomes associated with different rice subspecies and genotypes (Zhang et al., 2019; Ladha and Reddy, 2019).

3 Nitrogen management practices for improved nitrogen use efficiency

Agronomical practices play an important role in NUE improvement as they reduce N loss occur through volatilization, leaching and denitrification (Udvardi et al., 2021; Ladha et al., 2020). With advancements in technology, N fertilizers have been modified and customized for enhanced NUE. Experiments performed with graded levels of customized fertilizer have been shown to improve NUE in rice (Nagabovanalli Basavarajappa et al., 2021). To achieve maximum yield, farmers apply excess N fertilizers in fewer doses whereas plants need nutrition in small amounts in many doses. As a result, most of the applied fertilizers are not utilized by the plants bringing down the NUE. To overcome this problem from the fertilizer angle, slow-release fertilizers have been developed incorporating urease and/or nitrification inhibitors with some success for some crops in some countries where farmers could afford their high costs (Li et al., 2018). Recently, nanotechnology has been applied to further improve slow-release fertilizer formulations and its potential for NUE remains to be firmly established (Sahu et al., 2022). Biochar, a partial pyrolysis product of agricultural organic material, has been shown to improve crop efficiency by enhancing moisture retention, nutrient holding capacity of soil, nutrient availability and uptake. It has been shown that the application of biochar improved NUE up to 12% in rice (Liu et al., 2022d). Further, biochar-based N fertilizers such as biochar-coated urea improved NUE by reducing N loss from soil and increasing N uptake in oilseed rape (Jia et al., 2021). The NUE-enhancing properties of biochar-coated urea are yet to be established in rice. Bio-fertilizers, the beneficial microorganisms, are involved in the mineralization process to increase the availability of nutrients and enhance nutrient acquisition in plants. Diazotrophs convert reduced N_2 to ammonia, which is utilized by plants. Biological nitrogen fixation (BNF) helps in NUE improvement in cereals including rice (Ladha et al., 2022). Precision farming such as geographic information systems, remote sensing and variable rate applicators are the technological advancements for NUE improvements. Further, leaf colour chart, chlorophyll meter and drip fertigation help to provide N fertilizer in the right amount at the right time, which reduce N loss and also improve NUE in rice. A more detailed discussion of N management practices is beyond the scope of this chapter, due to its focus on molecular aspects in rice.

4 Biological intervention for nitrogen use efficiency improvement

The various N management practices operate within the inherent genetic capacity of the plants to use available N inputs, which needs to be improved in terms of uptake, utilization and remobilization efficiencies, minimally influenced by agro-climatic conditions (Mandal et al., 2018; Raghuram and Sharma, 2019; Sinha et al., 2020; Kumari and Raghuram, 2020; Moring et al., 2021 Madan et al., 2022). The continuous variability of NUE in the germplasm of every crop is a piece of firm evidence that it is a quantitative trait involving many genes regulating the underlying physiological, biochemical and molecular interactions. Genetic improvement efforts need germplasm diversity for phenotypic characterization/selection and/or identification of donor genotypes for breeding, quantitative trait loci, marker development or genome-wide association mapping. The availability of huge germplasm diversity in rice and advancements in its genomics, functional genomics and biotechnological tools provide immense opportunities to employ reverse genetics for NUE enhancement in rice (Mandal et al., 2018; Raghuram and Sharma, 2019).

4.1 Nitrogen use efficiency phenotypes

Phenotype is a set of observable and ideally measurable characteristics associated with an inherited trait such as plant NUE, whether morphological, physiological or biochemical, but stable in nature and genetically regulated. Plants of the same genotype interacting with the environment may manifest as different phenotypes, but different genotypes that vary in the extent of the same phenotypic traits are of particular interest in crop improvement for NUE. The NUE is a complex trait, and therefore the identification and characterization of the phenotype and the associated genomic regions/loci/genes is an important but unfinished task in most crops including cereals (Hawkesford and Griffiths, 2019), despite the recent progress in rice (Sharma et al., 2018, 2021; Kumari et al., 2021). The N influences plant growth traits throughout the developmental processes in rice, from germination to seed setting (Sharma et al., 2018, 2021). The root is the primary organ for N uptake, and it has been shown that either N deprivation or excess N causes significant changes in the root architecture of rice (Zhang et al., 2020a; Sharma et al., 2018). Fluctuation in N input causes poorly developed plants with altered tiller number and reduced grain yield (Wang et al., 2017). Further, it has been shown that yield components such as panicle development and grain filling are also influenced by N. Using yield as major differentiator between N response and NUE in rice, we found that out of 20 significantly N-responsive traits, only 6 were significantly correlated with NUE for nitrate or urea and 2 others were N-form specific (Sharma et al., 2021). They include germination, root length, flowering time, shoot length before and after harvest, total plant height, fresh biomass and dry biomass. Details of phenotypic traits are provided in Table 2. Such detailed characterization of the NUE phenotype will boost the identification of highly efficient NUE QTLs and/or the most important candidate genes, which can be used to expedite the NUE improvement programmes in rice.

4.2 Quantitative trail loci mapping, whole genome re-sequencing and genome-wide association study for nitrogen use efficiency candidates

Natural variations exist among the germplasms and their screening leads to the identification of robust and reproducible phenotypes as well as the isolation of alleles responsible for the natural phenotypes (Ogura and Busch, 2015). Identification of stable NUE QTLs, which are less influenced by genotype × environment interaction is the most important step for NUE improvement in

Table 2 Segregation and shortlisting of phenotypic parameters for N response, yield and NUE

No.	N-responsive parameters	Yield-related parameters - low N	N use efficiency parameters
1	[a,b,c]Days to germination (G)	[a,b,c]Days to germination (G)	[a,b,c]Days to germination (G)
2	[a, c]Chlorophyll content (V1)	[a,b,c]Shoot length before harvest (V7)	[a,b,c]Shoot length before harvest (V7)
3	[a,b,c]Green leaf number at veg. stage(V2)	[a]Total leaf number at flowering (V8)	[a,b,c]Fresh biomass (V11)
4	[b,c]Yellow leaf number at veg. stage (V3)	[a,b,c]Fresh biomass (V11)	[a,b,c]Shoot length aft. harvest (V12)
5	[a,b,c]Total leaf number at veg. stage (V4)	[a,b,c]Shoot length after harvest (V12)	[a,c] Root length (V13)
6	[a,b,c]Leaf width(V5)	[a,b,c]Root length(V13)	[a,b,c]Total plant height (V14)
7	[a,b,c]Stem thickness (V6)	[a,b,c]Total plant height (V14)	[a,b,c] Dry biomass (V15)
8	[a,b,c]Shoot length before harvest (V7)	[a,b,c]Dry biomass (V15)	[a,c] Days to flowering (R1)
9	[a,b,c]Total leaf number at flowering (V8)	[a,c]Days to flowering (R1)	
10	[a,b,c]Green leaf number at flowering (V9)		
11	[a,b,c]Yellow leaf number at flowering (V10)		
12	[a,b,c]Fresh biomass (V11)		
13	[a,b,c]Shoot length after harvest (V12)		
14	[a,b] Root length (V13)		
15	[a,b] Total plant height (V14)		
16	[a,b,c] Dry biomass (V15)		
17	[a,c]Days to flowering (R1)		
18	[a,c] Unfilled seed weight (R2)		
19	[a,c] Total seed weight (R3)		
20	[ns] Filled seed weight (R4)		
21	[ns] Panicle weight (R5)		
22	[ns]Filled seed number (R6)		
23	[ns] Total seed number (R7)		
24	[a] Unfilled seed number (R8)		
25	[a,b,c]Weight of panicle remains (R9)		

This table was adapted from our earlier publication (Sharma et al., 2021). Twenty of the 25 parameters were confirmed as N responsive based on the significance test and shown in the first column. Ten of these parameters shortlisted from feature selection and network analyses were subjected to correlation analysis with grain yield in normal and low N and those parameters that significantly correlated with yield at low N are shown in the second column. The NUE was calculated as yield per unit N input for all genotypes/treatments, and their correlation with each of these ten parameters was analysed. Those that significantly correlated with NUE are shown in the third column. The superscripted alphabet a/b/c over the parameter in the first column denotes parameter in combined nitrate and urea N, only nitrate N and only urea N.
G, germination parameter; ns, non-significant parameter; R, reproductive parameter; V, vegetative parameter.

rice. Many studies have been performed to identify the QTLs/genomic regions and or genes involved in NUE in rice (Table 3). Almost two decades ago, Yamaya et al. (2002) focused on N recycling (sink and source relation) and observed the overexpression of glutamate synthase (NADH-GOGAT from Japonica) in Kasalath, an indica cultivar, increased the grain weight (up to 80%). Since Indica cultivars contain lower NADH-GOGAT protein content in the sink organ as compared to Japonica, overexpression of Japonica NADH-GOGAT confirmed their role in sink regulation. Then, they used the protein content of GS1 (senescing leaf blade) and NADH-GOGAT (developing leaf blade), performed QTLs mapping and identified QTLs associated with N recycling and various agronomic traits. With the advancement of genome information and mapping tools, many NUE QTLs and associated candidate genes have been identified in rice (Table 3). Phenotypic variation is the primary requisite for QTLs identification. Most common phenotypic traits in QTL analyses were chlorophyll content, plant height, tiller number and traits associated with the reproductive phase such as panicle number, seeds and yield. In the context of these traits and various physiological parameters, the N content of plant/organ/tissue and N fertilizer input were used to define the various efficiencies including NUE. Phenotypic parameters used in earlier studies are not defined as phenotypes for NUE. Recently, comprehensive phenotypic analyses established the phenotypes for NUE in rice (Sharma et al., 2021). QTL identification using these defined NUE phenotypes may produce stable QTLs, which are not influenced by genotype × environment interaction. In many studies of QTL mapping, the identification of candidate genes/loci is hampered by a lack of completely sequenced genome information and a long timeline to accomplish the different steps involved in the experiments. In this context, GWAS have produced many NUE candidates within a short duration and less effort. The genes OsTCP19, OsNAC68 and OsMYB61 among others are the important candidates identified by GWAS in rice (Tables 3 and 4). Though GWAS produce candidate gene(s), their role in NUE must be validated in field condition using transgenic or genome-editing (CRISPR/Cas9) approaches.

4.3 Genetic manipulation of genes involved in nitrogen use efficiency

Functional characterization of a particular biological response may be achieved using forward and reverse genetics approaches. Studies using forward or reverse genetics have provided a wealth of information on NUE in crop plants (Zhang et al., 2020b; Neeraja et al., 2021; Liu et al., 2022a). Forward genetics begins with identified phenotypes and then the isolation of loci/

Table 3 List of QTLs and genomic regions for NUE in rice

Method	Traits	QTL information	Potential candidate gene/marker information	Reference
QTL mapping	EPN, GNPP	*qEPN-LN4/qEPN-HN3, qGNPP-LN3 qGNPP-LN6*	qEPN-LN4/qEPN-HN3 (SLR1, Os03g0707600), (OsbZIP59, Os07g0209800)	Liu et al. (2022b)
GWAS	Chlorophyll content (SPAD value)	*n-p8, qSdw8, n-r3, qNCP-3-1 Q6, Q31, qRW6 qYd6, qSdw6, Q31, qRW6*	q1 (LOC_Os08g07010, LOC_Os08g07620, LOC_Os08g07630, LOC_Os08g07660), q2 (LOC_Os03g10980, LOC_Os03g11030, LOC_Os03g11190, LOC_Os03g11250), q3 (LOC_Os05g44900, LOC_Os05g45020, LOC_Os05g45030, LOC_Os05g45140, LOC_Os05g45380, LOC_Os05g45410), q4 (LOC_Os06g43030, LOC_Os06g43090, LOC_Os06g43150, LOC_Os06g43304, LOC_Os06g43384), q5 (LOC_Os06g44250, LOC_Os06g44360, LOC_Os06g44690, LOC_Os06g44750, LOC_Os06g45020)	Li et al. (2022b)
GWAS	-	-	Cytosolic glutamine synthetase 1;1 (OsGS1;1, Os02g0735200)	Liu et al. (2022c)
WGR	Large-grain, yield and NUE	*Gs3 allele* (nonsense mutated GS3 gene)	Grain size 3 (GS3, Os03g0407400)	Yoon et al. (2021)
GWAS	NUE	*tld1-D*	Auxin-responsive GH3 gene (Os11g0528700)	Rakotoson et al. (2021)
GWAS	TRN and NUE	-	OsTCP19	Liu et al. (2021b)
GWAS	N absorption and utilization	-	OsNAC68	Xin et al. (2021)

GWAS	Plant height, tiller number, chlorophyll content and leaf length	-	MYB61 (Os01g0285300), MOC2 (Os01g0866400), OsOAT (Os03g0643300), OsCIPK04 (Os12g0603700)	Lv et al. (2021)
GWAS	NUE	-	OsNLP4 (Os09g0549450)	Yu et al. (2021)
QTL mapping	Grain nitrogen increase rate, agNUE	*qGIR-11, q-N8_agNUE-1, q-N8_agNUE-6, q-N18_agNUE-11*	qGIR-11 (chr 11, Start bp 4,068,737 and end bp 4,236,048), q-N8_agNUE-1 (Chr 1, Start bp 3,805,999 and end bp 4,740,546), q-N8_agNUE-6 (chr 6, Start bp 25,153,480 and end bp 25,272,880), q-N18_agNUE-11 (chr 11, Start bp 3,936,668 and end bp 4,083,717)	Bai et al. (2021)
QTL mapping	NUE	*qSBM1*	SBM1 (LOC_Os01g65120)	Xu et al. (2021)
QTL mapping	NDT and NUE	-	*The bins 16, 1301, 1465, 1486, 3464 and 6249 shared the QTLs or genes for NDT identified in this study and NUE*	Shen et al. (2021)
QTL mapping	NUE	*qNLA1* and *qNLA2*	MYB61 (LOC_Os01g18240)	Gao et al. (2020)
QTL mapping	NUE	*qDNR1*	DNR1 (LOC_Os01g08270)	Zhang et al. (2021)
QTL mapping	PNUE ANUE	-	Markers RM13181-RM13201 covering 467.706 kb has 42 candidates	Vishnukiran et al. (2020)
QTL mapping	NUE	*qCR2*	OsNR2 (Os08g0468700)	Gao et al. (2019)
GWAS	NUE	-	OsNPF6.1 (LOC_Os01g01360)	Tang et al. (2019)

(*Continued*)

Table 3 (*Continued*)

Method	Traits	QTL information	Potential candidate gene/marker information	Reference
Association mapping	Yield under low N	*qGYLN5-1* (Os05g0208000) *qGYLN7* (Os07g0617800) *qGYLN10* (Os10g0189600)	2-oxoglutarate/malate translocator (Os05g0208000), alanine aminotransferase (Os07g0617800) and pyridoxal phosphate-dependent transferase (Os10g0189600)	Rao et al. (2018)
QTL mapping	NUE	*qngr2*	NGR2/GRF4	Li et al. (2018)
QTL mapping	NUE	*qNUE6*	Peptide transporter PTR2 (Os06g0264500) and Asparagine synthetase (Os06g0265000)	Yang et al. (2017)
QTL mapping	NUE	*qNUE2.1* *qNUE4.1* *qNUE6.1* *qNUE6.2* *qNUE10.1* *qNUE10.2*	-	Zhou et al. (2017)
QTL mapping	aNUE pNUE agNUE	-	aNUE (RM055-RM3199 on chr3, RM433-RM230 on chr8) pNUE (RM287-RM209 on chr11) agNUE (RM055-RM3199 on chr3, RM433-RM230 on chr8)	Nguyen et al. (2016)
QTL mapping	Agricultural NUE	*qANAE4*	-	Dai et al. (2015)
QTL mapping	NUE	*qCHR10*	NRT1.1B/OsNPF6.5 (LOC_Os10g40600)	Hu et al. (2015)
QTL mapping	NUE	*TOND1*	TOND1 (LOC_Os12g43440)	Zhang et al. (2015)
QTL mapping	Grain yield per plant, spikelets per panicle, number of filled grains per panicle, number of panicles per plant	*qGYP-4, qSP-1a, qSP-3b, qFGP-1a, qFGP-3a, qFGP-8a, qPNP-3a*	qGYP-4 (RM273-RM241, ch4), qSP-1a (RM1-RM3746, chr1), qSP-3b (RM135-RM168, chr3), qFGP-1a (RM1-RM3746, chr1), qFGP-3a (RM135-RM168, chr 3) qFGP-8a (RM5556-RM310, chr8), qPNP-3a (RM135-RM168, chr1)	Yue et al. (2015)[a]

QTL mapping	NUE	*qNGR9*	Dense and erect panicle 1 (DEP1, Os09g0441900)	Sun et al. (2014)
QTL mapping	NDT and NUE	*qPE3, qPE7 and qPE11*	-	Wei et al. (2012)
QTL mapping	NUE	*qNUEn1, qNUEl11*	Four genomic regions, including C86-C2340 on chromosome 1, RZ599-R1738 on chromosome 2, RZ471-C1023 on chromosome 7 and R3203-RM20a on chromosome 11, were found to contain QTLs for both NUE and GY.	Wei et al. (2011)
QTL mapping	Relative panicle number and relative yield	*Qrpn-2a, Qrpn-2b, Qryd-7a, Qryd-10*	Relative panicle number [Qrpn-2a (RM211, chr 2), Qrpn-2b (RM322-RM424)], relative yield [Qryd-7a (RM542, chr7), Qryd-10 (RM467, chr10)]	Wang et al. (2009)
QTL mapping	NUE	*qNUE-3*	A QTL region between markers RZ678, RZ574 and RZ284 was associated with nitrogen use and yield.	Senthilvel et al. (2008)
QTL mapping	pNUE	*pnue9*	Marker interval S9093-S9073	Cho et al. (2007)
QTL mapping	Plant height, tillers number per plant, chlorophyll content (low and N0 conditions).	-	QTL for plant height (Rh4118-RM248 on chr7), QTLs for tillers number per plant (RM401-RM417 on chr4 and OSR31-RM55 on chr3), QTL for chlorophyll (RM324-RM26 on chr7)	Tong et al. (2006)
QTL mapping	Protein contents of GS1 (senescing leaf blade) and NADH-GOGAT (developing leaf blade)	-	Glutamine synthetase (GS1, Os02g0735200) and glutamate synthase (NADH-GOGAT, Os01g0681900)	Yamaya et al. (2002)

The QTLs were identified using multiple phenotypic traits, and we considered those associated with yield and/or NUE.
agNUE, agronomic nitrogen use efficiency; ANUE, agronomic nitrogen use efficiency; ARE, agricultural nitrogen-absorption efficiency; EPN, effective panicle number; GIR, grain nitrogen increase rate; GNPP, grain number per panicle; NUPE, nitrogen uptake efficiency; NUTE, nitrogen utilization efficiency; PNUE, physiological nitrogen use efficiency; TRN, tillering response to nitrogen.
[a]Considered those QTLs that explained maximum phenotypic variations in the studied cultivars.

Table 4 List of NUE genes known in rice

Gene symbol (RAP-DB Id)	Gene name	Transgenic nature	Subspecies	N-responsive/NUE phenotype	Direct or indirect role in NUE	Reference
N transport						
OsNRT2.3b (Os01g0704100)	High-affinity nitrate transporter 2.3	EMS mutant of OsNRT2.3b overexpression line	Japonica	EMS mutant had reduced growth time with improved nitrogen translocation efficiency, physiological and NUE	Direct	Chen et al. (2022)
OsAMT1;1 (Os04g0509600), OsAMT1;2 (Os02g0620600), OsAMT1;3 (Os02g0620500)	Ammonium transporter-1;1, and 1;2 and 1;3	Knockout triple mutant	Japonica	Decreased panicle number per plant and grain yield in the mutants.	Indirect	Konishi and Ma (2021)
OsNRT1.1B (Os03g0235900)	Nitrate transporter 1	Allelic variation	Indica vs Japonica	DNR1^{indica} allele promoted auxin-mediated grain yield and NUE in Indica.	Direct	Zhang et al. (2021)
OsNAR2.1 (Os02g0595900) and OsNRT2.3a (Os01g0704100)	Activator for NRT2 (OsNAR2.1) and high-affinity nitrate transporter 2.3 (OsNRT2.3a)	Co-overexpression	Japonica	Co-overexpression of both genes increased rice yield and NUE	Direct	Chen et al. (2020a)
OsAMT1;2 (Os02g0620600) and OsGOGAT1 (Os01g0681900)	Ammonium transporter-1;2 (OsAMT1;2) and NADH-glutamate synthase 1 (OsGOGAT1)	Activation tagging double mutant	Japonica	Concurrent activation of both the genes enhanced grain yield and NUE	Direct	Lee et al. (2020a)

OsLHT1 (Os08g0127100)	Lysine-histidine-like transporter 1	Knockout mutant	Japonica	Mutant showed decreased N uptake efficiency and physiological utilization efficiency	Direct	Guo et al. (2020)
NC4N	NC4N is a chimeric nitrate transporter, in which 2nd to 5th TMDs were replaced by the corresponding region of CHL1 transporter	NRT1.7p::NC4N::3	Japonica	Hyperactive chimeric nitrate transporter increased grain yield	Direct	Chen et al. (2020b)
OsAAP1 (Os07g0134000)	Amino acid permease 6	Overexpression and knockout (CRISPR)	Japonica	Overexpression increased NutE	Indirect	Ji et al. (2020)
OsDUR3 (Os10g0580400)	High-affinity urea transporter	Loss of function mutant	Japonica	Downregulation reduced yield	Indirect	Beier et al. (2019)
OsNRT1.1A/ OsNPF6.3 (Os08g0155400)	Nitrate transporter 1.1A	Overexpression and knockdown	Japonica	Overexpression increased NUE	Direct	Wang et al. (2018b)
OsNPF7.7 (Os10g0579600)	Peptide transporter 7.7 (both splice variants)	Overexpression	Japonica	Overexpression increased NUE	Direct	Huang et al. (2018)
OsAAP3 (Os06g0556000)	Amino acid permease 3	Overexpression and downregulation (CRISPR/Cas9)	Japonica	Downregulation enhanced NUE	Direct	Lu et al. (2018)
OsNAR2.1 (Os02g0595900)	Activator for NRT2	Overexpression	Japonica	Overexpression increased NUE	Direct	Chen et al. (2017)

(Continued)

Table 4 (*Continued*)

Gene symbol (RAP-DB Id)	Gene name	Transgenic nature	Subspecies	N-responsive/NUE phenotype	Direct or indirect role in NUE	Reference
OsPTR6/OsNPF7.3 (Os04g0597800)	Protein transporter 6/ peptide transporter family 7.3	Overexpression and knockdown	Japonica	Overexpression increased NUE	Direct	Fang et al. (2017)
OsNRT2.1 (Os02g0112100)	High-affinity nitrate transporter 2.1	Overexpression (pOsNAR2.1:OsNRT2.1)	Japonica	Overexpression increased NUE	Direct	Chen et al. (2016)
OsNRT2.3b (Os01g0704100)	High-affinity nitrate transporter 2.3	Overexpression	Japonica	Overexpression increased NUE	Indirect	Fan et al. (2016)
OsAMT1.3 (Os02g0620600)	Ammonium transporter 1-3	Overexpression	Japonica	Overexpression decreased yield at high N	Indirect	Bao et al. (2015)
OsPTR6/OsNPF7.3 (Os04g0597800)	Protein transporter 6/ peptide transporter family 7.3	Overexpression	Japonica	Overexpression decreased NUE	Direct	Fan et al. (2014)
OsAMT1;1 (Os04g0509600)	Ammonium transporter-1;1	Overexpression	Japonica	Overexpression increased NUE	Direct	Ranathunge et al. (2014)
OsPTR9 (Os06g0706400)	Protein transporter 9	Overexpression and knockdown mutant	Japonica	Overexpression increased NutE and yield	Direct	Fang et al. (2013)
OsNPF7.9 (Os02g0689900)	Nitrate transporter 1/ peptide transporter Family protein 7.9	Loss of function mutant	Japonica	Mutant showed decreased grain yield and NUE	Direct	Guan et al. (2022)
N assimilation						
GS1;3 (Os03g0712800)	Cytosolic glutamine synthetase	Knockout mutant	Japonica	Observed decrease in agronomic traits	Direct	Fujita et al. (2022)
HvAlaAT	Alanine aminotransferase 1	Overexpression	Japonica	Overexpression increased grain yield and NUE	Direct	Tiong et al. (2021)

TaGS1	Glutamine synthetase 1	Overexpression	Japonica	Overexpression increased grain yield and NUE	Direct	Wu et al. (2021)
OsGS1;2 (Os03g0223400) and OsGS2 (Os04g0659100)	Glutamine synthetase	Overexpression and RNAi lines	Japonica	Overexpression increased the tiller number	Direct	Wang et al. (2020b)
OsASN1 (Os03g0291500)	Asparagine synthetase 1	Overexpression and knockout mutants	Japonica	Overexpression increased grain yield, Pn and protein content	Indirect	Lee et al. (2020b)
OsNR2 (Os08g0468700)	Nitrate reductase 2	Allelic variation	Indica vs Japonica	Increased effective tiller number, grain yield and NUE in Indica.	Direct	Gao et al. (2019)
OsGS1;2 (Os03g0223400)	Cytosolic glutamine synthetase 1;2	Overexpression	Japonica	Overexpression increased utilization NUE	Direct	Brauer et al. (2011)
CsAlaAT2	Alanine aminotransferase	Overexpression	Japonica	Overexpression increased NUE	Direct	Sisharmini et al. (2019)
HvAlaAT	Alanine aminotransferase	Overexpression	Japonica	Overexpression increased NUE	Direct	Shrawat et al. (2008)
NADH-GOGAT (Os01g0681900)	NADH-dependent glutamate synthase	Overexpression	Indica	Overexpression increased N utilization and grain filling	Indirect	Yamaya et al. (2002)
Transcriptional regulators						
OsNhd1 (Os08g0157600)	N-mediated heading date-1	Knockout mutant	Japonica	Mutant showed higher N uptake efficiency (NUpE)	Direct	Li et al. (2022)

(*Continued*)

Table 4 *(Continued)*

Gene symbol (RAP-DB Id)	Gene name	Transgenic nature	Subspecies	N-responsive/NUE phenotype	Direct or indirect role in NUE	Reference
Ghd7 (Os07g0261200)-ARE1 (Os08g0224300)	Heading date 7 (Ghd7) and abnormal cytokinin response 1-1 repressor1 (ARE1)	Allelic variation	Japonica	Increased grain yield and NUE under low-N conditions	Direct	Wang et al. (2021b)
OsNAC68 (Os01g0675800)	NAC domain-containing protein 014	Overexpression	Japonica	Overexpression increased grain yield and NUE	Direct	Xin et al. (2021)
OsMYB305 (Os01g0637800)	Myb transcription factor	overexpression	Japonica	Overexpression increased N uptake and tiller number under low N	Indirect	Wang et al. (2020a)
ARE1 (Os08g0224300)	Abnormal cytokinin response 1-1 repressor1 (ARE1)	Loss of function mutant (allele)	Japonica	Knockdown of ARE1 increased NUE.	Direct	Wang et al. (2018a)
OsTBP2.1 (Os03g0657000)	TATA-box binding protein	Overexpression and knockout mutant	Japonica	Overexpression increased nitrogen uptake and yield	Direct	Zhang et al. (2022a)
Calcium-mediated signalling						
OsNLP3 (Os01g0236700)	Nodule inception (NIN)-like protein 3	Overexpression	Japonica	Overexpression increased grain yield and NUE	Direct	Zhang et al. (2022b)
OsNLP4 (Os09g0549450)	Nodule inception (NIN)-like protein 4	Overexpression and knockout mutants	Japonica	Overexpression increased tiller number, grain yield and NUE	Direct	Yu et al. (2021)

OsNLP1 (Os03g0131100)	Nodule inception (NIN)-like protein 1	Overexpression and knockout mutant	Japonica	Overexpression enhanced grain yield and NUE	Direct	Alfatih et al. (2020)
OsCIPK2 (Os07g0678600)	Calcineurin B-like interacting protein kinase 2	Overexpression	Japonica	Overexpression reduced PE, AE and RE at normal N condition	Direct	Khan et al. (2019)
G-protein-mediated signalling						
LW5/D1/RGA1 (Os05g0333200)	Leaf width 5/G-protein alpha subunit	Knockout mutant	Indica	Mutant showed increased N uptake efficiency	Indirect	Zhu et al. (2020)
DEP1 (Os09g0441900)	Dense and erect panicle 1	Overexpression of dep1 allele	Japonica	Overexpression increased agronomic traits including NUE	Direct	Zhao et al. (2019)
OsSPL14 (Os08g0509600)	Squamosa promoter-binding-like protein 14	Knockout mutants	Japonica	Knockouts of OsSPL14 showed reduced plant height and tiller number	Indirect	Jiang et al. (2020); Srikanth et al. (2016)
OsSPL18 (Os09g0507100)	Squamosa promoter-binding-like protein 18	Overexpression and knockout mutants	Japonica	OsSPL18 binds to the promoter of DEP1 (NUE QTL) and knockout mutant showed increased tiller number and reduced panicle length and grain number	Indirect	Yuan et al. (2019)
Protein modification and degradation						
OsATG8b (Os04g0624000)	Autophagy-associated gene 8B	Overexpression and knockdown mutant	Japonica	Overexpression conferred higher N recycling efficiency to grains	Direct	Zhen et al. (2021)

(*Continued*)

Table 4 *(Continued)*

Gene symbol (RAP-DB Id)	Gene name	Transgenic nature	Subspecies	N-responsive/NUE phenotype	Direct or indirect role in NUE	Reference
OsSAE1a (Os11g0497000)	SUMO-activating enzyme subunit 1	Knockdown mutants	Japonica	Down-regulation reduced panicle length and seed setting rate	Indirect	Wang et al. (2021d)
OsATG8a (Os07g0512200)	Autophagy-associated gene 8a	Overexpression	Japonica	Overexpression enhanced effective tiller numbers, NUpE and NUE	Direct	Yu et al. (2019)
OsATG8c (Os08g0191600)	Autophagy-associated gene 8c	Overexpression	Japonica	Overexpression increased yield, NUpE and NUE	Direct	Zhen et al. (2019)
OsPP2C9 (Os01g0846300)	Protein phosphatase 2C	Overexpression	Japonica	Overexpression increased NUE	Direct	Waqas et al. (2018)
OsBT (Os01g0908200)	BT1/BT2 ortholog	Loss of function mutant	Japonica	Loss of OsBT increased NUE	Direct	Araus et al. (2016)
Epigenetic regulation						
NGR5 (Os05g0389000)	Nitrogen-mediated tiller growth response 5	Loss of function mutant	Japonica	Mutant showed reduced tiller number, which was insensitive to change in N	Direct	Wu et al. (2020)
Stress						
OsLSD1.1 (Os08g0159500)	Lesion simulating disease 1	Knockout	Japonica	Mutant showed decrease in photosynthetic nitrogen use efficiency (PNUE)]	Direct	Xiang et al. (2021)

DREB1C (Os06g0127100)	Dehydration-responsive element-binding protein 1C	Overexpression lines and knockout mutants	Japonica	Overexpression lines showed increased yield and NUE due to enhanced nitrogen uptake and transport activity	Direct	Wei et al. (2022)
DST	Drought and salt tolerance	Loss of function mutant	Japonica	Mutant lines showed significant decrease in NUE	Direct	Han et al. (2022)
Miscellaneous						
PSF	Premature senescence of flag leaves	Gamma irradiation mutant	Indica	Increased N translocation efficiency in mutants	Indirect	Zakari et al. (2021)
OsENOD93-1 (Os06g0142350)	UDP-glucuronyl transferase early nodulin 93 gene 1	Overexpression	Japonica	Overexpression increased NUE	Direct	Bi et al. (2009)
OsPHR3 (Os02g0139000)	Phosphate starvation response 3	Overexpression	Japonica	Overexpression increased grain weight, grain yield and NUE	Direct	Sun et al. (2021)

HvAlaAT, *Hordeum vulgare* alanine aminotransferase; TaGS1, *Triticum aestivum* glutamine synthetase 1.

gene(s) responsible for observed phenotypes. N-responsive phenotypes have been used to identify the genes involved in NUE and recently discovered NUE phenotypes in rice (Sharma et al., 2021) would expedite the research on NUE. Traditional forward genetic approaches take longer time and, in many cases, lack the information on molecular basis and associated mechanisms responsible for observed phenotypes.

Reverse genetics begins with the known candidate gene and then generation of transgenic (loss or gain of function) plants to define the effect of manipulated gene(s) on the various traits/phenotypes. Advancements in genome sequencing information have provided lots of information about genes not only in the particular species but also in different species and germplasms. Therefore, reverse genetic approach permits tremendous possibility to develop and evaluate transgenic plants for NUE improvement under different N conditions. In the last two decades, the research on transgenic manipulation for NUE improvement has been focused on cereals including rice. We have covered the recent information on gene manipulation for NUE (Table 4) and also included information reported in earlier publications (Mandal et al., 2018; Raghuram and Sharma, 2019; Sinha et al., 2020; Kumari and Raghuram, 2020; Madan et al., 2022).

To identify the association of candidate genes with NUE, the most common targets for gene manipulations are N transporters (OsNPF7.9, OsNRT2.3b, OsAMT1;1, OsAMT1;2, OsAMT1;3), N assimilation enzymes (OsNR2, OsGS1;2, OsGS2, GS1;3, AlaAT) and genes involved in transcriptional regulations (OsTBP2.1, OsMYB305, OsNLP3, OsNLP4, OsNAC68, OsMYB305) as mentioned in Table 4 and similar family members in earlier publications (Mandal et al., 2018; Raghuram and Sharma, 2019; Sinha et al., 2020; Kumari and Raghuram, 2020; Madan et al., 2022). Identified NUE genes are associated with heading date (OsNhd1, Ghd7), autophagy pathway (OsATG8a, OsATG8b), epigenetic regulation (NGR5), stress (OsLSD1.1), calcium signalling (OsNLP1, 3 and 4) and other nutrient (OsPHR3) among others (Table 4).

4.4 Biological processes regulated by nitrogen use efficiency genes

The NUE genes mentioned in Table 3 were functionally categorized into various biological processes based on their validated biological functions known in the literature. The most prominent functional categories include N transport (37%), N assimilation (18%), transcriptional regulators (11%), protein modification and degradation (10%) and calcium-mediated signalling (7%) among others (Fig. 1). Details of biological processes are given later.

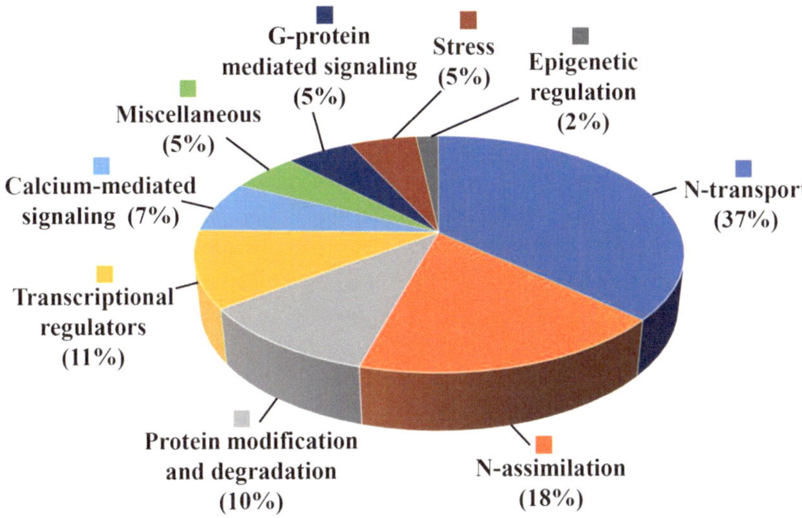

Figure 1 Pie diagram represents the functional categories of NUE genes known in rice. NUE genes were classified according to their function known in literature.

4.4.1 Nitrogen transport, nitrogen metabolism and nitrogen use efficiency

N uptake efficiency and subsequent utilization efficiency regulate NUE in plants. In rice, the majority of known NUE genes belong to the N transport (37%) and assimilation (18%) categories (Fig. 1). In rice, nitrate and ammonium are the two most common soil N forms, which are transported inside the cells using nitrate transporters (NRTs) and ammonium transporters (AMTs), respectively. Overexpression of NRTs such as OsNRT1;1A or AMTs such as OsAMT1;1 have been shown to improve the NUE, whereas their downregulation decreased NUE in rice (Table 4). This was not the case with OsAMT1.3, where its overexpression reduced yield at high N (Bao et al., 2015), which could be due to different transcriptional or translational regulation within the same gene families. Further, the triple mutants of OsAMTs (1;1, and 1;2 and 1;3) showed reduced NUE in rice (Konishi and Ma, 2021). Apart from nitrate and ammonium, amino acids present in the soil are acquired by the root through amino acid transporters (AATs). Amino acids are also generated through the reduction of nitrate via the GS-GOGAT pathway. In the context of remobilization, amino acids are considered the major source of N form and therefore play an important role in NUE. Downregulation of AATs such as amino acid permease 3 (Lu et al., 2018) and lysine-histidine-like transporter 1 (Guo et al., 2020) has been shown to reduce NUE in rice (Table 4). Considering the importance of AATs in N signalling, assimilation, metabolism and long-distance transport, limited information is available for AATs in rice compared to Arabidopsis. Further, the coordinated role of NRTs, AMTs and AATs in NUE is yet to be established.

During N assimilation, nitrate reductase (NR) converts nitrate into nitrite, which is then converted into ammonium by nitrite reductase (NiR) in plants. Ammonium goes through a series of enzymatic reactions and is involved in the generation of different amino acids The allelic variation of OsNR2 has demonstrated its role in NUE (Gao et al., 2019), whereas direct evidence supporting the role of NiR in rice NUE is yet to be established. Interestingly, manipulation of the promoter of NiR for the OsNLP4 transcription factor has improved yield at low N as compared to wild type in rice (Yu et al., 2021). Cytosolic and plastidic isoforms of glutamine synthetase (GS) are known in rice. Rice genome encodes three members of GS1 (GS1;1, GS1;2, and GS1;3) and overexpression of OsGS1;2 (Wang et al., 2020b) and GS1;3 (Fujita et al., 2022) improved NUE in rice (Table 4). The role of N-assimilation enzymes such as glutamate synthase (FD-GOGAT or NADH-GOGAT) in the context of other assimilatory enzymes and transporters is under progress (Li et al., 2022a) and needs deeper investigation for NUE improvement in rice.

4.4.2 Protein modification, degradation and nitrogen use efficiency

It has been established that post-translational modifications (PTMs) are involved in the regulation of protein stability, distribution and their function and therefore tightly control the biological responses in plants. PTMs play an important role in soil N uptake, assimilation and remobilization (Wang et al., 2021c). Phosphorylation is the most common PTM that occurs in plants. Phosphorylation of nitrate transporter (Ho et al., 2009) and ammonium transporter (Loque et al., 2007; Lanquar and Frommer, 2010) has been characterized, which suggests the importance of post-translational regulation to control the N uptake and translocation in plants. However, phosphorylation of urea transporter (DUR3) and phosphorylation-dependent urea uptake are not known in the plants.

The transporter NRT1.1 is known as a transceptor, i.e. it can act as a nitrate transporter and nitrate receptor in Arabidopsis. NRT1.1 works as a dual affinity transporter, i.e. it works as a low-affinity transporter under low nitrate availability, whereas phosphorylation of Thr101 amino acid by CIPK23 kinase converts NRT1.1 into a high-affinity nitrate transporter. Availability of high nitrate promotes dephosphorylation of Thr101 and NRT1.1 function as low-affinity nitrate transporter in Arabidopsis. Uptake of ammonium from soil is performed by ammonium transporters (AMTs) in plants. Time- and ammonium dose-dependent phosphorylation of T460 residue of Arabidopsis ammonium transporter AMT1.1 was observed (Loque et al., 2007). CIPK23 has been shown to phosphorylate the T460 residue of AMT1.1 in Arabidopsis but the loss of function of the cipk23 mutant provided evidence that additional kinases are involved in the phosphorylation of T460 residue in Arabidopsis (Straub et al., 2017). Addition of ammonium supply promoted T452 (similar to the T460 in Arabidopsis AMT1.1) phosphorylation in OsAMT1.2 by OsACTPK1 kinase in rice (Beier et al., 2018). More experimental pieces of evidence are needed to decode the transceptor function as in the case with Arabidopsis nitrate transporter NRT1.1. Based on the transceptor function of bacterial and fungal AMTs (Lanquar and Frommer, 2010), it was hypothesized that Arabidopsis AMTs may function as transceptor in plants. However, this hypothesis needs to be validated in plants.

N remobilization includes the transport of mainly nitrate and amino acids and therefore contributes an important step towards NUE improvement. Nitrogen limitation adaptation gene encodes a RING E3 ligase, which interacts and ubiquitinates with nitrate transporter AtNRT1.7 for its degradation via 26S proteasome pathway in Arabidopsis (Liu et al., 2017a). During N deficiency, ubiquitination controls the source-to-sink remobilization of leaf nitrate by regulating the protein abundance of AtNRT1.7 in Arabidopsis (Liu et al., 2017). Though the information is available on genes, proteins and enzymes involved in the ubiquitin-proteasome pathway, their role in rice NUE is not established.

Autophagy, an important component in the N remobilization, is an intracellular degradation process, which maintains cellular homeostasis under nutrient starvation condition by providing nutrients through the removal of damaged or aged proteins, or organelles in plants. It has been shown that overexpression of OsATG8a (Yu et al., 2019); OsATG8c (Zhen et al., 2019) increased NUE in rice (Table 4). These observations suggest that protein modifications have the unique potential to improve NUE and therefore we must explore these areas and identify candidate genes for improvement of NUE in rice.

4.4.3 Transcriptional regulators and nitrogen use efficiency

Transcription factors (TFs) bind to the cis-regulatory elements to regulate the expression of genes to control biological responses in plants. A transcriptional regulatory network (TRN) is formed by TFs and their interactions with target genes, and nitrate-regulated TRN has broadened molecular aspects of N transport and metabolism in Arabidopsis (Gaudinier et al., 2018). Despite the fact that such a comprehensive TRN is not known in crops, orthologs-based predicted N-responsive TRN revealed many TFs and their targets involved in N transport and metabolism in rice (Pathak et al., 2020; Sharma et al., 2022). Many TFs from various gene families have been characterized for their role in N response, but their role in NUE is limited in rice. For example, nitrate-regulated

1 (ANR1), a MADS-box TF, has been shown to regulate nitrate-dependent lateral root development in Arabidopsis (Zhang and Forde, 1998), whereas OsMADS25 regulates the nitrate-dependent architecture of the rice root system (Zhang et al., 2018). These TF family members have yet to be characterized for their role in the improvement of NUE in rice. Among TFs, NHD1, ARE1 and NAC68 are well-known TFs that regulate NUE in rice (Table 4).

4.4.4 Calcium signalling and nitrogen use efficiency

Nitrate-regulated primary responses include activation of many genes involved in myriads of pathways including calcium (Ca^{2+}) signalling in plant (Liu et al., 2020). Ca^{2+}-mediated signalling is one of the most evolutionarily conserved signalling, involved in the regulation of various physiological, stress, and transport processes in plants. Nitrate has been shown to induce Ca^{2+} ions level in cytosol and nucleus; however, such induction was not observed in nitrate transporter NRT1.1 mutants, suggesting nitrate-specific Ca^{2+} induction in Arabidopsis (Riveras et al., 2015). Calcium signalling involves Ca^{2+} ions sensor proteins such as calcineurin-B-like proteins (CBL) and calcium-dependent protein kinases (CPKs), which interact with and activate CBL-interacting protein kinases (CIPK) for the phosphorylation of target proteins for downstream signalling (Liu et al., 2021a).

Under low N conditions, CBL proteins (CBL1 and CBL9) interact with CIPK23 and form associated complexes (CBL1-CIPK23 and CBL9-CIPK23), which convert low-affinity nitrate transporter NPF6.3/NRT1.1 into high affinity after phosphorylation (Ho et al., 2009). The ABI2 protein, a specific protein phosphatase 2C, dephosphorylate NPF6.3/NRT1.1 transporter to antagonize CBL1/CBL9-CIPK23 complex phosphorylation responses in Arabidopsis (Leran et al., 2015). Other Ca^{2+} sensor proteins CPKs (CPK10, CPK30 and CPK32) are known to phosphorylate TF NIN-like protein-7 in the nucleus, which regulate a large set of genes to implement nitrate response in rice (Liu et al., 2017b). There are many Ca^{2+} signalling genes have been characterized for their N responses in Arabidopsis and rice, but only NLPs and CIPK2 have been shown to regulate NUE in rice (Table 4). The role of other core components of Ca^{2+} signalling in the regulation of rice NUE is yet to be established.

4.4.5 G-protein signalling and nitrogen use efficiency

Heterotrimeric G-proteins, which consist of three different subunits (α, β, and γ), have been implicated in the regulation of plant growth, development, stress and nutrient responses in rice (Cui et al., 2020; Pandey, 2019). Our earlier molecular studies, including genome-wide N responses, have demonstrated the role of G-proteins in the regulation of N responses in Arabidopsis (Chakraborty et al.,

2015a, 2019), maize (Raghuram et al., 1999) and rice (Ali et al., 2007; Pathak et al., 2021; Prasanna et al., 2023). To identify the genetic loci responsible for NUE, QTL mapping experiments were performed using rice genotypes with significant variation in the N-responsive traits, which led to the identification of a major QTL (qNGR9) for NUE in rice (Sun et al., 2014). Fine mapping of qNGR9 and genetic complementation experiments revealed a previously identified gene known as Dense and Erect Panicles 1 (DEP1). DEP1 is a G-protein gamma subunit, and the gain of function of *dep1* allele improved NUE in rice (Sun et al., 2014). Members of the gene family 'squamosa promoter binding protein-like' (OsSPL) have been shown to regulate growth and development in rice. For example, OsSPL14 regulates ideal plant architecture and it was differentially regulated under low N conditions (Srikanth et al., 2016). The OsSPL18 knockout mutant showed reduced panicle length and grain number while the number of tillers increased (Yuan et al., 2019). Further, OsSPL18 also binds to the promoter of DEP1, which controls NUE in rice. This clearly suggests their potential involvement in the regulation of NUE in rice.

In another study, EMS (ethyl methanesulfonate) mutants of indica rice were analysed, and the leaf width 5 (lw5) mutants were identified, which had broad width leaves and small grains (Zhu et al., 2020). Map-based cloning revealed the base substitution mutation in the Gα subunit (RGA1) leading to truncated RGA1 proteins in the mutants. This mutant showed improved photosynthetic and N utilization efficiency as compared to the wild-type plants (Zhu et al., 2020). Advancements in G-protein signalling research have expanded the signalling components in plants. The Arabidopsis G-protein interactome (Klopffleisch et al., 2011) and orthologs-based predicted rice G-protein interactome (Pathak et al., 2021) revealed many known and potential candidates associated with N responses/NUE in rice. As compared to the available information and identification of NUE QTL, this area is underexplored and needs much deeper investigation for the improvement of NUE.

4.4.6 Epigenetic regulation, stress and nitrogen use efficiency

Pie diagram shows that N transport and N assimilation are relatively more explored areas for NUE improvement, whereas biological processes such as 'Epigenetic regulation' and 'Stress' need much deeper investigation in the context of NUE (Fig. 1). Apart from transcriptional and post-transcriptional regulation, that epigenetic regulation also affects the expression of genes through different processes of DNA methylation and histone modifications in plants (Liang et al., 2020). It is known that N affects the DNA methylation status as well as histone modification dynamics in plants (Kou et al., 2011; Secco et al., 2017). The NR gene was methylated by chromomethylase 3, which was regulated by ammonium in Arabidopsis (Kim et al., 2015). It has been shown that histone

Table 5 List of intergenic markers developed for NUE genes in rice

Gene symbol (Id)	Gene name	Molecular marker name	Molecular marker type	Reference
SBM1 (LOC_Os01g65120)	Peptide transporter	SBM1 5U ID and SBM1 E3S	Indel and CAPS	Xu et al. (2021)
OsTCP19 (LOC_Os06g12230)	Teosinte branched/cycloidea/proliferating cell factors 19	TCP19 5U ID	Indel	Liu et al. (2021b)
OsNLP4 (LOC_Os09g37710)	Nodule inception (NIN)-like protein 4	NLP4 5U ID	Indel	Yu et al. (2021)
DNR1 (LOC_Os01g08270)	Aminotransferase	DNR1 I4 ID	Indel	Zhang et al. (2021)
NGR5 (LOC_Os05g32270)	Nitrogen-mediated tiller growth response 5	NGR5 5U ID and NGR5 I6 S	Indel and PARMS	Wu et al. (2020)
MYB61 (LOC_Os01g18240)	Myb transcription factor	MYB61 3U ID	Indel	Gao et al. (2020)
OsNAC42 (LOC_Os09g32040)	NAC domain-containing protein 042	NAC42 3U S	dCAPS	Tang et al. (2019)
OsNR2 (LOC_Os02g53130)	Nitrate reductase 2	NR2 E4 ID	Indel	Gao et al. (2019)
OsNPF6.1 (LOC_Os01g01360)	Nitrate transporter 1	NPF6.1 5U ID	Indel	Tang et al. (2019)
NGR2/GRF4 (LOC_Os02g47280)	Growth-regulating factor 4	NGR2 5U S and NGR2 I3 S	CAPS and PARMS	Li et al. (2018)
ARE1 (LOC_Os08g12780)	Abnormal cytokinin response 1-1 repressor1	ARE1 5U ID and ARE1 5U S	Indel and CAPS	Wang et al. (2018a)
NRT1.1B/OsNPF6.5 (LOC_Os10g40600)	Nitrate transporter 1	NRT1.1B I1 ID	Indel	Hu et al. (2015)
TOND1 (LOC_Os12g43440)	Tolerance of nitrogen deficiency 1	TOND1 5U ID	Indel	Zhang et al. (2015)
DEP1 (LOC_Os09g26999)	Dense and erect panicle 1	DEP1 E5 ID	Indel	Sun et al. (2014)

This adapted table was created based on the information provided in earlier publication (Li et al., 2022c).

H3 lysine 27 trimethylation (H3K27me3) regulates N-mediated tillering in rice (Wu et al., 2020). To promote N-mediated tillering, branching-inhibitory genes are inhibited after H3K27me3 modification by polycomb repressive complex 2, which was recruited by N-mediated tiller growth response 5 (NGR5) TF in rice (Wu et al., 2020). Despite available epigenetic regulation information, only NGR5 has been shown to regulate NUE in rice (Table 4). This clearly showed that the identification and manipulation of genes associated with epigenetic regulation may improve NUE in rice.

Crosstalk of N and stress signalling has been a relatively underexplored research area for crop improvement (Jangam and Raghuram, 2015). N and stress responses showed commonality in terms of physiological, biochemical and molecular changes in plants. For example, the roots are the primary organ to sense soil N fluctuation and drought condition. Photosynthesis, chlorophyll contents, etc. are affected under both conditions. Further, mapping of nitrate- or urea-responsive differentially expressed genes onto various biological pathways and processes revealed their involvement in abiotic and biotic stresses in rice (Pathak et al., 2020; Sharma et al., 2022; Mandal et al., 2022; Sharma et al., 2023). Few stress-associated genes have been characterized for their role in NUE (Table 4). For example, a recent study showed that the drought and salt tolerance (DST) TF regulates NUE by controlling the N assimilation gene OsNR1.2 in rice (Han et al., 2022). The loss of function of *dst* mutant showed reduced OsNR1.2 function and accordingly a significant reduction in NUE (Han et al., 2022). In another study, the screening of 118 potential TFs identified from different transcriptomic and metabolomics studies led to the identification of Dehydration-Responsive Element-Binding Protein 1C (OsDREB1C), which improved photosynthesis and NUE in rice (Wei et al., 2022). OsDREB1C is an AP2/ERF family TF, and its overexpression increased yield by up to 68%, including NUE in rice (Wei et al., 2022). These findings clearly show that unravelling the mechanisms controlling both N and stress responses and manipulating the identified candidate genes would provide tolerance to the plants and reduce the consumption of N fertilizers, contributing to sustainable agriculture.

4.5 Genetic and genomic strategies for manipulation of nitrogen use efficiency genes

Overexpression or mutation (knockdown or knockout) of these genes either increased or decreased the NUE in rice (Table 4). Interestingly, in few cases, two genes were manipulated to improve the NUE in rice (Table 4). For example, Chen et al. (2020a) demonstrated that transgenic plants co-expressing nitrate transporter (OsNRT2.3a) along with its interactor (OsNAR2.1) had increased NUE as compared to those single-gene transgenic rice plants overexpressing either OsNRT2.3a or OsNAR2.1 in rice. In this case, both genes are involved in the same transport pathway. In another example, two genes involved in two different pathways viz. ammonium transport (OsAMT1;2) and N-assimilation pathways (OsGOGAT1) were co-activated in rice using T-DNA activation tagging method (Lee et al., 2020a). Rice plants with double activation of ammonium transporter (OsAMT1;2) and NADH-glutamate synthase 1 (OsGOGAT1) showed improved grain yield under N-limiting condition. This clearly suggests that to deal with complex nature of NUE, we must simultaneously manipulate multiple targeted genes involved in same or different pathways for improved NUE in rice.

Allelic variations within subspecies and related germplasms have provided the natural resources to identify the target loci/genes responsible for the phenotypic variation including NUE (Table 2). A study conducted by Gao et al. (2019) showed that allelic variation in nitrate reductase 2 (OsNR2), a gene involved in N-assimilation pathway, is responsible for high NUE in Indica rice as compared to Japonica subspecies. They showed that Trp_{779} substitution in OsNR2 enzyme reduced its activity in Japonica as compared to Indica subspecies. In another study, a natural allelic variation was identified in nitrate transporter, which improved NUE by regulating auxin biosynthesis in rice (Zhang et al., 2021). An interesting work was performed by Chen et al. (2020a), where they first improved the NUE by overexpressing nitrate transporter (OsNRT2.3b) in rice. Next, they mutagenized the transgenic rice with EMS and isolated a transgenic plant having short duration with similar NUE. This study shows the integration of two methods, i.e. transgenic manipulation of a gene and screening of genetic variation (EMS mutagenesis) to develop a plant with multiple beneficial traits.

Improvement of NUE using chimeric protein is a unique methodology demonstrated by Chen et al. (2020b). They developed a chimeric nitrate transporter by replacing the transmembrane domain (TMD) region between second and fifth with CHL1 transporter. They observed increased NUE in Arabidopsis and grain yield in rice. This study provides an opportunity to develop synthetic chimeric protein (genotype-specific/inhibitory/activating domain) for NUE improvement. Consideration of synthetic promoter associated with stress and or nutrient responses would add another level of control to fine-tune the spatio-temporal regulation of chimeric protein for NUE improvement in rice.

4.6 Molecular markers for nitrogen use efficiency genes in rice

Molecular markers help in marker-assisted selection, which has been shown to improve the efficiency of rice cultivars. Research on NUE revealed many genes involved in NUE in rice. Recently, Li et al. (2022c) have developed the 18 intragenic molecular markers to track the coinheritance of 14 NUE genes studied in rice (Table 5). These are NRT1.1B, OsDEP1, OsNR2 and OsDNR1 among others, which are well established for their role in NUE in rice (Table 5). To develop the markers, 2 kb upstream region from start codon and 1 kb downstream region from stop codon were considered. While developing these intragenic markers, they observed structural variations in the 5′ UTR, which help to define genetic diversity across different germplasms. Identification of molecular markers associated with NUE is a necessary step for the wider application of discovered genes and to expedite the breeding programmes for NUE improvement.

5 Conclusion and future prospects

Improvement of NUE is one of the most important steps for sustainable agriculture, as well as for global sustainable N management. As a crop that feeds half the world and is one of the highest consumers of N fertilizers, rice is an important target crop for NUE. For improved NUE, either higher yields must be achieved using current N fertilizer inputs, or, even better, similar yields must be achieved using lower N inputs. Many studies have been performed to identify the molecular targets and associated signalling networks involved in rice NUE. Transgenic manipulation of selected molecular targets has improved yield and NUE in rice (Tables 3 and 4). Biological pathways primarily associated with manipulated NUE genes were N uptake, assimilation, translocation and remobilization. In the context of signalling events, N affects calcium-mediated signalling, hormone-dependent regulation, autophagy pathways and epigenetic regulation in rice (Table 4). Recent studies using other gene family members, such as calcium signalling associated NLP gene family, were characterized for NUE in rice. Further, NUE regulation through OsNGR5 opens an underexplored area of N and epigenetic regulation in plants.

During early studies, single-gene-based transgenic plants were used to understand the N response and NUE, but it appeared that this method does not provide complete information to improve NUE because of its complex nature. As per improved knowledge, manipulation of two or more genes involved in the same pathways has yielded better results in terms of NUE. For example, co-activation of nitrate transporter and its interactors provided increased NUE as compared to their single-gene transgenics (Chen et al., 2020). Since genotype and environment interaction affects NUE, we must also consider temporal and spatial gene regulatory networks controlling NUE in rice. Another underexplored area is gene splicing events and NUE. Alternative splicing of a gene produces different transcript variants, and it is possible that one or all splice variants may contribute to a particular trait or response. Liu et al. (2022c) have recently shown that glutamine synthase (*OsGS1;1*) has two splice variants viz. *OsGS1;1a* and *OsGS1;1b* and the high *OsGS1;1b* activity is associated with improved NUE in rice. This will help to shortlist/identify the specific transcript variants that actually contribute to NUE.

The development of 'omics' approaches enabled us to identify the temporal and spatial gene regulatory networks controlled by N in rice. However, such studies are limited in terms of comprehensive tissues-, time- and N dose-specific transcriptional regulation in diverse cultivars and not available for translational or post-translational regulations and metabolic regulation with multiple rice cultivars (genotypes). For reproducible and reliable conclusions of the 'omics' data, we must generate N-dose dependent, comprehensive temporal and spatial data sets in many rice genotypes contrasting for NUE in field condition. This will help to understand the genotype and environment interaction, and accordingly manipulation of candidate genes for their temporal and/or tissue-specific roles for NUE.

The development of rice cultivars with a synthetic control system for gene expression would be one of the most challenging tasks in the near future. We can modulate the function of specific gene(s) according to the soil N condition, environment and developmental stages of plants by spraying/supplying exogenous activator. This will also help to engineer the plant for dual or multiple regulatory system, i.e. regulation at transport for efficient N uptake from soil and regulation at the reproductive stage for improved yield at low N.

The number of QTLs and associated genes for NUE are increasing with time, but very few of them are used in the breeding programmes as the molecular markers for many QTLs and NUE genes validated in field condition are not available. In this context, Li et al. (2022c) have developed the molecular markers for validated 14 NUE genes in rice cultivars. In addition, CRISPR/Cas9 is the most important genome editing tool for precision breeding in plants. This method has the ability to generate transgene-free rice plants. After the identification of elite rice varieties with improved NUE and the responsible gene(s), CRISPR/Cas9 genome editing tool can be used to manipulate the genome to generate many novel germplasms, which contain homologous NUE gene(s) identified in elite rice varieties. This will help to expedite the breeding programmes for NUE improvement in rice.

6 Where to look for further information

Sutton et al. (2019), Raghuram et al. (2021), Sutton et al. (2021) and Raghuram et al. (2022a) provide detailed information on the scientific community, international collaborations, and the importance of global governance to regulate N policy and global climate change.

Liu et al. (2022a) and Hawkesford and Griffiths (2019) provide comprehensive information on the molecular and genetic aspects of N-related processes and NUE in crop plants.

Current progress related to N can be found on the following websites:

- International Nitrogen Management System (https://www.inms.international/).
- International Nitrogen Initiative (https://initrogen.org/).
- Global Partnership on Nutrient Management (http://www.nutrient-challenge.org/).

7 Acknowledgements

We thank the research grant from UKRI-GCRF South Asian Nitrogen Hub (SANH, NE/S009019/1), Science and Engineering Research Board (SERB, CRG/2021/007467) and Faculty Research Grant Scheme (GGSIPU/DRC/FRGS/2021/594/14, GGSIPU/DRC/FRGS/2022/1223/14). DKJ was supported by fellowships from the UKRI-GCRF SANH project and the CSIR Scientist's Pool Scheme 13(9243-A)/2023-POOL.

8 References

Abrol, Y. P., Adhya, T. K., Aneja, V. P., Raghuram, N., Pathak, H., Kulshrestha, U., Sharma, C. and Singh, B. (2017). *The Indian Nitrogen Assessment: Sources of Reactive Nitrogen, Environmental and Climate Effects, Management Options, and Policies*. Elsevier.

Alfatih, A., Wu, J., Zhang, Z. S., Xia, J. Q., Jan, S. U., Yu, L. H. and Xiang, C. B. (2020). Rice NIN-LIKE PROTEIN 1 rapidly responds to nitrogen deficiency and improves yield and nitrogen use efficiency. *Journal of Experimental Botany* 71(19): 6032–6042.

Ali, A., Sivakami, S. and Raghuram, N. (2007). Regulation of activity and transcript levels of NR in rice (Oryza sativa): roles of protein kinase and G-proteins. *Plant Science* 172(2): 406–413.

Araus, V., Vidal, E. A., Puelma, T., Alamos, S., Mieulet, D., Guiderdoni, E. and Gutiérrez, R. A. (2016). Members of BTB gene family regulate negatively nitrate uptake and nitrogen use efficiency in Arabidopsis thaliana and Oryza sativa. *Plant Physiology* 171(2): 1523–1532. DOI: 10.1104/pp.15.01731.

Bai, J., Piao, Z., Wan, C., Lee, G., Ruan, X., Luo, Z., Shi, F., Cong, X. and Yang, R. (2021). SLAF-based linkage map construction and QTL mapping of nitrogen use efficiency in rice (Oryza sativa L.). *Plant Molecular Biology Reporter* 39(4): 727–738.

Bao, A., Liang, Z., Zhao, Z. and Cai, H. (2015). Overexpressing of OsAMT1-3, a high affinity ammonium transporter gene, modifies rice growth and carbon-nitrogen metabolic status. *International Journal of Molecular Sciences* 16(5): 9037–9063.

Beier, M. P., Fujita, T., Sasaki, K., Kanno, K., Ohashi, M., Tamura, W., Konishi, N., Saito, M., Imagawa, F., Ishiyama, K., Miyao, A., Yamaya, T. and Kojima, S. (2019). The urea transporter DUR3 contributes to rice production under nitrogen-deficient and field conditions. *Physiologia Plantarum* 167(1): 75–89.

Beier, M. P., Obara, M., Taniai, A., Sawa, Y., Ishizawa, J., Yoshida, H., Tomita, N., Yamanaka, T., Ishizuka, Y., Kudo, S., Yoshinari, A., Takeuchi, S., Kojima, S., Yamaya, T. and Hayakawa, T. (2018). Lack of ACTPK1, an STY kinase, enhances ammonium uptake and use, and promotes growth of rice seedlings under sufficient external ammonium. *The Plant Journal* 93(6): 992–1006.

Bi, Y. M., Kant, S., Clarke, J., Gidda, S., Ming, F., Xu, J., Rochon, A., Shelp, B. J., Hao, L., Zhao, R., Mullen, R. T., Zhu, T. and Rothstein, S. J. (2009). Increased nitrogen-use efficiency in transgenic rice plants over-expressing a nitrogen-responsive early nodulin gene identified from rice expression profiling. *Plant, Cell and Environment* 32(12): 1749–1760.

Brauer, E. K., Rochon, A., Bi, Y. M., Bozzo, G. G., Rothstein, S. J. and Shelp, B. J. (2011). Reappraisal of nitrogen use efficiency in rice overexpressing glutamine synthetase1. *Physiologia Plantarum* 141(4): 361–372.

Briat, J. F., Gojon, A., Plassard, C., Rouached, H. and Lemaire, G. (2020). Reappraisal of the central role of soil nutrient availability in nutrient management in light of recent advances in plant nutrition at crop and molecular levels. *European Journal of Agronomy* 116: 126069.

Chakraborty, N., Kanyuka, K., Jaiswal, D. K., Kumar, A., Arora, V., Malik, A., Gupta, N., Hooley, R. and Raghuram, N. (2019). GCR1 and GPA1 coupling regulates nitrate, cell wall, immunity and light responses in Arabidopsis. *Scientific Reports* 9(1): 5838.

Chakraborty, N., Sharma, P., Kanyuka, K., Pathak, R. R., Choudhury, D., Hooley, R. and Raghuram, N. (2015a). G-protein alpha-subunit (GPA1) regulates stress, nitrate and phosphate response, flavonoid biosynthesis, fruit/seed development and substantially shares GCR1 regulation in A. thaliana. *Plant Molecular Biology* 89(6): 559–576.

Chen, J., Fan, X., Qian, K., Zhang, Y., Song, M., Liu, Y., Xu, G. and Fan, X. (2017). pOsNAR2.1:OsNAR2.1 expression enhances nitrogen uptake efficiency and grain yield in transgenic rice plants. *Plant Biotechnology Journal* 15(10): 1273-1283.

Chen, J., Liu, X., Liu, S., Fan, X., Zhao, L., Song, M., Fan, X. and Xu, G. (2020a). Co-overexpression of OsNAR2.1 and OsNRT2.3a increased agronomic nitrogen use efficiency in transgenic rice plants. *Frontiers in Plant Science* 11: 1245.

Chen, J., Wang, F., Lei, B., Qian, K., Wei, J. and Fan, X. (2022). Modulation of growth duration, grain yield and nitrogen recovery efficiency by EMS mutagenesis under OsNRT2.3b overexpression background in rice. *Agriculture* 12(6).

Chen, J., Zhang, Y., Tan, Y., Zhang, M., Zhu, L., Xu, G. and Fan, X. (2016). Agronomic nitrogen-use efficiency of rice can be increased by driving OsNRT2.1 expression with the OsNAR2.1 promoter. *Plant Biotechnology Journal* 14(8): 1705–1715.

Chen, K. E., Chen, H. Y., Tseng, C. S. and Tsay, Y. F. (2020b). Improving nitrogen use efficiency by manipulating nitrate remobilization in plants. *Nature Plants* 6(9): 1126–1135.

Cho, Y. I., Jiang, W., Chin, J. H., Piao, Z., Cho, Y. G., McCouch, S. and Koh, H. J. (2007). Identification of QTLs associated with physiological nitrogen use efficiency in rice. *Molecules and Cells* 23(1): 72–79.

Ciampitti, I. A., Briat, J. F., Gastal, F. and Lemaire, G. (2022). Redefining crop breeding strategy for effective use of nitrogen in cropping systems. *Communications Biology* 5(1): 823.

Ciampitti, I. A. and Lemaire, G. (2022). From use efficiency to effective use of nitrogen: a dilemma for maize breeding improvement. *Science of the Total Environment* 826: 154125.

Congreves, K. A., Otchere, O., Ferland, D., Farzadfar, S., Williams, S. and Arcand, M. M. (2021). Nitrogen use efficiency definitions of today and tomorrow. *Frontiers in Plant Science* 12: 637108.

Cui, Y., Jiang, N., Xu, Z. and Xu, Q. (2020). Heterotrimeric G protein are involved in the regulation of multiple agronomic traits and stress tolerance in rice. *BMC Plant Biology* 20(1): 90.

Dai, G. J., Cheng, S. H., Hua, Z. T., Zhang, M. L., Jiang, H. B., Feng, Y., Shen, X. H., Su, Y. A., He, N., Ma, Z. B., Ma, X. Q., Hou, S. G. and Wang, Y. R. (2015). Mapping quantitative trait loci for nitrogen uptake and utilization efficiency in rice (Oryza sativa L.) at different nitrogen fertilizer levels. *Genetics and Molecular Research* 14(3): 10404–10414.

Fan, X., Tang, Z., Tan, Y., Zhang, Y., Luo, B., Yang, M., Lian, X., Shen, Q., Miller, A. J. and Xu, G. (2016). Overexpression of a pH-sensitive nitrate transporter in rice increases crop yields. *Proceedings of the National Academy of Sciences of the United States of America* 113(26): 7118–7123.

Fan, X., Xie, D., Chen, J., Lu, H., Xu, Y., Ma, C. and Xu, G. (2014). Over-expression of OsPTR6 in rice increased plant growth at different nitrogen supplies but decreased nitrogen use efficiency at high ammonium supply. *Plant Science* 227: 1–11.

Fang, Z., Bai, G., Huang, W., Wang, Z., Wang, X. and Zhang, M. (2017). The rice peptide transporter OsNPF7.3 is induced by organic nitrogen, and contributes to nitrogen allocation and grain yield. *Frontiers in Plant Science* 8: 1338.

Fang, Z., Xia, K., Yang, X., Grotemeyer, M. S., Meier, S., Rentsch, D., Xu, X. and Zhang, M. (2013). Altered expression of the PTR/NRT1 homologue OsPTR9 affects nitrogen utilization efficiency, growth and grain yield in rice. *Plant Biotechnology Journal* 11(4): 446–458.

Fujita, T., Beier, M. P., Tabuchi-Kobayashi, M., Hayatsu, Y., Nakamura, H., Umetsu-Ohashi, T., Sasaki, K., Ishiyama, K., Murozuka, E., Kojima, M., Sakakibara, H., Sawa, Y., Miyao, A., Hayakawa, T., Yamaya, T. and Kojima, S. (2022). Cytosolic glutamine synthetase GS1;3 is involved in rice grain ripening and germination. *Frontiers in Plant Science* 13: 835835.

Gao, Y., Xu, Z., Zhang, L., Li, S., Wang, S., Yang, H., Liu, X., Zeng, D., Liu, Q., Qian, Q., Zhang, B. and Zhou, Y. (2020). MYB61 is regulated by GRF4 and promotes nitrogen utilization and biomass production in rice. *Nature Communications* 11(1): 5219.

Gao, Z., Wang, Y., Chen, G., Zhang, A., Yang, S., Shang, L., Wang, D., Ruan, B., Liu, C., Jiang, H., Dong, G., Zhu, L., Hu, J., Zhang, G., Zeng, D., Guo, L., Xu, G., Teng, S., Harberd, N. P. and Qian, Q. (2019). The indica nitrate reductase gene OsNR2 allele enhances rice yield potential and nitrogen use efficiency. *Nature Communications* 10(1): 5207.

Gaudinier, A., Rodriguez-Medina, J., Zhang, L., Olson, A., Liseron-Monfils, C., Bågman, A. M., Foret, J., Abbitt, S., Tang, M., Li, B., Runcie, D. E., Kliebenstein, D. J., Shen, B., Frank, M. J., Ware, D. and Brady, S. M. (2018). Transcriptional regulation of nitrogen-associated metabolism and growth. *Nature* 563(7730): 259–264.

Guan, Y., Liu, D. F., Qiu, J., Liu, Z. J., He, Y. N., Fang, Z. J., Huang, X. H. and Gong, J. M. (2022). The nitrate transporter OsNPF7.9 mediates nitrate allocation and the divergent nitrate use efficiency between indica and japonica rice. *Plant Physiology* 189(1): 215–229.

Guo, N., Gu, M., Hu, J., Qu, H. and Xu, G. (2020). Rice OsLHT1 functions in leaf-to-panicle nitrogen allocation for grain yield and quality. *Frontiers in Plant Science* 11: 1150.

Han, M. L., Lv, Q. Y., Zhang, J., Wang, T., Zhang, C. X., Tan, R. J., Wang, Y. L., Zhong, L. Y., Gao, Y. Q., Chao, Z. F., Li, Q. Q., Chen, G. Y., Shi, Z., Lin, H. X. and Chao, D. Y. (2022). Decreasing nitrogen assimilation under drought stress by suppressing DST-mediated activation of nitrate reductase 1.2 in rice. *Molecular Plant* 15(1): 167–178.

Hawkesford, M. J. and Griffiths, S. (2019). Exploiting genetic variation in nitrogen use efficiency for cereal crop improvement. *Current Opinion in Plant Biology* 49: 35-42.

Ho, C. H., Lin, S. H., Hu, H. C. and Tsay, Y. F. (2009). CHL1 functions as a nitrate sensor in plants. *Cell* 138(6): 1184–1194.

Hu, B., Wang, W., Ou, S., Tang, J., Li, H., Che, R., Zhang, Z., Chai, X., Wang, H., Wang, Y., Liang, C., Liu, L., Piao, Z., Deng, Q., Deng, K., Xu, C., Liang, Y., Zhang, L., Li, L. and Chu, C. (2015). Variation in NRT1.1B contributes to nitrate-use divergence between rice subspecies. *Nature Genetics* 47(7): 834–838.

Hu, D. W., Sun, Z. P., LI, T. L., Yan, H. and Zhang, H. (2014). Nitrogen nutrition index and its relationship with N use efficiency, tuber yield, radiation use efficiency, and leaf parameters in potatoes. *Journal of Integrative Agriculture* 13(5): 1008–1016.

Huang, W., Bai, G., Wang, J., Zhu, W., Zeng, Q., Lu, K., Sun, S. and Fang, Z. (2018). Two splicing variants of OsNPF7.7 regulate shoot branching and nitrogen utilization efficiency in rice. *Frontiers in Plant Science* 9: 300.

Jamora, N. and Ramaiah, V. (2022). Global demand for rice genetic resources. *CABI Agriculture and Bioscience* 3(1): 1-15.

Jangam, A. P. and Raghuram, N. (2015). Nitrogen and stress. *Elucidation of Abiotic Stress Signaling in Plants* 2: 323–339.

Ji, Y., Huang, W., Wu, B., Fang, Z. and Wang, X. (2020). The amino acid transporter AAP1 mediates growth and grain yield by regulating neutral amino acid uptake and reallocation in Oryza sativa. *Journal of Experimental Botany* 71(16): 4763–4777.

Jia, Y., Hu, Z., Ba, Y. and Qi, W. (2021). Application of biochar-coated urea controlled loss of fertilizer nitrogen and increased nitrogen use efficiency. *Chemical and Biological Technologies in Agriculture* 8(1): 1-11.

Jiang, M., He, Y., Chen, X., Zhang, X., Guo, Y., Yang, S., Huang, J. and Traw, M. B. (2020). CRISPR-based assessment of genomic structure in the conserved SQUAMOSA promoter-binding-like gene clusters in rice. *The Plant Journal* 104(5): 1301–1314.

Kabange, N. R., Lee, S. M., Shin, D., Lee, J. Y., Kwon, Y., Kang, J. W., Cha, J. K., Park, H., Alibu, S. and Lee, J. H. (2022). Multiple facets of nitrogen: From atmospheric gas to indispensable agricultural input. *Life* 12(8): 1272.

Khan, M. U., Li, P., Amjad, H., Khan, A. Q., Arafat, Y., Waqas, M., Li, Z., Noman, A., Islam, W., Wu, L., Zhang, Z. and Lin, W. (2019). Exploring the potential of overexpressed OsCIPK2 rice as a nitrogen utilization efficient crop and analysis of its associated rhizo-compartmental microbial communities. *International Journal of Molecular Sciences* 20(15): 3636.

Kim, J. Y., Kwon, Y. J., Kim, S. I., Kim, D. Y., Song, J. T. and Seo, H. S. (2015). Ammonium inhibits chromomethylase 3-mediated methylation of the Arabidopsis nitrate reductase gene NIA2. *Frontiers in Plant Science* 6: 1161.

Klopffleisch, K., Phan, N., Augustin, K., Bayne, R. S., Booker, K. S., Botella, J. R., Carpita, N. C., Carr, T., Chen, J. G., Cooke, T. R., Frick-Cheng, A., Friedman, E. J., Fulk, B., Hahn, M. G., Jiang, K., Jorda, L., Kruppe, L., Liu, C., Lorek, J., McCann, M. C., Molina, A.,

Moriyama, E. N., Mukhtar, M. S., Mudgil, Y., Pattathil, S., Schwarz, J., Seta, S., Tan, M., Temp, U., Trusov, Y., Urano, D., Welter, B., Yang, J., Panstruga, R., Uhrig, J. F. and Jones, A. M. (2011). Arabidopsis G-protein interactome reveals connections to cell wall carbohydrates and morphogenesis. *Molecular Systems Biology* 7: 532.

Konishi, N. and Ma, J. F. (2021). Three polarly localized ammonium transporter 1 members are cooperatively responsible for ammonium uptake in rice under low ammonium condition. *New Phytologist* 232(4): 1778–1792.

Kou, H. P., Li, Y., Song, X. X., Ou, X. F., Xing, S. C., Ma, J., Von Wettstein, D. and Liu, B. (2011). Heritable alteration in DNA methylation induced by nitrogen-deficiency stress accompanies enhanced tolerance by progenies to the stress in rice (Oryza sativa L.). *Journal of Plant Physiology* 168(14): 1685–1693.

Kumari, S. and Raghuram, N. (2020). Protein Phosphatases in N response and NUE in Crops. *Protein Phosphatases and Stress Management in Plants*: 233–244.

Kumari, S., Sharma, N. and Raghuram, N. (2021). Meta-analysis of yield-related and N-responsive genes reveals chromosomal hotspots, key processes and candidate genes for nitrogen-use efficiency in rice. *Frontiers in Plant Science* 12: 627955.

Ladha, J. K., Jat, M. L., Stirling, C. M., Chakraborty, D., Pradhan, P., Krupnik, T. J., Sapkota, T. B., Pathak, H., Rana, D. S., Tesfaye, K. and Gerard, B. (2020). Achieving the sustainable development goals in agriculture: the crucial role of nitrogen in cereal-based systems. *Advances in Agronomy* 163: 39–116.

Ladha, J. K., Kirk, G. J. D., Bennett, J., Peng, S., Reddy, C. K., Reddy, P. M. and Singh, U. (1998). Opportunities for increased nitrogen-use efficiency from improved lowland rice germplasm. *Field Crops Research* 56(1–2): 41–71.

Ladha, J. K., Pathak, H. P., Krupnik, T. J. Six, J. and Van Kessel, C. (2005). Efficiency of fertilizer nitrogen in cereal production: retrospect and prospect. *Advances in Agronomy* 87: 85–156.

Ladha, J. K., Peoples, M. B., Reddy, P. M., Biswas, J. C., Bennett, A., Jat, M. L. and Krupnik, T. J. (2022). Biological nitrogen fixation and prospects for ecological intensification in cereal-based cropping systems. *Field Crops Research* 283: 108541.

Ladha, J. K. and Reddy, P. M. (2019). The soil microbiome and crop nitrogen nutrition in anaerobic system: a case study in rice. In: Zeigler, R. S. (Ed.), *Sustaining Global Food Security*. CSIRO Publishing, pp. 278–302.

Ladha, J. K., Tirol-Padre, A., Reddy, C. K., Cassman, K. G., Verma, S., Powlson, D. S., van Kessel, C., de B Richter, D., Chakraborty, D. and Pathak, H. (2016). Global nitrogen budgets in cereals: a 50-year assessment for maize, rice and wheat production systems. *Scientific Reports* 6(1): 19355.

Lanquar, V. and Frommer, W. B. (2010). Adjusting ammonium uptake via phosphorylation. *Plant Signaling and Behavior* 5(6): 736–738.

Lee, S. (2021). Recent advances on nitrogen use efficiency in rice. *Agronomy* 11(4): 753.

Lee, S., Marmagne, A., Park, J., Fabien, C., Yim, Y., Kim, S. J., Kim, T. H., Lim, P. O., Masclaux-Daubresse, C. and Nam, H. G. (2020a). Concurrent activation of OsAMT1;2 and OsGOGAT1 in rice leads to enhanced nitrogen use efficiency under nitrogen limitation. *The Plant Journal* 103(1): 7–20.

Lee, S., Park, J., Lee, J., Shin, D., Marmagne, A., Lim, P. O., Masclaux-Daubresse, C., An, G. and Nam, H. G. (2020b). OsASN1 overexpression in rice increases grain protein content and yield under nitrogen-limiting conditions. *Plant and Cell Physiology* 61(7): 1309–1320.

Lemaire, G. and Ciampitti, I. (2020). Nitrogen use efficiency across genotype-by-environment-by-management scenarios: a review. *Plants* 9: 1-8.

Lemaire, G., Jeuffroy, M. H. and Gastal, F. (2008). Diagnosis tool for plant and crop N status in vegetative stage: theory and practices for crop N management. *European Journal of Agronomy* 28(4): 614-624.

Leran, S., Edel, K. H., Pervent, M., Hashimoto, K., Corratgé-Faillie, C., Offenborn, J. N., Tillard, P., Gojon, A., Kudla, J. and Lacombe, B. (2015). Nitrate sensing and uptake in Arabidopsis are enhanced by ABI2, a phosphatase inactivated by the stress hormone abscisic acid. *Science Signaling* 8(375): ra43-ra43.

Li, J. Y., Wang, J. and Zeigler, R. S. (2014). The 3,000 rice genomes project: new opportunities and challenges for future rice research. *GigaScience* 3: 2047-17X.

Li, K., Zhang, S., Tang, S., Zhang, J., Dong, H., Yang, S., Qu, H., Xuan, W., Gu, M. and Xu, G. (2022a). The rice transcription factor Nhd1 regulates root growth and nitrogen uptake by activating nitrogen transporters. *Plant Physiology* 189(3): 1608-1624.

Li, Q., Lu, X., Wang, C., Shen, L., Dai, L., He, J., Yang, L., Li, P., Hong, Y., Zhang, Q., Dong, G., Hu, J., Zhang, G., Ren, D., Gao, Z., Guo, L., Qian, Q., Zhu, L. and Zeng, D. (2022b). Genome-wide association study and transcriptome analysis reveal new QTL and candidate genes for nitrogen-deficiency tolerance in rice. *The Crop Journal* 10(4): 942-951.

Li, P., Li, Z., Liu, X. Zhang, H., Wang, Q., Li, N., Ding, H. and Yao, F. (2022c). Development and application of intragenic markers for 14 nitrogen-use efficiency genes in rice (Oryza sativa L.). *Frontiers in Plant Science* 13: 891860.

Li, S., Tian, Y., Wu, K., Ye, Y., Yu, J., Zhang, J., Liu, Q., Hu, M., Li, H., Tong, Y., Harberd, N. P. and Fu, X. (2018). Modulating plant growth-metabolism coordination for sustainable agriculture. *Nature* 560 (7720): 595-600.

Li, T., Zhang, W., Yin, J., Chadwick, D., Norse, D., Lu, Y., Liu, X., Chen, X., Zhang, F., Powlson, D. and Dou, Z. (2018). Enhanced-efficiency fertilizers are not a panacea for resolving the nitrogen problem. *Global Change Biology* 24(2): e511-e521.

Liang, Z., Riaz, A., Chachar, S., Ding, Y., Du, H. and Gu, X. (2020). Epigenetic modifications of mRNA and DNA in plants. *Molecular Plant* 13(1): 14-30.

Liu, K. H., Diener, A., Lin, Z., Liu, C. and Sheen, J. (2020). Primary nitrate responses mediated by calcium signalling and diverse protein phosphorylation. *Journal of Experimental Botany* 71(15): 4428-4441.

Liu, K. H., Niu, Y., Konishi, M., Wu, Y., Du, H., Sun Chung, H., Li, L., Boudsocq, M., McCormack, M., Maekawa, S., Ishida, T., Zhang, C., Shokat, K., Yanagisawa, S. and Sheen, J. (2017b). Discovery of nitrate-CPK-NLP signalling in central nutrient-growth networks. *Nature* 545(7654): 311-316.

Liu, L., Gao, H., Li, S., Han, Z. and Li, B. (2021a). Calcium signaling networks mediate nitrate sensing and responses in Arabidopsis. *Plant Signaling & Behavior* 16(10): 1938441.

Liu, Q., Wu, K., Song, W., Zhong, N., Wu, Y. and Fu, X. (2022a). Improving crop nitrogen use efficiency toward sustainable green revolution. *Annual Review of Plant Biology* 73(1): 523-551.

Liu, W., Sun, Q., Wang, K., Du, Q. and Li, W. X. (2017a). Nitrogen Limitation Adaptation (NLA) is involved in source-to-sink remobilization of nitrate by mediating the degradation of NRT1.7 in Arabidopsis. *New Phytologist* 214(2): 734-744.

Liu, X., Jiang, H., Yang, J., Han, J., Jin, M., Zhang, H., Chen, L., Chen, S. and Teng, S. (2022b). Comprehensive QTL analyses of nitrogen use efficiency in indica rice. *Frontiers in Plant Science* 13: 992225.

Liu, X., Tian, Y., Chi, W., Zhang, H., Yu, J., Chen, G., Wu, W., Jiang, X., Wang, S., Lin, Z., Xuan, W., Ye, J., Wang, B., Liu, Y., Sun, Z., Xu, D., Wang, C. and Wan, J. (2022c). Alternative splicing of OsGS1;1 affects nitrogen-use efficiency, grain development, and amylose content in rice. *The Plant Journal* 110(6): 1751–1762.

Liu, Y., Li, H., Hu, T., Mahmoud, A., Li, J., Zhu, R., Jiao, X. and Jing, P. (2022d). A quantitative review of the effects of biochar application on rice yield and nitrogen use efficiency in paddy fields: a meta-analysis. *Science of the Total Environment* 830: 154792.

Liu, Y., Wang, H., Jiang, Z., Wang, W., Xu, R., Wang, Q., Zhang, Z., Li, A., Liang, Y., Ou, S., Liu, X., Cao, S., Tong, H., Wang, Y., Zhou, F., Liao, H., Hu, B. and Chu, C. (2021b). Genomic basis of geographical adaptation to soil nitrogen in rice. *Nature* 590(7847): 600–605.

Loque, D., Lalonde, S., Looger, L. L., von Wirén, N. and Frommer, W. B. (2007). A cytosolic trans-activation domain essential for ammonium uptake. *Nature* 446(7132): 195–198.

Lu, K., Wu, B., Wang, J., Zhu, W., Nie, H., Qian, J., Huang, W. and Fang, Z. (2018). Blocking amino acid transporter OsAAP3 improves grain yield by promoting outgrowth buds and increasing tiller number in rice. *Plant Biotechnology Journal* 16(10): 1710–1722.

Lv, Y., Ma, J., Wang, Y., Wang, Q., Lu, X., Hu, H., Qian, Q., Guo, L. and Shang, L. (2021). Loci and natural alleles for low-nitrogen-induced growth response revealed by the genome-wide association study analysis in rice (Oryza sativa L.). *Frontiers in Plant Science* 12: 770736.

Madan, B., Malik, A. and Raghuram, N. (2022). Crop nitrogen use efficiency for sustainable food security and climate change mitigation. *Plant Nutrition and Food Security in the Era of Climate Change*: 47–72.

Mandal, V. K., Jangam, A. P., Chakraborty, N. and Raghuram, N. (2022). Nitrate-responsive transcriptome analysis reveals additional genes/processes and associated traits viz. height, tillering, heading date, stomatal density and yield in japonica rice. *Planta* 255(2): 42.

Mandal, V. K., Sharma, N. and Raghuram, N. (2018). Molecular targets for improvement of crop nitrogen use efficiency: current and emerging options. *Engineering Nitrogen Utilization in Crop Plants*: 77–93.

Móring, A., Hooda, S., Raghuram, N., Adhya, T. K., Ahmad, A., Bandyopadhyay, S. K., Barsby, T., Beig, G., Bentley, A. R., Bhatia, A., Dragosits, U., Drewer, J., Foulkes, J., Ghude, S. D., Gupta, R., Jain, N., Kumar, D., Kumar, R. M., Ladha, J. K., Mandal, P. K., Neeraja, C. N., Pandey, R., Pathak, H., Pawar, P., Pellny, T. K., Poole, P., Price, A., Rao, D. L. N., Reay, D. S., Singh, N. K., Sinha, S. K., Srivastava, R. K., Shewry, P., Smith, J., Steadman, C. E., Subrahmanyam, D., Surekha, K., Venkatesh, K., Varinderpal-Singh, Uwizeye, A., Vieno, M. and Sutton, M. A. (2021). Nitrogen challenges and opportunities for agricultural and environmental science in India. *Frontiers in Sustainable Food Systems* 5(13): 505347.

Nagabovanalli Basavarajappa, P., Shruthi, L. M., Lingappa, M., Kadalli, G. G. and Goudra Mahadevappa, S. (2021). Nutrient requirement and use efficiency of rice (Oryza sativa L.) as influenced by graded levels of customized fertilizer. *Journal of Plant Nutrition* 44(19): 2897–2911.

Neeraja, C. N., Barbadikar, K. M., Mangrauthia, S. K., Rao, P. R., Subrahmanayam, D. and Sundaram, R. M. (2021). Genes for NUE in rice: a way forward for molecular breeding and genome editing. *Plant Physiology Reports* 26(4): 587–599.

Nguyen, H. T. T., Dang, D. T., Van Pham, C. and Bertin, P. (2016). QTL mapping for nitrogen use efficiency and related physiological and agronomical traits during the vegetative phase in rice under hydroponics. *Euphytica* 212(3): 473–500.

Ogura, T. and Busch, W. (2015). From phenotypes to causal sequences: using genome wide association studies to dissect the sequence basis for variation of plant development. *Current Opinion in Plant Biology* 23: 98–108.

Pandey, S. (2019). Heterotrimeric G-protein signaling in plants: conserved and novel mechanisms. *Annual Review of Plant Biology* 70: 213–238.

Pathak, R. R., Ahmad, A., Lochab, S. and Raghuram, N. (2008). Molecular physiology of plant nitrogen use efficiency and biotechnological options for its enhancement. *Current Science* 94(11): 1394–1403.

Pathak, R. R., Jangam, A. P., Malik, A., Sharma, N., Jaiswal, D. K. and Raghuram, N. (2020). Transcriptomic and network analyses reveal distinct nitrate responses in light and dark in rice leaves (Oryza sativa indica var. Panvel1). *Scientific Reports* 10(1): 12228.

Pathak, R. R., Mandal, V. K., Jangam, A. P., Sharma, N., Madan, B., Jaiswal, D. K. and Raghuram, N. (2021). Heterotrimeric G-protein alpha subunit (RGA1) regulates tiller development, yield, cell wall, nitrogen response and biotic stress in rice. *Scientific Reports* 11(1): 2323.

Prasanna, J. A., Mandal, V. K., Kumar, D., Chakraborty, N. and Raghuram, N. (2023). Nitrate-responsive transcriptome analysis of rice RGA1 mutant reveals the role of G-protein alpha subunit in negative regulation of nitrogen-sensitivity and use efficiency. *Plant Cell Reports* 42(12): 1–24.

Raghuram, N., Aziz, T., Kant, S., Zhou, J. and Schmidt, S. (2022a). Editorial: nitrogen use efficiency and sustainable nitrogen management in crop plants. *Frontiers in Plant Science* 13: 862091.

Raghuram, N., Chandok, M. R. and Sopory, S. K. (1999). Light regulation of nitrate reductase gene expression in maize involves a G-protein. *Molecular Cell Biology Research Communications* 2(2): 86–90.

Raghuram, N. and Sharma, N. (2019). Improving crop nitrogen use efficiency. *Comprehensive Biotechnology* 4: 211–220.

Raghuram, N., Sutton, M. A., Jeffery, R., Ramachandran, R. and Adhya, T. K. (2021). From South Asia to the world: embracing the challenge of global sustainable nitrogen management. *One Earth* 4(1): 22–27.

Rakotoson, T., Dusserre, J., Letourmy, P., Frouin, J., Ratsimiala, I. R., Rakotoarisoa, N. V., Cao, T., Vom Brocke, K., Ramanantsoanirina, A., Ahmadi, N. and Raboin, L. (2021). Genome-wide association study of nitrogen use efficiency and agronomic traits in upland rice. *Rice Science* 28(4): 379–390.

Ranathunge, K., El-kereamy, A., Gidda, S., Bi, Y. M. and Rothstein, S. J. (2014). AMT1;1 transgenic rice plants with enhanced NH4+ permeability show superior growth and higher yield under optimal and suboptimal NH4+ conditions. *Journal of Experimental Botany* 65(4): 965–979.

Rao, I. S., Neeraja, C. N., Srikanth, B., Subrahmanyam, D., Swamy, K. N., Rajesh, K., Vijayalakshmi, P., Kiran, T. V., Sailaja, N., Revathi, P., Rao, P. R., Rao, L. V. S., Surekha, K., Babu, V. R. and Voleti, S. R. (2018). Identification of rice landraces with promising yield and the associated genomic regions under low nitrogen. *Scientific Reports* 8(1): 9200.

Riveras, E., Alvarez, J. M., Vidal, E. A., Oses, C., Vega, A. and Gutiérrez, R. A. (2015). The calcium ion is a second messenger in the nitrate signaling pathway of Arabidopsis. *Plant Physiology* 169(2): 1397–1404.

Sahu, T. K., Kumar, M., Kumar, N. Chandrakar, T. and Singh, T. P. (2022). Effect of Nano urea application on growth and productivity of rice (Oryza sativa L.) under mid land situation of Bastar region. *Journal of Pharmaceutical Innovation* 11: 185–187.

Secco, D., Whelan, J., Rouached, H. and Lister, R. (2017). Nutrient stress-induced chromatin changes in plants. *Current Opinion in Plant Biology* 39: 1–7.

Senthilvel, S., Vinod, K. K., Malarvizhi, P. and Maheswaran, M. (2008). QTL and QTL × environment effects on agronomic and nitrogen acquisition traits in rice. *Journal of Integrative Plant Biology* 50(9): 1108–1117.

Sharma, N., Jaiswal, D. K., Kumari, S. Das, G.K., Panda, S., Anandan, A. and Raghuram, N. (2023). Genome-wide urea response in rice genotypes contrasting for nitrogen use efficiency. *International Journal of Molecular Sciences*24(7): 6080.

Sharma, N., Kumari, S., Jaiswal, D. K. and Raghuram, N. (2022). Comparative transcriptomic analyses of nitrate-response in rice genotypes with contrasting nitrogen use efficiency reveals common and genotype-specific processes, molecular targets and nitrogen use efficiency-candidates. *Frontiers in Plant Science* 13: 881204.

Sharma, N., Sinha, V. B., Gupta, N., Rajpal, S., Kuchi, S., Sitaramam, V., Parsad, R. and Raghuram, N. (2018). Phenotyping for nitrogen use efficiency: rice genotypes differ in N-responsive germination, oxygen consumption, seed urease activities, root growth, crop duration, and yield at low N. *Frontiers in Plant Science* 9: 1452.

Sharma, N., Sinha, V. B., Prem Kumar, N. A., Subrahmanyam, D., Neeraja, C. N., Kuchi, S., Jha, A., Parsad, R., Sitaramam, V. and Raghuram, N. (2020). Nitrogen use efficiency phenotype and associated genes: roles of germination, flowering, root/shoot length and biomass. *Frontiers in Plant Science* 11: 587464.

Shen, C., Chen, K., Cui, Y., Chen, J., Mi, X., Zhu, S., Zhu, Y., Ali, J., Ye, G., Li, Z. and Xu, J. (2021). QTL Mapping and favorable allele mining of nitrogen deficiency tolerance using an interconnected breeding population in rice. *Frontiers in Genetics* 12: 616428.

Shrawat, A. K., Carroll, R. T., DePauw, M., Taylor, G. J. and Good, A. G. (2008). Genetic engineering of improved nitrogen use efficiency in rice by the tissue-specific expression of alanine aminotransferase. *Plant Biotechnology Journal* 6(7): 722–732.

Singh, U., Ladha, J. K., Castillo, E. G., Punzalan, G., Tirol-Padre, A. and Duqueza, M. (1998). Genotypic variation in nitrogen use efficiency in medium- and long-duration rice. *Field Crops Research* 58(1): 35–53.

Sinha, V. B., Jangam, A. P. and Raghuram, N. (2020). Biological determinants of crop nitrogen use efficiency and biotechnological avenues for improvement. *Just Enough Nitrogen*: 157–171.

Sisharmini, A., Apriana, A., Khumaida, N., Trijatmiko, K. R. and Purwoko, B. S. (2019). Expression of a cucumber alanine aminotransferase2 gene improves nitrogen use efficiency in transgenic rice. *Journal, Genetic Engineering and Biotechnology* 17(1): 9.

Srikanth, B., Rao, I. S., Surekha, K., Subrahmanyam, D., Voleti, S. R. and Neeraja, C. N. (2016). Enhanced expression of OsSPL14 gene and its association with yield components in rice (Oryza sativa) under low nitrogen conditions. *Gene* 576(1 Pt 3): 441–450.

Straub, T., Ludewig, U. and Neuhauser, B. (2017). The kinase CIPK23 inhibits ammonium transport in Arabidopsis thaliana. *Plant Cell* 29(2): 409–422.

Sun, H., Qian, Q., Wu, K., Luo, J., Wang, S., Zhang, C., Ma, Y., Liu, Q., Huang, X., Yuan, Q., Han, R., Zhao, M., Dong, G., Guo, L., Zhu, X., Gou, Z., Wang, W., Wu, Y., Lin, H. and Fu, X. (2014). Heterotrimeric G proteins regulate nitrogen-use efficiency in rice. *Nature Genetics* 46(6): 652–656.

Sun, Y., Hu, Z., Wang, X., Shen, X., Hu, S., Yan, Y., Kant, S., Xu, G., Xue, Y. and Sun, S. (2021). Overexpression of OsPHR3 improves growth traits and facilitates nitrogen use efficiency under low phosphate condition. *Plant Physiology and Biochemistry* 166: 712–722.

Sutton, M. A., Bleeker, A., Howard, C., Erisman, J. W., Abrol, Y. P., Bekunda, M., Datta, A., Davidson, E., De Vries, W., Oenema, O. and Zhang, F. S. (2013). *Our Nutrient World. The Challenge to Produce More Food & Energy with Less Pollution*. Centre for Ecology and Hydrology (CEH), Edinburgh UK, on behalf of the Global Partnership on Nutrient Management (GPNM) and the International Nitrogen Initiative (INI).

Sutton, M. A., Howard, C. M., Kanter, D. R., Lassaletta, L., Móring, A., Raghuram, N. and Read, N. (2021). The nitrogen decade: mobilizing global action on nitrogen to 2030 and beyond. *One Earth* 4(1): 10–14.

Sutton, M. A., Raghuram, N., Adhya, T. K., Baron, J., Cox, C., de Vries, W., Hicks, K., Howard, C., Ju, X., Kanter, D. and Masso, D. (2019). The nitrogen fix: from nitrogen cycle pollution to nitrogen circular economy-frontiers 2018/19: emerging issues of environmental concern chapter. *Frontiers: Emerging Issues of Environmental Concern* 4: 2018/19.

Tang, W., Ye, J., Yao, X., Zhao, P., Xuan, W., Tian, Y., Zhang, Y., Xu, S., An, H., Chen, G., Yu, J., Wu, W., Ge, Y., Liu, X., Li, J., Zhang, H., Zhao, Y., Yang, B., Jiang, X., Peng, C., Zhou, C., Terzaghi, W., Wang, C. and Wan, J. (2019). Genome-wide associated study identifies NAC42-activated nitrate transporter conferring high nitrogen use efficiency in rice. *Nature Communications* 10(1): 5279.

Tiong, J., Sharma, N., Sampath, R., MacKenzie, N., Watanabe, S., Metot, C., Lu, Z., Skinner, W., Lu, Y., Kridl, J., Baumann, U., Heuer, S., Kaiser, B. and Okamoto, M. (2021). Improving nitrogen use efficiency through overexpression of alanine aminotransferase in rice, wheat, and barley. *Frontiers in Plant Science* 12: 628521.

Tong, H. H., Mei, H. W., Yu, X. Q., Xu, X. Y., Li, M. S., Zhang, S. Q. and Luo, L. J. (2006). Identification of related QTLs at late developmental stage in rice (Oryza sativa L.) under two nitrogen levels. *Yi Chuan Xue Bao* 33(5): 458–467.

Udvardi, M., Below, F. E., Castellano, M. J., Eagle, A. J., Giller, K. E., Ladha, J. K., Liu, X., Maaz, T. M., Nova-Franco, B., Raghuram, N., Robertson, G. P., Roy, S., Saha, M., Schmidt, S., Tegeder, M., York, L. M. and Peters, J. W. (2021). A research road map for responsible use of agricultural nitrogen. *Frontiers in Sustainable Food Systems* 5: 660155.

UNEA-4 UNEP (2019). *United Nations Environment Assembly of the United Nations Environment Programme, Fourth Session, Nairobi, 11–15 March 2019. Resolution on Sustainable Nitrogen Management* UNEP/EA.4/L.16. https://wedocs.unep.org/bitstream/handle/20.500.11822/28478/English.pdf?sequence=3&isAllowed=y

Vishnukiran, T., Neeraja, C. N., Jaldhani, V., Vijayalakshmi, P., Raghuveer Rao, P., Subrahmanyam, D. and Voleti, S. R. (2020). A major pleiotropic QTL identified for yield components and nitrogen content in rice (Oryza sativa L.) under differential nitrogen field conditions. *PLoS ONE* 15(10): e0240854.

Wang, D., Xu, T., Yin, Z., Wu, W., Geng, H., Li, L., Yang, M., Cai, H. and Lian, X. (2020a). Overexpression of OsMYB305 in rice enhances the nitrogen uptake under low-nitrogen condition. *Frontiers in Plant Science* 11: 369.

Wang, F. and Peng, S.-b. (2017). Yield potential and nitrogen use efficiency of China's super rice. *Journal of Integrative Agriculture* 16(5): 1000–1008.

Wang, F., Yoshida, H. and Matsuoka, M. (2021a). Making the 'Green Revolution' truly green: improving crop nitrogen use efficiency. *Plant and Cell Physiology* 62(6): 942–947.

Wang, Q., Nian, J., Xie, X., Yu, H., Zhang, J., Bai, J., Dong, G., Hu, J., Bai, B., Chen, L., Xie, Q., Feng, J., Yang, X., Peng, J., Chen, F., Qian, Q., Li, J. and Zuo, J. (2018a). Genetic variations in ARE1 mediate grain yield by modulating nitrogen utilization in rice. *Nature Communications* 9(1): 735.

Wang, Q., Su, Q., Nian, J., Zhang, J., Guo, M., Dong, G., Hu, J., Wang, R., Wei, C., Li, G., Wang, W., Guo, H. S., Lin, S., Qian, W., Xie, X., Qian, Q., Chen, F. and Zuo, J. (2021b). The Ghd7 transcription factor represses ARE1 expression to enhance nitrogen utilization and grain yield in rice. *Molecular Plant* 14(6): 1012–1023.

Wang, R., Qian, J., Fang, Z. and Tang, J. (2020b). Transcriptomic and physiological analyses of rice seedlings under different nitrogen supplies provide insight into the regulation involved in axillary bud outgrowth. *BMC Plant Biology* 20(1): 197.

Wang, W., Hu, B., Yuan, D., Liu, Y., Che, R., Hu, Y., Ou, S., Liu, Y., Zhang, Z., Wang, H., Li, H., Jiang, Z., Zhang, Z., Gao, X., Qiu, Y., Meng, X., Liu, Y., Bai, Y., Liang, Y., Wang, Y., Zhang, L., Li, L., Sodmergen, Jing, H., Li, J. and Chu, C. (2018b). Expression of the nitrate transporter gene OsNRT1.1A/OsNPF6.3 confers high yield and early maturation in rice. *The Plant Cell* 30(3): 638–651.

Wang, W., Li, A., Zhang, Z. and Chu, C. (2021c). Posttranslational modifications: regulation of nitrogen utilization and signaling. *Plant and Cell Physiology* 62(4): 543–552.

Wang, X., Jain, A., Pei, W., Hu, Z., Sun, S. and Huang, X. (2021d). Knockdown of OsSAE1a affects acquisition and mobilization of nitrogen, and growth and development of rice. *Plant Growth Regulation* 94(3): 221–231.

Wang, Y., Lu, J., Ren, T., Hussain, S., Guo, C., Wang, S., Cong, R. and Li, X. (2017). Effects of nitrogen and tiller type on grain yield and physiological responses in rice. *AoB Plants* 9(2): plx012.

Wang, Y., Sun, Y.-J., Chen, D.-Y. and Si-Bin, Y. U. (2009). Analysis of quantitative trait loci in response to nitrogen and phosphorus deficiency in rice using chromosomal segment substitution lines. *Acta Agronomica Sinica* 35(4): 580–587.

Waqas, M., Feng, S., Amjad, H., Letuma, P., Zhan, W., Li, Z., Fang, C., Arafat, Y., Khan, M. U., Tayyab, M. and Lin, W. (2018). Protein phosphatase (PP2C9) induces protein expression differentially to mediate nitrogen utilization efficiency in rice under nitrogen-deficient condition. *International Journal of Molecular Sciences* 19(9): 2827.

Wei, D., Cui, K., Pan, J., Ye, G., Xiang, J., Nie, L. and Huang, J. (2011). Genetic dissection of grain nitrogen use efficiency and grain yield and their relationship in rice. *Field Crops Research* 124(3): 340–346.

Wei, D., Cui, K., Ye, G., Pan, J., Xiang, J., Huang, J. and Nie, L. (2012). QTL mapping for nitrogen-use efficiency and nitrogen-deficiency tolerance traits in rice. *Plant and Soil* 359(1–2): 281–295.

Wei, S., Li, X., Lu, Z., Zhang, H., Ye, X., Zhou, Y., Li, J., Yan, Y., Pei, H., Duan, F., Wang, D., Chen, S., Wang, P., Zhang, C., Shang, L., Zhou, Y., Yan, P., Zhao, M., Huang, J., Bock, R., Qian, Q. and Zhou, W. (2022). A transcriptional regulator that boosts grain yields and shortens the growth duration of rice. *Science* 377(6604): eabi8455.

Winiwarter, W., Amon, B., Bodirsky, B. L., Friege, H., Geupel, M., Lassaletta, L. and Raghuram, N. (2022). Focus on reactive nitrogen and the UN sustainable development goals. *Environmental Research Letters* 17(5): 050401.

Wu, D., Li, Y., Cao, Y., Hu, R., Wu, X., Zhang, W., Tao, W., Xu, G., Wang, X. and Zhang, Y. (2021). Increased glutamine synthetase by overexpression of TaGS1 improves grain yield and nitrogen use efficiency in rice. *Plant Physiology and Biochemistry* 169: 259-268.

Wu, K., Wang, S., Song, W., Zhang, J., Wang, Y., Liu, Q., Yu, J., Ye, Y., Li, S., Chen, J., Zhao, Y., Wang, J., Wu, X., Wang, M., Zhang, Y., Liu, B., Wu, Y., Harberd, N. P. and Fu, X. (2020). Enhanced sustainable green revolution yield via nitrogen-responsive chromatin modulation in rice. *Science* 367(6478): p.eaaz2046.

Xiang, J., Qian, K., Zhang, Y., Chew, J., Liang, J., Zhu, J., Zhang, Y. and Fan, X. (2021). OsLSD1.1 is involved in the photosystem II reaction and affects nitrogen allocation in rice. *Plant Physiology and Biochemistry* 166: 246-257.

Xin, W., Wang, J., Li, J., Zhao, H., Liu, H., Zheng, H., Yang, L., Wang, C., Yang, F., Chen, J. and Zou, D. (2021). Candidate gene analysis for nitrogen absorption and utilization in japonica rice at the seedling stage based on a genome-wide association study. *Frontiers in Plant Science* 12: 670861.

Xu, J., Shang, L., Wang, J., Chen, M., Fu, X., He, H., Wang, Z., Zeng, D., Zhu, L., Hu, J., Zhang, C., Chen, G., Gao, Z., Zou, W., Ren, D., Dong, G., Shen, L., Zhang, Q., Li, Q., Guo, L., Qian, Q. and Zhang, G. (2021). The SEEDLING BIOMASS 1 allele from indica rice enhances yield performance under low-nitrogen environments. *Plant Biotechnology Journal* 19(9): 1681-1683.

Yamaya, T., Obara, M., Nakajima, H., Sasaki, S., Hayakawa, T. and Sato, T. (2002). Genetic manipulation and quantitative-trait loci mapping for nitrogen recycling in rice. *Journal of Experimental Botany* 53(370): 917-925.

Yang, X., Xia, X., Zhang, Z., Nong, B., Zeng, Y., Xiong, F., Wu, Y., Gao, J., Deng, G. and Li, D. (2017). QTL mapping by whole genome re-sequencing and analysis of candidate genes for nitrogen use efficiency in Rice. *Frontiers in Plant Science* 8: 1634.

Yoon, D. K., Suganami, M., Ishiyama, K., Kagawa, T., Tanaka, M., Nagao, R., Takagi, D., Ishida, H., Suzuki, Y., Mae, T., Makino, A. and Obara, M. (2022). The gs3 allele from a large-grain rice cultivar, Akita 63, increases yield and improves nitrogen-use efficiency. *Plant Direct* 6(7): e417.

Yu, J., Xuan, W., Tian, Y., Fan, L., Sun, J., Tang, W., Chen, G., Wang, B., Liu, Y., Wu, W., Liu, X., Jiang, X., Zhou, C., Dai, Z., Xu, D., Wang, C. and Wan, J. (2021). Enhanced OsNLP4-OsNiR cascade confers nitrogen use efficiency by promoting tiller number in rice. *Plant Biotechnology Journal* 19(1): 167-176.

Yu, J., Zhen, X., Li, X., Li, N. and Xu, F. (2019). Increased autophagy of rice can increase yield and nitrogen use efficiency (NUE). *Frontiers in Plant Science* 10: 584.

Yuan, H., Qin, P., Hu, L., Zhan, S., Wang, S., Gao, P., Li, J., Jin, M., Xu, Z., Gao, Q., Du, A., Tu, B., Chen, W., Ma, B., Wang, Y. and Li, S. (2019). OsSPL18 controls grain weight and grain number in rice. *Journal of Genetics and Genomics* 46(1): 41-51.

Yue, F., Rong-rong, Z., Ze-chuan, L., Li-yong, C., Xing-hua, W. and Shi-hua, C. (2015). Quantitative trait locus analysis for rice yield traits under two nitrogen levels. *Rice Science* 22(3): 108-115.

Zakari, S. A., Zaidi, S. H. R., Sunusi, M. and Dauda, K. D. (2021). Nitrogen deficiency regulates premature senescence by modulating flag leaf function, ROS homeostasis, and intercellular sugar concentration in rice during grain filling. *Journal, Genetic Engineering and Biotechnology* 19(1): 177.

Zhang, G., Xu, N., Chen, H., Wang, G. and Huang, J. (2018). OsMADS25 regulates root system development via auxin signalling in rice. *The Plant Journal* 95(6): 1004-1022.

Zhang, H. and Forde, B. G. (1998). An Arabidopsis MADS box gene that controls nutrient-induced changes in root architecture. *Science* 279(5349): 407–409.

Zhang, J., Liu, Y. X., Zhang, N., Hu, B., Jin, T., Xu, H., Qin, Y., Yan, P., Zhang, X., Guo, X., Hui, J., Cao, S., Wang, X., Wang, C., Wang, H., Qu, B., Fan, G., Yuan, L., Garrido-Oter, R., Chu, C. and Bai, Y. (2019). NRT1.1B is associated with root microbiota composition and nitrogen use in field-grown rice. *Nature Biotechnology* 37(6): 676–684.

Zhang, S., Zhu, L., Shen, C., Ji, Z., Zhang, H., Zhang, T., Li, Y., Yu, J., Yang, N., He, Y., Tian, Y., Wu, K., Wu, J., Harberd, N. P., Zhao, Y., Fu, X., Wang, S. and Li, S. (2021). Natural allelic variation in a modulator of auxin homeostasis improves grain yield and nitrogen use efficiency in rice. *Plant Cell* 33(3): 566–580.

Zhang, Y., Iqbal, M. F., Wang, Y., Qian, K., Xiang, J., Xu, G. and Fan, X. (2022a). OsTBP2.1, a TATA-binding protein, alters the ratio of OsNRT2.3b to OsNRT2.3a and improves rice grain yield. *International Journal of Molecular Sciences* 23(18): 10795.

Zhang, Y., Tan, L., Zhu, Z., Yuan, L., Xie, D. and Sun, C. (2015). TOND1 confers tolerance to nitrogen deficiency in rice. *Plant Journal: For Cell and Molecular Biology* 81(3): 367–376.

Zhang, Y.-J., Xu, J.-N., Cheng, Y.-D., Wang, C., Liu, G. and Yang, J. (2020a). The effects of water and nitrogen on the roots and yield of upland and paddy rice. *Journal of Integrative Agriculture* 19(5): 1363–1374.

Zhang, Z., Gao, S. and Chu, C. (2020b). Improvement of nutrient use efficiency in rice: current toolbox and future perspectives. *TAG. Theoretical and Applied Genetics* 133(5): 1365–1384.

Zhang, Z. S., Xia, J. Q., Alfatih, A., Song, Y., Huang, Y. J., Sun, L. Q., Wan, G. Y., Wang, S. M., Wang, Y. P., Hu, B. H., Zhang, G. H., Qin, P., Li, S. G., Yu, L. H., Wu, J. and Xiang, C. B. (2022b). Rice NIN-LIKE PROTEIN 3 modulates nitrogen use efficiency and grain yield under nitrate-sufficient conditions. *Plant, Cell and Environment* 45(5): 1520–1536.

Zhao, B. (2014). Determining of a critical dilution curve for plant nitrogen concentration in winter barley. *Field Crops Research* 160: 64–72.

Zhao, M., Zhao, M., Gu, S., Sun, J., Ma, Z., Wang, L., Zheng, W. and Xu, Z. (2019). DEP1 is involved in regulating the carbon-nitrogen metabolic balance to affect grain yield and quality in rice (Oriza sativa L.). *PLoS ONE* 14(3): e0213504.

Zhen, X., Li, X., Yu, J. and Xu, F. (2019). OsATG8c-mediated increased autophagy regulates the yield and nitrogen use efficiency in rice. *International Journal of Molecular Sciences* 20(19): 4956.

Zhen, X., Zheng, N., Yu, J., Bi, C. and Xu, F. (2021). Autophagy mediates grain yield and nitrogen stress resistance by modulating nitrogen remobilization in rice. *PLoS ONE* 16(1): e0244996.

Zhou, Y., Tao, Y., Tang, D., Wang, J., Zhong, J., Wang, Y., Yuan, Q., Yu, X., Zhang, Y., Wang, Y., Liang, G. and Dong, G. (2017). Identification of QTL associated with nitrogen uptake and nitrogen use efficiency using high throughput genotyped CSSLs in rice (Oryza sativa L.). *Frontiers in Plant Science* 8: 1166.

Zhu, Y., Li, T., Xu, J., Wang, J., Wang, L., Zou, W., Zeng, D., Zhu, L., Chen, G., Hu, J., Gao, Z., Dong, G., Ren, D., Shen, L., Zhang, Q., Guo, L., Hu, S., Qian, Q. and Zhang, G. (2020). Leaf width gene LW5/D1 affects plant architecture and yield in rice by regulating nitrogen utilization efficiency. *Plant Physiology and Biochemistry* 157: 359–369.

Chapter 4

Improving the effective use of nitrogen on major field crops across the globe: a new paradigm

Ignacio A. Ciampitti, Kansas State University, USA; and Gilles Lemaire*, Honorary Director of Research, INRAE, 86600 Lusignan, France*

1 Introduction

Historically, crop improvement has been highly dependent on the adoption of technology to express a high yield under favorable environments, either via breeding genotypes for a high potential yielding capacity and/or via removing yield-limiting factors. From a crop breeding perspective, this yield improvement reinforced dependency on nitrogen (N) fertilization of modern genotypes as it has been well documented for cereals such as maize (*Zea mays* L.) (Ciampitti and Vyn, 2012), wheat (*Triticum aestivum* L.) (Cormier et al., 2013),

* Corresponding authors: ciampitti@ksu.edu; gilles.lemaire.inra@gmail.com.

http://dx.doi.org/11.19103/AS.2024.0135.23

and oilseed rape (Stahl et al., 2019). Historically, yields have improved roughly two-fold while the use of N fertilizers have increased seven-fold during the last half-century (Tilman et al., 2002), resulting in a decrease in overall nutrient use efficiency (yield per unit of input use).

Another strategy for crop improvement is to improve the capacity of the crop to acquire limiting soil N resources more efficiently and then maintain their yielding capacity. Although several breeding programs have identified potential traits for directly targeting crop N uptake improvement (e.g. root size, transporters, morphology), most progress in N over time has been achieved unintentionally by selecting for yielding capacity. As suggested by Sadras and Richards (2014), indirect selection for yield can be taken as the benchmark for any other direct approach targeting a specific plant trait. Cormier et al. (2016) have already identified the lack of breeding programs directly targeting the improvement of N acquisition capacity. In the same review, the authors compared N gains over time for modern relatives of older wheat varieties, with improvements ranging from 0.3% to 1.2% gain per year (Ortiz-Monasterio et al., 1997; Sylvester-Bradley and Kindred, 2009; Cormier et al., 2013). Previous studies have focused on exploring genotypic variation for N, for example, in rice (*Oryza sativa* L.) under different environmental conditions (Sing et al., 1998; Tirol Padre et al., 1996; Ladha et al., 1998). However, an approach to more adequately compare genotypes, considering changes in crop N status at similar biomass under different environments and management (G × E × M), has not been formalized. Therefore, providing foundational knowledge in order to guide future crop improvement, more directly targeting the effective use of N, is a research priority.

2 The traditional paradigm of nitrogen use efficiency for crop improvement

Nitrogen use efficiency (NUE) has been defined as the marginal response of crop yield (ΔY) to any additional increase in N supply (ΔN), as explained in Box 1. Enhancing NUE is an elusive and complex goal but necessary to secure the sustainability of current food production systems (Hirel et al., 2007). It has been well documented for many major field crops that NUE increases with reduced N supply and crop N status (Gastal et al., 2014; Ciampitti and Vyn, 2012), according to the law of 'diminishing returns' (Mitscherlich, 1924). With the purpose of studying NUE, Moll et al. (1982) defined N uptake efficiency (NuptE) as the increase in plant N uptake per unit of N supply, and N utilization efficiency (NutE) as the increase in yield per unit of plant N uptake (Box 1). For major field crops producing grains or tubers, these two terms, NuptE and NutE, need to be integrated with the overall ability of the plant to allocate carbon (C) to the harvested organs, expressed as harvest index (HI). From a soil-plant system perspective, NUE is a more relevant metric for evaluating the overall efficiency of a productive system, rather than to characterize the rate of crop improvement (Fig. 1).

Box 1: Why NUE is an elusive breeding target for crop improvement

Nitrogen use efficiency (NUE) as a traditional N metric refers to a crop system, i.e. the interaction of plants in a community with soil-weather under a management system.

For a better interpretation, NUE values are often decomposed (Moll et al., 1982):

NUE(Nf) = $\Delta Y/\Delta Nf$ (1) Y = yield, Nf = N fertilizer, Δ = delta

N uptake efficiency (also referred to as 'N recovery efficiency'); NupE as the increment of N uptake by crop (ΔNup) per unit of ΔNf:

NupE = $\Delta Nup/\Delta Nf$ (2)

N utilization efficiency (also referred to as 'physiological N efficiency'); NutE as the increment in crop mass (ΔW) per unit of ΔNup:

NutE = $\Delta W/\Delta Nup$ (3)

Harvest Index (HI) is the increment in ΔY to increment in ΔW:

HI = $\Delta Y/\Delta W$ (4)

Therefore NUE = NupE × NutE × HI (5)

Equation (5) must be completed by taking into account grain N per unit of grain weight harvested, that is the grain N concentration (%Ng):

NUE = NupE × NutE × HI × %Ng (6)

NUE is not necessary for analyzing crop N performance

Regardless of the expression of NUE (Eq. 5 or 6), this metric faces several challenges as a relevant indicator for analyzing the intrinsic plant performance for effective use of N, including the following:

- *Nf application also affects endogenous soil N supply (Ns). So ΔNf does not represent the increment in N available for plant (Na).*
- *The Na remains a virtual entity, inaccessible to any measurement or evaluation in field conditions owing to its high spatial and temporal variation at micro-scale rhizosphere level (Briat* et al., *2020). The Na depends on both (co-regulated by) soil and plant factors.*
- *The NupE reflects both (i) the capacity of plant to modify Na via its own N rhizospheric activity and (ii) the intrinsic plant N uptake capacity, i.e. the plant N uptake at similar plant growth. NupE of crop species and genotypes must be compared at similar crop mass otherwise the bigger crop mass would have the higher NupE, which is very trivial.*
- *Larger W promotes a greater NutE relative to small W, as W increases more proportionally (in an allometric relationship)*

relative to crop N uptake. Thus, NutE must also be compared at similar W when evaluating different scenarios.

- *Increases in NutE mirror declines in plant N concentration, leading to grain N dilution as already documented in many cereals due to grain yield improvement.*
- *Trade-off between NupE and NutE (Lemaire and Ciampitti, 2020): a plant with high NupE would be considered as having low responsiveness to N fertilization (low NutE) and then classified as a low NUE plant but with a reduced environmental N footprint relative to a low NupE and high NutE.*

The traditional paradigm of NUE, with NuptE and NutE, identifies improvements in N but changes in biomass and yield potential can mislead the outcomes of this N metric. This approach does not permit the evaluation of different G × E × M scenarios, overlooking changes in crop biomass to provide a fair comparison of actual N changes (Fig. 2) (Lemaire and Ciampitti, 2020). In addition, crop N uptake is co-regulated by both soil N supply and crop N uptake capacity (mainly determined by plant growth and root absorption) (Briat et al., 2020). Thus, the NUE approach results in a biased assessment of changes

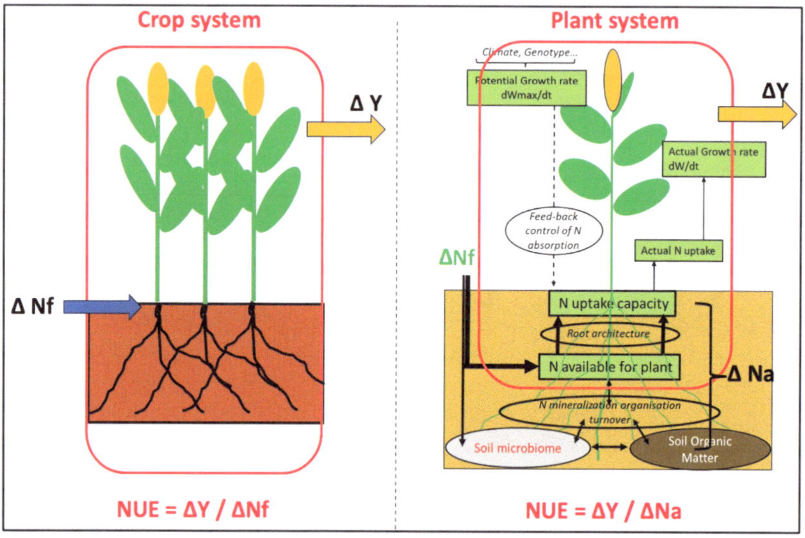

Figure 1 Schematic representation of the nitrogen (N) use efficiency (NUE) concept on a well-defined system, crop and plant systems. For the 'crop system' (plants and soil), NUE can be expressed as an increment in crop yield (ΔY) per unit of increment of applied N fertilizer (ΔNf). For the 'plant system', NUE must be calculated as an increment in crop yield (ΔY) per unit of increment in N available for the plant (ΔNa).

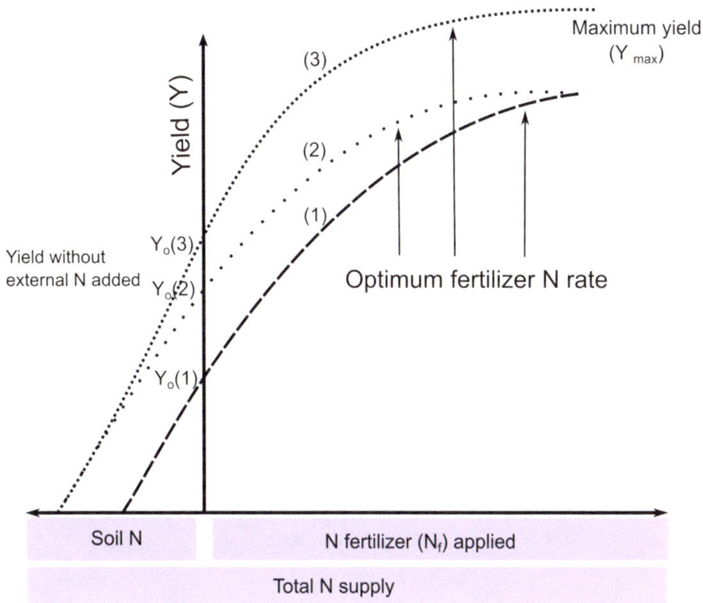

Figure 2 Response of crop yield (Y) to fertilizer N rates (Nf) according to variations in soil N. Optimum Nf depends on the total soil N and the maximum Y (Ymax), reflected in different lines: (1) reduced soil N but with high Nf; (2) high soil N, both with similar Y; and (3) high soil N and Ymax, with similar total N supply (2). Source: redrawn from Lemaire and Ciampitti (2020).

in yield under varying N supply, resulting in significant uncertainty in terms of defining optimal fertilizer N needs (Correndo et al., 2021).

3 A more functional nitrogen indicator for evaluating different genotype-environment-management scenarios

The concept of co-regulation of crop N uptake by both soil N availability and plant growth rate (Devienne-Barret et al., 2000; Lemaire et al., 2008; Briat et al., 2020) indicates that the rate of plant N uptake depends not only on the soil N supply but is also feedback-regulated by plant N demand over time. Therefore, any change in the crop growth rate would have an impact on crop N uptake, even if soil N remains unchanged. This feedback regulation is expressed via the allometric relationship between crop N uptake (N) and crop biomass (W) as proposed by Greenwood et al. (1990), Eq. 1:

$N = aW^b$ (1), a and b = coefficients from the N dilution curve

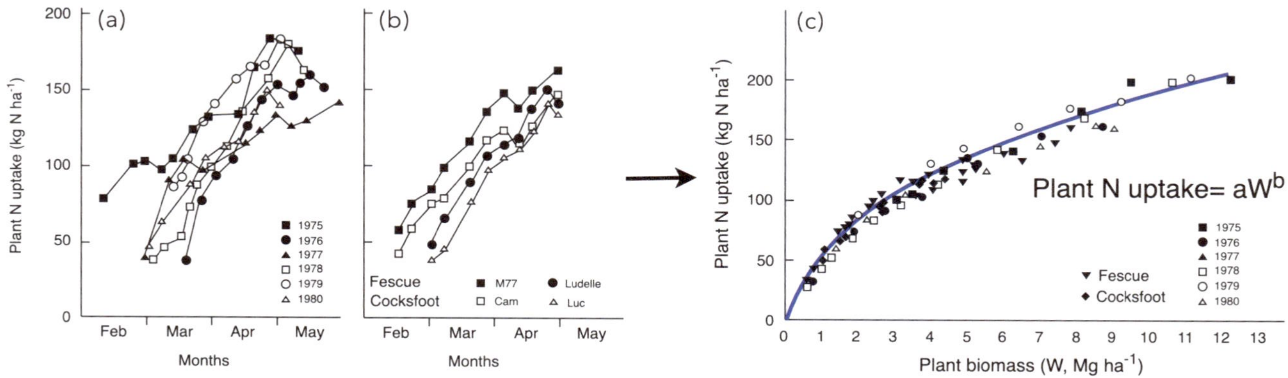

Figure 3 Plant N uptake at varying months of the growing season, years, and crop species, fescue and cocksfoot (panels a and b), and overall plant N uptake relationship with changes in plant biomass. Source: data redrawn from Lemaire and Salette (1984a,b).

Thus, the quantity of N uptake per unit of W declines as W increases, resulting in an overall decline in N/W, reflected as the N dilution curve. In summary, the overall allometric relationship between plant N uptake and W remains constant regardless of the sampling time, environments (years), or genotypes (similar species-cultivar) under non-limiting N conditions (see example in Fig. 3; Lemaire and Salette, 1984a,b).

A critical N-W allometry can be obtained when soil N matches plant N demand, i.e. the minimum crop N uptake for achieving maximum crop W (W_{max}). This critical N-W allometry trajectory can be written as follows:

$$N_c = a_c(W_{max})^b \quad (2)$$

As shown by several authors (Greenwood et al., 1990; Lemaire et al., 2008; Justes et al., 1994; Plénet and Lemaire, 2000), the critical N uptake in Equation (2) can be determined for each crop species as being relatively stable across genotype-environment conditions (see Ciampitti et al., 2021a,b; Fernandez et al., 2021). The utilization of the critical N curves assists in the calculation of the crop N nutrition index (NNI). This N index is obtained as the ratio between the observed N uptake (N_{act}) by a crop having reached a given biomass (W_{act}) and the corresponding critical N uptake (N_c) as calculated from Equation (2):

$$NNI = N_{act}/N_c \quad (3)$$

Determination of NNI allows the diagnosis of the crop's N nutrition status via quantification of the degree of satisfaction of the crop's N demand according to Lemaire et al. (2008). When NNI is below 1, there are limitations in soil N and the crop is unable to take up the quantity of N required to match its growth capacity. For example, NNI equal to 0.8 indicates that 20% of the crop N demand has not been satisfied by N supply, reflecting the intensity of crop N deficiency under a specific environment (Fig. 4).

In a recent review, Lemaire and Ciampitti (2020) reported that the use of NUE (herein defined as yield to N supply ratio) for comparison across different G × E × M scenarios remains a highly elusive concept (see Box 1). This can be explained by at least two main reasons: (1) NUE increases as the crop biomass (W) increases either with age, and/or genotype and/or environment, resulting in the N dilution process and (2) NUE decreases asymptotically with crop N nutrition status (NNI). Therefore, NNI is a foundational concept to compare crop performances, taking into account W, across different G × E × M scenarios (see Box 2). This NNI approach can help analyze quantitatively and directly the 'crop N deficiency' as the reduction in crop mass W (−ΔW) or in grain yield Y (−ΔY) when NNI < 1 by using the N deficiency Index (NDI), as calculated by 1-NNI.

In an earlier publication, Ciampitti et al. (2022) identified two major potential strategies for increasing overall effective use of N by major field crops

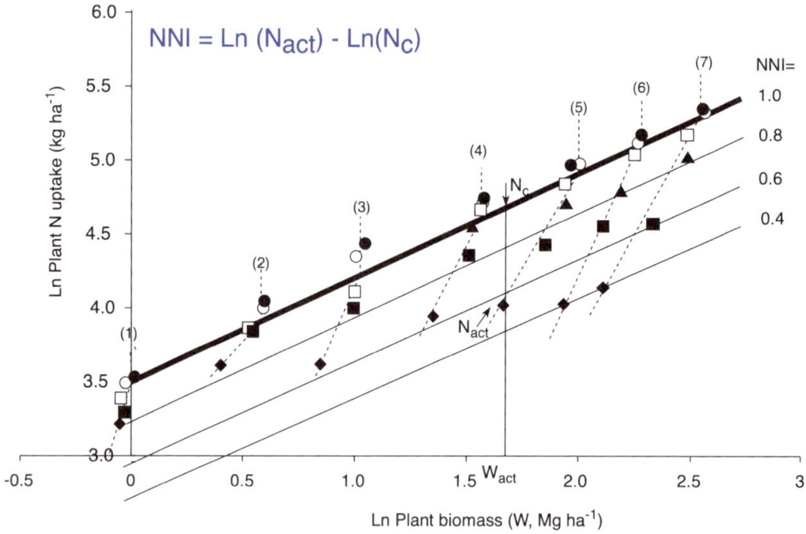

Figure 4 Ln-Ln relationship between plant nitrogen (N) uptake and biomass (W) from the fourth leaf (4th) (1) until silage harvest stage (7) for a maize crop receiving different N fertilization rate: dark diamond (no N), dark square (30 kg N ha^{-1}), dark triangle (80 kg N ha^{-1}), open square (130 kg N ha^{-1}), dark circle (180 kg N ha^{-1}), open circle (230 kg N ha^{-1}), and dark circle (280 kg N ha^{-1}). Dotted lines (1) to (7) represent the plant N uptake-to-plant W response under different fertilizer N rates for each stage. The bold line represents the critical N uptake curve (Nc) as determined for maize by Plénet and Lemaire (2000): $N = 36.4\ W^{0.66}$ corresponding to the N nutrition index (NNI) equal to 1. As soil N availability becomes limited for plant N uptake, this leads to a corresponding decrease in NNI. N_{act} = actual plant N concentration, N_c = critical N concentration, and W_{act}= current biomass. Source: data reinterpreted from Plénet and Lemaire (2020).

(see Box 2). The first, and current strategy, breeding for high yield under high N fertilization, increases maximum yield but with a risk of high dependency on external N inputs. The current strategy, first scenario, has been documented as a major gain in the traditionally defined NUE for major cereals, diluting protein concentration and maximizing NutE (for maize, Ciampitti and Vyn, 2012; for sorghum, Demarco et al., 2022; for wheat, Bogard et al., 2010; Gaju et al., 2011). The second scenario can be presented as an increase in intrinsic N uptake capacity, enhancing the ability of the plant to uptake N under low N supply (mainly via changes in NuptE). Potential improvements in HI automatically improved NUE via increases in NHI, resulting in better N remobilization (Ladha et al., 1998; Ladha et al., 2020; Fernández et al., 2022). However, the pathway of increasing intrinsic plant N uptake capacity with better reproductive N utilization has yet to be explored in order to more directly and effectively increase yields and effective use of N.

Box 2: NNI as a relevant phenotyping tool for crop N improvement

The nitrogen nutrition index (NNI) is the ratio between the actual crop N uptake 'Nupa', i.e. the quantity of N accumulated in crop and the critical N uptake 'Nupc', i.e. the minimum crop N uptake necessary for achieving maximum crop mass (Wmax) as determined by the critical N uptake curve: Nupc = a (Wmax)^b.

NNI expresses the overall plant N status, values =1 or >1 indicate that plant N demand is satisfied or luxury N conditions, and values <1 indicate a plant N deficiency. For example, an NNI of 0.6 reflects that only 60% of the total plant N demand has been satisfied, estimating the intensity of a N deficiency scenario.

As a phenotyping tool, the determination of this N metric at multiple times during the vegetative period is both labor- and cost-intensive. Thus, a recent study defined the most promising sampling time related to yield definition of major cereal field crops as the moment around flowering (Rodriguez et al., 2023).

The following figure represents: (A) the linear + plateau relationship between crop mass (W) and NNI at flowering; (B) the quantity of N fertilizer (Nf) applied for, allowing the crop to reach different NNI values. Without Nf application, the plant can reach a given value NNI_0 according to soil N supply capacity. Increasing Nf will be needed in order to achieve greater values of NNI until satisfying the crop N demand, value = 1 for Wmax.

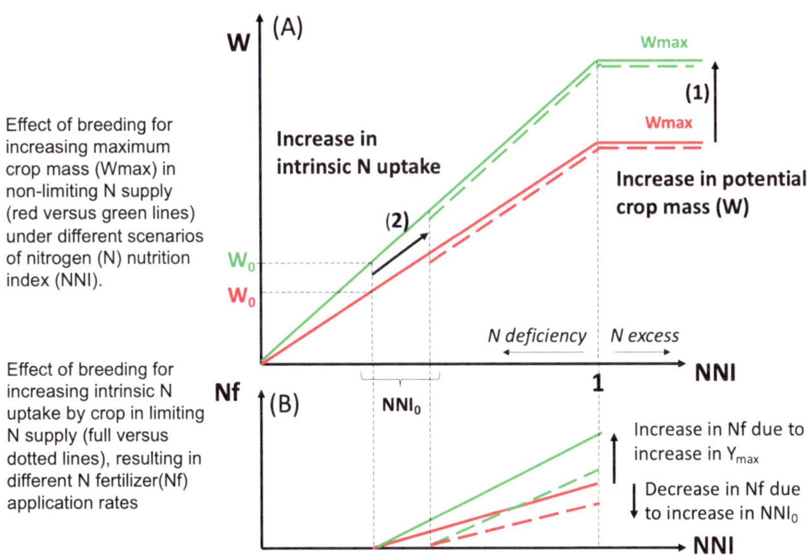

Two strategies for improving plant N performance are presented:

(1) A breeding progress in Wmax (from red to green lines) in (A), with a need to increase Nf, making the crop more sensitive to N deficiencies (B).

(2) A breeding progress in intrinsic N uptake capacity of plant, reaching a greater W with a higher level of NNI_0. In these conditions, the same Wmax value can be reached with a lower Nf application rate (dotted versus full lines, B).

4 Interactions of plant nitrogen status with other nutrients and water

Other limiting factors can be taken into account when investigating the interaction of plant N status with other nutrients, including those explored for N with sulfur (Ciampitti et al., 2013; Salvagiotti et al., 2017; Carciochi et al., 2020; Gomez et al., 2021) and with phosphorus (Sadras, 2006; Bélanger et al., 2017; Fontana et al., 2021). The law of the optimum proposed by Liebscher (1895) provides an overview on which the yield response to the availability of a nutrient is also dependent on the availability of other factors. This concept promotes the study of interactions between factors, rather than the opposite view of studying them individually. More recently this concept has been discussed by Lemaire et al. (2021) in order to bring back the study of interactions (so-called co-limitations) between nutrients and other resources such as radiation, temperature, and water (Sinclair and Park, 1993; Cossani and Sadras, 2018; Carciochi et al., 2020; Kunrath et al., 2020; Sadras et al., 2022). More recent studies into investigating the universality of N dilution curves for different crops (Ciampitti et al., 2021a,b; Fernandez et al., 2021; Fu et al., 2023), the effect of water deficit (Ciampitti et al., 2022) on those curves, and defining the saturated water accumulation curve (Zhao et al., 2022) for field crops will further improve nutrient diagnoses. It is evident that future frameworks considering co-limitation of different plant resources are needed in order to advance more universal and scientific input management in our complex agricultural farming systems.

5 Potential use of the nitrogen nutrition index across genotype-environment-management scenarios

For multiple field crops, critical N dilution curves have been established with a high adoption of the Bayesian method in the literature published in the last few years (Makowski et al., 2020). The challenge when comparing G × E × M scenarios lies mainly in situations in which a critical N dilution curve has not been established by a crop. Therefore, it is critical to evaluate potential alternatives to distinguish the different scenarios. Greenwood et al. (1990)

demonstrated that a large proportion of the variation in plant N concentration (%N) was accounted for a N dilution model using a 'b' coefficient (−0.5) with the 'a' coefficient changing depending on the crop species (C3 greater than C4). The same authors also demonstrated that a small improvement in the model was clear when both 'a' and 'b' coefficient vary among crop species. However, it was not until fairly recently that a new discussion on dataset characteristics for determining critical N curves was published (Fernandez et al., 2022b), when a new analysis from multiple C3 and C4 species uncovered more similar values for 'b' rather than 'a' coefficients (greater for C3 relative to C4 crop species). To follow up this outcome, the dataset synthesized by Ciampitti et al. (2022) was reutilized to demonstrate this point (Fig. 5).

In this new analysis, we present evidence that the 'b' coefficient could be fixed to 2/3 as a relative NNI approach to compare different G × E × M scenarios among crop species. Similar intercepts for both C3 versus C4 species are only linked to the variation of NNI within the dataset for each crop species. This value of 2/3 for 'b' coefficient will correspond to the theoretical value if we consider that N accumulation scales to the crop leaf area while W should scale to canopy volume (see Lemaire and Gastal, 1997).

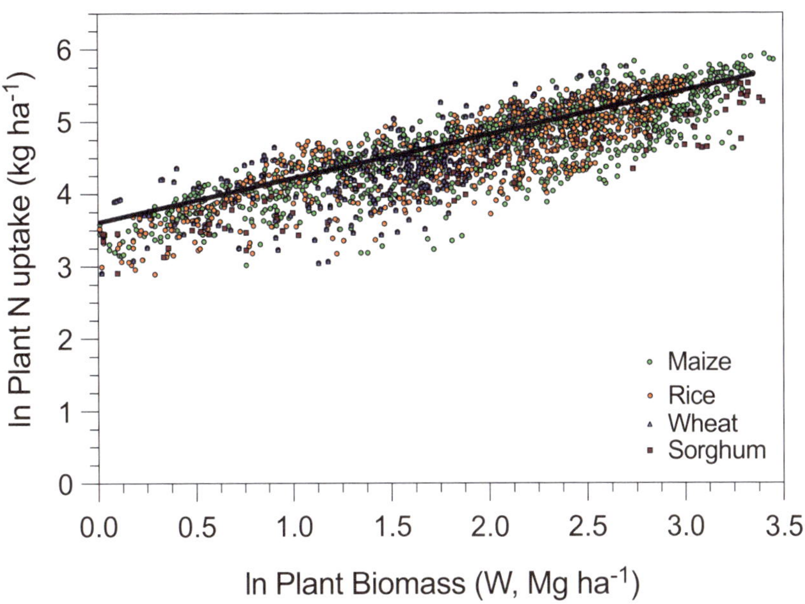

Figure 5 Ln-Ln relationship between crop nitrogen (N) uptake and biomass (W) for different crop species; observations corresponded to C3 (rice and wheat) and to C4 (maize and sorghum) crop species. Equations for each crop species: Maize = 3.45 + 0.61X; Sorghum = 3.18 + 0.61X; Rice = 3.36 + 0.66X; and Wheat = 3.36 + 0.67X. The black line represents all observations across crop species with a N nutrition index (NNI) equal to 1 (sufficiency), equation = 3.61 + 0.61X (R^2 = 0.95, n = 507). Source: dataset from Ciampitti et al. (2022).

Based on this brief analysis, it seems less plausible to breed for increased NutE at similar W or yield target. However, as highlighted previously, a major avenue for increasing the effective use of N for field crops is to increase NupE via NNI approach. The other potential change can be linked to an overall increase in the N harvest index (NHI). The HI is a relevant goal for crop improvement and plant breeding. The NHI is a major component determining the overall grain N concentration and target protein (Sinclair, 1998). The overall concept of increasing HI has been proposed as a strategy for short-term improvement in tropical grain legumes (Lawn, 1989). However, one of the potential limiting factors for increasing HI is the overall partitioning in N to the reproductive organs, NHI and with more importance in protein-rich grain/seed crops. For cereals, the grain HI and NHI are more closely and linearly related with N concentration in both vegetative and reproductive plant tissues relative to this ratio for leguminous (Lawn, 1989; Lemaire and Ciampitti, 2000). Therefore, depending on the crop species, achieving a similar HI can produce different

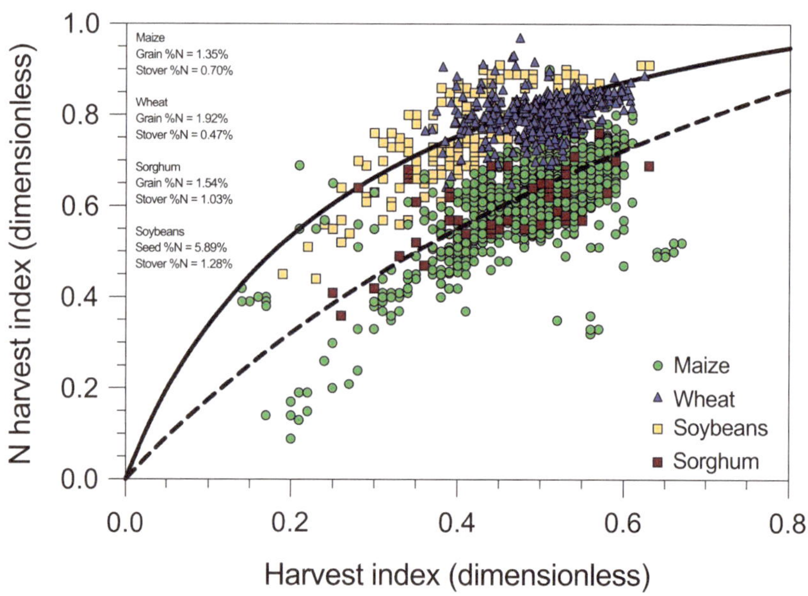

Figure 6 Relationship between nitrogen harvest index (NHI) and grain harvest index (HI) for different field crops - maize, soybeans, and wheat - reporting the overall average N concentration for the reproductive (grain/seed) and vegetative (stover) plant fractions. The full black line represents a model with greater N concentration in reproductive organs (soybeans and wheat), and the dashed line portrays a model with similar N concentration in both grain and stover (maize and sorghum). Source: data obtained from Ciampitti and Vyn, 2012 (for maize), Tamagno et al. (2017) (for soybeans), Ciampitti and Prasad, 2016 (for sorghum) and Bogard et al., 2010 and Gaju et al., 2011 (for wheat). Figure redrawn from Lemaire and Ciampitti (2020).

NHI based on the overall protein target level. For example, both soybeans and wheat crops will need to achieve greater NHI to maintain their protein levels as the HI increases, relative to cereals such as maize and sorghum (Fig. 6). This relationship between NHI and HI is a direct effect of N concentration in reproductive organs (excess or dilution), with more linear increments in cereals relative to legumes. Thus, in order to change the overall NHI, high HI can be a potential direct breeding aim.

From a growth cycle perspective, up to the point of flowering the N-W allometry (Fig. 3) is mainly governed by the ratio of leaf area to crop W (Lemaire et al., 2008). After flowering, during reproductive growth, N is remobilized in a more direct relationship based on the overall total accumulated at flowering (for maize and sorghum; Ciampitti et al., 2013; Ciampitti and Prasad, 2016). A similar N dilution occurs in the harvestable organs during the crop's reproductive phase (Plénet and Lemaire, 2000). This phenomenon has more recently been demonstrated by Zhao et al. (2020) for maize and wheat crops (Fig. 7), accounting for variations in final grain weight. A recent review by Rodriguez et al. (2023) synthesized information across three major crop species (maize, rice, and wheat), demonstrating a clear relationship between crop NNI status around flowering time and grain number determination. After flowering,

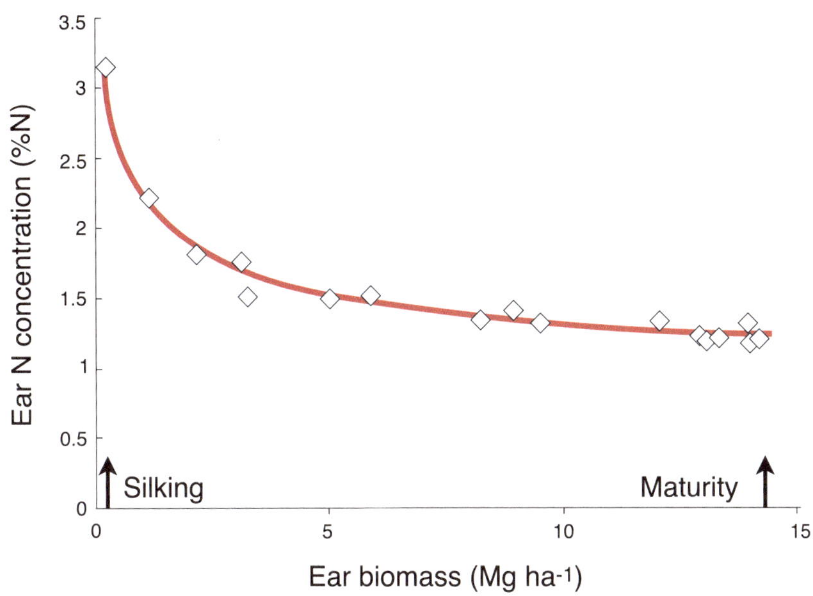

Figure 7 Curve for critical N dilution in the ear organ for maize crop determined from silking to maturity: %Ear N concentration (%N) = 2.3 ear biomass$^{-0.25}$. Source: data redrawn from Plénet and Lemaire (2000).

the utilization of reproductive NNI is a clear indicator of the plant's N status and a potential proxy for better defining grain weight and the target level to achieve high protein values at harvest time.

6 A retrospective analysis of true gains of nitrogen over time for major field crops

Previous studies have reported changes in NUE (as the ratio yield to N supply), mainly linked to yield gains over time and an overall N dilution process, protein reductions (e.g. for maize, Ciampitti and Vyn, 2012; for sorghum, Ciampitti and Prasad, 2016). For example, for maize, modern hybrids present a more consistent increase in crop W and N uptake with fertilizer N rates (Ciampitti and Vyn, 2012) (Fig. 8). The overall greater NutE reported by the same authors for modern maize hybrids was mainly the result of grain N dilution associated with higher crop W. Lastly, from a yield perspective for different cereals, major gains have been reported due to changes in HI and biomass over time, but changes in intrinsic N capacity are yet to be documented (not reflected in maize hybrids for the 'New Era'; Fig. 8).

A new synthesis-analysis focused on four major field crop species (barley, maize, sorghum, and wheat) and their overall changes in NNI over time are presented as a key output of this chapter (Fig. 9). This evaluation concluded that historical yield gains were achieved maintaining similar crop NNI (not different from zero for all slopes) over time (Fig. 9). These findings reveal the lack of an effective change in intrinsic N uptake capacity and in the ability to reduce the dependency of these cereal crops on N fertilization, as already demonstrated by Ciampitti and Lemaire (2022).

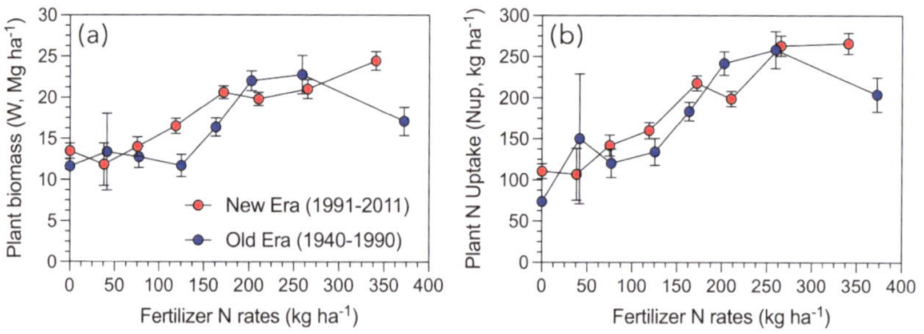

Figure 8 Changes in plant biomass (panel A) and plant N uptake (panel B), all at end of the season for maize crop, relative to different fertilizer N rate levels for materials classified as 'Old Era' data from publications ranging from 1940 to 1990 and 'New Era' for maize hybrids obtained from studies published from 1991 to 2011. Source: data collected from Ciampitti and Vyn (2012).

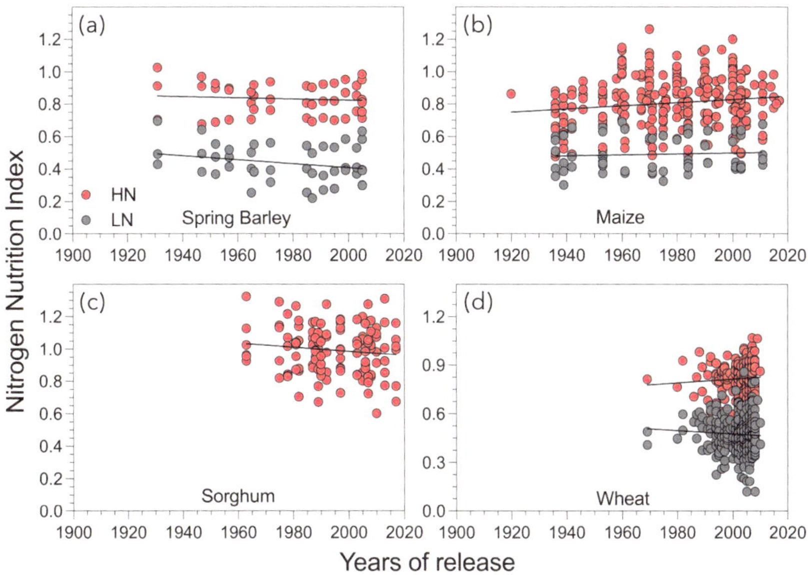

Figure 9 Nitrogen nutrition index (NNI) over time for major cereal crops - spring barley (Bingham et al., 2012), maize (Chen et al., 2014, 2015; Emmett et al., 2018; Mueller et al., 2019; Woli et al., 2016; Fernandez et al., 2022c), sorghum (Demarco et al., 2022), and wheat (Cormier et al., 2013). LN = low fertilizer N rate, HN = high fertilizer N rate. All fitted models presented slopes that did not differ from zero, without showing a consistent (positive or negative) over time.

7 A path forward for enhancing nitrogen uptake and improve yields

To enhance plant resiliency via development of a 'N reserve' pool, the hypothesis formulated by Plénet and Lemaire (2000) was that a crop NNI value above 1 was needed prior to flowering (Fig. 10). This pathway for increasing yields directly targeting changes in plant N status should be pursued as a strategy to increase crop resiliency under adverse conditions during the crop's reproductive period (Fig. 10). In a different crop improvement strategy, a recent study published by Fernandez et al. (2022a) presented new evidence that enhancing vegetative N allocation to the leaves located in the lower part of the canopy can help increase N remobilization following flowering when (1) soil N availability becomes exhausted and/or (2) N uptake capacity of the root declines (senescence).

8 Using the nitrogen nutrition index for a holistic approach to defining crop nitrogen status

The approach of using crop NNI as a diagnosis tool focuses on the concept of comparing the relative yield (or crop W) versus instantaneous NNI determined

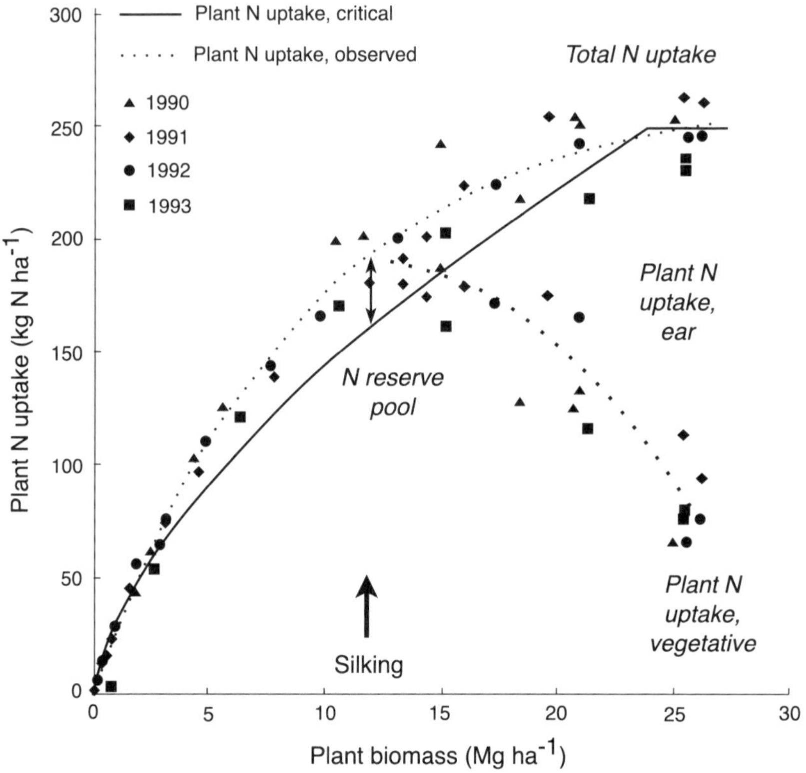

Figure 10 Plant nitrogen (N) uptake relative to plant biomass over time, from vegetative to maturity for maize crop. Source: figure redrawn from Plénet and Lemaire (2000).

at different points during the season (Fig. 11). The strength of the association has a physiological basis due to the relevancy of NNI as a predictor of grain number (when comparison is determined until late vegetative period), one of the major components for attainable yield within each environment (Plénet and Cruz, 1997). However, crop NNI at the end of the vegetative period does not take into account potential fluctuations in the N status of the harvestable organs during the reproductive period (Zhao et al., 2020). In addition, more recent studies have explored the importance of plant N status following flowering not only to maximize yield but also to evaluate the impact on other quality components such as protein levels (Zhao et al., 2016, 2020; Ata-Ul-Karim et al., 2017; Feng et al., 2022). Thus, it is critical to assess crop NNI at different timings when focusing on improving the overall N status of the crop. From the perspective of yield components, gains in weight of harvestable organs or in their %N must be obtained at a target number of grains/seeds, otherwise, the trade-off between grains/seeds number and weight (and with %N) can mask true changes. In summary, utilizing vegetative NNI up until flowering as a diagnosis of grain

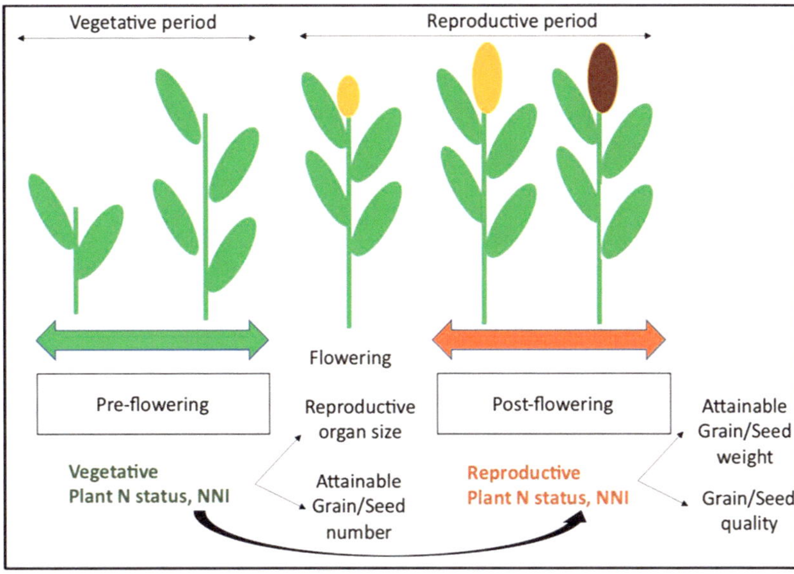

Figure 11 Theoretical framework for the relationship between plant nitrogen (N) status and relative yield (yield to maximum yield ratio). This approach emphasizes the importance of plant N status for defining critical yield components and the target for different timing on phenotyping depending on the objective for trait crop improvement. The dark arrow indicates that the reproductive plant N status diagnosis must be subordinated to the previous vegetative plant N status diagnosis.

number, and reproductive NNI (N status in harvestable organs) as a reflection of the grain/seed weight (and target protein levels), provides a comprehensive sequential approach to assessing N for major field crops (Fig. 11).

9 Conclusion and future trends

This chapter provides an overview of the key concepts to understand current progress on effectively managing N and crop improvement over time for major field crops. The important change with this new concept is the move away from the classical approach of NUE measuring crop response to N fertilizer application to the new approach of NNI for estimating the capacity of crop to obtain soil N and increase N storage per unit of biomass. Foundational concepts on the development of critical N dilution curves to understand the minimum requirement of N to maximize plant growth are critical for the development of practical diagnostic tools to assist farmers, agronomists, researchers, and breeders to better manage this nutrient.

From a crop breeding perspective, the focus of a more holistic assessment of plant N status will serve as a foundation for phenotyping in terms of specific plant traits. A future pathway for increasing plant N uptake capacity under different conditions is needed as a target to maintain current yield gains while reducing dependency on N fertilization. Lastly, the use of this N metric needs to take into account potential limitations of other nutrient and resources (such as water, radiation, etc.) in order to obtain a robust diagnosis method for more precisely guiding N recommendations in our complex farming systems.

10 Acknowledgements

Contribution no. 23-304-B from the Kansas Agricultural Experiment Station.

11 References

Ata-Ul-Karim, S. T., Zhu, Y., Cao, Q., Rehmani, M. I. A., Cao, W. X. and Tang, L. (2017). In-season assessment of grain protein and amylose content in rice using critical nitrogen dilution curve. *Eur J Agron* 90:139–151. doi: 10.1016/j.eja.2017.08.001.

Bélanger, G., Ziadi, N., Lajeunesse, J., Jouany, C., Virkajarvi, P., Sinaj, S. and Nyiraneza, J. (2017). Shoot growth and phosphorus-nitrogen relationship of grassland swards in response to mineral phosphorus fertilization. *Field Crops Res* 204:31–41.

Bingham, I. J., Karley, A. J., White, P. J., Thomas, W. T. B. and Russell, J. R. (2012). Analysis of improvements in nitrogen use efficiency associated with 75 years of spring barley breeding. *Eur J Agron* 42:49–58.

Bogard, M., Allard, V., Brancourt-Hulmel, M., Heumez, E., Machet, J. M., Jeuffroy, M. H., Gate, P., Martre, P. and Le Gouis, J. (2010). Deviation from the grain protein concentration-grain yield negative relationship is highly correlated to post-anthesis N uptake in winter wheat. *J Exp Bot*. https://doi.org/10.1093/jxb/erq238.

Briat, J. F., Gojon, A., Plassard, C., Rouached, H. and Lemaire, G. (2020). Reappraisal of the central role of soil nutrient availability in nutrient management in light of recent advances in plant nutrition at crop and molecular levels. *Eur J Agron* 116. https://doi.org/10.1016/j.eja.2020.126069.

Carciochi, W. D., Salvagiotti, F., Pagani, A., Reussi Calvo, N. I., Eyherabide, M., Sainz Rozas, H. R. and Ciampitti, I. A. (2020). Nitrogen and sulfur interaction on nutrient use efficiencies and diagnostic tools in maize. *Eur J Agron* 116:126045.

Chen, K., Kumudini, S. V., Tollenaar, M. and Vyn, T. J. (2015). Plant biomass and nitrogen partitioning changes between silking and maturity in newer versus older maize hybrids. *Field Crops Res* 183:315–328.

Chen, Y., Xiao, C., Chen, X., Li, Q., Zhang, J., Chen, F., Yuan, L. and Mi, G. (2014). Characterization of the plant traits contributed to high grain yield and high grain nitrogen concentration in maize. *Field Crops Res* 159:1–9.

Ciampitti, I., van Versendaal, E., Rybecky, J. F., Lacasa, J., Fernandez, J., Makowski, D. and Lemaire, G. (2022). A global dataset to parametrize critical nitrogen dilution curves for major crop species. *Sci Data* 9(1):277.

Ciampitti, I. A. and Lemaire, G. (2022). From use efficiency to effective use of nitrogen : A dilemma for maize breeding improvement. *Sci Total Environ* 826:154125.

Ciampitti, I. A. and Prasad, P. V. V. (2016). Historical synthesis-analysis of changes in grain nitrogen dynamics in sorghum. *Front Plant Sci* 7. https://doi.org/10.3389/fpls.2016.00275.

Ciampitti, I. A. and Vyn, T. J. (2012). Physiological perspectives of changes over time in maize yield dependency on nitrogen uptake and associated nitrogen use efficiency: a review. *Field Crops Res* 43:48–67.

Ciampitti, I. A., Briat, J. F., Gastal, F. and Lemaire, G. (2022). Redefining crop breeding strategy for effective use of nitrogen in cropping systems. *Commun Biol* 5(1):823. https://doi.org/10.1038/s42003-022-03782-2.

Ciampitti, I. A., Camberato, J. J., Murrell, S. T. and Vyn, T. J. (2013). Maize nutrient accumulation and partitioning in response to plant density and nitrogen rate: I. Macronutrients. *Agron J* 105(3):783–795.

Ciampitti, I. A., Fernandez, J., Tamagno, S., Zhao, B., Lemaire, G. and Makowski, D. (2021a). Does the critical N dilution curve for maize crop vary across genotype x environment x management scenarios?–A Bayesian analysis. *Eur J Agron* 123:126202.

Ciampitti, I. A., Makowski, D., Fernandez, J., Lacasa, J. and Lemaire, G. (2021b). Does water availability affect the critical N dilution curves in crops? A case study for maize, wheat, and tall fescue crops. *Field Crops Res* 273:108301.

Cormier, F., Faure, S., Dubreuil, P., Heumez, E., Beauchêne, K., Lafarge, S., Praud, S. and Le Gouis, J. (2013). A multi-environmental study of recent breeding progress on nitrogen use efficiency in wheat (Triticum aestivum L.). *Theor Appl Genet* 126(12):3035–3048.

Cormier, F., Foulkes, J., Hirel, B., Gouache, D., Moënne-Loccoz, Y. and Le Gouis, J. (2016). Breeding for increased nitrogen-use efficiency: a review for wheat (T. aestivum L.). *Plant Breed* 135(3):255–278.

Correndo, A. A., Tremblay, N., Coulter, J. A., Ruiz-Diaz, D., Franzen, D., Nafziger, E., Prasad, V., Rosso, L. H. M., Steinke, K., Du, J., Messina, C. D. and Ciampitti, I. A. (2021). Unraveling uncertainty drivers of the maize yield response to nitrogen: A Bayesian and machine learning approach. *Ag Forest Meteorol* 311:108668.

Cossani, C. M. and Sadras, V. O. (2018). Water-nitrogen co-limitation in grain crops. *Adv Agron* 50:231–274.

Demarco, P. A., Mayor, L., Messina, C., Prasad, P. V. V., Morris, G. P. and Ciampitti, I. A. (2022). Retrospective study in U.S. commercial sorghum breeding: III. Nitrogen internal efficiency. *Crop Sci*. https://doi.org/10.1002/csc2.20763.

Devienne-Barret, F., Justes, E., Machet, J. M. and Mary, B. (2000). Integrated control of nitrate uptake by crop growth rate and soil nitrate availability under field conditions. *Ann Bot* 86(5):995–1005.

Emmett, B. D., Buckley, D. H., Smith, M. E. and Drinkwater, L. E. (2018). Eighty years of maize breeding alters plant nitrogen acquisition but not rhizosphere bacterial community composition. *Plant Soil* 431(1-2):53–69.

Feng, W., Li, X., Dong, H., Qin, Y., Sun, M., Shao, J., Zheng, C. and Li, P. (2022). Fruits-based critical nitrogen dilution curve for diagnosing nitrogen status in cotton. *Front Plant Sci* 13:801968.

Fernandez, J. A., Lemaire, G., Belanger, G., Gastal, F., Makowski, D. and Ciampitti, I. A. (2021). Revisiting the critical nitrogen dilution curve for tall fescue: A quantitative synthesis. *Eur J Agron* 131:126380.

Fernandez, J. A., Habben, J. E., Schussler, J. R., Masek, T., Weers, B., Bing, J. and Ciampitti, I. A. (2022a). *zmm28* transgenic maize increases both N uptake- and N utilization-efficiencies. *Commun Biol* 5(1):555. https://doi.org/10.1038/s42003-022-03501-x.

Fernandez, J. A., van Versendaal, E., Lacasa, J., Makowski, D., Lemaire, G. and Ciampitti, I. A. (2022b). Dataset characteristics for the determination of critical nitrogen dilution curves: from past to new guidelines. *Eur J Agron* 139:126568.

Fernández, J. A., Messina, C. D., Salinas, A., Prasad, P. V. V., Nippert, J. B. and Ciampitti, I. A. (2022). Kernel weight contribution to yield genetic gain of maize: a global review and US case studies. *J Exp Bot* 73(11):3597–3609.

Fontana, M., Bélanger, G., Histe, J., Ziadi, N., Elfouki, S., Bragazza, L., Liebisch, F. and Sinaj, S. (2021). Critical plant phosphorus for winter wheat assessed for long-term field experiments. *Eur J Agron* 126:126263.

Fu, Z., Zhang, R., Zhang, J., Zhang, K., Cao, Q., Tian, Y., Zhu, Y., Cao, W. and Liu, X. (2023). Potential of establishing the universal critical nitrogen dilution curve for japonica rice. *Plant Phenom* 5(0036). https://doi.org/10.34133/plantphenomics.0036.

Gaju, O., Allard, V., Martre, P., Snape, J. W., Heumez, E., LeGouis, J., Moreau, D., Bogard, M., Griffiths, S., Orford, S., Hubbart, S. and Foulkes, M. J. (2011). Identification of traits to improve the nitrogen-use efficiency of wheat genotypes. *Field Crops Res* 123(2):139–152.

Gastal, F., Lemaire, G., Durand, J.-L. and Louarn, G. (2014). Quantifying crop responses to nitrogen and avenues to improve nitrogen-use efficiency. In *Crop Physiology, Application for Genetic Improvement and Agronomy*; Sadras, V. O. and Calderini, D. F., Eds.; Academia Press Oxford, Oxford, pp. 161–206. ISBN 978-0-12-4171046.

Gomez, F. M., Prystupa, P., Boero, J. J. and Gutierrez, F. H. (2021). Sulfur partitioning and grain concentration differed from that for nitrogen in malting barley. *Field Crops Res* 263:108053. Available at: https://www.sciencedirect.com/science/article/pii/S037842902031337X.

Greenwood, D. J., Lemaire, G., Gosse, G., Cruz, P., Draycott, A. and Neeteson, J. J. (1990). Decline in percentage N of C3 and C4 crops with increasing plant mass. *Ann Bot* 66(4):425–436.

Hirel, B., Le Gouis, J., Ney, B. and Gallais, A. (2007). The challenge of improving nitrogen use efficiency in crop plants: towards a more central role of genetic variability and quantitative genetics within integrated approaches. *J Exp Bot* 58(9):2369–2387.

Justes, E., Mary, B., Meynard, J.-M., Machet, J.-M. and Thelier-Huche, L. (1994). Determination of a critical nitrogen dilution curve for winter wheat crops. *Ann Bot* 74(4):397–407.

Kunrath, T. R., Lemaire, G., Texeira, E., Brown, H. E., Ciampitti, I. A. and Sadras, V. O. (2020). Allometric relationships between nitrogen uptake and transpiration to untangle interactions between nitrogen supply and drought in maize and sorghum. *Eur J Agron* 120:126145.

Ladha, J. K., Kirk, G. J. D., Bennett, J., Peng, S., Reddy, C. K., Reddy, P. M. and Singh, U. (1998). Opportunities for increased nitrogen use efficiency from improved lowland rice germplasm. *Field Crops Res* 56(1–2):41–71.

Ladha, J. K., Jat, M. L., Stirling, C. M., Chakraborty, D., Pradhan, P., Krupnik, T. J., Sapkota, T. B., Pathak, H., Rana, D. S., Tesfaye, K. and Gerard, B. (2020). Achieving the sustainable development goals in agriculture: the crucial role of nitrogen in cereal-based systems. *Adv Agron* 163:39–116.

Lawn, R. J. (1989). Agronomic and physiological constraints to the productivity of tropical grain legumes and prospects for improvement. *Exp Agric* 25(4):509–528.

Lemaire, G. and Ciampitti, I. (2020). Crop mass and N status as prerequisite covariables for unraveling nitrogen use efficiency across genotype-by-environment-by-management scenarios: a review. *Plants* 9(10):1309.

Lemaire, G. and Salette, J. (1984a). Relation entre dynamique decroissance et dynamique de prélèvement d'azote par un peuplement de graminées fourragères. 1 - Etude de l'effet du milieu. *Agronomie* 4:423-430.
Lemaire, G. and Salette, J. (1984b). Relation entre dynamique de croissance et dynamique de prélèvement d'azote par un peuplement de graminées fourragères. 2 - etude de la variabilité entre génotypes. *Agronomie* 4:431-436.
Lemaire, G. and Gastal, F. (1997). N uptake and distribution in plant canopies. In *Diagnosis on the Nitrogen Status in Crops*; Lemaire, G., Ed.; Springer, Heidelberg, Germany, pp. 3-43.
Lemaire, G., Jeuffroy, M. H. and Gastal, F. (2008). Diagnosis tool for plant and crop N status in vegetative stage. Theory and practices for crop N management. *Eur J Agron* 28(4):614-624.
Lemaire, G., Jeuffroy, M. H. and Gastal, F. (2008). Diagnosis tool for plant and crop N status in vegetative stage. Theory and practices for crop N management. *Eur J Agron* 28(4):614-624.
Lemaire, G., Tang, L., Bélanger, G., Zhu, Y. and Jeuffroy, M. (2021). Forward new paradigms for crop mineral nutrition and fertilization towards sustainable agriculture. *Eur J Agron* 125:126248.
Liebscher, G. (1895). Untersuchungen über die Bestimmung Des Düngerbedürfnisses Der Ackerböden und Kulturpflanzen. *J Landwirtsch* 43:49-216.
Makowski, D., Zhao, B., Ata-Ul-Karim, S. T. and Lemaire, G. (2020). Analyzing uncertainty in critical nitrogen dilution curves. *Eur J Agron* 118. https://doi.org/10.1016/j.eja.2020.126076.
Mitscherlich, E. A. (1924). *Die. Bestimung des Düngerbedürjnisses des Bodens*. Paul Barey, Berlin, p. 100.
Moll, R. H., Kamprath, E. J. and Jackson, W. A. (1982). Analysis and interpretation of factors which contribute to efficiency of nitrogen utilization. *Agron J* 74(3):562-564.
Mueller, S. M., Messina, C. D. and Vyn, T. J. (2019). Simultaneous gains in grain yield and nitrogen efficiency over 70 years of maize genetic improvement. *Sci Rep* 9(1):9095.
Ortiz-Monasterio, J. I., Satre, K. D., Rajaram, S.. and McMahon, M. (1997). Genetic progress in wheat yield and nitrogen use efficiency under four nitrogen rates. *Crop Sci* 37(3):898-904.
Plénet, D. and Cruz, P. (1997). The nitrogen requirement of major agricultural crops. Maize and sorghum. In *Diagnosis of the Nitrogen Status in Crops*; Lemaire, G., Ed.; Springer-Verlag, Heidelberg, pp. 93-106.
Plénet, D. and Lemaire, G. (2000). Relationships between dynamics of nitrogen uptake and dry matter accumulation in maize crops. *Plant Soil* 216:65-82.
Sadras, V. O. (2006). The N:P stoichiometry of cereal, grain legume and oilseed crops. *Field Crops Res* 95(1):13-29.
Sadras, V. O. and Richards, R. A. (2014). Improvement of crop yield in dry environments: benchmarks, levels of organization and role of nitrogen. *J Exp Bot* 65(8):1981-1995.
Sadras, V. O., Giordano, N., Correndo, A., Cossani, C. M., Ferreyra, J. M., Caviglia, O. P., Coulter, J. A., Ciampitti, I. A. and Lollato, R. P. (2022). Temperature-driven developmental modulation of yield response to nitrogen in wheat and maize. *Front Agron* 4:903340.
Salvagiotti, F., Prystupa, P., Ferraris, G., Couretot, L., Magnano, L., Dignani, D. and Gutiérrez BGutiérrez-Boemoem, F. H. (2017). N:P:S stoichiometry in grains and physiological attributes associated with grain yield in maize as affected by phosphorus and sulfur nutrition. *Field Crops Res* 203:128-138.

Sinclair, T. R. (1998). Historical changes in harvest index crop N accumulation. *Crop Sci* 38(3):638–643.

Sinclair, T. R. and Park, W. I. (1993). Inadequacy of the Liebig limiting-factor paradigm for explaining varying crop yields. *Agron J* 85(3):742–746.

Stahl, A., Vollrath, P., Samans, B., Frisch, M., Wittkop, B. and Snowdon, R. J. (2019). Effect of breeding on nitrogen use efficiency-associate traits in oilseed rape. *J Exp Bot* 70(6):1969–1986.

Singh, U., Ladha, J. K., Castillo, E. G., Punzalan, G., Tirol-Padre, A. and Duqueza, M. (1998). Genotypic variation in nitrogen use efficiency in medium-and long duration rice. *Field Crops Res* 58(1):35–53.

Sylvester-Bradley, R. and Kindred, D. R. (2009). Analysing nitrogen responses of cereals to prioritize routes to the improvement of nitrogen use efficiency. *J Exp Bot* 60(7):1939–1951.

Tamagno, S., Balboa, G. R., Assefa, Y., Kovács, P., Casteel, S. N., Salvagiotti, F., García, F. O., Stewart, W. M. and Ciampitti, I. A. (2017). Nutrient partitioning and stoichiometry in soybean: A synthesis-analysis. *Field Crops Res* 200:18–27.

Tilman, D., Cassman, K. G., Matson, P. A., Naylor, R. and Polasky, S. (2002). Agricultural sustainability and intensive production practices. *Nature* 418(6898):671–677.

Tirol Padre, A., Ladha, J. K., Singh, U., Laureles, E., Punzalan, G. and Akita, S. (1996). Grain yield performance of rice genotypes at sub-optimal levels of soil N as affected by N uptake and utilization efficiency. *Field Crops Res* 46(1–3):127–143.

Woli, K. P., Boyer, M. J., Elmore, R. W., Sawyer, J. E., Abendroth, L. J. and Barker, D. W. (2016). Corn era hybrid response to nitrogen fertilization. *Agron J* 108(2):473–486.

Zhao, B., Ata-Ul-Karim, S. T., Yao, X., et al. (2016). A new curve of critical nitrogen concentration based on spike dry matter for winter wheat in Eastern China. *PLOS ONE*. https://doi.org/10.1371/journal.pone.0164545.

Part 2

Monitoring and optimising nitrogen use

Chapter 5

Developments in proximal sensors to detect crop nitrogen status

Brenda S. Tubana and Daniel Forestieri, Louisiana State University, USA

1 Introduction

Considerable research effort has been focused on developing best field practices to improve crop nitrogen use efficiency given the vital role of nitrogen in plant health and crop productivity. Nitrogen is the most limiting nutrient in crop production, mainly due to its dynamic nature in the environment and its vulnerability to losses from soil. This well-known attribute of nitrogen highlights the need for efficient management practices for nitrogen fertilizer in crop production. Nitrogen fertilizer is considered one of the most expensive agricultural inputs due to the high nitrogen requirements of plants. This can also include the costs associated with the environmental consequences of nitrogen fertilizer mismanagement (e.g. leaching into watercourses). Among the technologies developed to improve the management of nitrogen in the soil, non-invasive evaluation via proximal sensing represents a major turning point. This approach allows rapid and regular assessment of crop nitrogen status.

Traditionally, crop nitrogen status has been determined based on the direct measurement of plant nitrogen concentration and subsequently estimating plant nitrogen needs (Binford et al., 1992; Fox et al., 1989; Hong et al., 1990; Justes et al., 1997; Lemaire and Gastal, 1997; Magdoff et al., 1990; Sims et al.,

http://dx.doi.org/10.19103/AS.2024.0135.07

1995). The current laboratory procedure to determine plant tissue nitrogen is the dry combustion method whereas, historically, titration and colorimetric were popular methods for quantification of nitrogen concentration in digested plant samples. These methods are accurate for evaluating crop nitrogen status and generally correlate well with crop yield. However, they are time-consuming, require a trained operator and can be expensive (Argenta et al., 2004).

Such limitations have prompted the development and adoption of rapid, automated techniques that can produce near real-time results. Some of these techniques rely on visual assessment of crop nitrogen status (Furuya, 1987). This application is highly practical as leaf/canopy greenness can be related directly to soil nitrogen (Schröder et al., 2000). The evaluation of plant health using leaf colour began long before the development of leaf colour charts to more precisely measure plant nitrogen status, beginning in the late nineteenth and early twentieth centuries when scientists reported observable changes in leaf spectra of stressed plants (Coblentz, 1912; Shull, 1929; Sorby, 1873).

To avoid measurement bias from individuals making subjective assessments using leaf colour charts, sensors have been developed for rapid and near real-time detection of stressed plants (Niak et al., 2017). An important step forward was the use of spectral ratio reflection for rapid quantification of the colour of intact plant leaves (Inada, 1963). From the 1970s, instruments were developed to more precisely characterize and quantify leaf spectra (Jackson et al., 1986; Myers, 1983). The objective of this chapter is to describe the principles, types and use of proximal sensors for assessing crop nitrogen status and their use in developing decision tools for the effective management of nitrogen.

2 Principles of proximal sensing

The electromagnetic spectrum ranges from gamma rays (shortest wavelength) to radio waves (longest wavelength) (Fig. 1). The human eye can only detect wavelengths from about 400 nm (violet) to 700 nm (red). Sensors are able to collect imagery at different frequencies and wavelengths that are invisible to the human eye. Proximal sensing involves receiving information by being in contact with or close to the crop (typically under 2 m) (Thompson et al., 2017; Usha and Singh, 2013). Common layers such as red, green, blue and near-infrared can be measured and combined to assess the health of plants, track plant growth and consequently determine management of agricultural inputs.

The electromagnetic radiation that strikes leaf surfaces can be reflected, absorbed, scattered and transmitted (Fig. 2). The wavebands within the electromagnetic spectrum that are relevant for evaluating plant health include the visible and infrared. Photosynthetic pigments and cell structure primarily influence leaf spectral reflectance from visible and near-infrared wavelengths

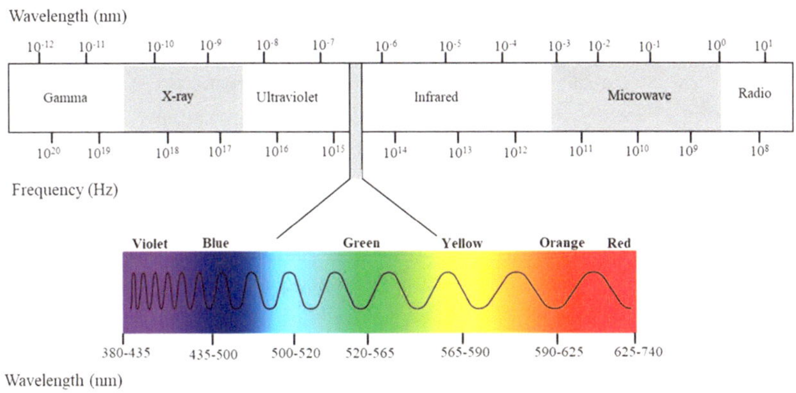

Figure 1 The electromagnetic spectrum with expanded visible light waveband.

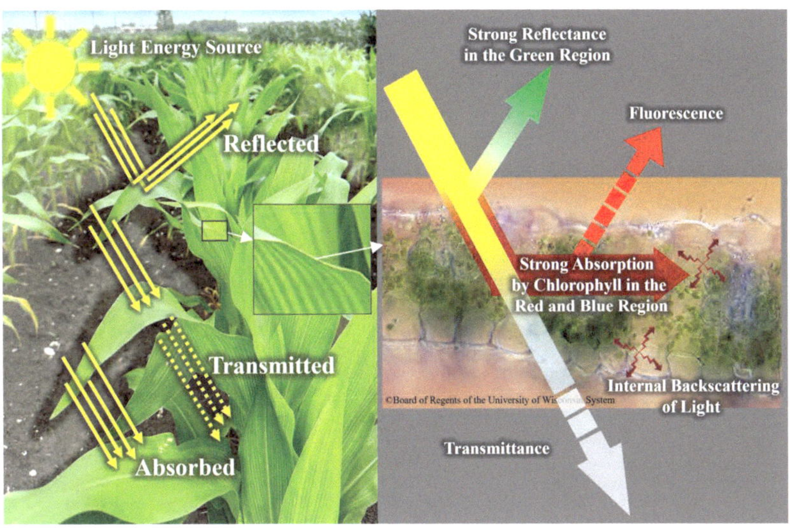

Figure 2 Diagram representation of the fate of light energy as it strikes leaf surfaces (leaf cross section photo credit: Board of Regents of the University of Wisconsin Systems).

(Zhu et al., 2020), and these plant traits are easily altered under unfavourable and stressful growing conditions (Mishra et al., 2017). Many sensors are therefore designed to capture leaf reflectance, transmittance and fluorescence from the visible and near-infrared regions of the electromagnetic spectrum as a way to assess plant health.

Chlorophylls are the plant pigments involved in harvesting sunlight energy. This energy is then converted to the energy storage molecules adenosine triphosphate (ATP) and nicotine adenine dinucleotide phosphoric acid (NADPH), via the light reaction process of photosynthesis. Nitrogen is vital for the biosynthesis of chlorophylls. Nitrogen is assimilated by plants to produce amino acids, the building blocks of protein, and is therefore essential for metabolism, growth and biomass production. Nitrogen can therefore significantly impact plant foliage greenness, density and mass. As each molecule of chlorophyll contains four atoms of nitrogen, the relationship between the chlorophyll content of the plant and leaf nitrogen content is well established (Schepers et al., 1996; Cartelat et al., 2005; Samborski et al., 2009). Stress-induced reduction in plant chlorophyll content corresponds to increased reflectance readings near 700 nm and decreased reflectance readings in the 530–630 nm range (Lichtenthaler et al., 1986; Niak et al., 2017).

The leaf colour chart was the first non-invasive tool developed to monitor crop nitrogen status via measuring variations in leaf colour (Fig. 3). Sensors capable of capturing reflectance, transmittance and fluorescence from visible and near-infrared radiation can now use these colour variations to assess crop nitrogen status. There are three types of nitrogen-sensitive proximal sensors (Padilla et al., 2018):

- Chlorophyll meters;
- Fluorescence-based flavonol meters; and
- Reflectance sensors.

These sensors provide indices to measure light (reflectance, transmittance and fluorescence) and the presence of compounds sensitive to crop nitrogen content.

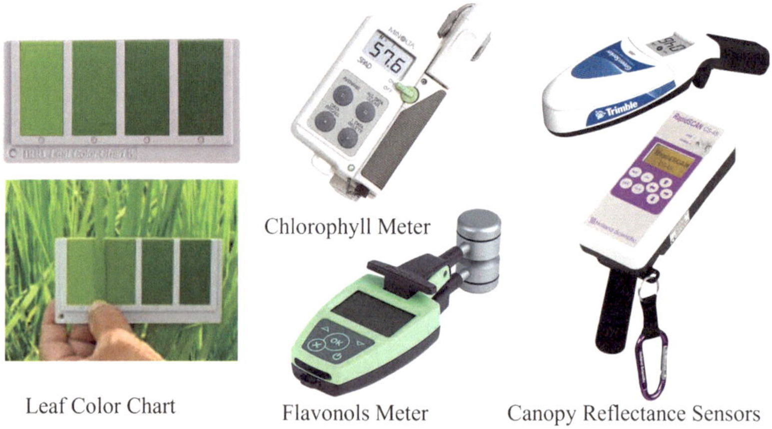

Figure 3 Leaf color chart for nitrogen management in rice (credit: IRRI Knowledge Bank), fluorescence-based flavonols meter, transmittance-based chlorophyll meter, and GreenSeeker and RapidScan CS-45 canopy reflectance sensors.

3 Types of proximal sensors

3.1 Leaf colour chart

The leaf colour chart is a tool extensively used by farmers worldwide for visual and subjective assessment of nitrogen fertilizer requirements of plants (Carter and Knapp, 2001). Leaf colour charts remain a practical tool for small-scale crop producers in particular. They enable real-time assessment of the nitrogen requirement of a plant and are thus an important component of decision support systems (or DSS) for nitrogen management (Johnson and Huggins, 1999). However, leaf colour charts are sensitive to sunlight variability and interpretative errors by users who necessarily make subjective judgements in trying to match a leaf colour to the chart.

A leaf colour chart is composed of a range of green plastic chips or panels that cover a scale from deficient (yellowish green) to excessive (dark green) leaf nitrogen content (Witt et al., 2005) (Fig. 3). The range of green colours is created based on the wavelength characteristic of leaves and calibrated to leaf nitrogen content. This tool measures leaf greenness by visually comparing light reflection from the surface of the leaves to the panel (Yang et al., 2003). Several types of leaf colour charts (initially for use in rice cultivation) have been developed following the release of the first Japanese prototype (Furuya, 1987). The main types are categorized based on the number of panels:

- The 4–6 panel type developed by the International Rice Research Institute (IRRI) (Philippines); and
- The 8 panel type was developed by Zhejiang Agricultural University (China) and the University of California Davis (USA).

The range of green colours varies among these types of leaf colour chart due to regional differences in (rice) cultivars. The variations in leaf greenness among rice cultivars have been addressed by using a different panel on a leaf colour chart as the critical colour for the application of nitrogen.

Depending on the shades of the green panel, nitrogen fertilization can be skipped if a reading of 4 or above is collected at panicle initiation in rice (Gill et al., 2023). The critical leaf colour values range from 3 for light green foliage (aromatic varieties) to 4 for other varieties (Thind and Gupta, 2010). Ladha et al. (2005) also described different critical values depending on the planting method, with critical leaf colour chart values of 4 for transplanted rice and a value of 3 for wet-seeded rice. The leaf colour charts can be used in other crops, in addition to rice, and the colour panels in leaf colour charts have been adjusted for multiple crops as shown in the Nitrogen Parameters website (https://www.nitrogenparameters.com/index.html). Singh et al. (2020) indicated that growers in South Asia have successfully used leaf colour charts to increase nitrogen use efficiency in cereals on a field-specific basis.

3.2 Chlorophyll meters

Chlorophyll meters were the first group of nitrogen-sensitive proximal sensors, indirectly estimating the relative chlorophyll content per unit of a leaf surface area (Monje and Bugbee, 1992) (Table 1). Francis and Piekielek (1999) found that chlorophyll meters can detect the onset of nitrogen stress in plants before it becomes visible to the human eye. Most chlorophyll meters are handheld and clipped onto or placed close to the leaf surface, allowing a fast, easy, on-site measurement of leaf nitrogen content (Watanabe, 1980).

Chlorophyll meters estimate the relative leaf chlorophyll content, or greenness, by measuring the absorbance and transmittance of radiation by the leaf at the red and near-infrared bands. Chlorophyll molecules absorb red radiation and transmit near-infrared radiation (Fox et and Walthall, 2008; Cerovic et al., 2012); hence, the amount of red light absorbed by the leaf surface increases with chlorophyll content (Hu et al., 2011; Schepers et al., 1996). This is the principal basis of designing this type of sensor, known as transmittance-based chlorophyll meters.

Padilla et al. (2018) reviewed the use of chlorophyll meters and other proximal optical sensors in vegetable production. Commercial chlorophyll meter designs differ in a number of ways, including:

Table 1 Types of proximal sensors for leaf transmittance and fluorescence measurements

Sensor type	Device	Manufacturer	Light output type	Wavelengths (nm)
Chlorophyll	SPAD-502	Konica Minolta (Tokyo, Japan)	Transmittance	650 and 940
	N-tester	Yara International (Oslo, Norway)	Transmittance	650 and 960
	atLeaf+	FT Green (Wilmington, DE, USA)	Transmittance	660 and 940
	MC-100 Chlorophyll Concentration Meter	Apogee Instruments (Logan, UT, USA)	Transmittance	653 and 931
	CCM-200 Chlorophyll Content Meter Plus	Opti-Sciences (Hudson, NH, USA)	Transmittance	653 and 931
	DUALEX	Force-A (Orsay, France)	Transmittance	710 and 850
	MULTIPLEX LI-600	Force-A (Orsay, France)	Fluorescence	516, 685 and 735
	LI-600N	LI-COR Biosciences (NE, USA)	Fluorescence	680-780
Flavonols	DUALEX	Force-A (Orsay, France)	Fluorescence	375, 650
	MULTIPLEX	Force-A (Orsay, France)	Fluorescence	590, 735, 985

- Measuring principle (i.e. transmittance versus fluorescence);
- The wavelengths used;
- Measurement units; and
- The calibration equations used to convert electrical signals into chlorophyll content estimates.

Several commercially available transmittance-based chlorophyll meters include:

- The Soil Plant Analysis Development, or SPAD-502, meter (Konica Minolta, Japan) (Fig. 3); and
- N-tester (Yara International, Norway).

More recent, low-cost examples include:

- The atLEAF+ sensor (FT Green LLC, USA); and
- The MC-100 Chlorophyll Concentration Meter (Apogee Instruments Inc., USA).

These chlorophyll meters use radiation at 650 ± 10 nm (red) and 930–960 nm (far-red). DUALEX (Force-A, France) is also a transmittance-based chlorophyll meter that uses wavelengths at 710 and 850 nm.

The MULTIPLEX (Force-A, France) is a fluorescence-based chlorophyll meter using wavelengths of 516, 685 and 735 nm. The following section provides more information on fluorescence techniques. The many models of chlorophyll meters complicate comparisons in how accurately they measure plant chlorophyll content (Parry et al., 2014; Taskos et al., 2015; Zhu et al., 2012). For this reason, the interpretation and corresponding (nitrogen) recommendations should be specific to the type of chlorophyll meter.

3.3 Fluorescence meters

Another type of proximal sensing technique to evaluate crop nitrogen status is through the estimation of fluorescent molecules in plants using fluorescence-based meters or fluorometers (Table 1). There are only a few components in plant leaves that fluoresce, including chlorophyll and flavonols. Fluorescent molecules absorb and emit light at specific wavelengths, typically at longer wavelengths, mainly blue-green fluorescence (400–600 nm), with chlorophyll fluorescence in the 650–800 nm range (Krause and Weis, 1984; Lang et al., 1992). The two most commonly used methods for collecting plant fluorescence data are (Lawson and Vialet-Chabrand, 2018):

- Pulse-amplitude modulations of light being measured; and
- Continuous illumination.

Fluorometers consist mainly of a light source, detector, light guide and devices for wavelength selection and signal processing (Tremblay et al., 2007). Devices used in the pulse-amplitude method comprise a pulse-measuring light source, an actinic light source and a saturating light. In contrast, light is not pulsed in the continuous illumination method (Brooks and Niyogi, 2011).

The basic principle behind the fluorescence technique lies in the relationship between the rate of the fluorescent molecule-dependent biochemical process (e.g. photosynthesis) and the dissipation of fluorescence (Lichtenthaler and Rinderle, 1988), measured as a spectrum from a single point in time or a change over time termed chlorophyll fluorescence kinetics (Lei et al., 2016). Fluorescence data can be expressed in ratios for the evaluation of plant stress. Fluorescence ratios at red/far-red and blue/green amplitude can differentiate nitrogen deficiency, leaf rust and powdery mildew stresses in plants (Burling et al., 2011). The chlorophyll fluorescence emission ratio of red and far-red radiation has been used to estimate plant chlorophyll content (Buschmann, 2007; Lichtenthaler et al., 1986; Tremblay et al., 2012; Zhang et al., 2012). The chlorophyll fluorescence emission technique was first developed to assess the photosynthetic performance of plants based on light energy distribution (Strasser et al., 2004). Buschmann (2007) noted that the ratio of red to far-red chlorophyll fluorescence decreases with chlorophyll content, which is attributed to the re-absorption of red chlorophyll fluorescence within the leaf.

The fluorescence technique can also estimate the content of plant flavonols, a class of polyphenolic compounds that build up in leaves under conditions of lower nitrogen supply (Bragazza and Freeman, 2007; Liu et al., 2010). Leaf flavonols are thus inversely related to leaf chlorophyll content (Cartelat et al., 2005). A similar principle to the chlorophyll fluorescence technique thus applies to measuring relative flavonols content (Agati et al., 2005; Bilger et al., 1997; Cerovic et al., 2002).

Flavonols content is estimated based on the emission of fluorescence in the red to far-red wavebands by mesophyll chlorophyll when the leaf is hit by red and ultraviolet radiation (Buschmann, 2007; Krause and Weis, 1991; Maxwell and Johnson, 2000). Flavonols in the leaf epidermis can absorb ultraviolet radiation, subsequently reducing the emission of far-red chlorophyll fluorescence under ultraviolet excitation but not during red excitation. Far-red chlorophyll fluorescence under red and ultraviolet excitation are then used to estimate leaf flavonols content (Cartelat et al., 2005; Cerovic et al., 2002; Goulas et al., 2004).

The two most commonly used flavonols meters are:

- DUALEX (Fig. 3); and
- MULTIPLEX.

Just like chlorophyll meters, these measure a dimensionless value that has a strong association with measured (using wet chemistry) flavonols (Ben Ghozlen et al., 2010; Cerovic et al., 2012; Tremblay et al., 2012). DUALEX is a leaf-clip instrument that takes flavonols reading from a small section of a leaf, whereas the MULTIPLEX is a handheld proximal-type sensor that takes flavonols reading 10 cm from the leaf (no contact) with a field of view of 4–8 cm in diameter (Debuisson et al., 2010; Zhang et al., 2012). The MULTIPLEX sensor has also been used as an indicator of plant physiological stress by using multiple excitation light sources (ultraviolet, blue, green and red). MULTIPLEX sensors measure anthocyanin content, flavonol content, chlorophyll content and chlorophyll fluorescence, which are associated with nitrogen deficiency due to decreased leaf chlorophyll and increased leaf phenolic content (Ben Ghozlen et al., 2010).

3.4 Reflectance sensors

In proximal reflectance sensors, the crop nitrogen status is obtained from canopy reflectance readings at specific wavelengths, commonly those in the visible, red-edge and near-infrared regions of electromagnetic radiation (Hatfield et al., 2008; Knipling, 1970; Ollinger, 2011; Peñuelas et al., 1994) (Table 2). Proximal reflectance sensors are positioned relatively close to the crop (at 0.4–3 m from the canopy). This generates more extensive (canopy) information about a plant for each scan as opposed to a single (leaf) index value that a chlorophyll or flavonols meter can take (Richardson et al., 2002). On a given sampling area, reflectance sensors can thus take more readings in each scan than a chlorophyll meter, thus potentially, reducing variability (Han et al., 2002).

Plants contain pigments that absorb and reflect specific wavelengths of visible and near-infrared radiations (Fig. 4). Photosynthetically active pigments, such as chlorophyll, strongly absorb red light leaving very low reflectance and nearly zero transmittance at this wavelength. The reflectance and transmittance of near-infrared radiation (700–1350 nm) can also relate to the amount of living cells in leaf tissues (Bell et al., 2004). Reflectance sensors can therefore provide information on both colour (greenness) and plant biomass.

According to Knipling (1970), plants normally absorb approximately 90% of the visible radiation (390 nm to 750 nm) and reflect about 50% of the near-infrared (750–1300 nm) radiation. This means that the reflectance of red and near-infrared radiation on a healthy plant is low and high, respectively. Similarly, the degree of absorbance and reflectance varies with crop nitrogen content, such that nitrogen-deficient crops reflect more visible and reflect less near-infrared than nitrogen-sufficient crops (Peñuelas et al., 1994; Schepers et al., 1996).

Table 2 Types of proximal canopy reflectance sensors for plant health monitoring and for nitrogen management

Device	Manufacturer	Light source	Wavelengths (nm)	Angle	Operation height	Data output
MSR5/87/16R	CropScan Inc. (Rochester, MN, USA)	Passive	460, 510, 560, 610, 660, 710, 760, 810	Nadir	0.8–1.5 m	Leaf area index and biomass estimation
CropSpec	Topcon Positioning Systems, Inc. (Livermore, CA, USA)	Passive	730-740, 800810	Oblique	Tractor cab	Biomass index and N recommendations
Spectral Reflectance Sensor	METER Group, Inc. (Pullman, WA, USA)	Passive	532, 570, 650, 810	Hemispherical	0.8–2.5 m	Biomass index and phenotyping
OptRX Crop Sensor	Ag Leader Technology (Ames, IA, USA)	Active	670, 728, 775	Nadir	0.2–2.5 m	Vegetation index and nitrogen recommendations
N-Sensor ALS	Yara International (Oslo, Norway)	Active	670, 730, 780	Nadir	Tractor cab	Biomass index and nitrogen recommendations
Crop Circle ACS 430	Holland Scientific (Lincoln, NE, USA)	Active	670, 730, 780	Nadir	0.2–2.5 m	Crop biomass and nitrogen uptake
Crop Circle ACS 470	Holland Scientific (Lincoln, NE, USA)	Active	450, 550, 650, 670, 730, 760	Nadir	0.2–2.5 m	Crop biomass and nitrogen uptake
RapidScan CS-45	Holland Scientific (Lincoln, NE, USA)	Active	670, 730, 780	Nadir	0.2–2.5 m	Vegetation index and nitrogen recommendations
GreenSeeker	Trimble Inc. (Sunnyvale, CA, USA)	Active	650, 770	Nadir	0.6–1.6 m	Yield potential and nitrogen responsiveness
GreenSeeker Handheld	Trimble Inc. (Sunnyvale, CA, USA)	Active	660, 780	Nadir	0.6–1.6 m	Yield potential and nitrogen responsiveness

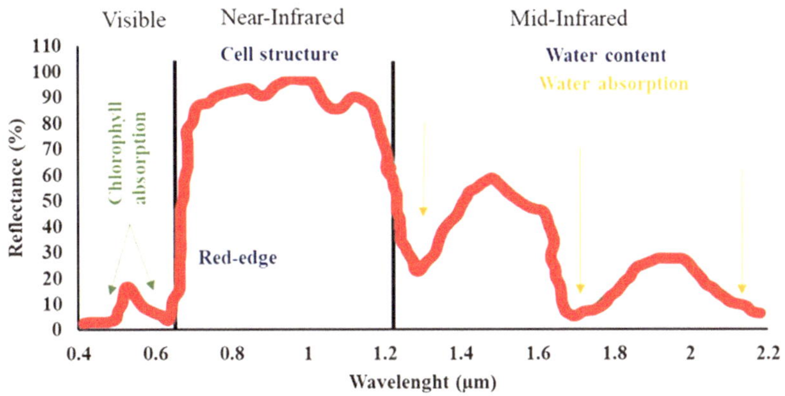

Figure 4 Plant canopy spectral reflectance curve with absorption features of leaf pigments, cell structure, and water.

The wavelengths used in reflectance sensors are selected based on their sensitivity to changing chlorophyll status, foliage density and biomass, with the nitrogen health status of a plant affecting all of these. These plant traits often correspond to four narrow bands: 675 nm (red absorption maxima), 905 nm (near-infrared reflectance peak), 720 nm (mid-portion of the red edge) and 550 nm (green reflectance maxima) (Gianquinto et al., 2011; Read et al., 2022; Thenkabail et al., 2002; Ulissi et al., 2011).

However, Daughtry et al. (2000) reported that background reflectance has a compounding effect on the assessment of plant nitrogen status based on single-band canopy reflectance. This complication means that many commercially available reflectance sensors are built using two or more wavebands. Through linear combinations or transformation processes, reflectance readings from multiple wavebands are expressed as spectral vegetation indices in ratios or normalized forms. Vegetation indices that combine reflectance from 2 to 3 wavelengths manifest a stronger association with crop biophysical properties at a higher level of precision as opposed to using single-band canopy reflectance (Bannari et al., 1995; Harrell et al., 2011; Lofton et al., 2012; Scotford and Miller, 2005; Teal et al., 2006; Xue et al., 2004).

Among the long list of vegetation indices developed for crop nitrogen status monitoring, the normalized difference vegetation index, or NDVI, is probably the most widely used (Rouse, Jr, et al., 1974; Sellers, 1985). The NDVI uses the relationship of visible low reflectance (strong absorption of the red light of chlorophyll) and low absorption (high reflectance) in the near-infrared region, using the formula: NDVI = (ρNIR – ρRED)/(ρNIR + ρRED). Values for NDVI range between –1 ānd +1. By contrast, unhealthy vegetation absorbs more near-infrared light and reflects more visible light (Fig. 5).

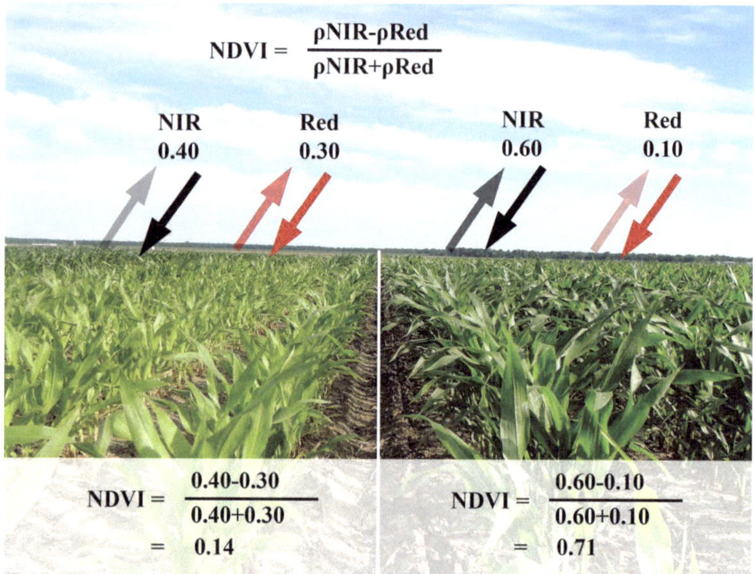

Figure 5 Hypothetical NDVI values of unhealthy (left) and health (right) corn plants at V10 growth stage.

The simple ratio, or ratio vegetation index (RVI) (Jordan, 1969), is another vegetation index that is computed as the ratio of reflectance at the near-infrared and red wavebands. Simple ratios and many normalized indices are measured directly using the crop canopy reflectance, and the readings generated from the sensors are based on the spectral mixture analysis of green vegetation, soil and shade in each field of view (Bateson and Curtiss, 1996; Roberts et al., 1993; Smith et al., 1994). When the field of view is exclusively occupied by green vegetation at plant canopy closure, these vegetation indices saturate and lose sensitivity to changing crop biophysical attributes. Some vegetation indices can distinguish vegetation from soil such as the soil-adjusted vegetation index or SAVI (Huete, 1988). The SAVI is essentially NDVI except for the added parameter L in the equation to account for the brightness of the soil background. There are over 40 spectral vegetation indices (Bannari et al., 1995; Ollinger, 2011); in addition to NDVI, RVI, NDRE, SR and SAVI, there are several different canopy reflectance vegetation indices that can be used to monitor crop nitrogen status (Table 3).

Reflectance sensors are mainly grouped into two based on the light source:

- Passive; and
- Active.

Table 3 Summary of the most common canopy reflectance vegetation indices for monitoring plant health status

Vegetation index	Equation	Author
SR (Simple Ratio)	NIR/Red	Birth and McVey (1968)
NDVI (Normalized Difference Vegetation Index)	(NIR - Red)/(NIR + Red)	Rouse Jr et al. (1975)
GNDVI (Green Normalized Difference Vegetation Index)	(NIR - Green) + (NIR + Green)	Ma et al. (1996)
GVI (Green Ratio Vegetation Index)	NIR/Green	Birth and McVey (1968)
RVI (Red Ratio Vegetation Index)	NIR/Red	Birth and McVey (1968)
GRVI (Green-Red Vegetation Index)	(Green - Red)/(Green + Red)	Tucker (1979)
SAVI (Soil-Adjusted Difference Vegetation Index)	(NIR - Red)/(NIR + Red + L)*(1 + L)	Huete (1988)
RDVI (Renormalized Difference Vegetation Index)	$\sqrt{(NIR - Red)/(NIR + Red)}$	Roujean and Breon (1995)
OSAVI (Optimized SAVI)	$\left[(NIR/Red) - 1\right] / \sqrt{1 + (NIR/Red)}$	Rondeaux et al. (1996)
MSR (Modified SR)	(NIR - Red)/(NIR + Red + 0.16)	Chen (1996)
CI (Chlorophyll Index)	(NIR/Red) - 1	Gitelson et al. (2003)
CVI (Chlorophyll Vegetation Index)	(NIR/Green)*(Red/Green)	Vincini et al. (2008)
REI (Red-Edge Index)	NIR/Red-Edge	Vogelmann et al. (1993)
RENDVI (Red-Edge Normalized Difference Vegetation Index)	(NIR - Red - Edge)/(NIR + Red - Edge)	Gitelson and Merzlyak (1994)
Ratio RENDVI/NDVI	RENDVI/NDVI	Varco et al. (2013)

Passive sensors are highly dependent on ambient illumination (Heege and Reusch, 1996; Stone et al., 1996) and are built with two sets of photodetectors that measure the incident radiation above the crop canopy and the radiation reflected from the canopy. Active sensors were subsequently developed to address the issue of ambient illumination effects on reflectance readings (Beck and Vyse, 1995). Active sensors have a light source that operates at two or three specific wavebands, typically a combination of one-two visible and near-infrared radiation. The light source is modulated, allowing the detector system of the sensors to capture the reflected radiation from their own light source, thus eliminating the possible interference of reflectance from the ambient source

(Solari et al., 2008). Active sensors can thus be used in any irradiance conditions and have many applications under field conditions (Fitzgerald, 2010).

Despite the known versatility of active sensors, passive reflectance sensor types remain an option in the market. The Yara passive N-sensor (Yara International ASA, Oslo, Norway) operates on two diode array spectrometers, one for measuring ambient light and one for crop light reflectance at 450 and 900 nm (Zillman et al., 2006). The Yara N-sensor can be mounted on a tractor roof and used for on-the-go monitoring of crop nitrogen status and applying the appropriate amount of nitrogen fertilizer.

Commercially available active sensors include:

- Crop Circle (Holland Scientific, Lincoln, NE);
- GreenSeeker (Trimble); and
- N-Sensor ALS.

GreenSeeker operates in both red (650 nm) and near-infrared (780 nm) parts of the spectrum, whereas Crop Circle uses ACD Polysource light covering the amber (590 nm) and near-infrared (880 nm) using a single LED light source. A more advanced Crop Circle model has been developed operating in six spectral bands (450, 550, 650, 670, 730 and >760 nm).

There are various models of Crop Circle and GreenSeeker sensors. Some are simple, cheap, handheld and cannot store data (Fig. 3). Other more expensive models have data storage capacity and can therefore be used for continuous data recording. They are often mounted on tractors and connected to GPS systems for field mapping. They are primarily used in this configuration for automatic, variable-rate applications of fertilizers.

4 Proximal sensor applications to estimate crop nitrogen requirement

Use of proximal sensors requires that crops should be established and have reached their active vegetative growth stage. The objective is to obtain near real-time crop information which can then be used immediately to derive a recommendation for nitrogen application and subsequent fertilization. It has taken many years of research to build a database to establish a nitrogen fertilizer recommendation system based on sensor readings using different devices, such as chlorophyll meters and reflectance sensors. The process involved calibrating the sensor readings with biological parameters such as plant nitrogen content, biomass and yield, then generating nitrogen status interpretations and recommendations for the rate of fertilizer application (Table 2).

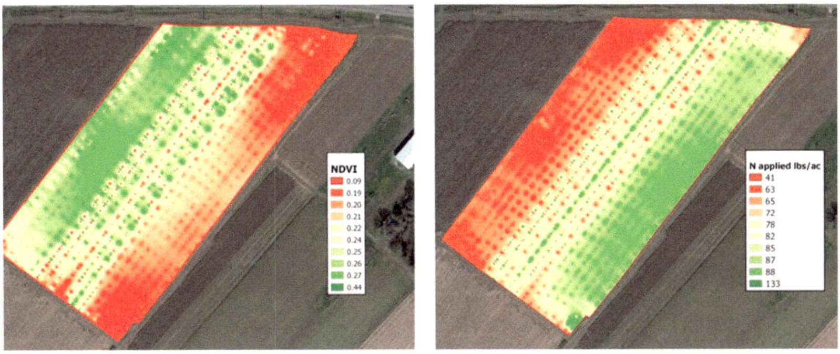

Figure 6 Applied nitrogen fertilizer (left) and NDVI (right) maps of a sugarcane field in Napoleonville, Louisiana.

Figure 6 provides an example of NDVI data collected from a sugarcane field in Louisiana (USA) before nitrogen fertilization. The photo illustrates the locations in the field with low NDVI values receiving a higher amount of nitrogen and vice versa. Variable-rate technology allows nitrogen application at different rates across a field by matching vegetation indices and yield potential with a specific input rate.

Peng et al. (1995) and Singh and Ali (2020) observed differences in rice grain yield and biomass production by delaying nitrogen application over different weeks after transplanting of rice. They used a leaf colour chart and a SPAD meter for site-specific nitrogen management of transplanted rice. Ali et al. (2021) also found that nitrogen fertilizer efficiency and wheat grain yield increased when both the atLeaf meter and leaf colour chart were used compared with the grower's existing sensor-based nitrogen management. Singh and Ali (2020) found that sensor-guided fertilizer nitrogen applications using the leaf colour chart, SPAD meter and reflectance sensor have improved wheat yield and nitrogen fertilizer use efficiency applied at the tillering stage. Access to cutting-edge technologies, fertilizer prices, cultivars and best management practices to understand both soil spatial and temporal variability are the main challenges that growers in developing countries face when establishing site-specific nutrient management programmes. This highlights the need to fully understand different systems and sensors to determine in-season nitrogen application requirements (Peng et al., 2010; Singh et al., 2018).

4.1 Chlorophyll meters

The literature contains a vast amount of information describing the relationship between chlorophyll readings and leaf nitrogen content, with the relationship between chlorophyll readings and maize leaf nitrogen content being linear

(Chapman and Barreto, 1997; Bullock and Anderson, 1998; Hodgen et al., 2006; Schepers et al., 1992). Studies have shown partial saturation of chlorophyll meter readings in plants/leaves with high nitrogen and chlorophyll content (Markwell et al., 1995; Monje and Bugbee, 1992; Parry et al., 2014). The partial saturation response does not occur in all plants; hence, linearity between chlorophyll meter readings and plant/leaf nitrogen content is retained (Cartelat et al., 2005; Li et al., 2013a; Tremblay et al., 2012).

However, this linear relationship is not the case for other plants where partial saturation does occur; as a result, chlorophyll readings tend to plateau with leaf nitrogen content as reported for barley, wheat, oats and rice (Chubachi and Oikawa, 1986; Debaeke et al., 2006; Giunta et al., 2002; Peltonen et al., 1995; Peng et al., 1993). Others recognized the need to adjust chlorophyll readings to overcome non-linearity and generate reasonable crop nitrogen status evaluation using, for example, specific leaf weight (Peng et al., 1993) and a normalization procedure through the introduction of chlorophyll readings from a fully fertilized reference strip in the same field (Debaeke et al., 2006; Ortuzar-Iragorri et al., 2005).

Chlorophyll meter readings have been used in a decision tool for topdress nitrogen fertilization (Maiti et al., 2004; Turner and Jund, 1991; Turner and Jund, 1994; Wells et al., 1993). One approach, called the sufficiency index, takes the ratio of chlorophyll readings of a field in question and a well-nitrogen-fertilized strip and, once the ratio value falls below the established threshold, a supplemental application of nitrogen is recommended. The threshold value for rice is set at 0.9 and, below this value, a topdress nitrogen rate of 30 kg ha^{-1} is recommended (Hussain et al., 2000). For maize, a sufficiency index below 0.95 is deemed nitrogen deficient and a side-dress nitrogen rate of approximately 35 kg ha^{-1} is recommended (Francis and Piekielek, 1999; Piekielek et al., 1995; Varvel et al., 1997).

A more popular approach is based on applying topdress/side-dress nitrogen when the chlorophyll reading is less than a set critical reading (Argenta et al., 2004; Balasubramanian et al., 1999; Singh et al., 2002; Maiti et al., 2004). An example of interpretation based on a set critical level for maize is Argenta et al. (2004). Here, the established critical chlorophyll readings taken by SPAD meter are 45.4, 52.1, 55.3 and 58.0 at 3–4, 6–7, 10–11 expanded leaves, and silking stages, respectively. Both approaches are effective, but the sufficiency index is less popular due to the yearly requirement for the establishment of a well-nitrogen-fertilized reference strip for each field or management zone. Varvel et al. (2007) subsequently developed an algorithm for maize nitrogen recommendations using a chlorophyll meter-based sufficiency index with the nitrogen rate estimated based on an established relationship between nitrogen rate and sufficiency index at three vegetative stages of maize.

4.2 Fluorescence meters

The effective use of flavonols meters relies on eliminating, as much as possible, factors affecting flavonols production in plants other than nitrogen supply. Studies have shown a clear response of flavonols content of plants to water stress, radiation and salinity stress (Ksouri et al., 2007; Fortier et al., 2010; Tattini et al., 2004). While the flavonols content of a plant also varies within a season and with changing solar irradiance (Fortier et al., 2010; Tattini et al., 2004), daytime has no apparent effect on flavonols estimated by fluorescence-based meters (Cartelat et al., 2005; Tremblay et al., 2012; Zhang et al., 2012).

The use of fluorescence-based flavonols meters is still in an early adoption stage for assessing crop nitrogen status. To date, the use and application of flavonols meters for monitoring nitrogen status have been tested in grain crops such as maize (Longchamps and Khosla, 2014; Quemada et al., 2014; Tremblay et al., 2007; Zhang et al., 2012), wheat (Cartelat et al., 2005; Tremblay et al., 2010) and rice (Li et al., 2013b). Flavonols meters can be utilized in two ways (Cartelat et al., 2005; Tremblay et al., 2012):

- Use of flavonols estimates as a stand-alone indictor of crop nitrogen status
- Use of flavonols estimates in combination with chlorophyll readings to compute the Nitrogen Balance Index or NBI. Both chlorophyll and flavonol contents respond to nitrogen supply in opposing ways; thus, the value of NBI computed as the ratio of chlorophyll to flavonols content theoretically increases with nitrogen supply.

Unlike proximal reflectance sensors, soil does not interfere with the signals that handheld devices, such as chlorophyll and flavonols meters, receive. However, chlorophyll and flavonols readings are made at the leaf level, usually on multiple leaves to obtain a good representation of a given field or plot. Some practices can be implemented to obtain reliable data, such as the correct selection of leaf to take readings from. Selection is based on leaf age or position in the canopy, quality (avoid damaged leaves), side (adaxial versus abaxial, or both) and within-leaf position (near base versus tip versus mid-section). However, the need for a representative sample to ensure representative readings across a plt is a significant drawback, in addition to the fact that leaf-level measurements limit the integration of other crop parameters such as depth or height. The ability to collect readings at the crop canopy level has been the main advantage of proximal reflectance sensors over chlorophyll and flavonols meters. These sensors save labour and time and can take readings from more plants and more extensive areas than chlorophyll and flavonols meters in the same amount of time (Blackmer et al., 1994; Han et al., 2002).

4.3 Reflectance sensors

Reflectance sensors take canopy reflectance readings at two to three specific wavebands with the output as a reflectance reading from each waveband, pre-computed vegetation indices from the reflectance reading or both. Between these two outputs, the vegetation indices are often used for assessing crop nitrogen status, either based on absolute sufficiency values or in combination with reference plots.

The absolute sufficiency values rely on calibration studies to build databases to establish the interpretation of sensor measurements into categories of deficient, sufficient and excess nitrogen (Thompson et al., 2017). Yield and indicators of crop nitrogen status are used as response variables in the calibration process. This requites reference plots involving establishment of well-nitrogen-fertilized plots within the field of interest. To obtain a crop nitrogen assessment, the sensor measurement from the field of interest is divided by the measurement taken from the reference plot. This is also known as the Nitrogen Sufficiency Index or NSI (Debaeke et al., 2006; Piekielek et al., 1995). An NSI value <1 indicates deficiency, whereas a value of 1 indicates sufficiency. No value >1 is possible because this NSI approach assumes that excess nitrogen uptake does not occur (Holland and Schepers, 2013; Zhu et al., 2012).

The NSI is the opposite of the nitrogen Response Index or RI (Lofton et al., 2012; Mullen et al., 2003; Tubaña et al., 2012). Johnson and Raun (2003) introduced the concept of RI with yield as the parameter. The nitrogen RI based on sensor readings is determined by taking the ratio of vegetation index (e.g. NDVI) between the reference plot and field of interest such that an RI value >1 indicates the nitrogen supply is inadequate, whereas 1 or <1 means soil supply is adequate. This approach also provides an estimate of yield increase that can be attributed to nitrogen. For example, an RI value of 1.32 means that: (i) the supply of nitrogen is insufficient since RI < 1 and (ii) the potential increase in yield due to nitrogen is 0.32/1 or 32%.

5 Integration of crop data to develop algorithms to predict nitrogen requirements

Fluorescence spectra are less affected by crop biomass but respond significantly to different nitrogen levels at the blue, green, red and near-infrared wavelengths (Bredemeier and Schmidhalter, 2005; Heege et al., 2008; Langsdorf et al., 2000; McMurtrey et al., 1994). These sensors are less affected by soil background, even at the early growth stage of a crop, making them a powerful potential tool for crop nitrogen content sensing to offset limitations with proximal reflectance sensors.

There have, however, been few studies on the potential of these types of proximal sensors for predicting crop variables sensitive to nitrogen. The normalized SPAD index was evaluated by Debaeke et al. (2006) as a predictor of the yield and protein content of durum wheat cultivars. Huang et al. (2019) reported the use of a fluorescence canopy sensor to predict rice nitrogen uptake and yield. Ali et al. (2014) also used a chlorophyll meter to predict the yield of dry direct-seeded rice in northwestern India. The result showed chlorophyll readings using a SPAD meter were a better yield predictor than leaf colour chart readings. Guo et al. (2020) established the relationship between the modified blue vegetation index and chlorophyll estimates using a SPAD meter to predict maize yield. A derivative of chlorophyll readings, i.e. the canopy chlorophyll density-based index, was also used for yield prediction in rice (Liu et al., 2017).

A significant amount of research has focused on predicting nitrogen-responsive crop variables using reflectance sensors. Active sensors such as GreenSeeker and Crop Circle have established algorithms to estimate nitrogen requirements for a wide range of crops based on predicted yield potential and within-season crop response to nitrogen fertilization (Cao et al., 2016; Farid et al., 2022; Harrell et al., 2011; Larson et al., 2020; Lofton et al., 2012a; Moges et al., 2007; Teal et al., 2006; Tubaña et al., 2012; Yao et al., 2012). Biomass yield and nitrogen uptake are variables that influence crop nitrogen requirements. Cao et al. (2013) and Harrell et al. (2011) have demonstrated the potential of both GreenSeeker and Crop Circle in predicting these variables. GreenSeeker NDVI was used in wheat as a predictor of biomass, grain yield and grain protein (Freeman et al., 2003; Raun et al., 2001; Stone et al., 1996). Phillips et al. (2004) used NDVI as a predictor of winter wheat tiller density. Lofton et al. (2012a) used NDVI to predict cane tonnage and sugar yield, while Amaral et al. (2015) evaluated the performance of Crop Circle and GreenSeeker canopy sensors to assess sugarcane biomass and nitrogen status. These reflectance sensors are used for non-destructive measurement of canopy information relevant for estimating crop nitrogen requirements, while others are used for phenotyping (Table 2).

Canopy reflectance-predicted yield is the most commonly used baseline information for generating in-season nitrogen fertilizer recommendations. The works of Raun et al. (2003) in wheat, Tubaña et al. (2012) in rice and Lofton et al. (2012b) in sugarcane have demonstrated that the estimate of in-season crop response to nitrogen application or RI (described in the previous section) is a good indicator of nitrogen supply from the soil at the time sensor readings are taken. A nitrogen prescription/application recommendation is generated as the difference between the crop nitrogen demand, the product of predicted yield, the nitrogen content of the harvested part of the crop and the RI-estimated soil nitrogen supply. The basis of using these two components independently to estimate crop nitrogen requirements has been reviewed by Raun et al. (2010). Crop yield level and nitrogen response are unrelated. In years with above-average growing environments, high yield potential does not proportionately increase the crop's response to nitrogen application.

6 Conclusion and future trends

The development of proximal sensors is one of the major turning points in precision management of nitrogen in crop production following discovery of leaf spectra as a potential basis for plant health monitoring in the late nineteenth and early twentieth centuries. The non-invasive, rapid and near real-time acquisition of crop nitrogen status using proximal sensors paved the way to developing more powerful and practical techniques to generate nitrogen prescriptions for crops, with many of them configured for automatic, variable-rate application. The successful use of proximal sensors, from transmittance- and fluorescence-based meters to passive and active reflectance sensors, relies on building strong databases to: (i) calibrate sensor readings to crop variables that are nitrogen sensitive and (ii) generate nitrogen status interpretations and recommendations. As technology progresses, smaller and more portable yet powerful and multifunctional sensors will be built. This means that the data collection and calibration process should continue to bring meaningful interpretation from these sensors and ultimately develop nitrogen assessment and decision tools.

7 Where to look for further information

A standard introduction to proximal sensing and its application to plant nitrogen health monitoring is the article of F.M. Padilla entitled 'Proximal optical sensors for nitrogen management of vegetable crops: a review' (see Padilla et al., 2018). Another good source is the special issue of Remote Sensing 'Remote sensing for precision nitrogen management' by Miao et al. (2020).

Articles published in Precision Agriculture (https://link.springer.com/journal/11119) and the proceedings of the Annual International Conference on Precision Agriculture (https://www.ispag.org/icpa) are good sources of information to keep up to date with the trends on proximal sensing in crop nitrogen management. Events, webinars, and symposium organized by professional societies like the International Society of Precision Agriculture (https://www.ispag.org/Events), American Society of Agronomy - Precision Agriculture Systems Community (https://www.agronomy.org/membership/sections-communities/agronomic-production-systems), and Society of Precision Agriculture Australia (https://www.spaa.com.au/resources-spaa/spaa-event

-proceedings/) continue to provide viewpoints of scientists, growers, extension personnel, practitioners, and among others on the subject.

8 References

Agati, G., Pinelli, P., Ebner, S. C., Romani, A., Cartelat, A. and Cerovic, Z. G. (2005). Nondestructive evaluation of anthocyanins in olive (*Olea europaea*) fruits by in situ chlorophyll fluorescence spectroscopy. *J. Agric. Food Chem.* 53(5):1354–1363.

Ali, A. M., Ibrahim, S. M., Hassany, W. M., El-Sadek, A. N. and Bijay-Singh, (2021). Fixed-time corrective dose fertilizer nitrogen management in wheat using atLeaf meter and leaf color chart. *Exp. Agri.* 57(4):232–243.

Ali, A. M., Thind, H. S., Sharma, S. and Varinderpal-Singh, (2014). Prediction of dry direct-seeded rice yields using chlorophyll meter, leaf color chart and GreenSeeker optical sensor in northwestern India. *Field Crops Res.* 161:11–15.

Amaral, L. R., Molin, J. P., Portz, G., Finazzi, F. B. and Cortinove, L. (2015). Comparison of crop canopy reflectance sensors used to identify sugarcane biomass and nitrogen status. *Precision Agric.* 16(1):15–28.

Argenta, G., Silva, P. R. F. D. and Sangoi, L. (2004). Leaf relative chlorophyll content as an indicator parameter to predict nitrogen fertilization in maize. *Cien. Rural* 34(5):1379–1387.

Balasubramanian, V., Morales, A. C., Cruz, R. T. and Abdulrahman, S. (1999). On-farm adaptation of knowledge-intensive nitrogen management technologies for rice systems. *Nutr. Cyc. Agroeco* 53:59–69.

Bannari, A., Morin, D., Bonn, F. and Huete, A. R. (1995). A review of vegetation indices. *Remote Sens. Rev.* 13(1–2):95–120.

Bateson, C. A. and Curtiss, B. (1996). A method for manual end member selection and spectral unmixing. *Remote Sens. Environ.* 55:255–269.

Beck, J. and Vyse, T. (1995). Structure and method usable for differentiating a plant from soil in a field. U.S. Patent no. 5389781. Issued 14 Feb. 1995.

Bell, G. E., Howell, B. M., Johnson, G. V., Raun, W. R., Solie, J. B. and Stone, M. L. (2004). Optical sensing of turfgrass chlorophyll content and tissue nitrogen. *HortScience* 39(5):1130–1132.

Ben Ghozlen, N., Cerovic, Z. G., Germain, C., Toutain, S. and Latouche, G. (2010). Non-destructive optical monitoring of grape maturation by proximal sensing. *Sensors* 10(11):10040–10068.

Bilger, W., Veit, M., Schreiber, L. and Schreiber, U. (1997). Measurement of leaf epidermal transmittance of UV radiation by chlorophyll fluorescence. *Physiol. Plant.* 101Z:754–763.

Binford, G. D., Blackmer, A. M. and Cerrato, M. E. (1992). Relationship between corn yields and soil nitrate in late spring. *Agron. J.* 84(1):53–59.

Birth, G. S. and McVey, G. R. (1968). Measuring the color of growing turf with a reflectance spectrophotometer. *Agron. J.* 60(6):640–643.

Blackmer, T. M., Schepers, J. S. and Varvel, G. E. (1994). Light reflectance compared with other nitrogen stress measurements in corn leaves. *Agron. J.* 86(6):934–938.

Bragazza, L. and Freeman, C. (2007). High nitrogen availability reduces polyphenol content in Sphagnum peat. *Sci. Total Environ.* 377(2–3):439–443.

Bredemeier, C. and Schmidhalter, U. (2005). Laser-induced chlorophyll fluorescence sensing to determine biomass and nitrogen uptake of winter wheat under controlled environment and field condition. In: Proceedings of the 5th European Conference on Precision Agriculture, pp. 273–280. Wageningen Academic Publishers, Wageningen.

Brooks, M. D. and Niyogi, K. K. (2011). Use of a pulse-amplitude modulated chlorophyll fluorometer to study the efficiency of photosynthesis in Arabidopsis plants. In: Jarvis, R. P. (Ed.) *Chloroplast Research in Arabidopsis: Methods in Molecular Biology*, Volume 755, pp. 299–310. Humana Press, Totowa, NJ. ISBN 978-1-61779-236-6.

Bullock, D. G. and Anderson, D. S. (1998). Evaluation of the Minolta SPAD-502 chlorophyll meter for nitrogen management in corn. *J. Plant Nutr.* 21(4):741–755.

Burling, K., Hunsche, M. and Noga, G. (2011). Use ofblue-green and chlorophyll fluorescence measurements for differentiation between nitrogen deficiency and pathogen infection in winter wheat. *J. Plant Physiol.* 168(14):1641–1648.

Buschmann, C. (2007). Variability and application of the chlorophyll fluorescence emission ratio red/far-red of leaves. *Photosynth. Res.* 92(2):261–271.

Cao, Q., Miao, Y., Shen, J., Yu, W., Yuan, F., Cheng, S., Huang, S., Wang, H., Yang, W. and Liu, F. (2016). Improving in-season estimation of rice yield potential and responsiveness to topdressing nitrogen application with crop circle active crop canopy sensor. *Precis. Agric.* 17(2):136–154.

Cartelat, A., Cerovic, Z. G., Goulas, Y., Meyer, S., Lelarge, C., Prioul, J.-L., Barbottin, A., Jeuffroy, M.-H., Gate, P., Agati, G. and Moya, I. (2005). Optically assessed contents of leaf polyphenolics and chlorophyll as indicators of nitrogen deficiency in wheat (*Triticum aestivum* L.). *Field Crops Res.* 91(1):35–49.

Carter, G. A. and Knapp, A. K. (2001). Leaf optical properties in higher plants: linking spectral characteristics to stress and chlorophyll concentration. *Am. J. Bot.* 88(4):677–684.

Cerovic, Z. G., Masdoumier, G., Ghozlen, N. B. and Latouche, G. (2012). A new optical leaf-clip meter for simultaneous non-destructive assessment of leaf chlorophyll and epidermal flavonoids. *Physiol. Plant.* 146(3):251–260.

Cerovic, Z. G., Ounis, A., Cartelat, A., Latouche, G., Goulas, Y., Meyer, S. and Moya, I. (2002). The use of chlorophyll fluorescence excitation spectra for the non-destructive in situ assessment of UV-absorbing compounds in leaves. *Plant Cell Environ.* 25:1663–1676.

Chapman, S. C. and Barreto, H. J. (1997). Using a chlorophyll meter to estimate specific leaf nitrogen of tropical maize during vegetative growth. *Agron. J.* 89(4):557–562.

Chen, J. M. (1996). Evaluation of vegetation indices and a modified simple ratio for boreal applications. *Can. J. Remote. Sens.* 25:295–309.

Chubachi, T. I. A. and Oikawa, T. (1986). The diagnosis of nitrogen nutrition of rice plants (Sasanishiki) using chlorophyll meter. *Jap. J. Soil Sci. Plant Nutr.* 57:190–193.

Coblentz, W. W. (1912). The diffuse reflecting power of various substances. *Bull. Bur. Stand.* 9:283–325.

Daughtry, C. S. T., Walthall, C. L., Kim, M. S., de Colstoun, E. B. and McMurtrey, J. E. (2000). Estimating corn leaf chlorophyll concentration from leaf and canopy reflectance. *Remote Sens. Environ.* 74:229–239.

Debaeke, P., Rouet, P. and Justes, E. (2006). Relationship between the normalized SPAD index and the nitrogen nutrition index: application to durum wheat. *J. Plant Nutr.* 29(1):75–92.

Debuisson, S., Germain, C., Panigai, L., Le Moigne, M., Fadaili, E. M., Evain, S. and Cerovic, Z. G. (2010). Using Multiplex® and GreenSeeker™ to manage spatial variation in vine vigor in champagne. In: Proceedings of the 10th International Conference Precision Agriculture. Denver, CO, 18 July 2010.

Farid, H. U., Khan, Z. M., Anjum, M. N., Shakoor, A. and Qureshi, H. S. (2022). Precision nitrogen management for cotton using GreenSeeker handheld crop sensors. *Environ. Sci. Proc.* 23:12.

Fitzgerald, G. J. (2010). Characterizing vegetation indices derived from active and passive sensors. *Int. J. Remote Sens.* 31(16):4335-4348.

Fortier, E., Desjardins, Y., Tremblay, N., Bélec, C. and Côté, M. (2010). Influence of irrigation and nitrogen fertilization on broccoli polyphenolics concentration. *Acta Hortic.* 856:55-62.

Fox, R. H., Roth, G. W., Iversen, K. V. and Piekielek, W. P. (1989). Comparison of soil and tissues nitrate tests for predicting soil nitrogen availability to corn. *Agron. J.* 81(6):971-974.

Fox, R. H. and Walthall, C. L. (2008). Crop monitoring technologies to assess nitrogen status. In: Schepers, J. S. and Raun, W. R. (Eds) *Nitrogen in Agricultural Systems, Agronomy Monograph No. 49*, pp. 647-674. American Society of Agronomy, Crop Science Society of America, Soil Science Society of America, Madison, WI.

Francis, D. D. and Piekielek, W. P. (1999). Assessing crop nitrogen needs with chlorophyll meters. *Site-Specific Management Guidelines SSMG-12*. Potash & Phosphate Institute. Reference 99082/Item#10-1012.

Freeman, K. W., Raun, W. R., Johnson, G. V., Mullen, R. W., Stone, M. L. and Solie, J. B. (2003). Late-season prediction of wheat grain yield and grain protein. *Commun. Soil Sci. Plant Anal.* 34(13-14):1837-1852.

Furuya, S. (1987). Growth diagnosis of rice plants by means of leaf color. *Jpn Agric. Res. Quart.* 20:147-153.

Gianquinto, G., Orsini, F., Sambo, P. and Paino D'Urzo, M. P. (2011). The use of diagnostic optical tools to assess nitrogen status and to guide fertilization of vegetables. *Horttechnology* 21(3):287-292.

Gill, A., Rajput, A., Sharma, A. and Devi, O. R. (2023). Soil farming techniques for climate resilient agriculture. *Indian J. Fertil* 10:14-27.

Gitelson, A. and Merzlyak, M. N. (1994). Spectral reflectance changes associated with autumn senescence of *Aesculus hippocastanum* L. and *Acer platanoides* L. leaves: spectral features and relation to chlorophyll estimation. *J. Plant Physiol.* 143(3):286-292.

Gitelson, A. A., Gritz, Y. and Merzlyak, M. N. (2003). Relationships between leaf chlorophyll content and spectral reflectance and algorithms for non-destructive chlorophyll assessment in higher plant leaves. *J. Plant Physiol.* 160(3):271-282.

Giunta, F., Motzo, R. and Deidda, M. (2002). SPAD readings and associated leaf traits in durum wheat, barley and triticale cultivars. *Euphytica* 125(2):197-205.

Goulas, Y., Cerovic, Z. G., Cartelat, A. and Moya, I. (2004). Dualex: a new instrument for field measurements of epidermal ultraviolet absorbance by chlorophyll fluorescence. *Appl. Opt.* 43(23):4488-4496.

Han, S., Hendrickson, L. L. and Ni, B. (2002). Comparison of satellite and aerial imagery for detecting leaf chlorophyll content in corn. *Trans. ASAE* 45:1229.

Harrell, D., Tubaña, B., Walker, T. and Phillips, S. (2011). Estimating rice grain yield potential using normalized difference vegetation index. *Agro. J.* 103:1717-1723.

Hatfield, J. L., Gitelson, A. A., Schepers, J. S. and Walthall, C. L. (2008). Application of spectral remote sensing for agronomic decisions. *Agron. J.* 100(S3):S117-S131.

Heege, J. J. and Reusch, S. (1996). Sensor for on the go control of site-specific nitrogen top dressing. ASAE Paper no. 961018. ASAE, St. Joseph, MI.

Heege, H. J., Reusch, S. and Thiessen, E. (2008). Prospects and results for optical systems for site-specific on-the-go control of nitrogen-top-dressing in Germany. *Precis. Agric.* 9(3):115–131.

Hodgen, P., Solari, F., Schepers, J. S., Ferguson, R. and Shanahan, J. (2006). Variability of chlorophyll meter readings of individual maize plants. In: *Annual Meetings Abstracts [CD-ROM]*.ASA, CSSA and SSSA, Madison, WI.

Holland, K. H. and Schepers, J. S. (2013). Use of a virtual-reference concept to interpret active crop canopy sensor data. *Precis. Agric.* 14(1):71–85.

Hong, S. D., Fox, R. H. and Piekielek, W. P. (1990). Field evaluation of several chemical indexes of soil nitrogen availability. *Plant Soil* 123(1):83–88.

Hu, J., He, D. and Yang, P. (2011). Study on plant nutrition indicator using leaf spectral transmittance for nitrogen detection. *Adv. Inf. Commun. Technol.* 347:504–513.

Huang, S., Miao, Y., Yuan, F., Cao, Q., Ye, H., Lenz-Wiedemann, V. I. S. and Bareth, G. (2019). In-season diagnosis of rice nitrogen status using proximal fluorescence canopy sensor at different growth stages. *Remote Sens.* 11(16):1847.

Huete, A. R. (1988). A soil-adjusted vegetation index (SAVI). *Remote Sens. Environ.* 25(3):295–309.

Hussain, F., Bronson, K. F., Yadvinder, S., Singh, B. and Peng, S. (2000). Use of chlorophyll meter sufficiency indices for nitrogen management of irrigated rice in Asia. *Agron. J.* 92(5):875–879.

Inada, K. (1963). Studies on a method for determining the deepness of green and color chlorophyll content of intact crop leaves and its practical applications: 1. Principle for estimating the deepness of green color and chlorophyll content of whole leaves. *Jpn. J. Crop Sci.* 32(2):157–162.

Jackson, R. D., Pinter, P. J., Reginato, R. J. and Idso, S. B. (1986). Detection and evaluation of plant stresses for crop management decisions. *IEEE Trans. Geosci. Remote. Sens.* 1:99–106.

Johnson, G. A. and Huggins, D. R. (1999). Knowledge-based decision support strategies: linking spatial and temporal components within site-specific weed management. *J. Crop Prod.* 2(1):225–238.

Johnson, G. V. and Raun., W. R. (2003). Nitrogen response index as a guide to fertilizer management. *J. Plant Nutr.* 26(2):249–262.

Jordan, C. R. (1969). Derivation of leaf-area index from quality of light on the forest floor. *Ecology* 50:663–666.

Justes, E., Jeuffroy, M. H. and Mary, B. (1997). Wheat, barley, and durum wheat. In: *Diagnosis of the Nitrogen Status in Crops*, pp. 73-91. Springer, Berlin.

Knipling, E. B. (1970). Physical and physiological basis for the reflectance of visible and near-infrared radiation from vegetation. *Remote Sens. Environ.* 1(3):155–159.

Krause, G. H. and Weis, E. (1984). Chlorophyll fluorescence as a tool in plant physiology: II. Interpretation of fluorescence signals. *Photosynth. Res.* 5(2):139–157.

Krause, G. H. and Weis, E. (1991). Chlorophyll fluorescence and photosynthesis: the basics. *Annu. Rev. Plant Physiol. Plant Mol. Biol.* 42(1):313–349.

Ladha, J. K., Pathak, H., Krupnik, T. J., Six, J. and Van Kessel, C. (2005). Efficiency of fertilizer nitrogen in cereal production: retrospect and prospects. *Agron.* 1:85–156.

Lang, M., Siel, P., Braunová, Z. and Lichtenthaler, H. K. (1992). Investigations of the blue-green fluorescence emission of plant leaves. *Bot. Acta* 105(6):435–440.

Langsdorf, G., Buschmann, C., Sowinska, M., Babani, F., Mokry, M., Timmermann, F. and Lichtenthaler, H. K. (2000). Multicolour fluorescence imaging of sugar beet leaves with different nitrogen status by flash lamp UV-excitation. *Photosynthetica* 38(4):539–551.

Larson, J. A., Stefanini, M., Yin, X., Boyer, C. N., Lambert, D. M., Zhou, X. V., Tubana, B. S., Scharf, P., Varco, J. J., Dunn, D. J., Savoy, H. J. and Buschermohle, M. J. (2020). Effects of landscape, soils, and weather on yields, nitrogen use, and profitability with sensor-based variable rate nitrogen management in cotton. *Agronomy* 10(12):1858.

Lawson, T. and Vialet-Chabrand, S. (2018). Chlorophyll fluorescence imaging. In: Covsho, E. (Ed.) *Photosynthesis: Methods in Molecular Biology*, Volume 1770, pp. 121–140. Springer, New York. ISBN 978-1-4939-7785-7.

Lei, R., Du, Z., Qiu, Y. and Zhu, S. (2016). The detection of hydrogen peroxide involved in plant virus infection by fluorescence spectroscopy: detection of hydrogen peroxide in plant by fluorescence spectroscopy. *Luminescence* 31(5):1158–1165.

Lemaire, G. and Gastal, F. (1997). N uptake and distribution in plant canopies. In: *Diagnosis of the Nitrogen Status in Crops*, pp. 3–43. Springer, Berlin.

Li, H., Zhao, C., Huang, W. and Yang, G. (2013). Non-uniform vertical nitrogen distribution within plant canopy and its estimation by remote sensing: a review. *Field Crops Res.* 142:75–84.

Li, J. W., Zhang, J. X., Zhao, Z., Lei, X. D., Xu, X. L., Lu, X. X., Weng, D. L., Gao, Y. and Cao, L. K. (2013a). Use of fluorescence-based sensors to determine the nitrogen status of paddy rice. *J. Agric. Sci.* 151(6):862–871.

Lichtenthaler, H. K., Buschmann, C., Rinderle, U. and Schmuck, G. (1986). Application of chlorophyll fluorescence in ecophysiology. *Radiat. Environ. Biophys.* 25(4):297–308.

Lichtenthaler, H. K. and Rinderle, U. (1988). The role of chlorophyll fluorescence in the detection of stress conditions in plants. *CRC Crit. Rev. Anal. Chem.* 19:S29–S85.

Liu,W., Zhu, D. W., Liu, D. H., Geng, M. J., Zhou,W. B., Mi,W. J., Yang, T. W. and Hamilton, D. (2010). Influence of nitrogen on the primary and secondary metabolism and synthesis of flavonoids in *Chrysanthemum morifolium* Ramat. *J. Plant Nutr.* 33(2):240–254.

Liu, X., Zhang, K., Zhang, Z., Cao, Q., Lv, Z., Yuan, Z., Tian, Y., Cao, W. and Zhu, Y. (2017). Canopy chlorophyll density-based index for estimating nitrogen status and predicting yield in rice. *Front. Plant Sci.* 8:1829.

Lofton, J., Tubaña, B. S., Kanke, Y., Teboh, J. and Viator, H. (2012a). Predicting sugarcane response to nitrogen using a canopy reflectance-based response index value. *Agron. J.* 104(1):106–113.

Lofton, J., Tubana, B. S., Kanke, Y., Teboh, J., Viator, H. and Dalen, M. (2012b). Estimating sugarcane yield potential using an in-season determination of normalized difference vegetative index. *Sensor* 12(6):7529–7547.

Longchamps, L. and Khosla, R. (2014). Early detection of nitrogen variability in maize using fluorescence. *Agron. J.* 106(2):511–518.

Ma, B. L., Morrison, M. J. and Dwyer, L. M. (1996). Canopy light reflectance and field greenness to assess nitrogen fertilization and yield of maize. *Agron. J.* 88(6):915–920.

Magdoff, F. R., Jokela, W. E., Fox, R. H. and Griffin, G. F. (1990). A soil test for nitrogen availability in the northeastern United States. *Commun. Soil Sci. Plant Anal.* 21(13–16):1103–1115.

Maiti, D., Das, D. K., Karak, T. and Banerjee, M. (2004). Management of nitrogen through the use of leaf color chart (LCC) and soil plant analysis development (SPAD) or chlorophyll meter in rice under irrigated ecosystem. *Sci.WorldJ.* 4:838–846.

Markwell, J., Osterman, J. C. and Mitchell, J. L. (1995). Calibration of the Minolta SPAD-502 leaf chlorophyll meter. *Photosynth. Res.* 46(3):467–472.

Maxwell, K. and Johnson, G. N. (2000). Chlorophyll fluorescence–a practical guide. *J. Exp. Bot.* 51(345):659–668.

McMurtrey, J. E., III, Chappelle, E. W., Kim, M. S., Meisinger, J. J. and Corp, L. A. (1994). Distinguishing nitrogen fertilization levels in field corn (*Zea mays* L.) with actively induced fluorescence and passive reflectance measurements. *Remote Sens. Environ.* 47(1):36–44.

Mishra, P., Asaari, M. S. M., Herrero-Langreo, A., Lohumi, S., Diezma, B. and Scheunders, P. (2017). Close range hyperspectral imaging of plants: a review. *Biosyst. Eng.* 164:49–67. DOI: 10.1016/j.biosystemseng.2017.09.009.

Moges, S. M., Girma, K., Teal, R. K., Freeman, K. W., Zhang, H., Arnall, D. B., Holtz, S. L., Tubaña, B. S., Walsh, O., Chung, B. and Raun, W. R. (2007). In-season estimation of grain sorghum yield potential using a handheld optical sensor. *Arch. Agron. Soil Sci.* 53(6):617–628.

Monje, O. A. and Bugbee, B. (1992). Inherent limitations of nondestructive chlorophyll meters: a comparison of two types of meters. *HortScience* 27(1):69–71.

Mullen, R. W., Freeman, K. W., Raun, W. R., Johnson, G. V., Stone, M. L. and Solie, J. B. (2003). Identifying an in-season response index and the potential to increase wheat yield with nitrogen. *Agron. J.* 95(2):347–351.

Myers, V. I. (1983). Remote sensing application in agriculture. In: *Manual of Remote Sensing*. Volume II, Colwell, R. N. (Ed.), pp. 2111–2228. American Society of Photogrammetry, Fall Church, VA.

Niak, H. D., Zhang, J., Lofquist, A., Assefa, T., Sarkar, S., Ackerman, D., Singh, A., Singh, A. K. and Ganapathysubramanian, B. A. (2017). A real-time phenotyping framework using machine learning for plant stress severity in soybean. *Plant Meth.* 13:23.

Ollinger, S. V. (2011). Sources of variability in canopy reflectance and the convergent properties of plants. *New Phytol.* 189(2):375–394.

Ortuzar-Iragorri, M. A., Alonso, A., Castellón, A., Besga, G., Estavillo, J. M. and Aizpurua, A. (2005). N-tester use in soft winter wheat: evaluation of nitrogen status and grain yield prediction. *Agron. J.* 97(5):1380–1389.

Padilla, F. M., Gallardo, M., Peña-Fleitas, M. T., de Souza, R. and Thompson, R. B. (2018). Proximal optical sensors for nitrogen management of vegetable crops: a review., *Sensors* 18(7):2083. DOI: 10.3390/s18072083.

Parry, C., Blonquist, J. M. and Bugbee, B. (2014). In situ measurement of leaf chlorophyll concentration: analysis of the optical/absolute relationship. *Plant Cell Environ.* 37(11):2508–2520.

Peltonen, J., Virtanen, A. and Haggren, E. (1995). Using a chlorophyll meter to optimize nitrogen fertilizer application for intensively-managed small-grain cereals. *J. Agron. Crop Sci.* 174(5):309–318.

Peng, S., Buresh, R. J., Huang, J., Zhong, X., Zou, Y., Yang, J., Wang, G., Liu, Y., Hu, R., Tang, Q., Cui, K., Zhang, F. and Dobermann, A. (2010). Improving nitrogen fertilization in rice by site-specific N management: a review. *Agron. Sustain. Dev.* 30(3):649–656.

Peng, S., Cassman, K. G. and Kropff, M. J. (1995). Relationship between leaf photosynthesis and nitrogen content of field-grown rice in the tropics. *Crop Sci.* 35(6):1627–1630.

Peng, S., Garcıa, F. V., Laza, R. C. and Cassman, K. G. (1993). Adjustment for specific leaf weight improves chlorophyll meter's estimate of rice leaf nitrogen concentration. *Agron. J.* 85(5):987–990.

Peñuelas, J., Gamon, J. A., Fredeen, A. L., Merino, J. and Field, C. B. (1994). Reflectance indices associated with physiological changes in nitrogen- and water-limited sunflower leaves. *Remote Sens. Environ.* 48(2):135–146.

Phillips, S. B., Keahey, D. A., Warren, J. G. and Mullins, G. L. (2004). Estimating winter wheat tiller density using spectral reflectance sensors for early-spring, variable-rate nitrogen applications. *Agron. J.* 96(3):591–600.

Piekielek, W. P., Fox, R. H., Toth, J. D. and Macneal, K. E. (1995). Use of a chlorophyll meter at the early dent stage of corn to evaluate nitrogen sufficiency. *Agron. J.* 87(3):403–408.

Quemada, M., Gabriel, J. and Zarco-Tejada, P. (2014). Airborne hyperspectral images and ground-level optical sensors as assessment tools for maize nitrogen fertilization. *Remote Sens.* 6(4):2940–2962.

Raun, W. R., Solie, J. B., Johnson, G. V., Stone, M. L., Lukina, E. V., Thomason, W. E. and Schepers, J. S. (2001). In-season prediction of potential grain yield in winter wheat using canopy reflectance. *Agron. J.* 93(1):131–138.

Raun, W. R., Solie., J. B. and Stone, M. L. (2010). Independence of yield potential and crop nitrogen status. *Prec. Agric.* DOI 10.1007/s111119-010-9196-z.

Read, J. J., Tarpley, L., McKinion, J. M. and Reddy, K. R. (2022). Narrow-waveband reflectance ratios for remote estimation of nitrogen status in cotton. *J. Environ. Qual.* 31(5):1442–1452.

Richardson, A. D., Duigan, S. P. and Berlyn, G. P. (2002). An evaluation of non-invasive methods to estimate foliar chlorophyll content. *New Phytol.* 153(1):185–194.

Roberts, D. A., Smith, M. O. and Adams, J. B. (1993). Green vegetation non-photosynthetic vegetation and soils in AVIRIS data. *Remote Sens. Environ.* 44(2–3):255–269.

Rondeaux, G., Steven, M. and Baret, F. (1996). Optimization of soil-adjusted indices. *Remote Sens. Environ.* 55:95–107.

Roujean, J. L. and Breon, F. M. (1995). Estimating PAR absorbed by vegetation from bidirectional reflectance measurements. *Remote Sens. Environ.* 51(3):375–384.

Rouse, Jr., J. W., Hass, R. H., Schell, J. A. and Deering, D. W. (1974). Monitoring vegetation systems in the great plains with ERTS. In: Third ERTS Symposium, NASA SP-351 1:309–314.

Samborski, S. M., Tremblay, N. and Fallon, E. (2009). Strategies to make use of plant sensors-based diagnostic information for nitrogen recommendations. *Agron. J.* 101(4):800–816.

Schepers, J. S., Blackmer, T. M., Wilhelm, W. W. and Resende, M. (1996). Transmittance and reflectance measurements of corn leaves from plants with different nitrogen and water supply. *J. Plant Physiol.* 148(5):523–529.

Schepers, J. S., Francis, D. D., Vigil, M. and Below, F. E. (1992). Comparison of corn leaf nitrogen concentration and chlorophyll meter readings. *Commun. Soil Sci. Plant Anal.* 23(17–20):2173–2187.

Schröder, J. J., Neeteson, J. J., Oenema, O. and Struik, P. C. (2000). Does the crop or the soil indicate how to save nitrogen in maize production? Reviewing the state of the art. *Field Crops Res.* 66(2):151–164.

Scotford, I. M. and Miller, P. C. H. (2005). Applications of spectral reflectance techniques in northern European cereal production: a review. *Biosyst. Eng.* 90(3):235–250.

Sellers, P. J. (1985). Canopy reflectance, photosynthesis and transpiration. *Int. J. Remote Sens.* 6(8):1335–1372.

Shull, C. A. (1929). A spectrophotometric study of reflection of light from leaf surfaces. *Bot. Gaz* 87(5):583–607.

Sims, J. T., Vasilas, B. L., Gartley, K. L., Milliken, B. and Green, V. (1995). Evaluation of soil and plant nitrogen tests for maize on manured soils of the Atlantic Coastal Plain. *Agron. J.* 87(2):213–222.

Singh, B. and Ali, A. M. (2020). Using handheld chlorophyll meters and canopy reflectance sensors for fertilizer nitrogen management in cereals in small farms in developing countries. *Sensors* 20(4):1127.

Singh, B., Singh, V., Singh, Y., Kumar, A., Sharma, S., Thind, H. S., Choudhary, O. P. and Vashistha, M. (2018). Site-specific fertilizer nitrogen management in irrigated wheat using chlorophyll meter (SPAD meter) in the north-western India. *J. Indian Soc. Soil Sci.* 66(1):53–65.

Singh, B., Singh, V. and Ali, A. M. (2020). Site-specific fertilizer nitrogen management in cereals in South Asia. *Sustain. Agric. Res.* 39:137–178.

Singh, B., Singh, Y., Ladha, J. K., Bronson, K. F., Balasubramanian, V., Singh, J. and Khind, C. S. (2002). Chlorophyll meter and leaf color chart-based nitrogen management for rice and wheat in northwestern India. *Agron. J.* 94(4):821–829.

Smith, M. O., Adams, J. B. and Sabol, D. E. (1994). Spectral mixture analysis – new strategies for the analysis of multispectral data. In: Hill, J. and Megier, J. (Eds) *Imaging Sectrometry – A Tool for Environment Observations*, pp. 125–143. Kluwer Academic Publishers, Boston, MA.

Solari, F., Shanahan, J., Ferguson, R., Schepers, J. and Gitelson, A. (2008). Active sensor reflectance measurements of corn nitrogen status and yield potential. *Agron. J.* 100(3):571–579.

Sorby, H. C. (1873). XXII. On comparative vegetable chromatology. *Proc. R. Soc. Lond.* 21:442–483.

Stone, M. L., Solie, J. B., Raun, W. R., Whitney, R. W., Taylor, S. L. and Ringer, J. D. (1996). Use of spectral radiance for correcting in-season fertilizer nitrogen deficiencies in winter wheat. *Trans. ASAE* 39:1623–1631.

Stone, M. L., Solie, J. B., Whitney, R. W., Raun, W. R. and Lees, H. L. (1996). Sensors for detection of nitrogen in winter wheat. SAE Paper no. 96-1757. Society of Automotive Engineers, Warrendale, PA.

Strasser, R. J., Tsimilli-Michael, M. and Srivastava, A. (2004). Analysis of the chlorophyll a fluorescence transient. In: Papageorgiou, G. C. and Govindjee (Eds) *Chlorophyll a Fluorescence: A Signature of Photosynthesis*, 321–362. Springer, New York.

Taskos, D. G., Koundouras, S., Stamatiadis, S., Zioziou, E., Nikolaou, N., Karakioulakis, K. and Theodorou, N. (2015). Using active canopy sensors and chlorophyll meters to estimate grapevine nitrogen status and productivity. *Precis. Agric.* 16(1):77–98.

Tattini, M., Galardi, C., Pinelli, P., Massai, R., Remorini, D. and Agati, G. (2004). Differential accumulation of flavonoids and hydroxycinnamates in leaves of Ligustrum vulgare under excess light and drought stress. *New Phytol.* 163(3):547–561.

Teal, R. K., Tubaña, B., Girma, K., Freeman, K. W., Arnall, D. B., Walsh, O. and Raun, W. R. (2006). In-season prediction of corn grain yield potential using normalized difference vegetation index. *Agron. J.* 98(6):1488–1494.

Thenkabail, P. S., Smith, R. B. and De Pauw, E. (2002). Evaluation of narrowband and broadband vegetation indices for determining optimal hyperspectral wavebands for agricultural crop characterization. *Photogramm. Eng. Remote Sens.* 68:607–621.

Thind, H. S. and Gupta, R. K. (2010). Need based nitrogen management using the chlorophyll meter and leaf color chart in rice and wheat in South Asia: a review. *Nutr. Cycl. Agroecosystems* 88:361–380.

Thompson, R. B., Tremblay, N., Fink, M., Gallardo, M. and Padilla, F. M. (2017). Tools and strategies for sustainable nitrogen fertilisation of vegetable crops. In: Tei, F., Nicola, S. and Benincasa, P. (Eds) *Advances in Research on Fertilization Management in Vegetable Crops*, pp. 11–63. Springer, Berlin.

Tremblay, N., Wang, Z. and Bélec, C. (2007). Evaluation of the Dualex for the assessment of corn nitrogen status. *J. Plant Nutr.* 30(9):1355–1369.

Tremblay, N., Wang, Z. and Bélec, C. (2010). Performance of DUALEX in spring wheat for crop nitrogen status assessment, yield prediction and estimation of soil nitrate content. *J. Plant Nutr.* 33(1):57–70.

Tremblay, N., Wang, Z. and Cerovic, Z. G. (2012). Sensing crop nitrogen status with fluorescence indicators: a review. *Agron. Sustain. Dev.* 32(2):451–464.

Tubaña, B. S., Harrell, D. L., Walker, T., Teboh, J., Lofton, J. and Kanke, Y. (2012). In-season canopy reflectance-based estimation of rice yield response to nitrogen. *Agron. J.* 104(6):1604–1611.

Tucker, C. J. (1979). Red and photographic infrared linear combinations for monitoring vegetation. *Remote Sens. Environ.* 8(2):127–150.

Turner, F. T. and Jund, M. F. (1991). Chlorophyll meter to predict nitrogen topdress requirement for semidwarf rice. *Agron. J.* 83(5):926–928.

Turner, F. T. and Jund, M. F. (1994). Assessing the nitrogen requirements of rice crops with a chlorophyll meter. *Aus. J. Expert. Agric.* 34:1001–1005.

Ulissi, V., Antonucci, F., Benincasa, P., Farneselli, M., Tosti, G., Guiducci, M., Tei, F., Costa, C., Pallottino, F.,Pari, L. and Menesatti, P. (2011). Nitrogen concentration estimation in tomato leaves by VIS-NIR non-destructive spectroscopy. *Sensors* 11(6):6411–6424.

Usha, K. and Singh, B. (2013). Potential applications of remote sensing in horticulture–a review. *Sci. Hortic. (Amst.)* 153:71–83.

Varco, J. J., Fox, A. A., Raper, T. B. and Hubbard, K. J. (2013). Development of sensor-based detection of crop nitrogen status for utilization in variable rate nitrogen fertilization. In: Precision Agriculture 2013–Papers Presented at the 9th European Conference on Precision Agriculture, ECPA 2013, pp. 145–150. Wageningen Academic Publisher, Lleida.

Varvel, G. E., Schepers, J. S. and Francis, D. D. (1997). Ability for in-season correction of nitrogen deficiency in corn using chlorophyll meters. *Soil Sci. Soc. Am. J.* 61(4):1233–1239.

Varvel, G. E., Wilhelm, W. W., Shanahan, J. F. and Schepers, J. S. (2007). An algorithm for corn nitrogen recommendations using a chlorophyll meter-based sufficiency index. *Agron. J.* 99(3):701–706.

Vincini, M., Frazzi, E. and D'Alessio, P. (2008). Abroad-band leaf chlorophyll vegetation index at the canopy scale. *Precis. Agric.* 9(5):303–319.

Vogelmann, J. E., Rock, B. N. and Moss, D. M. (1993). Red edge spectral measurements from sugar maple leaves. *Int. J. Remote Sens.* 14(8):1563–1575.

Watanabe, S. (1980). Development of a digital chlorophyll meter: I. Structure and performance. *Jpn. J. Crop Sci.* 49:89–90.

Wells, B. R., Norman, R. J. and Helms, R. S. (1993). Use of plant area measurements to estimate mid-season nitrogen rates for rice. Rice Research Studies 1992. University of Arkansas Research Series No. 431:137-141.

Witt, C., Pasuquin, J. M. C. A., Mutters, R. and Buresh, R. J. (2005). New leaf color chart for effective nitrogen management in rice. *Better Crops* 89:36-39.

Xue, L., Cao, W., Luo, W., Dai, T. and Zhu, Y. (2004). Monitoring leaf nitrogen status in rice which canopy spectral reflectance. *Agron. J.* 96(1):135-142.

Yang, W. H., Peng, S., Huang, J., Sanico, A. L., Buresh, R. J. and Witt, C. (2003). Using leaf color charts to estimate leaf nitrogen status of rice. *Agron. J.* 95(1):212-217.

Yao, Y., Miao, Y., Huang, S., Gao, L., Ma, X., Zhao, G., Jiang, R., Chen, X., Zhang, F., Yu, K., Gnyp, M. L., Bareth, G., Liu, C., Zhao, L., Yang, W. and Zhu, H. (2012). Active canopy sensor-based precision N management strategy for rice. *Agron. Sustain. Dev.* 32(4):925-933.

Zhang, Y., Tremblay, N. and Zhu, J. (2012). A first comparison of Multiplex® for the assessment of corn nitrogen status. *J. Food Agric. Environ.* 10:1008-1016.

Zhu, J., He, W., Yao, J., Yu, Q., Xu, C., Huang, H., Mhae, B. and Jandug, C. (2020). Spectral reflectance characteristics and chlorophyll content estimation model of Uqeros aquifolioides leaves at different altitudes in Sejila Mountain. *Appl. Sci.* 10:3636.

Zhu, J., Tremblay, N. and Liang, Y. (2012). Comparing SPAD and atLEAF values for chlorophyll assessment in crop species. *Can. J. Soil Sci.* 92(4):645-648.

Zillman, E., Graeff, S., Link, J., Batchelor, W. D. and Claupein, W. (2006). Assessment of cereal nitrogen requirements derived by optical on the go sensors on heterogeneous soils. *Agron. J.* 98(3):682-690.

Chapter 6

Synchronizing nitrogen fertilizer application to crop nitrogen needs

Bijay-Singh[1], Punjab Agricultural University, India; R. J. Buresh, Independent Researcher, USA; and S. Peng, Huazhong Agricultural University, China

1 Introduction

Fertilizer nitrogen (N) is applied to crops and cropping systems to ensure sufficient food production for the global population. However, indiscriminate fertilizer use can have extensive negative impacts on the environment. For example, on a global basis, fertilizer N constitutes the single largest source of nitrous oxide, a potent and long-lived greenhouse gas (GHG) (Winiwarter et al., 2018) which contributes to pervasive nitrate pollution of surface and ground water bodies (Bijay-Singh and Craswell, 2021). It is therefore essential to limit the excessive application of fertilizer N in agricultural systems. However, to

[1] Corresponding author e-mail: bijaysingh20@hotmail.com.

http://dx.doi.org/10.19103/AS.2024.0135.10

avoid food insecurity, N input via biological or industrial sources will have to be increased in agroecosystems that exhibit poor crop yields due to an insufficient supply of N to crop plants. Cropping systems would benefit from a more precise needs-based N application to improve fertilizer N use efficiency (FNUE) and sustain soil organic matter and soil fertility (Udvardi et al., 2021).

FNUE, defined as the proportion or percentage of fertilizer N recovered in the aboveground crop biomass during the growing season, is governed by complex N transformations in the soil–plant system, as well as soil and fertilizer management practices. Achieving synchrony between applied N from all sources, including fertilizers, and crop N demand, in addition to considering the temporal and spatial variability of available soil N, is crucial to optimize yield, profit and FNUE in crop production, as well as minimizing N-related environmental degradation (Cassman et al., 2002; Shanahan et al., 2008; Cassman and Dobermann, 2022). Strategies to achieve greater synchrony also involve enhancing the efficiency of applied N as well as N derived from soil organic matter and biological N fixation (Crews and Peoples, 2005; Ladha et al., 2022).

To improve FNUE in crop production systems, Cassman et al. (2002) proposed an N synchrony framework for managing fertilizer N using strategies or technologies that lead to 'just-in-time N supply', i.e. the dosage is not excessive or deficient throughout the crop growth season. These technologies include:

- Using split applications of fertilizer N;
- Precision agriculture tools such as chlorophyll metres and canopy reflectance sensors; and
- Fertilizer source modification.

They have been designed to concomitantly increase crop yields and FNUE. When crops absorb more N and convert it into economic yield, the amount of N that is vulnerable to loss to the environment is reduced. This chapter explores the causes of asynchrony between N supply and crop demand and evaluates the strategies and technologies that have been developed to synchronize fertilizer N application to crop N requirements.

2 Why there is asynchrony between crop nitrogen demand, soil nitrogen supply and fertilizer nitrogen application

Nitrogen asynchrony refers to the availability of N in the soil not being closely aligned in space and time with N demands of crop plants. Synchrony is influenced by the N uptake pattern of crop plants, fertilizer N management, environmental factors dictating net N mineralization in the soil and their interactions. When N availability in the soil is insufficient to meet crop N demands, productivity is limited and crop yields are increased by supplying N via fertilizer or manure. However, the widespread use of fertilizer N may lead to asynchrony if N availability outpaces or lags behind crop N demand. Asynchrony with excessive N may lead to high yield levels not limited by N availability but result in losses of N to the environment (Crews and Peoples, 2005; Ladha et al., 2020).

2.1 Crop nitrogen demand

Crop N demand is the amount of N taken up by the crop when, under the given environmental conditions, N supply to the crop ceases to be a limiting factor for growth. Although there is a great deal of uncertainty surrounding the amount of N a crop needs to absorb from the soil to ensure growth and new tissue synthesis (Grindlay, 1997), it is known that N uptake by the crop is less than the crop N demand when mineral N available in the soil for uptake by roots is insufficient (van Keulen et al., 1989; Justes et al., 1994). Crop biomass yield and the physiological requirements for tissue N greatly influence crop N demand (Gastal and Lemaire, 2002). Biomass yield and thus crop N demand are also influenced by crop management practices and weather prevailing during the crop growing period. Therefore, year-to-year variations in weather can lead to large differences in crop N demand. Similarly, variations in solar radiation and temperature in irrigated crops and changing rainfall patterns (amount and distribution) in rainfed crops influence yield potential and crop N demand. The actual crop yields obtained by farmers are also adversely influenced by suboptimal nutrient supply, weed competition, damage from insects and diseases and interaction of weather and management, which can result in large year-to-year and field-to-field variations in crop N demand (Shanahan et al., 2008).

As shoot growth in crop plants determines N demand and potential N uptake (Grindlay, 1997), N content on a weight basis is generally used as the index of crop N nutrition. At any point during the growth of food crops, there is a critical N content below which growth is reduced. During early crop growth stages, when large amounts of organic N compounds needed for photosynthesis and plant growth are synthesized, N content on a weight basis is relatively high. As critical N content differs with the age of the crop, potential N requirement is the sum of the N needed for new growth and the N required to overcome the existing deficit. At any crop stage when the N content falls below the critical level, N deficiency occurs leading to asynchrony between N supply and N demand. At any point during crop growth, the rate of N uptake is defined by a number of factors including: crop N demand, N uptake capacity and the N supply (Singh et al., 2001).

2.2 Soil nitrogen supply

A substantial amount of N is mineralized from soil organic matter during a crop growing season, but due to the diversity and complexity of processes controlling both net N mineralization and N uptake by plants, it is challenging to ensure their synchronization. The temporal drivers of dynamics of net N mineralization may be related to, but not directly aligned with, the drivers of N uptake by crop plants (Grandy et al., 2012). In addition, our understanding of the dynamics of N mineralization-immobilization processes is inadequate to comprehend in terms of N synchronization with crop N uptake and demand (Fierer et al., 2009).

Only a portion of fertilizer N applied in a cropping season directly reaches the soil mineral N pool within the root zone of crops where it can be taken up by crop plants. Some applied N can be lost from the soil-plant system, and some remaining fertilizer N is converted into organic forms and becomes a part of the soil organic matter referred to as soil N (Bijay-Singh and Sapkota, 2023). Thus, soil N contains portions of fertilizer N applied over several previous years and mineralization-immobilization turnover, controlled by soil temperature and moisture, releases a small portion of soil N to the mineral N pool during a crop growing season. Additionally, asymbiotically fixed N, biologically fixed N, N added through organic sources and N deposited by rainfall as components of soil N contribute to spatial and temporal variability in soil N supply (Ladha et al., 2016). Substantial variations in crop N demand as well as soil N supply even within a field (Scharf et al., 2005) are also linked with the spatial distribution of soil water content and soil water-holding capacity (Schmidt et al., 2007).

As more than 50% of the crop N uptake typically comes from soil N (Ladha et al., 2005; Gardner and Drinkwater, 2009), which has been enriched via N applied through fertilizers and other sources over previous years, considerable asynchrony between crop N demand and soil N supply is expected. As the application of split doses of fertilizer N supplements the contribution of soil N to N uptake by a crop, it may add to N synchrony. Asynchrony between crop N demand and soil N supply is also determined by the extent of loss of soil N from the soil-plant system. Factors that influence the loss of soil N via leaching, denitrification, ammonia volatilization and run-off include inherent soil fertility, soil texture, rate and time of application of external N sources, soil pH, soil temperature, soil moisture, soil oxygen level, root length density, N uptake potential per unit length of root and crop N demand (Ladha et al., 1998; Singh et al., 2001).

2.3 Fertilizer nitrogen application

Uniform fertilizer N rates, formulated from crop response data averaged over large geographic areas with a similar climate and landforms, when applied to spatially variable landscapes lead to considerable asynchrony between crop N demand and N supply from soils and fertilizers. Uniform fertilizer N application rates may be justified for ease and convenience of farmers (Lambert et al., 2006; Hong et al., 2007), but these discount the fact that the supply of N from the soil and fertilizer, crop N uptake and crop response to N are not spatially uniform (Inman et al., 2005). In developing countries, fertilizer N is generally managed in the form of blanket recommendations consisting of uniform fertilizer N rates over vast areas. As blanket recommendations are designed to produce optimum yields in all the fields in the region, these lead to over-application of fertilizer N in many fields (Bijay-Singh et al., 2020). Large pre-plant application of fertilizer N, as in the US maize belt region (Cassman et al., 2002) or in the intensive wheat–maize system in China (Cui et al., 2008a,b), also result in poor synchronization with crop N demand. Rice farmers in Asia often split apply fertilizer N with a relatively large proportion of the total N applied as a basal dose and at tillering. Excessive early N can lead to luxuriant vegetative growth, which can result in loss of yield (Peng et al., 2010) associated with increased lodging (Pan et al., 2019) and diseases such as sheath blight (Tang et al., 2007). Farmers in South Asia gain a sense of comfort against yield loss by applying a basal dose of fertilizer N (Varinderpal-Singh et al., 2010). These large pre-plant or basal N applications lead to high soil inorganic N levels prior to the rapid N uptake phase of the crops resulting in poor N synchrony (Shanahan et al., 2008). In Spain, Lopez-Bellido et al. (2005) observed that utilization of applied fertilizer N by wheat was only 14.1% when applied at sowing compared with 54.8% when top-dressed at initiation of the stem elongation stage of the crop.

Fertilizer N rates are frequently not formulated based on soil N supply and crop N demand, as estimated by the response of crop yield to applied N, and are only 'yield-based'. A yield goal is set and multiplied by a constant factor to estimate fertilizer N requirement that is essentially based on removal of N by the crop at harvest, N associated with biomass production and estimated FNUE (Meisinger and Randall, 1991). Such 'mass-balance' approaches to determine fertilizer N rates are highly simplified and cannot lead to synchronization between N supply and demand. Fertilizer N rates should be determined by eliminating uncertainties in soil N supply and crop N demand. Average fertilizer N rates for maize (*Zea mays* L.) could be reduced an estimated 20–40% by eliminating uncertainty in soil N supply and an estimated 3–10% by eliminating uncertainty in potential crop N demand (Lobell, 2007).

3 Improving fertilizer nitrogen use efficiency through synchronizing fertilizer nitrogen application with soil nitrogen and crop nitrogen demand

The concept of synchrony between N supply, from fertilizer and soil, and crop N demand is at the core of FNUE. Improving N synchrony in crop production systems can be challenging, but it is currently one of the most important objectives of agricultural research. It is very difficult to achieve N synchrony in most cereal grain crops because during the typical 90–100 day growing season there are only 30–40 days mid-season that crop plants rapidly take up N and accumulate biomass at a significant rate (Udvardi et al., 2021). Osterholz et al. (2017) recorded an N uptake rate by maize as high as 5 kg N ha^{-1} day^{-1} for 3–4 weeks, which quickly became negligible. The application of the majority of fertilizer N or mineralization of large amounts of soil N during favourable soil temperatures and moisture not coinciding with the period of rapid N uptake by the crop will result in N asynchrony. The asynchrony between availability of N and crop N demand results in low FNUE (Robertson and Vitousek, 2009).

Spatial synchrony is as important as temporal synchrony for synchronizing crop N demand with fertilizer N supply and soil N availability. Crop N demand varies across landscapes in a given cropping season and from year to year at a single location (Basso et al., 2019), and its prediction at the time of deciding fertilizer N application, and also considering possible future weather scenarios, is not easy. It is a challenging task to manage fertilizer N during crop production in a way that N is never more or less than the crop N demand during the crop growth period. Over the past few decades, scientists have been experimenting with different ways to improve N synchrony to increase the efficiency of N fertilizer use. Several management options are now available to increase FNUE via the improved matching of N supply with crop N demand. These can be grouped into two broad categories. One category of options includes spatially and temporally unambiguous fertilizer management approaches based on adjusting timing and the rate of fertilizer N to nearly match the crop N uptake pattern and demand. These include a range of innovative precision N management strategies relying on dynamics of leaf chlorophyll and other crop vigour-related measures. Some site-specific N management strategies utilize tools such as on-the-go soil and crop sensors that are capable of sensing crop N status in 'real-time' and deliver spatially variable fertilizer N applications as per crop requirement. Site-specific N management also involves the use of precision agricultural techniques such as GPS mapping to create detailed maps of field variability and to apply fertilizer in a targeted and precise manner (Shanahan et al., 2008). The second category of options to enhance FNUE through improved N synchrony revolves around the development of fertilizers and chemical inhibitors that regulate the release of N to match the N uptake pattern by the crop plants. Commonly identified as enhanced efficiency fertilizers (EEFs), these are formulations with coatings of materials such as polymers that resist immediate solubilization of fertilizer N or contain urease or nitrification inhibitors. Innovative fertilizers are being developed by taking into consideration the physiological needs of crops rather than starting with chemistry of the N fertilizer (Bindraban et al., 2020).

4 Innovative nitrogen management strategies: in-season monitoring of crop nitrogen and/or biomass status

Nitrogen synchronization in crop production involves dynamic adjustment of fertilizer use to optimally bridge the deficit between N demands of the crop and the indigenous N supply defined as the N from soil and all sources other than fertilizer. It can be achieved in a site-specific manner and only when the crop is growing in the field. The effects of soil, climate and management conditions on N supply and demand are reflected by leaf N status, which can be used for guiding fertilizer N input to achieve N synchrony. As plant leaves are the principal organs that transform solar radiation into chemical energy via photosynthesis and supply N-containing photosynthetic products for the development of green plant biomass (Kusano et al., 2011; Makino, 2011; Bianculli et al., 2016; Bailey and Leegood, 2016), leaf greenness or chlorophyll content and crop biomass reflect soil N supply as well as N requirement for optimum growth. A low chlorophyll content or lesser green colour of leaves along with reduced biomass production are associated with an insufficient supply of N (Fageria et al., 2011; Zhao et al., 2005). Growing plants can be used in different ways, at any time during the cropping season, to assess the status of soil N supply and crop N demand, and strategies that have evolved over recent decades can be classified in three distinct categories and are described in the following subsections. In general, different approaches revolve around comparing the leaf N status and biomass of crops to be fertilized with that of crops grown under no N limiting conditions (Shanahan et al., 2008).

4.1 Visual assessments using calibration reference plots

As developed by Raun et al. (2005), calibration reference plot technology for improved in-season N fertilizer management in cereal crops is based on visually inspecting the growth and greenness in a group of plots (each measuring 1 m^2) established by the pre-plant application of a range of fertilizer N levels that are superimposed on top of the farmer's N practice. The N topdress rate is determined as the fertilizer N level among the group of plots showing maximum biomass production. In case more than one plot in the group of calibration reference plots shows near maximum biomass production, the lowest pre-plant N rate among these constitutes the amount of additional N to be topdressed to achieve optimum grain yield. The visual inspection of reference plots to determine optimum fertilizer N rate is performed prior to, but very close to, topdressing of the fertilizer.

The strategy, as developed by Raun et al. (2005), not only provides a method for determining the topdress N application rate that accounts for N mineralization and atmospheric deposition but also visual interpretation of N mineralized in the soil plus atmospheric N deposition during planting to the time when fertilizer N is applied mid-season. Although using calibration reference plots to determine the mid-season fertilizer N topdress rate is not based on prediction of yield potential as in some other strategies (such as Raun et al. (2002), Buresh and Witt (2007)), it does take into account the temporal and spatial variability among production fields and provides a simple method to estimate the extent of N available to the crop via mineralization, rainfall or atmospheric deposition; it also determines mid-season fertilizer N needs of the crop based on the growing conditions encountered up until the interpretation of the calibration reference plots.

As reported by Shanahan et al. (2008), a programmable applicator has been developed to establish calibration reference plots by applying pre-plant rates ranging from 0 to 300 kg N ha^{-1}, in progressive incremental rates as low as 10 kg N ha^{-1} over user defined distances. This strategy is highly advantageous as reference plots are created within the field, which has direct educational value for the farmer in terms of recognizing signs of N stress or adequacy and taking timely action for the need-based application of fertilizer N to the crop.

4.2 Leaf nitrogen sensing using a chlorophyll metre and leaf colour chart

Approximately 80% of N in the leaves of crop plants is contained in chloroplasts in the parenchyma cells of the leaf mesophyll. In the chloroplasts, about 50% of this N is associated with photosynthetic proteins involved in harvesting sunlight and leaf chlorophyll content is linked to N supply (Peng et al., 1995). Due to the close association between leaf greenness and leaf chlorophyll content (Madeira et al., 2000), the N status and greenness of the topmost fully expanded leaf of the growing crop plants at any time reflect the N demand as it dictates the photosynthetic rate and biomass production (Peng et al., 1996; Buresh, 2007). Thus, leaf greenness, assessed via comparison with shades of green colour on a leaf colour chart (LCC) or leaf chlorophyll content as measured by a chlorophyll metre, can be used as indicators of the N requirements of field crops. Although the hand-held Minolta Soil Plant Analysis Development (SPAD) 502®, known as a SPAD metre, is the most commonly used chlorophyll metre, atLeaf (Zhu et al., 2012) and Yara (Aranguren et al., 2019) also produce chlorophyll metres, which quickly measure light transmittance or absorbance to estimate leaf greenness, in terms of the relative quantity of chlorophyll, in a non-destructive manner. Smartphone apps have also been developed for the non-destructive estimation of leaf chlorophyll (Vesali et al., 2015) and leaf colour (Intaravanne and Sumriddetchkajorn, 2015; Paleari et al., 2019; Tao et al., 2020).

The LCC estimates leaf greenness by visually comparing light reflection from the leaf surface and different panels with varying shades of green embedded on a plastic strip or chart (Yang et al., 2003). It has proved to be an efficient and economical substitute to chlorophyll metres for poor and marginal farmers in developing countries. To deal with field-to-field and year-to-year differences in fertilizer N demand and N supply, and consequential N asynchrony, chlorophyll content and greenness of the topmost fully expanded leaves of crop plants can serve as a non-destructive method to guide within-season N management in field crops. The use of chlorophyll metres or LCC to sense leaf N can help avoid over- and under-application of fertilizer N and help achieve optimum crop yields and minimal N-related environmental impact.

Chlorophyll metres and LCCs can be used with real-time (RT) and fixed-time adjustable-dose (FT) N management, which can significantly contribute towards increasing FNUE by improving N synchrony in crop production (Peng et al., 2006; Wang et al., 2007). RT N management was developed and evaluated in the 1990s (Peng et al., 1996) and the application time of N is not pre-set in this strategy. The time and number of N applications are dynamically adjusted to crop N demand by applying fertilizer N when the N content of the topmost fully expanded leaves, determined non-destructively by a chlorophyll metre or LCC, falls below a critical level (Tao et al., 1990; Peng et al., 1996; Balasubramanian et al., 1999; Yang et al., 2003). For rice, leaf N status is monitored every 7–10 days from approximately 2 weeks after planting. A fixed N rate between 20 kg N ha^{-1} and 45 kg N ha^{-1} is applied when the SPAD or LCC value falls below a critical threshold determined for different varietal groups in a region by conducting experiments prior to commencing RT N management. FT was developed during 1997–2000 (Wang et al., 2001). Its key components are estimating crop N requirement before the start of the cropping season, applying an optimal amount of basal N, splitting the remaining fertilizer N for application at critical crop growth stages, and then adjusting these in-season N topdressings upward or downward based on chlorophyll metre or LCC readings (Dobermann et al., 2002; Peng et al., 2006; Witt et al., 2007). In FT N management, the time and number of fertilizer N applications are fixed but the rate of each N application is defined by N synchrony between soil N and crop N demand across season and location.

RT N management using the LCC and SPAD metre has been evaluated for rice in a large number of on-farm locations in South and Southeast Asia

Table 1 Average grain yield, total N uptake, recovery efficiency of N (RE_N), agronomic N use efficiency (AE_N) and PFP_N for RT N management, FT adjustable-dose N management and FFP in rice across four sites in China and 2 years (2001 and 2002)

Treatment	Yield (t ha^{-1})	N uptake (kg ha^{-1})	RE_N (%)	AE_N (kg grain kg^{-1} N)	PFP_N (kg grain kg^{-1} N)
FFP	7.2 b	194 a	48 b	3.6 b	35 c
RT	7.4 a	140 c	66 a	14.5 a	123 a
FT	7.7 a	165 b	66 a	11.8 a	72 b

Within a column, means followed by different letters (a, b and c) are significantly different at the 0.05 probability level according to Tukey's multiple comparison test. Modified from Peng et al. (2006).

since 1998 (Balasubramanian et al., 2000). Peng et al. (2006) evaluated both RT and FT N management of farmers' fertilizer practice (FFP) in four major rice-growing provinces in China (Table 1) and observed that while yield and agronomic and recovery efficiencies of rice with RT were on par with FT, fertilizer N application with RT was significantly less than FT. As a result, the partial factor productivity of N (PFP_N) with RT was higher than FT. When compared with blanket recommendations for fertilizer N, the same rice yield level could generally be achieved with RT but with 20–30% less fertilizer N application (Peng et al., 1996; Balasubramanian et al., 1999, 2000). However, this approach for increasing FNUE via improved N synchrony has not become popular with Chinese farmers primarily due to the increased labour required to implement RT.

Over last two decades, numerous studies have been carried out in South Asia, to evaluate RT N management as a possible strategy to manage fertilizer N in rice and maize as per crop N demand and to increase FNUE (Bijay-Singh et al., 2020). RT N management does not work with wheat because fertilizer N application to wheat must coincide with irrigation events at critical growth stages of the crop; but in rice and maize N can be applied whenever the leaf N status falls below a threshold level. While some researchers have used a SPAD metre, more studies are based on using LCC as it is an inexpensive alternative to a chlorophyll metre. Identifying the threshold N status of the topmost fully exposed leaf constitutes the starting point to practise RT and it was found to vary for different varietal groups of rice between shades 3 and 5 using an LCC and between 32 and 37.5 on a SPAD metre. For example, in the Indo-Gangetic Plains of India, Shukla et al. (2004) found LCC shades 3, 4 and 5 to be the thresholds for basmati, inbred and hybrid rice cultivars, respectively. In the Philippines, for rice cultivar IR72 SPAD, values of 35 and 32 proved to be the critical leaf N status when grown in dry and wet seasons, respectively. Bijay-Singh et al. (2020) reviewed the outcome of seven studies on SPAD metre-based RT carried out in South Asia. In all the studies, significant increases in different measures of FNUE were observed compared with when fertilizer N was managed following blanket recommendations or FFP.

The chlorophyll metre has also been used for RT following a threshold sufficiency index based on chlorophyll readings. The sufficiency index is defined as the ratio of the chlorophyll metre readings of the test field to that of a well-fertilized reference plot (Varvel et al., 1997). This approach has the advantage of being self-calibrating for different regions, seasons and cultivars. Its usefulness in increasing FNUE by need-based fertilizer N application has been demonstrated in maize by Varvel et al. (1997) and in rice by Hussain et al. (2000), Bijay-Singh et al. (2006) and Bijay-Singh (2008). Experiments conducted by Balasubramanian et al. (1999) and Bijay-Singh et al. (2002) further revealed that by avoiding a basal fertilizer N dose if a grain yield of 3 t ha^{-1} or more was obtained in no-N control plots and RT N management started 2 weeks after transplanting, improved N synchrony between fertilizer N and crop N demand and, in turn, high FNUE were observed. Decision support systems based on SPAD metre readings have also been developed to guide in-season fertilizer N application in field crops (Goffart and Olivier, 2004). After reviewing the research on fertilizer N management using a SPAD metre in maize in the USA, Shanahan et al. (2008) concluded that a chlorophyll metre can help maintain yield with less fertilizer N than single pre-plant application, but it is a tool for maintaining an adequate but not over-supply of N to the crop.

Results from numerous studies on LCC-based RT in rice and maize carried out in South Asia (Bijay-Singh, 2022) can be categorized in two major groups:

- Same grain yield with reduced total fertilizer N application than FFP or blanket recommendation; and
- Increased grain yield with the same or reduced total fertilizer N application than FFP or blanket recommendation.

In a third group, with a few studies conducted on rice in India, increased yield with no change in total applied N was achieved with a fixed dose of fertilizer N applied by adjusting the time of split applications as per LCC-based RT N management (Bijay-Singh, 2022). In all the studies in the three groups, FNUE indices increased significantly (wherever studied) emphasizing the role of improve N synchrony. In 50 LCC-based RT on-farm experiments conducted by Varinderpal-Singh et al. (2007), the schedule of application time of fertilizer N doses as summarized in Table 2 revealed that while the first dose was applied in all the experiments, the second and third doses were applied only in 90% and 50% of the experiments. As the organic carbon content of the soils of the experiments varied from 0.2% to 0.9%, soil N supply was expected to be low leading to the first LCC-guided fertilizer N dose generally being applied within 40 days of transplanting of rice. The second dose was most often applied

Table 2 Summary of times of application of fertilizer N doses (30 kg N ha^{-1} each) to transplanted rice following RT N management in 50 on-farm experiments conducted in northwestern India by Varinderpal-Singh et al. (2007)

	Percentage of farms		
Days after transplanting of rice	First dose	Second dose	Third dose
Less than 20	44	4	-
21-40	54	42	6
41-55	2	36	16
More than 55	-	8	28
	100	90	50

between 21 days and 55 days after the transplanting of rice. Depending upon crop N demand, the third LCC-guided dose was applied in half of the experiments but in 28% of the experiments it was 55 days after the transplanting of rice. It demonstrates that to achieve N synchrony between crop N demand and N supply, LCC-based RT guides the application of fertilizer N dose even 55 days after transplanting.

Several variants of FT N management are available in terms of critical LCC/chlorophyll metre values and percentage of total fertilizer N to be applied at different critical growth stages of rice (Dobermann et al., 2002; Peng et al., 2006, 2010; Wang et al., 2007; Witt et al., 2007). Bijay-Singh et al. (2012) conducted a series of experiments on FT N management on irrigated rice in northwestern India and found improved congruence between N supply and crop N demand was achieved by applying a dose of N at the transplanting of rice before applying fertilizer N doses adjustable based on leaf N status expressed via LCC measurements at maximum tillering, panicle initiation and flowering initiation stages of the crop. Following this approach, the optimum rice grain yield and high N use efficiency were observed but with less fertilizer N application than the blanket recommendation. Similar results were reported by Ali et al. (2015) using LCC and SPAD-metre-assisted FT N management in dry direct seeded rice grown in northwestern India.

As fertilizer N application to wheat is linked with irrigation events, improved N synchrony and N application as per crop N demand is only possible by following the FT approach. In the Indo-Gangetic Plains in India, where some studies on FT have been carried out in recent years (Bijay-Singh et al., 2020), blanket recommendation for fertilizer N management is to apply in two split doses, half at sowing and the remaining half along with first irrigation at the crown root initiation stage about 3 weeks after sowing. It should lead to a very high degree of N asynchrony as all the N is applied within 3 weeks of sowing wheat. Varinderpal-Singh et al. (2012) conducted a series of experiments and formulated an FT strategy which consists of applying a fixed small dose of fertilizer N at the sowing of the crop and another fixed dose at the crown root initiation or first irrigation stage when the leaves are yet too small and then applying an N dose adjusted as per leaf N status, estimated by LCC or a chlorophyll metre, at the maximum tillering stage of the crop coinciding with the second irrigation event. The work of Islam et al. (2014) and Bijay-Singh et al. (2002, 2013) convincingly proved the role of applying N to maintain leaf N status at the maximum tillering stage to achieve a high yield of wheat. Thus, the FT strategy formulated by Varinderpal-Singh et al. (2012) resulted in high yield levels, reduced total fertilizer N use as well as improved agronomic and recovery efficiencies of fertilizer N in wheat whether used with an LCC or a SPAD metre (Bijay-Singh et al., 2018). It was applicable to both *Triticum aestivum* and *T. turgidum* genotypes (Varinderpal-Singh et al., 2017).

4.3 Canopy reflectance sensors for matching fertilizer nitrogen management and crop nitrogen demand

Leaves appear green because chlorophyll absorbs most of the light in the blue (around 450 nm) and red (around 670 nm) parts of the spectrum for photosynthesis but reflects green light (around 550 nm). Reflectance measurement of the visible spectrum from the crop canopy can therefore assess crop N requirement as a positive relationship exists between leaf greenness or chlorophyll content and crop N status. In addition, because reflectance of as much as 60% incident near infrared radiation (NIR) (around 780 nm) is determined by the structure of mesophyll tissues, measured spectral reflectance from the canopy in the red and NIR bands could estimate plant biomass as well as leaf N content (Ma et al., 1996). As the crop canopy develops, reflectance of the visible spectrum is reduced, while NIR reflectance increases (Major et al., 2004). Solie et al. (1996) used canopy reflectance measurements to vary fertilizer N rates. In 2001, this led to the commercial development of the GreenSeeker active sensor to variably apply fertilizer N based on optical sensing (Mulla, 2013). Later, Crop Circle, another active crop sensor, was developed (Holland et al., 2004) which used green and NIR bands to estimate N stress and fertilizer N needs of the crop.

Output from both the sensors is expressed as the Normalized Difference Vegetation Index (NDVI), which can be translated into fertilizer N rate using a suitable algorithm. Raun et al. (2001, 2002) used in-season NDVI measurements to estimate the potential yield from the early season crop growth rate and developed the concept of the response index as the ratio of NDVI values in the test crop and in a well-fertilized reference strip. The algorithm, based on potential yield and response index, can estimate fertilizer N requirement based on crop N demand and account for both spatial (field-to-field as well as within-field) and temporal (season-to-season) variability in N supply and crop performance. It is now the commonly used algorithm for fertilizer N management in cereals based on canopy reflectance sensors (Bijay-Singh and Ali, 2020). Crop reflectance sensors were initially developed to manage fertilizer N application rates on-the-go by mounting on tractors. In recent years, pole-mounted and hand-held versions of both GreenSeeker and Crop Circle have also become available to aid in-season fertilizer N decisions.

Crop N demand-based fertilizer N management using canopy reflectance sensors has shown promising results with cereal crops in many studies (Fox and Walthall, 2008; Hatfield et al., 2008; Scotford and Miller, 2005). Traditionally, sensors as tools for need-based fertilizer N application have been tested on wheat (Thomason et al., 2011; Basso et al., 2016; Samborski et al., 2016; Bijay-Singh et al., 2017; Bronson et al., 2017) and maize (Dellinger et al., 2008; Solari et al., 2008; Kitchen et al., 2010; Schmidt et al., 2011). In recent years, several studies have focused on vegetable crops (Padilla et al., 2018). However, in transplanted rice, the exposed water background interferes with reflectance measurements. When interference was reduced to a negligible level, due to the extent of canopy development around the panicle initiation stage of a rice crop, several sensor-based evaluations of fertilizer N management showed promising results in rice (Nguyen et al., 2008; Xue and Yang, 2008; Bajwa et al. 2010; Tubaña et al., 2012; Yao et al., 2012; Ali et al., 2014; Xue et al., 2014; Bijay-Singh et al., 2015; Rehman et al., 2019, 2022) but only for applying N after the panicle initiation stage of the crop. Due to the availability of pole-mounted and hand-held versions of the sensors, several studies on sensor-based N management for smallholder farms have become available from developing countries such as India and China (Bijay-Singh and Ali, 2020).

5 Innovative nitrogen management strategies: site-specific nutrient management using split applications

As farmers' N fertilizer practices frequently do not synchronize N supply with N demand, the crop yield and FNUE of farmers can be increased by adjusting the times and amounts of split N applications even without the use of tools for estimating crop N status. The use of split N applications better synchronized with crop N demand can be particularly attractive to smallholders with sufficient labour for multiple fertilizer applications. A critical component of site-specific nutrient management (SSNM), as developed for rice by the International Rice Research Institute (IRRI) and then evolved to impact nutrient management for numerous crops, is fixed timing for split doses of fertilizer N that improves N synchronization even without adjustments for leaf N status.

Starting in the mid-1990s, the IRRI developed, evaluated and disseminated an SSNM approach for rice that included split fertilizer N applications better matching crop requirements (Dobermann et al., 2002, 2004; Buresh et al.,

2019). From 2005, the SSNM principles arising from a decade of research on rice were used to develop improved N management in maize and wheat (Buresh, 2010). The SSNM approach evolved from the QUEFTS (Quantitative Evaluation of the Fertility of Tropical Soils) model (Janssen et al., 1990). SSNM used the direct relationship between crop yield and plant N uptake at maturity to determine total plant N required for a target yield (Witt et al., 1999). Plant N uptake in N omission plots was used to determine the indigenous supply of N from sources other than fertilizer, and fertilizer N rate for a target yield was calculated using an estimated increase in plant N from applied N (i.e. the difference between total N uptake for the target yield and indigenous N supply) and an estimated FNUE (Witt and Dobermann, 2004). The calculation of total fertilizer N rate was subsequently simplified using estimated yield gain from applied N (i.e. the difference between target yield and yield without applied N) and a target agronomic efficiency for fertilizer N (AE_N) (Buresh and Witt, 2007). The total fertilizer N for the cropping season was then distributed at critical crop growth stages to match the needs of the crop for fertilizer N. N management based on crop need, as refined and promoted within the SSNM approach for rice before 2010, included both dynamic in-season N adjustment using the LCC or chlorophyll metre and static split applications using fixed times with fixed proportions of the total fertilizer N (Dobermann et al., 2004; Witt et al., 2007). Key principles that emerged for rice were to: (1) apply only a moderate amount of fertilizer N to young rice within 14 days after transplanting or 21 days after sowing, when plant growth and need for supplemental N are minimal and (2) manage fertilizer N to ensure sufficient N supply at the critical growth stages of mid-tillering and panicle initiation.

SSNM research in China revealed that rice farmers over-applied fertilizer N during the early vegetative phase (Wang et al., 2001; Hu et al., 2007) resulting in luxuriant tillering and biomass production, which increased the likelihood of yield loss due to lodging, diseases and insect pests (Zhong et al., 2010). On-farm trials with 107 farmers across six provinces and multiple years found a significant 5% yield increase and a near doubling of AE_N with an average reduction of 62 kg N ha^{-1} mostly during the early vegetative phase (Peng et al., 2010). In 2007, these findings helped launch the 'Three Controls Technology' (3CT) in Guangdong Province, which increased rice yields by 5–10% through more synchronous application of approximately 20% less fertilizer N than used by farmers (Zhong et al., 2010). In 3CT, a fertilizer N rate was determined before the growing season based on expected crop response to N and then fixed proportions of the total N were distributed at fixed growth stages using 40% as basal, 20% at mid-tillering, 30% at panicle initiation and 10% at heading. The increase in yield was attributed to reduced production of unproductive tillers that, in turn, reduced crop susceptibility to lodging and produced a more open crop canopy, which was less susceptible to diseases and insect pests thereby reducing the need for fungicides and insecticides. The term 'three controls' helped highlight to farmers the ability of the technology to control N fertilizer use, unproductive tillers and diseases and insects. 3CT has been widely adopted by rice farmers in southern China (Wang et al., 2017; Wehmeyer et al., 2020). In 2019, the adoption of 3CT covered 2.5 million hectares planted to rice across four provinces, and the increase in farmers' yield with 3CT averaged 0.6 t ha $^{-1}$ (Xuhua Zhong, personal communication, February 2023).

With the advancement of information technology in recent years, several web- and mobile phone-based decision support tools have been developed to disseminate SSNM-based recommendations customized for a farmer's specific crop-growing conditions. These tools use site and crop information available from a farmer or local extension worker to calculate, before a cropping season, a location-specific recommendation for N, P and K fertilizer management, which includes rates and times for split applications of N fertilizer. In 2008, IRRI in collaboration with national partners developed Nutrient Manager, a web-based decision support tool (Buresh et al., 2012), which was upgraded in 2013 to Rice Crop Manager (RCM) (Buresh et al., 2019). RCM synchronizes N supply with demand without the need for estimating leaf N status. Fertilizer N is split into three fixed-time applications using principles outlined by Witt et al. (2007) that ensure only 25–33% of total N is applied within 2 weeks of crop establishment, the remaining N is divided between tillering and panicle initiation, and the N applied at tillering is less than the N applied at panicle initiation (Buresh et al., 2019). The occurrence of tillering and panicle initiation depend on growth duration of a rice cultivar and age of seedlings for transplanted rice. Therefore, RCM uses farmer-provided information on their crop establishment practice (i.e. transplanting, wet seeding or dry seeding), rice cultivar and seedling age for transplanted rice to provide farmers with 5-day windows, expressed as the number of days after crop establishment, for N application at tillering and panicle initiation (Buresh et al., 2019). Banayo et al. (2018a,b) showed that RCM reduced farmers' N use in the first 15 days after transplanting and increased farmers' N use at panicle initiation, thereby contributing to more spikelets per m^2 and higher yield. Country-specific versions of RCM have been released across Asia. In the Philippines, 2.66 million RCM recommendations were provided from 2013 to early 2021; and in India, 250 000 recommendations were provided from 2017 to early 2021 (Chivenge et al., 2021). RCM in the Philippines was incorporated into a digital rice advisory service developed through a partnership of IRRI with national organizations, and from 2022 this advisory service including RCM has fully transitioned to the national agricultural system in the Philippines (https://rcm.da.gov.ph/home).

Nutrient Expert is another decision support tool for maize, wheat, rice and other crops that calculates an N, P and K fertilizer recommendation using an SSNM approach and uses split N applications to synchronize N supply with crop demand (Kumar et al., 2012; Pampolino et al., 2012; Xu et al., 2017). Nutrient Expert for maize recommends two or three splits for N fertilizer (Pampolino et al., 2012). Nutrient Expert for rice recommends three N splits that distribute fertilizer N using fixed times and proportions with 30% as basal, 35% at tillering and 35% at panicle initiation (Xu et al., 2017). Nutrient Expert for wheat includes a basal N application and then the option for farmers to select additional N applications that fit into their irrigation scheduling or growing conditions (Kumar et al., 2012; Zhang et al., 2018). Numerous studies have shown higher grain yield and higher PFP_N for maize, wheat and rice with SSNM-based nutrient management – recommended with tools such as Nutrient Expert and RCM – than with FFP (Chivenge et al., 2021). The reported benefits from SSNM relative to FFP arise from differences in P and K management and N rate as well as differences in the distribution of N during crop growth. Some of the reported benefits with SSNM could arise from improved N synchronization associated with changes in the distribution of N during crop growth. However, evaluations of SSNM seldom enable the quantification of benefits arising solely from improved N synchronization.

6 Innovative nitrogen management strategies: management zones

To improve congruence of fertilizer N supply with crop N demand, the concept of delineating the spatial variability in a field into management zones to variably apply N and improve FNUE has been advocated (Franzen et al., 2002; Ferguson et al., 2003). It is a soil-based approach in which management zones are identified as smaller homogeneous areas occurring within an individual field and stable over time. Due to homogeneous attributes and soil conditions, similar results with respect to yield potential of crops and FNUE are expected in a management zone. As N management is optimized within individual management zones, spatial variability in terms of N synchrony may be taken care of to a large extent but the management zone approach cannot deal with year-to-year temporal variability in N supply and crop N demand. Another limitation of this approach is that management zones are generally delineated based on spatial variability in attributes such as topography, soil colour, electrical conductivity, soil maps, intensive soil sampling and previous-year yield maps, but these only partially determine the crop yield potential and even less consistently characterize spatial variation in crop N requirement (Schepers et al., 2004; Lambert et al., 2006). Thus, the management zone approach alone may not prove adequately efficient to variably apply fertilizer N to crops because weather-mediated variability in crop N demand cannot be accounted for (Shanahan et al., 2008).

7 Innovative nitrogen management strategies: enhanced efficiency fertilizers

Introduced for commercial use in recent decades, EEFs aim to improve synchrony between N supply from the fertilizer and uptake by the crop and, in turn, offer the potential to enhance FNUE. EEFs consist of formulations based on slowing the rate of N release from granules of conventional fertilizers by coating with polymers or delaying N transformations in the soil associated with the activity of urease enzyme and nitrifying microbes. Results of numerous studies conducted with EEFs all over the world vary widely due to differences in cropping systems, climate or soil conditions (Li et al., 2018). Confounding factors make it difficult to determine which EEF product will work efficiently under a given set of soil and climatic conditions.

Researchers have recently used meta-analysis tools (Linquist et al., 2013; Abalos et al., 2014; Qiao et al., 2015; Xia et al., 2017; Li et al., 2018) to assess the integrative efficacy of one or more types of EEFs in different crops and regions. While Abalos et al. (2014) and Qiao et al. (2015) restricted their analysis to nitrification and/or urease inhibitors, Linquist et al. (2013) analysed the work on EEFs only on rice and Xia et al. (2017) based the meta-analysis on publications originating from China. Li et al. (2018) conducted a comprehensive meta-analysis based on studies conducted across the world on all major types of EEFs (polymer-coated fertilizers, nitrification inhibitors, urease inhibitors and both urease and nitrification inhibitors combined) and published during 1980–2016. It revealed that while yield and FNUE responses to polymer-coated fertilizers were 5% and 26%, respectively, in rice, there was little impact on wheat and maize yields. The efficacy of polymer-coated fertilizers was greatly influenced by climate and soil conditions. Polymer-coated fertilizers boosted yield by 4–7% and FNUE by 14–15% under favourable temperature and moisture conditions. The positive effects were more prominent under irrigated cropping systems and in neutral (pH 6 to 8) soils (Li et al., 2018). Urease inhibitors exhibited positive responses in yield and FNUE across all cropping systems, but irrigated crops in warmer regions showed greater yield and N use efficiency. In rice-paddy systems (n = 100), average yield and FNUE improvements were 9% and 29%, respectively. Vegetable crops showed little response to the application of nitrification inhibitors with fertilizer N; wheat and maize responded marginally but the response of rice was modest – average yield increased by 6% and FNUE improved by 11%. Nitrification inhibitors were more effective in irrigated than under rainfed conditions and in soils with a pH above 8 (Li et al., 2018). The combined application of urease and nitrification inhibitors along with fertilizer N (n = 133) increased yield by 11% and FNUE by 33% in grassland ecosystems. No improvement in yield or FNUE was observed in regions receiving less than 800 mm rainfall. Thus, crops grown in irrigated fields were benefitted (yield and

FNUE) more from treating fertilizer N with double inhibitors than those grown under rainfed conditions (Li et al., 2018).

Due to the high cost of EEFs, their adoption by farmers requires evidence of consistent performance even under favourable soil, crop and climatic conditions along with relevant information on economic advantages. Recently, Lam et al. (2022) showed that benefits of EEFs due to avoided social costs of fertilizer N-related degradation of the environment considerably outweigh their costs. However, relying on EEFs alone to significantly enhance N synchrony may not be possible without efficiently improving typical fertilizer N management practices. Yield responses to EEFs may be site specific as indicated from the meta-analyses conducted by Linquist et al. (2013), Abalos et al. (2014) and Thapa et al. (2016). The greatest opportunity to achieve N synchrony between N supply and crop N demand potentially lies in combining EEFs with knowledge-based fertilizer N management practices.

8 Conclusion and future trends

Due to the high N requirement of fast-growing crops in intensive agroecosystems, mismatches can occur between supply of N in the soil and demand of N by the crop. Variability in available soil N across regions and climatic zones is both temporal and spatial, and there is huge potential for asynchrony between N supply and crop N demand. Nitrogen synchrony is further enhanced when N-rich fertilizers are used to supply N to crops but are not managed adequately. Mineral fertilizers can also contribute to N asynchrony via their effect on N transformations in the soil, and crops may experience an excess or a shortage of N. The concept of N synchrony is at the core of improving FNUE in agroecosystems. When crop N needs are closely aligned with N supply in space and time, crop yield levels are high and losses of N to the environment are reduced.

Approaches to improve N synchrony in farmers' fields have been formulated during the last 2–3 decades to improve FNUE, particularly in cereal crops. There is considerable scope to increase yields and FNUE of farmers simply through the use of an SSNM approach to N management that targets an achievable high yield. Such an approach first estimates total N fertilizer requirement using the expected crop response to applied N and a target high N agronomic efficiency achievable with good nutrient management. This approach then distributes the fertilizer N among several split applications based on best-management practices arising from past agronomic research. The experience with *Three Control Technology* in China illustrates the scope for large-scale increases in yield and FNUE in areas with excessive N fertilizer use by simply reducing fertilizer N rates to better match anticipated crop response to applied N and then adjusting the distribution of N in split applications to better match in-season crop N demand.

Since the N status of young leaves and crop biomass reflect N supply to roots in the soil, the monitoring of leaf N status using a chlorophyll metre or leaf colour chart and biomass of crops can also be used to increase N synchrony. With fixed time N management, fertilizer N is applied at fixed critical stages of the crop, but the amount is adjusted upward or downwards based on leaf N status. With real-time N management, fertilizer N doses are applied whenever the leaf N status is below a threshold value. The total amount of applied N is consequently decided in real time. Canopy reflectance sensors have also shown promise to improve N synchrony. These are being used in developed countries as a variable-rate fertilizer technology, but in small farms in developing countries hand-held versions of sensors are becoming popular. However, large-scale adoption by large-scale mechanized farmers in the form of on-the-go fertilizer application is not increasing possibly because the investment needed in equipment, software, data and labour exceed the increased benefits in yields and N saving. On the other hand, a low level of mechanization is proving a barrier for medium and small farmers in developing countries (Lowenberg-DeBoer and Erickson, 2019). Although the use of low-tech tools such as the leaf colour chart is slowly increasing in developing countries, suitable extension strategies are grossly lacking to address the complexity of implementing N synchrony for smallholders.

Improving N synchrony by following management zone approach has limited scope as it cannot consider temporal variations in soil N supply and crop N demand. Similarly, EEFs can help improve N synchrony only under specific soil, crop and climatic conditions, and their high cost can prove a barrier to their widespread use. There is need to explore combining EEFs with knowledge-based site-specific fertilizer management strategies to achieve high yields and high FNUE through improved N synchrony. Controlled-release coatings currently in use in EEFs have been designed based on the predicted crop N demand. These cannot respond to N needs of the crop plants in real time or take into account microbial interactions in the soil–plant system. As reviewed by Lam et al. (2022), new economical EEFs are already being designed using enzyme inhibitors with modifiable chemical structures and biodegradable coatings that respond to plant rhizosphere signalling molecules.

Despite proven benefits of improved N management in terms of increasing productivity and rational use of fertilizer N, the adoption rate by farmers is lower than expected (Cassman and Dobermann, 2022). There is need for a paradigm shift in optimization of site-specific management because it is possibly more complex for modern high-dimensional cropping systems. Saikai et al. (2020) have shown that machine learning and artificial intelligence approaches can be applied in developing optimized and tailored site-specific management solutions. Because most smallholder farmers do not have access to knowledge-based agricultural solutions, basic smartphone technology can be effectively used to improve in-person agricultural extension efforts in spreading improved N management practices provided by the smartphone apps and ensuring messages are scientifically sound. According to Fabregas et al. (2019), the rapid spread of GPS-enabled smartphones to smallholder farmers would not only help in spreading actionable advice but also utilize feedback mechanisms for testing and improvement of N management technologies.

Smartphone- and web-based decision tools, such as RCM and Nutrient Expert, enable rapid calculation and deployment of customized, science-based guidelines that enable smallholder farmers to improve synchronization of N fertilizer. However, these tools do not reduce the need for extension and crop advisory services. Rather, the effective use of such tools necessitates increased timeliness, technical expertise and professionalism in extension and crop advisory services to help ensure knowledge-intensive information is received, understood and implemented by farmers in a timely and appropriate fashion. An assessment of RCM dissemination has indicated that benefits to farmers have been constrained by inabilities of extension services to ensure timely delivery to farmers of time-sensitive crop management information such as the dates and amounts for each split application of N fertilizer. Timely delivery of information can be particularly problematic for locations without reliable internet access and for remote locations lacking consistent contact with extension professionals or crop advisors; however, phone messaging services can be used to deploy customized reminders that farmers receive just-in-time before each recommended N application.

Integration of smartphone- and web-based decision tools for site-specific N management into the digital agricultural platforms of national agricultural systems can help ensure their sustainable financial support and operation. Examples include the existing integration in the Philippines of the RCM decision tool into a much larger RCM advisory service that is maintained and operated by the Department of Agriculture (https://rcm.da.gov.ph/home), and in Indonesia the country-specific version of RCM, known locally by the acronym LKP, is part of an agricultural information system (Santoso et al., 2023). Digital crop advisory services could provide a platform to interface digital decision tools that provide customized N management guidelines before a cropping season with digital systems that compile results from in-season monitoring to estimate crop N status. When proximal or remote sensing indicates appreciable deviations from anticipated crop N status, a fixed-time N dose recommended before the cropping season could be correspondingly adjusted upward or downward. Phone messaging services embedded within the crop advisory service could be used to inform farmers in a timely manner of an adjusted N dose.

9 Where to look for further information

Bijay-Singh et al. (2020) and Bijay-Singh (2022) provide detailed information on site-specific N management. Buresh et al. (2019) explains how the timing of fertilizer N for rice can be adjusted for crop establishment method, growth duration of a rice cultivar and seedling age for transplanted rice. Cassman and Dobermann (2022) and Udvardi et al. (2021) critically analyse bottlenecks and lack in adoption by farmers of more synchronous N management.

Further reading about the contribution of canopy reflectance sensors to improve N synchrony and fertilizer N use efficiency is provided by Padilla et al. (2018) and Mulla (2013) for developed countries and Bijay-Singh and Ali (2020) for smallholder farmers in developing countries.

The latest and comprehensive information about the effectiveness of EEFs in improving N use efficiency through increased matching between N release from fertilizer and crop N uptake can be obtained from the meta-analyses carried out by Li et al. (2018) and Xia et al. (2017) and a recent review by Lam et al. (2022).

10 References

Abalos, D., Jeffery, S., Sanz-Cobena, A., Guardia, G. and Vallejo, A. (2014). Meta-analysis of the effect of urease and nitrification inhibitors on crop productivity and nitrogen use efficiency, *Agric. Ecosyst. Environ.* 189, 136–144. https://doi.org/10.1016/j.agee.2014.03.036.

Ali, A. M., Thind, H. S., Sharma, S. and Varinderpal-Singh (2014). Prediction of dry direct-seeded rice yields using chlorophyll meter, leaf color chart and GreenSeeker optical sensor in northwestern India, *Field Crops Res.* 161, 11–15. https://doi.org/10.1016/j.fcr.2014.03.001.

Ali, A. M., Thind, H. S., Sharma, S. and Yadvinder-Singh (2015). Site-specific nitrogen management in dry direct-seeded rice using chlorophyll meter and leaf colour chart, *Pedosphere* 25(1), 72–81. https://doi.org/10.1016/S1002-0160(14)60077-1.

Aranguren, M., Castellón, A. and Aizpurua, A. (2019). Crop sensor-based in-season nitrogen management of wheat with manure application, *Remote Sens.* 11(9), 1094. https://doi.org/10.3390/rs11091094.

Bailey, K. J. and Leegood, R. C. (2016). Nitrogen recycling from the xylem in rice leaves: dependence upon metabolism and associated changes in xylem hydraulics, *J. Exp. Bot.* 67(9), 2901–2911. http://doi.org/10.1093/jxb/erw132.

Bajwa, S. G., Mishra, A. R. and Norman, R. J. (2010). Canopy reflectance response to plant nitrogen accumulation in rice, *Precis. Agric.* 11(5), 488–506. https://doi.org/10.1007/s11119-009-9142-0.

Balasubramanian, V., Morales, A. C., Cruz, R. T. and Abdulrachman, S. (1999). On-farm adaptation of knowledge-intensive nitrogen management technologies for rice systems, *Nutr. Cycl. Agroecosyst.* 53(1), 59–69. https://doi.org/10.1023/A:1009744605920.

Balasubramanian, V., Morales, A. C., Thiyagarajan, T. M., Nagarajan, R., Babu, M., Abdulrachman, S. and Hai, L. H. (2000). Adoption of the chlorophyll meter (SPAD) technology for real-time N management in rice: a review, *Int. Rice Res. Newslett* 25(1), 4–8.

Banayo, N. P. M. C., Bueno, C. S., Haefele, S. M., Desamero, N. V. and Kato, Y. (2018a). Site-specific nutrient management enhances sink size, a major yield constraint in rainfed lowland rice, *Field Crops Res.* 224, 76–79. https://doi.org/10.1016/j.fcr.2018.05.006.

Banayo, N. P. M. C., Haefele, S. M., Desamero, N. V. and Kato, Y. (2018b). On-farm assessment of site-specific nutrient management for rainfed lowland rice in the Philippines, *Field Crops Res.* 220, 88–96. https://doi.org/10.1016/j.fcr.2017.09.011.

Basso, B., Fiorentino, C., Cammarano, D. and Schulthess, U. (2016). Variable rate nitrogen fertilizer response in wheat using remote sensing, *Precis. Agric.* 17(2), 168–182. https://doi.org/10.1007/s11119-015-9414-9.

Basso, B., Shuai, G., Zhang, J. and Robertson, G. P. (2019). Yield stability analysis reveals sources of large-scale nitrogen loss from the U.S. midwest, *Sci. Rep.* 9(1), 5774. https://doi.org/10.1038/s41598-019-42271-1.

Bianculli, M. L., Aguirrezábal, L. A. N., Irujo, G. A. P. and Echarte, M. M. (2016). Contribution of incident solar radiation on leaves and pods to soybean seed weight and composition, *Eur. J. Agron.* 77, 1–9. http://doi.org/10.1016/j.eja.2016.03.002.

Bijay-Singh (2008). Crop demand-driven site-specific nitrogen applications in rice (*Oryza sativa*) and wheat (*Triticum aestivum*): some recent advances, *Indian J. Agron.* 53, 157–166.

Bijay-Singh (2022). Leaf colour chart: an incredible tool for field-specific management of fertilizer nitrogen in cereals in South Asia, *Agric. Res. J.* 59(4), 583-601. https://doi.org/10.5958/2395-146X.2022.00087.4.

Bijay-Singh and Ali, A. M. (2020). Using hand-held chlorophyll meters and canopy reflectance sensors for fertilizer nitrogen management in cereals in small farms in developing countries, *Sensors* 20(4), 1127. https://doi.org/10.3390/s20041127.

Bijay-Singh and Craswell, E. (2021). Fertilizers and nitrate pollution of surface and ground water: an increasingly pervasive global problem, *SN Appl. Sci.* 3(4), 1–24. https://doi.org/10.1007/s42452-021-04521-8.

Bijay-Singh, Gupta, R. K., Yadvinder-Singh, Gupta, S. K., Singh, J., Bains, J. S. and Vashishta, M. (2006). Need-based nitrogen management using leaf color chart in wet direct seeded rice in north western India, *J. New Seeds* 8(1), 35–47. https://doi.org/10.1300/J153v08n01_03.

Bijay-Singh and Sapkota, T. B. (2023). The effects of adequate and excessive application of mineral fertilizers on the soil. In: Goss, M. J. and Oliver, M. A. (Eds). *Encyclopedia of Soils in the Environment* (2nd edn.), Elsevier Ltd. https://doi.org/10.1016/B978-0-12-822974-3.00051-3.

Bijay-Singh, Varinderpal-Singh and Ali, A. M. (2020). Site-specific fertilizer nitrogen management in cereals in South Asia, *Sust. Agric. Rev.* 39, 137–178. https://doi.org/10.1007/978-3-030-38881-2_6.

Bijay-Singh, Varinderpal-Singh, Purba, J., Sharma, R. K., Jat, M. L., Yadvinder-Singh, Thind, H. S., Gupta, R. K., Chaudhary, O. P., Chandna, P., Khurana, H. S., Kumar, A., Jagmohan-Singh, Uppal, H. S., Uppal, R. K., Vashistha, M. and Gupta, R. (2015). Site-specific nitrogen management in irrigated transplanted rice (*Oryza sativa*) using an optical sensor, *Precis. Agric.* 16, 455–475. https://doi.org/10.1007/s11119-015-9389-6.

Bijay-Singh, Varinderpal-Singh, Yadvinder-Singh, Thind, H. S., Kumar, A., Choudhary, O. P., Gupta, R. K. and Vashistha, M. (2017). Site-specific fertilizer nitrogen management using optical sensor in irrigated wheat in the north-western India, *Agric. Res.* 6(2), 159–168. https://doi.org/10.1007/s40003-017-0251-0.

Bijay-Singh, Varinderpal-Singh, Yadvinder-Singh, Thind, H. S., Kumar, A., Gupta, R. K., Kaul, A. and Vashistha, M. (2012). Fixed-time adjustable dose site-specific fertilizer nitrogen management in transplanted irrigated rice (*Oryza sativa* L.) in South Asia, *Field Crops Res.* 126, 63–69. https://doi.org/10.1016/j.fcr.2011.09.007.

Bijay-Singh, Varinderpal-Singh, Yadvinder-Singh, Thind, H. S., Choudhary, O. P., Gupta, R. K., Choudhary, O. P., Gupta, R. K. and Vashistha, M. (2013). Supplementing fertilizer nitrogen application to irrigated wheat at maximum tillering stage using chlorophyll meter and optical sensor, *Agric. Res.* 2(1), 81–89. https://doi.org/10.1007/s40003-013-0053-y.

Bijay-Singh, Varinderpal-Singh, Yadvinder-Singh, Kumar, A., Sharma, S., Thind, H. S., Choudhary, O. P. and Vashistha, M. (2018). Site-specific fertilizer nitrogen management in irrigated wheat using chlorophyll meter (SPAD meter) in the north-western India, *J. Indian Soc. Soil Sci.* 66(1), 53–65. https://doi.org/10.5958/0974-0228.2018.00006.3.

Bijay-Singh, Yadvinder-Singh, Ladha, J. K., Bronson, K. F., Balasubramanian, V., Singh, J. and Khind, C. S. (2002). Chlorophyll meter and leaf color chart-based nitrogen management for rice and wheat in northwestern India, *Agron. J.* 94(4), 821–829. https://doi.org/10.2134/agronj2002.8210.

Bindraban, P. S., Dimkpa, C. O., White, J. C., Franklin, F. A., Melse-Boonstra, A., Koele, N., Pandey, R., Rodenburg, J., Senthilkumar, K., Demokritou, P. and Schmidt, S. (2020). Safeguarding human and planetary health demands a fertilizer sector transformation, *Plants People Planet* 2(4), 302–309. https://doi.org/10.1002/ppp3.10098.

Bronson, K. F., White, J. W., Conley, M. M., Hunsaker, D. J., Thorp, K. R., French, A. N., Mackey, B. E. and Holland, K. H. (2017). Active optical sensors in irrigated durum wheat: nitrogen and water effects, *Agron. J.* 109(3), 1060–1071. https://doi.org/10.2134/agronj2016.07.0390.

Buresh, R. J. (2007). Fertile progress, *Rice Today* 6(3), 32–33. Available at: https://ricetoday.irri.org/fertile-progress.

Buresh, R. J. (2010). Nutrient best management practices for rice, maize, and wheat in Asia, 19th World Congress of Soil Science, Soil Solutions for a Changing World, 1–6 August 2010, Brisbane, Australia. Published on DVD. Available at: https://iuss.org/19th%20WCSS/Symposium/pdf/0792.pdf.

Buresh, R. J., Castillo, R. L., Torre, J. C. D., Laureles, E. V., Samson, M. I., Sinohin, P. J. and Guerra, M. (2019). Site-specific nutrient management for rice in the Philippines: calculation of field-specific fertilizer requirements by Rice Crop Manager, *Field Crops Res.* 239, 56–70. https://doi.org/10.1016/j.fcr.2019.05.013.

Buresh, R. J., Castillo, R., van den Berg, M. and Gabinete, G. (2012). Nutrient management decision tool for small-scale rice and maize farmers, *Technical Bulletin* 190, Food and Fertilizer Technology Center, Taipei, Taiwan. Available at: https://km.fftc.org.tw/article/1978, Accessed 25 February 2023.

Buresh, R. J. and Witt, C. (2007). Site-specific nutrient management. In: *Fertilizer Best Management Practices*, International Fertilizer Industry Association (IFA), Paris, pp. 47–55. Available at: https://www.fertilizer.org/images/Library_Downloads/2007_IFA_FBMP%20Workshop_Brussels.pdf#page=54.

Cassman, K. G. and Dobermann, A. (2022). Nitrogen and the future of agriculture: 20 years on, *Ambio* 51(1), 17–24. https://doi.org/10.1007/s13280-021-01526-w.

Cassman, K. G., Dobermann, A. and Walters, D. T. (2002). Agroecosystems, nitrogen-use efficiency, and nitrogen management, *Ambio* 31(2), 132–140. https://doi.org/10.1579/0044-7447-31.2.132.

Chivenge, P., Saito, K., Bunquin, M. A., Sharma, S. and Dobermann, A. (2021). Co-benefits of nutrient management tailored to smallholder agriculture, *Glob. Food Secur.* 30, 100570. https://doi.org/10.1016/j.gfs.2021.100570.

Crews, T. E. and Peoples, M. B. (2005). Can the synchrony of nitrogen supply and crop demand be improved in legume and fertilizer-based agroecosystems? A review, *Nutr. Cycl. Agroecosyst.* 72(2), 101–120. https://doi.org/10.1007/s10705-004-6480-1.

Cui, Z. L., Chen, X. P., Zhang, F. S., Miao, Y. X. and Li, J. L. (2008a). On-farm evaluation of the improved soil Nmin-based nitrogen management for summer wheat in North China Plain, *Agron. J.* 100, 517–525. https://doi.org/10.2134/agronj2007.0194.

Cui, Z. L., Miao, Y. X., Zhang, F. S. and Chen, X. P. (2008b). Soil nitrate-N levels required for high yield maize production in the North China Plain, *Nutr. Cycl. Agroecosyst.* 82(2), 187–196. https://doi.org/10.1007/s10705-008-9180-4.

Dellinger, A. E., Schmidt, J. P. and Beegle, D. B. (2008). Developing nitrogen fertilizer recommendations for corn using an active sensor, *Agron. J.* 100(6), 1546–1552. https://doi.org/10.2134/agronj2007.0386.

Dobermann, A., Witt, C. and Dawe, D. (Eds). (2004). *Increasing Productivity of Intensive Rice Systems through Site-Specific Nutrient Management*, Science Publishers, Inc., and International Rice Research Institute, Enfield/Los Baños.

Dobermann, A., Witt, C., Dawe, D., Abdulrachman, S., Gines, H. C., Nagarajan, R., Satawathananont, S., Son, T. T., Tan, P. S., Wang, G. H., Chien, N. V., Thoa, V. T. K., Phung, C. V., Stalin, P., Muthukrishnan, P., Ravi, V., Babu, M., Chatuporn, S., Sookthongsa, J., Sun, Q., Fu, R., Simbahan, G. C. and Adviento, M. A. A. (2002). Site-specific nutrient management for intensive rice cropping systems in Asia, *Field Crops Res.* 74(1), 37–66. https://doi.org/10.1016/S0378-4290(01)00197-6.

Fabregas, R., Kremer, M. and Schilbach, F. (2019). Realizing the potential of digital development: the case of agricultural advice, *Science* 366(6471), 1328. https://doi.org/10.1126/science.aay3038.

Fageria, N. K., Baligar, V. C. and Jones, C. A. (2011). *Growth and Mineral Nutrition of Field Crops* (3rd edn.), CRC Press, Boca Raton, FL.

Ferguson, R. B., Lark, R. M. and Slater, G. P. (2003). Approaches to management zone definition for use of nitrification inhibitors, *Soil Sci. Soc. Am. J.* 67(3), 937–947. https://doi.org/10.2136/sssaj2003.9370.

Fierer, N., Grandy, A. S., Six, J. and Paul, E. A. (2009). Searching for unifying principles in soil ecology, *Soil Biol. Biochem.* 41(11), 2249–2256. https://doi.org/10.1016/j.soilbio.2009.06.009.

Fox, R. H. and Walthall, C. L. (2008). Crop monitoring technologies to assess nitrogen status. In: Schepers, J. S. and Raun, W. R. (Eds). *Nitrogen in Agricultural Systems*, Agronomy Monograph No. 49, American Society of Agronomy, Crop Science Society of America, Soil Science Society of America, Madison, WI, pp. 647–674.

Franzen, D. W., Hopkins, D. H., Sweeney, M. D., Ulmer, M. K. and Halvorson, A. D. (2002). Evaluation of soil survey scale for zone development of site-specific nitrogen management, *Agron. J.* 94(2), 381–389. https://doi.org/10.2134/agronj2002.3810.

Gardner, J. B. and Drinkwater, L. E. (2009). The fate of nitrogen in grain cropping systems: a meta-analysis of ^{15}N field experiments, *Ecol. Appl.* 19(8), 2167–2184. https://doi.org/10.1890/08-1122.1.

Gastal, F. and Lemaire, G. (2002). N uptake and distribution in crops: an agronomical and ecophysiological perspective, *J. Exp. Bot.* 53(370), 789–799. https://doi.org/10.1093/jexbot/53.370.789.

Goffart, J.-P. and Olivier, M. (2004). Management of N-fertilization of the potato crop using total N-advice soft-ware and in-season chlorophyll-meter measurements. In: MacKerron, D. K. L. and Haverkort, A. J. (Eds). *Decision Support Systems in Potato Production*, Wageningen Academic Publishers, Wageningen, pp. 69–83.

Grandy, A., Kallenbach, C., Loecke, T. D., Snapp, S. S. and Smith, R. G. (2012). The biological basis for nitrogen management in agroecosystems. In: Cheeke, T. E., Coleman, D. C. and Wall, D. H. (Eds). *Microbial Ecology in Sustainable Agroecosystems*, CRC Press, Boca Raton, FL, pp. 113–132.

Grindlay, D. J. C. (1997). Towards an explanation of crop nitrogen demand based on the optimization of leaf nitrogen per unit leaf area, *J. Agric. Sci.* 128(4), 377–396. https://doi.org/10.1017/S0021859697004310.

Hatfield, J. L., Gitelson, A. A., Schepers, J. S. and Walthall, C. L. (2008). Application of spectral remote sensing for agronomic decisions, *Agron. J.* 100(S3), S117–S131. https://doi.org/10.2134/agronj2006.0370c.

Holland, K. H., Schepers, J. S., Shanahan, J. F. and Horst, G. L. (2004). Plant canopy sensor with modulated polychromatic light. In: Proceedings of the 7th International Conference on Precision Agriculture, Minneapolis, MN, USA, 25–28 July 2004.

Hong, N., Scharf, P. C., Davis, J. G., Kitchen, N. R. and Sudduth, K. A. (2007). Economically optimal nitrogen rate reduces soil residual nitrate, *J. Environ. Qual.* 36(2), 354–362. https://doi.org/10.2134/jeq2006.0173.

Hu, R., Cao, J., Huang, J., Peng, S., Huang, J., Zhong, X., Zou, Y., Yang, J. and Buresh, R. J. (2007). Farmer participatory testing of standard and modified site-specific nitrogen management for irrigated rice in China, *Agric. Syst.* 94(2), 331–340. https://doi.org/10.1016/j.agsy.2006.10.002.

Hussain, F., Bronson, K. F., Yadvinder-Singh, Bijay-Singh and Peng, S. (2000). Use of chlorophyll meter sufficiency indices for nitrogen management of irrigated rice in Asia, *Agron. J.* 92(5), 875–879. https://doi.org/10.2134/agronj2000.925875x.

Inman, D., Khosla, R., Westfall, D. G. and Reich, R. (2005). Nitrogen uptake across site specific management zones in irrigated corn production systems, *Agron. J.* 97(1), 169–176. https://doi.org/10.2134/agronj2005.0169.

Intaravanne, Y. and Sumriddetchkajorn, S. (2015). Android-based rice leaf color analyzer for estimating the needed amount of nitrogen fertilizer, *Comput. Electron. Agric.* 116, 228–233. https://doi.org/10.1016/j.compag.2015.07.005.

Islam, M. R., Haque, K. S., Akter, N. and Karim, M. A. (2014). Leaf chlorophyll dynamics in wheat based on SPAD meter reading and its relationship with grain yield, *Sci. Agric.* 4(1), 13–18. https://doi.org/10.15192/PSCP.SA.2014.4.1.1318.

Janssen, B. H., Guiking, F. C. T., van der Eijk, D., Smaling, E. M. A., Wolf, J. and van Reuler, H. (1990). A system for quantitative evaluation of the fertility of tropical soils (QUEFTS), *Geoderma* 46(4), 299–318. https://doi.org/10.1016/0016-7061(90)90021-Z.

Justes, E., Mary, B., Meynard, J.-M., Machet, J.-M. and Thelier-Huche, L. (1994). Determination of a critical nitrogen dilution curve for winter wheat crops, *Ann. Bot.* 74(4), 397–407. https://doi.org/10.1006/anbo.1994.1133.

Kitchen, N. R., Sudduth, K. A., Drummond, S. T., Scharf, P. C., Palm, H. L., Roberts, D. F. and Vories, E. D. (2010). Ground-based canopy reflectance sensing for variable-rate nitrogen corn fertilization, *Agron. J.* 102(1), 71–84. https://doi.org/10.2134/agronj2009.0114.

Kumar, A., Majumdar, K., Jat, M. L., Pampolino, M., Kamboj, B. R., Bishnoi, D. K., Kumar, V. and Johnston, A. M. (2012). Evaluation of nutrient expert™ for wheat, *Better Crops, South Asia* 6(1), 27–29.

Kusano, M., Fukushima, A., Redestig, H. and Saito, K. (2011). Metabolomic approaches toward understanding nitrogen metabolism in plants, *J. Exp. Bot.* 62(4), 1439–1453. http://doi.org/10.1093/jxb/erq417.

Ladha, J. K., Jat, M. L., Stirling, C. M., Chakraborty, D., Pradhan, P., Krupnik, T. J., Sapkota, T. B., Pathak, H., Rana, D. S., Tesfaye, K. and Gerard, B. (2020). Achieving the sustainable development goals in agriculture: the crucial role of nitrogen in cereal-based systems, *Adv. Agron.* 163, 39–116. https://doi.org/10.1016/bs.agron.2020.05.006.

Ladha, J. K., Kirk, G. J. D., Bennett, J., Peng, S., Reddy, C. K., Reddy, P. M. and Singh, U. (1998). Opportunities for increased nitrogen-use efficiency from improved lowland rice germplasm, *Field Crops Res.* 56(1–2), 41–71. https://doi.org/10.1016/S0378-4290(97)00123-8.

Ladha, J. K., Pathak, H., Krupnik, T. J., Six, J. and van Kessel, C. (2005). Efficiency of fertilizer nitrogen in cereal production: retrospects and prospects, *Adv. Agron.* 87, 85–156. https://doi.org/10.1016/S0065-2113(05)87003-8.

Ladha, J. K., Peoples, M. B., Reddy, P. M., Biswas, J. C., Bennett, A., Jat, M. L. and Krupnik, T. J. (2022). Biological nitrogen fixation and prospects for ecological intensification in cereal-based cropping systems, *Field Crops Res.* 283, 108541. https://doi.org/10.1016/j.fcr.2022.108541.

Ladha, J. K., Tirol-Padre, A., Reddy, C. K., Cassman, K. G., Verma, S., Powlson, D. S., van Kessel, C., de B Richter, D., Chakraborty, D. and Pathak, H. (2016). Global nitrogen budgets in cereals: a 50-year assessment for maize, rice and wheat production systems, *Sci. Rep.* 6, 19355. https://doi.org/10.1038/srep19355.

Lam, S. K., Wille, U., Hu, H. W., Caruso, F., Mumford, K., Liang, X., Pan, B., Malcolm, B., Roessner, U., Suter, H., Stevens, G., Walker, C., Tang, C., He, J. Z. and Chen, D. (2022). Next-generation enhanced-efficiency fertilizers for sustained food security, *Nat. Food* 3(8), 575–580. https://doi.org/10.1038/s43016-022-00542-7.

Lambert, D. M., Lowenberg-DeBoer, J. and Malzer, G. L. (2006). Economic analysis of spatial-temporal patterns in corn and soybean response to nitrogen and phosphorus, *Agron. J.* 98(1), 43–54. https://doi.org/10.2134/agronj2005.0005.

Li, T., Zhang, W., Yin, J., Chadwick, D., Norse, D., Lu, Y., Liu, X., Chen, X., Zhang, F., Powlson, D. and Dou, Z. (2018). Enhanced efficiency fertilizers are not a panacea for resolving the nitrogen problem, *Glob. Change Biol.* 24(2), e511–e521. https://doi.org/10.1111/gcb.13918.

Linquist, B. A., Liu, L., van Kessel, C. and van Groenigen, K. J. (2013). Enhanced efficiency nitrogen fertilizers for rice systems: meta-analysis of yield and nitrogen uptake, *Field Crops Res.* 154, 246–254. https://doi.org/10.1016/j.fcr.2013.08.014.

Lobell, D. B. (2007). The cost of uncertainty for nitrogen fertilizer management: a sensitivity analysis, *Field Crops Res.* 100(2–3), 210–217. https://doi.org/10.1016/j.fcr.2006.07.007.

Lopez-Bellido, L., Lopez-Bellido, R. J. and Redondo, R. (2005). Nitrogen efficiency in wheat under rainfed Mediterranean conditions as affected by split nitrogen application, *Field Crops Res.* 94(1), 86–97. https://doi.org/10.1016/j.fcr.2004.11.004.

Lowenberg-DeBoer, J. and Erickson, B. (2019). Setting the record straight on precision agriculture adoption, *Agron. J.* 111(4), 1552–1569. https://doi.org/10.2134/agronj2018.12.0779.

Ma, B. L., Morrison, M. J. and Dwyer, L. M. (1996). Canopy light reflectance and field greenness to assess nitrogen fertilization and yield of maize, *Agron. J.* 88(6), 915–920. https://doi.org/10.2134/agronj1996.00021962003600060011x.

Madeira, A. C., Mendonca, A., Ferriera, M. E. and de Taborda, M. L. (2000). Relationship between spectroradiometric and chlorophyll measurements in green beans, *Commun. Soil Sci. Plant Anal.* 31(5–6), 631–643. http://doi.org/10.1080/00103620009370465.

Major, D. J., Baumeister, R., Touré, A. and Zhao, S. (2004). Methods of measuring and characterizing the effects of stresses on leaf and canopy signatures. In: VanToai, T., Major, D., McDonald, M., Schepers, J. and Tarpley, (Eds), *Digital Imaging and Spectral Techniques: Applications to Precision Agriculture and Crop Physiology*, American Society of Agronomy, Minneapolis, MN, pp. 81–93. https://doi.org/10.2134/asaspecpub66.c8.

Makino, A. (2011). Photosynthesis, grain yield, and nitrogen utilization in rice and wheat, *Plant Physiol.* 155(1), 125–129. http://doi.org/10.1104/pp.110.165076.

Meisinger, J. J. and Randall, G. W. (1991). Estimating nitrogen budgets for soil-crop systems. In: Follett, R. F., Keeney, D. R. and Cruse, R. M. (Eds). *Managing Nitrogen for Groundwater Quality and Farm Profitability*, Soil Science Society of America, Madison, WI, 85–124.

Mulla, D. J. (2013). Twenty five years of remote sensing in precision agriculture: key advances and remaining knowledge gaps, *Biosyst. Eng.* 114(4), 358–371. https://doi.org/10.1016/j.biosystemseng.2012.08.009.

Nguyen, H. T., Lee, K. G. and Lee, B. W. (2008). Recommendation of nitrogen topdressing rates at panicle initiation stage of rice using canopy reflectance, *J. Crop Sci. Biotech.* 11, 141–150.

Osterholz, W. R., Rinot, O., Liebman, M. and Castellano, M. J. (2017). Can mineralization of soil organic nitrogen meet maize nitrogen demand?, *Plant Soil* 415(1–2), 73–84. https://doi.org/10.1007/s11104-016-3137-1.

Padilla, F. M., Gallardo, M., Peña-Fleitas, M. T., De Souza, R. and Thompson, R. B. (2018). Proximal optical sensors for nitrogen management of vegetable crops: a review, *Sensors* 18(7), 2083. https://doi.org/10.3390/s18072083.

Paleari, L., Movedi, E., Vesely, F. M., Thoelke, W., Tartarini, S., Foi, M., Boschetti, M., Nutini, F. and Confalonieri, R. (2019). Estimating crop nutritional status using smart apps to support nitrogen fertilization: a case study on paddy rice, *Sensors* 19(4), 981. https://doi.org/10.3390/s19040981.

Pampolino, M., Majumdar, K., Jat, M. L., Satyanarayana, T., Kumar, A., Shahi, V. B., Gupta, N. and Singh, V. (2012). Development and evaluation of nutrient expert for wheat in South Asia, *Better Crops* 96(3), 29–31.

Pan, J., Zhao, J., Liu, Y., Huang, N., Tian, K., Shah, F., Liang, K., Zhong, X. and Liu, B. (2019). Optimized nitrogen management enhances lodging resistance of rice and its morpho-anatomical, mechanical, and molecular mechanisms, *Sci. Rep.* 9(1), 20274. https://doi.org/10.1038/s41598-019-56620-7.

Peng, S., Buresh, R. J., Huang, J., Yang, J., Zou, Y., Zhong, X., Wang, G. and Zhang, F. (2006). Strategies for overcoming low agronomic nitrogen use efficiency in irrigated rice systems in China, *Field Crops Res.* 96(1), 37–47. https://doi.org/10.1016/j.fcr.2005.05.004.

Peng, S., Buresh, R. J., Huang, J., Zhong, X., Zou, Y., Yang, J., Wang, G., Liu, Y., Hu, R., Tang, Q., Cui, K., Zhang, F. and Dobermann, A. (2010). Improving nitrogen fertilization in rice by site-specific N management: a review, *Agron. Sustain. Dev.* 30(3), 649-656. https://doi.org/10.1051/agro/2010002.

Peng, S., Cassman, K. G. and Kropff, M. J. (1995). Relationship between leaf photosynthesis and nitrogen content of field-grown rice in the tropics, *Crop Sci.* 35(6), 1627-1630. https://doi.org/10.2135/cropsci1995.0011183X003500060018x.

Peng, S., Garcia, F. V., Laza, R. C., Sanico, A. L., Visperas, R. M. and Cassman, K. G. (1996). Increased N-use efficiency using a chlorophyll meter on high-yielding irrigated rice, *Field Crops Res.* 47(2-3), 243-252. https://doi.org/10.1016/0378-4290(96)00018-4.

Qiao, C., Liu, L., Hu, S., Compton, J. E., Greaver, T. L. and Li, Q. (2015). How inhibiting nitrification affects nitrogen cycle and reduces environmental impacts of anthropogenic nitrogen input, *Glob. Change Biol.* 21(3), 1249-1257. https://doi.org/10.1111/gcb.12802.

Raun, W. R., Solie, J. B., Johnson, G. V., Stone, M. L., Lukina, E. V., Thomason, W. E. and Schepers, J. S. (2001). In-season prediction on potential grain yield in winter wheat using canopy reflectance, *Agron. J.* 93(1), 131-138. https://doi.org/10.2134/agronj2001.931131x.

Raun, W. R., Solie, J. B., Johnson, G. V., Stone, M. L., Mullen, R. W., Freeman, K. W., Thomason, W. E. and Lukina, E. V. (2002). Improving nitrogen use efficiency in cereal grain production with optical sensing and variable rate application, *Agron. J.* 94(4), 815-820. https://doi.org/10.2134/agronj2002.8150.

Raun, W. R., Solie, J. B., Stone, M. L., Zavodny, D. L., Martin, K. L. and Freeman, K. W. (2005). Automated calibration stamp technology for improved in-season nitrogen fertilization, *Agron. J.* 97(1), 338-342. https://doi.org/10.2134/agronj2005.0338a.

Rehman, T. H., Borja Reis, A. F., Akbar, N. and Linquist, B. A. (2019). Use of normalized difference vegetation index to assess N status and predict grain yield in rice, *Agron. J.* 111(6), 2889-2898. https://doi.org/10.2134/agronj2019.03.0217.

Rehman, T. H., Lundy, M. E. and Linquist, B. A. (2022). Comparative sensitivity of vegetation indices measured via proximal and aerial sensors for assessing N status and predicting grain yield in rice cropping systems, *Remote Sens.* 14(12), 2770. https://doi.org/10.3390/rs14122770.

Robertson, G. P. and Vitousek, P. M. (2009). Nitrogen in agriculture: balancing the cost of an essential resource, *Annu. Rev. Environ. Resour.* 34(1), 97-125. https://doi.org/10.1146/annurev.environ.032108.105046.

Saikai, Y., Patel, V. and Mitchell, P. D. (2020). Machine learning for optimizing complex site-specific management, *Comput. Electron. Agric.* 174, 105381. https://doi.org/10.1016/j.compag.2020.105381.

Samborski, S. M., Gozdowski, D., Stępień, M., Walsh, O. S. and Leszczyńska, E. (2016). On-farm evaluation of an active optical sensor performance for variable nitrogen application in winter wheat, *Eur. J. Agron.* 74, 56-67. https://doi.org/10.1016/j.eja.2015.11.020.

Santoso, A. B., Girsang, S. S., Raharjo, B., Pustika, A. B., Hutapea, Y., Kobarsih, M., Suprihatin, A., Manurung, E. D., Siagian, D. R., Hanapi, S., Purba, T., Parhusip, D., Budiarti, S. W., Wanita, Y. P., Hatmi, R. U., Girsang, M. A., Haloho, L., Waluyo, Suparwoto, Yustisia and Sudarmaji (2023). Assessing the challenges and opportunities of agricultural information systems to enhance farmers' capacity and target rice production in Indonesia, *Sustainability* 15(2), 1114. https://doi.org/10.3390/su15021114.

Scharf, P. C., Kitchen, N. R., Sudduth, K. A., Davis, J. G., Hubbard, V. C. and Lory, J. A. (2005). Field-scale variability in optimal nitrogen fertilizer rate for corn, *Agron. J.* 97(2), 452-461. https://doi.org/10.2134/agronj2005.0452.

Schepers, A. R., Shanahan, J. F., Liebig, M. A., Schepers, J. S., Johnson, S. H. and Luchiari Jr., A. (2004). Appropriateness of management zones for characterizing spatial variability of soil properties and irrigated corn yields across years, *Agron. J.* 96(1), 195-203. https://doi.org/10.2134/agronj2004.1950.

Schmidt, J., Beegle, D., Zhu, Q. and Sripada, R. (2011). Improving in-season nitrogen recommendations for maize using an active sensor, *Field Crops Res.* 120(1), 94-101. https://doi.org/10.1016/j.fcr.2010.09.005.

Schmidt, J. P., Hong, N., Dellinger, A., Beegle, D. B. and Lin, H. (2007). Hillslope variability in corn response to nitrogen linked to in-season soil moisture redistribution, *Agron. J.* 99(1), 229-237. https://doi.org/10.2134/agronj2006.0187.

Scotford, I. M. and Miller, P. C. H. (2005). Applications of spectral reflectance techniques in northern European cereal production: a review, *Biosyst. Eng.* 90(3), 235-250. https://doi.org/10.1016/j.biosystemseng.2004.11.010.

Shanahan, J. F., Kitchen, N. R., Raun, W. R. and Schepers, J. S. (2008). Responsive in-season nitrogen management for cereals, *Comput. Electron. Agric.* 61(1), 51-62. https://doi.org/10.1016/j.compag.2007.06.006.

Shukla, A. K., Ladha, J. K., Singh, V. K., Dwivedi, B. S., Balasubramanian, V., Gupta, R. K., Sharma, S. K., Singh, Y., Pathak, H., Pandey, P. S., Padre, A. T. and Yadav, R. L. (2004). Calibrating the leaf color chart for nitrogen management in different genotypes of rice and wheat in a systems perspective, *Agron. J.* 96(6), 1606-1621. https://doi.org/10.2134/agronj2004.1606.

Singh, U., Giller, K. E., Palm, C. A., Ladha, J. K. and Breman, H. (2001). Synchronizing N release from organic residues: opportunities for integrated management of N, *TheScientificWorld* 1(Suppl. 2), 880-886. https://doi.org/10.1100/tsw.2001.361.

Solari, F., Shanahan, J., Ferguson, R., Schepers, J. and Gitelson, A. (2008). Active sensor reflectance measurements of corn nitrogen status and yield potential, *Agron. J.* 100(3), 571-579. https://doi.org/10.2134/agronj2007.0244.

Solie, J. B., Raun, W. R., Whitney, R. W., Stone, M. L. and Ringer, J. D. (1996). Optical sensor based field element size and sensing strategy for nitrogen application, *Trans. ASAE* 39, 1983-1992. https://doi.org/10.13031/2013.27700.

Tang, Q., Peng, S., Buresh, R. J., Zou, Y., Castilla, N. P., Mew, T. W. and Zhong, X. (2007). Rice varietal difference in sheath blight development and its association with yield loss at different levels of N fertilization, *Field Crops Res.* 102(3), 219-227. https://doi.org/10.1016/j.fcr.2007.04.005.

Tao, M., Ma, X., Huang, X., Liu, C., Deng, R., Liang, K. and Qi, L. (2020). Smartphone-based detection of leaf color levels in rice plants, *Comput. Electron. Agric.* 173, 105431. https://doi.org/10.1016/j.compag.2020.105431.

Tao, Q. N., Fang, P., Wu, L. H. and Zhou, W. (1990). Study of leaf color diagnosis of nitrogen nutrition in rice plants, *Soils* 22, 190-193.

Thapa, R., Chatterjee, A., Awale, R., McGranahan, D. A. and Daigh, A. (2016). Effect of enhanced efficiency fertilizers on nitrous oxide emissions and crop yields: a meta-analysis, *Soil Sci. Soc. Am. J.* 80(5), 1121-1134. https://doi.org/10.2136/sssaj2016.06.0179.

Thomason, W. E., Phillips, S. B., Davis, P. H., Warren, J. G., Alley, M. M. and Reiter, M. S. (2011). Variable nitrogen rate determination from plant spectral reflectance in soft red winter wheat, *Precis. Agric.* 12(5), 666–681. https://doi.org/10.1007/s11119-010-9210-5.

Tubaña, B. S., Harrell, D. L., Walker, T., Teboth, J., Lofton, J. and Kanke, Y. (2012). In-season canopy reflectance-based estimation of rice yield response to nitrogen, *Agron. J.* 104(6), 1604–1611. https://doi.org/10.2134/agronj2012.0214.

Udvardi, M., Below, F. E., Castellano, M. J., Eagle, A. J., Giller, K. E., Ladha, J. K., Liu, X., Maaz, T. M., Nova-Franco, B., Raghuram, N., Robertson, G. P., Roy, S., Saha, M., Schmidt, S., Tegeder, M., York, L. M. and Peters, J. W. (2021). A research road map for responsible use of agricultural nitrogen, *Front. Sustain. Food Syst.* 5, 660155. https://doi.org/10.3389/fsufs.2021.660155.

Van Keulen, H., Goudriaan, J. and Seligman, N. G. (1989). Modelling the effects of nitrogen on canopy development and crop growth. In: Russel, G., Marshall, B. and Jarvis, P. G. (Eds). *Plant Canopies, Their Growth, Form and Function*, Cambridge University Press, Cambridge, pp. 83–104.

Varinderpal-Singh, Bijay-Singh, Yadvinder-Singh, Thind, H. S., Buttar, G. S., Kaur, S. and Bhowmik, A. (2017). Site-specific fertilizer nitrogen management for timely sown irrigated wheat (*Triticum aestivum* L. and *Triticum turgidum* L. spp. durum) genotypes, *Nutr. Cycl. Agroecosyst.* 109(1), 1–16. https://doi.org/10.1007/s10705-017-9860-z.

Varinderpal-Singh, Bijay-Singh, Yadvinder-Singh, Thind, H. S. and Gupta, R. K. (2010). Need based nitrogen management using the chlorophyll meter and leaf colour chart in rice and wheat in South Asia: a review, *Nutr. Cycl. Agroecosyst.* 88(3), 361–380. https://doi.org/10.1007/s10705-010-9363-7.

Varinderpal-Singh, Bijay-Singh, Yadvinder-Singh, Thind, H. S., Kumar, A. and Vashistha, M. (2012). Establishment of threshold leaf colour greenness for need-based fertilizer nitrogen management in irrigated wheat (*Triticum aestivum* L.) using leaf colour chart, *Field Crops Res.* 130, 109–119. https://doi.org/10.1016/j.fcr.2012.02.005.

Varinderpal-Singh, Yadvinder-Singh, Bijay-Singh, Baldev-Singh, Gupta, R. K., Jagmohan-Singh, Ladha, J. K. and Balasubramanian, V. (2007). Performance of site-specific nitrogen management for irrigated transplanted rice in northwestern India, *Arch. Agron. Soil Sci.* 53(5), 567–579. https://doi.org/10.1080/03650340701568971.

Varvel, G. E., Schepers, J. S. and Francis, D. D. (1997). Ability for in-season correction of nitrogen deficiency in corn using chlorophyll meters, *Soil Sci. Soc. Am. J.* 61(4), 1233–1239. https://doi.org/10.2136/sssaj1997.03615995006100040032x.

Vesali, F., Omid, M., Kaleita, A. and Mobli, H. (2015). Development of an android app to estimate chlorophyll content of corn leaves based on contact imaging, *Comput. Electron. Agric.* 116, 211–220. https://doi.org/10.1016/j.compag.2015.06.012.

Wang, G. H., Dobermann, A., Witt, C., Sun, Q. Z. and Fu, R. X. (2001). Performance of site-specific nutrient management for irrigated rice in south-east China, *Agron. J.* 93(4), 869–878. https://doi.org/10.2134/agronj2001.934869x.

Wang, G., Zhang, Q. C., Witt, C. and Buresh, R. J. (2007). Opportunities for yield increases and environmental benefits through site-specific nutrient management in rice systems of Zhejiang Province, China, *Agric. Syst.* 94(3), 801–806. https://doi.org/10.1016/j.agsy.2006.11.006.

Wang, H., Hu, R., Chen, X., Zhong, X., Zheng, Z., Huang, N. and Xue, C. (2017). Reduction in nitrogen fertilizer use results in increased rice yields and improved environmental protection, *Int. J. Agric. Sustain.* 15(6), 681–692. https://doi.org/10.1080/14735903.2017.1398627.

Wehmeyer, H., de Guia, A. H. and Connor, M. (2020). Reduction of fertilizer use in South China - impacts and implications on smallholder rice farmers, *Sustainability* 12(6), 2240. https://doi.org/10.3390/su12062240.

Winiwarter, W., Höglund-Isaksson, L., Klimont, Z., Schöpp, W. and Amann, M. (2018). Technical opportunities to reduce global anthropogenic emissions of nitrous oxide, *Environ. Res. Lett.* 13(1), 014011. https://doi.org/10.1088/1748-9326/aa9ec9.

Witt, C., Buresh, R. J., Peng, S., Balasubramanian, V. and Dobermann, A. (2007). Nutrient management. In: Fairhurst, T. H., Witt, C., Buresh, R. J., and Dobermann, A. (Eds). *Rice: A Practical Guide to Nutrient Management*, International Rice Research Institute/International Potash Institute, Los Baños, Philippines/Singapore, pp. 1–45.

Witt, C. and Dobermann, A. (2004). Toward a decision support system for site-specific nutrient management. In: Dobermann, A., Witt, C. and Dawe, D. (Eds). *Increasing Productivity of Intensive Rice Systems through Site-Specific Nutrient Management*, Science Publishers, Enfield, NH and International Rice Research Institute (IRRI), Los Baños, pp. 358–395.

Witt, C., Dobermann, A., Abdulrachman, S., Gines, H. C., Guanghuo, W., Nagarajan, R., Satawatananont, S., Thuc Son, T., Sy Tan, P., Van Tiem, L., Simbahan, G. C. and Olk, D. C. (1999). Internal nutrient efficiencies of irrigated lowland rice in tropical and subtropical Asia, *Field Crops Res.* 63(2), 113–138. https://doi.org/10.1016/S0378-4290(99)00031-3.

Xia, L. L., Lam, S. K., Chen, D. L., Wang, J. Y., Tang, Q. and Yan, X. Y. (2017). Can knowledge-based N management produce more staple grain with lower greenhouse gas emission and reactive nitrogen pollution? A meta-analysis, *Glob. Change Biol.* 23(5), 1917–1925. https://doi.org/10.1111/gcb.13455.

Xu, X., He, P., Yang, F., Ma, J., Pampolino, M. F., Johnston, A. M. and Zhou, W. (2017). Methodology of fertilizer recommendation based on yield response and agronomic efficiency for rice in China, *Field Crops Res.* 206, 33–42. https://doi.org/10.1016/j.fcr.2017.02.011.

Xue, L., Li, G., Qin, X., Yang, L. and Zhang, H. (2014). Topdressing nitrogen recommendation for early rice with an active sensor in South China, *Precis. Agric.* 15(1), 95–110. https://doi.org/10.1007/s11119-013-9326-5.

Xue, L. H. and Yang, L. Z. (2008). Recommendations for nitrogen fertilizer topdressing rates in rice using canopy reflectance spectra, *Biosyst. Eng.* 100(4), 524–534. https://doi.org/10.1016/j.biosystemseng.2008.05.005.

Yang, W. H., Peng, S., Huang, J., Sanico, A. L., Buresh, R. J. and Witt, C. (2003). Using leaf color charts to estimate leaf nitrogen status of rice, *Agron. J.* 95(1), 212–217. https://doi.org/10.2134/agronj2003.212.

Yao, Y., Miao, Y., Huang, S., Gao, L., Ma, X., Zhao, G., Jiang, R., Chen, X., Zhang, F., Yu, K., Gnyp, M. L., Bareth, G., Liu, C., Zhao, L., Yang, W.. and Zhu, H. (2012). Active canopy sensor-based precision N management strategy for rice, *Agron. Sustain. Dev.* 32(4), 925–933. https://doi.org/10.1007/s13593-012-0094-9.

Zhang, J. J., He, P., Xu, X. P., Ding, W. C., Ullah, S., Wang, Y. L., Jia, L. L., Cui, R. Z., Wang, H. T. and Zhou, W. (2018). Nutrient Expert improves nitrogen efficiency and environmental benefits for winter wheat in China, *Agron. J.* 110(2), 696–706. https://doi.org/10.2134/agronj2017.05.0291.

Zhao, D. L., Reddy, K. R., Kakani, V. G. and Reddy, V. R. (2005). Nitrogen deficiency effects on plant growth, leaf photosynthesis, and hyperspectral reflectance properties of sorghum, *Eur. J. Agron.* 22(4), 391–403. http://doi.org/10.1016/j.eja.2004.06.005.

Zhong, X., Peng, S., Huang, N., Buresh, R. J., Tian, K. and Singleton, G. R. (2010). The development and extension of "three controls" technology in Guangdong, China. In: Palis, F. G., Singleton, G. R., Casimero, M. C. and Hardy, B. (Eds). *Research to Impact: Case Studies for Natural Resource Management for Irrigated Rice in Asia*, International Rice Research Institute, Los Baños, pp. 221–232. Available at: http://books.irri.org/getpdf.htm?book=9789712202599.

Zhu, J., Tremblay, N. and Liang, Y. (2012). Comparing SPAD and atLEAF values for chlorophyll assessment in crop species, *Can. J. Soil Sci.* 92(4), 645–648. https://doi.org/10.4141/cjss2011-100.

Chapter 7

Developments in the use of enhanced efficiency nitrogen fertilizers

Shu Kee Lam, Baobao Pan, Xia Liang, Arvin R. Mosier and Deli Chen, The University of Melbourne, Australia

1 Introduction

Humans have substantially accelerated the world's creation of reactive nitrogen (Nr: all species of nitrogen (N) except N_2) and transformed the global N cycle. Accompanying the rising demands for food and energy, on both an absolute and per capita basis, are substantial increases in anthropogenic production of Nr, from 29 Tg N year^{-1} in the 1860s to 267 Tg N year^{-1} in the 2010s (Tian et al., 2022). Among the total anthropogenic Nr inputs to global terrestrial ecosystems, N fertilizer inputs on cropland and pasture increased from 18 Tg N year^{-1} to 105 Tg N year^{-1} from the 1960s to the 2010s. These fertilizers have dramatically increased global agricultural productivity and by now fed around half of humanity (Galloway et al., 2008; Erisman et al., 2013).

Although the essential role of Nr in agricultural production is undeniable, the unintended consequences of the excessive Nr inputs have been seen in present-day environment on all spatial scales: local, regional, national and global. Once Nr has escaped into the natural environment through the pathways of ammonia volatilization (NH_3), nitrous oxide (N_2O) emission, nitrate (NO_3^-) leaching and runoff (Fig. 1), it cascades through atmospheric, terrestrial, aquatic pools and causes multiple effects during its lifetime (Galloway et al., 2003).

http://dx.doi.org/10.19103/AS.2024.0135.11

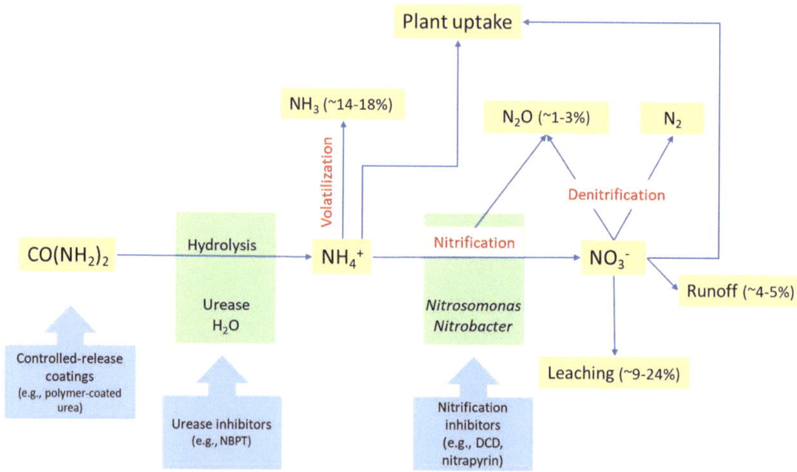

Figure 1 Nr loss pathways and how EEFs reduce risk of Nr losses. Values in the bracket represent the % loss from conventional synthetic N fertilizer input at a global scale (Lin et al., 2001; Bouwman et al., 2002; Davidson 2009; Zhou and Butterbach-Bahl et al., 2014; Pan et al., 2016; Wang et al., 2017; Xia et al., 2018; Wei et al., 2021).

Based on global data synthesis and model estimation, an average of around 14–18% of applied synthetic fertilizer N is lost as NH_3, 1–3% as N_2O, 9–24% via leaching and 4–5% via runoff from global croplands (Fig. 1; Lin et al., 2001; Bouwman et al., 2002; Davidson, 2009; Zhou and Butterbach-Bahl, 2014; Pan et al., 2016; Wang et al., 2017; Xia et al., 2018; Wei et al., 2021). In the atmosphere, Nr contributes to smog formation, stratospheric ozone depletion and enhanced greenhouse effect. In terrestrial ecosystems, Nr leads to soil acidification, forest dieback, and biodiversity loss. In marine and freshwater ecosystems, Nr contributes to groundwater pollution, ocean acidification, and eutrophication, causing algal blooms and biodiversity loss (Sutton et al., 2013). In human society, the high levels of Nr in water and air have been directly and indirectly connected with human ailments, diseases and allergies which influence human health and welfare (Townsend et al., 2003).

The Nr loss from agricultural systems represents a substantial cost to farmers, the agricultural industry and the environment. In the European Union (EU), the Nr-related damage has been estimated at €75 billion to €485 billion per year (van Grinsven et al., 2013). Optimizing Nr use to satisfy the society's growing food demand while minimizing its negative impacts is one of the grand challenges we face in the twenty-first century.

2 Use of enhanced efficiency fertilizers

Reducing Nr losses and improving N use efficiency (NUE) in agriculture can bring about productivity and profitability benefits. The NUE of synthetic and organic fertilizers can be improved by careful planning and implementation of best management practices (BMPs) such as the 4R Nutrient Stewardship strategies - the 'right' rate, source, timing and placement of N fertilizers to match Nr supply with crop requirements and to minimize Nr losses from agriculture (Sutton et al., 2013). For example, deep placement of urea is widely recognized as an effective mitigation measure of NH_3 volatilization (e.g. Pan et al., 2016; Xia et al., 2017).

A better synchrony of fertilizer N supply with crop demand could also be achieved through improved fertilizer technologies, particularly enhanced efficiency fertilizers (EEFs). Efficiency-enhancing fertilizer products include fertilizers stabilized with urease inhibitors (UIs) (Section 2.1), nitrification inhibitors (NIs) (Section 2.2) and controlled-release coatings (Section 2.3). EEFs are designed to increase nutrient availability to crops and reduce nutrient losses to the environment, but they are not the panacea for addressing the challenge of Nr use (Li et al., 2018). Adoption of EEFs must be coupled with implementation of knowledge-based fertilizer management practices (e.g. the 4R's) to maximize the economic, environmental and social benefits of fertilizer technologies.

2.1 Urease inhibitors

Urea is the most widely used form of N fertilizer, accounting for more than half of global N fertilizer use in 2011-20 (IFASTAT, 2023). Once dissolved in water, urea is converted to ammonium bicarbonate through the process of urea hydrolysis within a few days upon application, and is readily lost as NH_3 gas if not incorporated into the soil (Kissel et al., 1988). This process is catalysed by the enzyme urease and involves proton consumption. It increases the soil pH in the surrounding of urea granules (Ernst and Massey, 1960; Terman, 1980) and shifts the equilibrium more towards NH_3 gas from NH_4^+.

Conditions that favour NH_3 volatilization from urea application include surface application, light rainfall (<10 mm) and/or irrigation shortly after application, open crop canopies, high temperature, high soil pH and low cation exchange capacity of soils (White et al., 2002; Pan et al., 2016). NH_3 emission from agricultural sources has contributed significantly to air pollution, soil acidification, water eutrophication, biodiversity loss and damage to human health (Fowler et al., 2013; Galloway et al., 2008).

UIs are chemical compounds coated on or incorporated in urea or urea-containing fertilizers that can reduce the rate of hydrolysis and NH_3 volatilization. Specifically, UIs suppress the activity of enzyme urease over a certain period of time (Fig. 2). Thus, the efficiency of urea and of N fertilizers

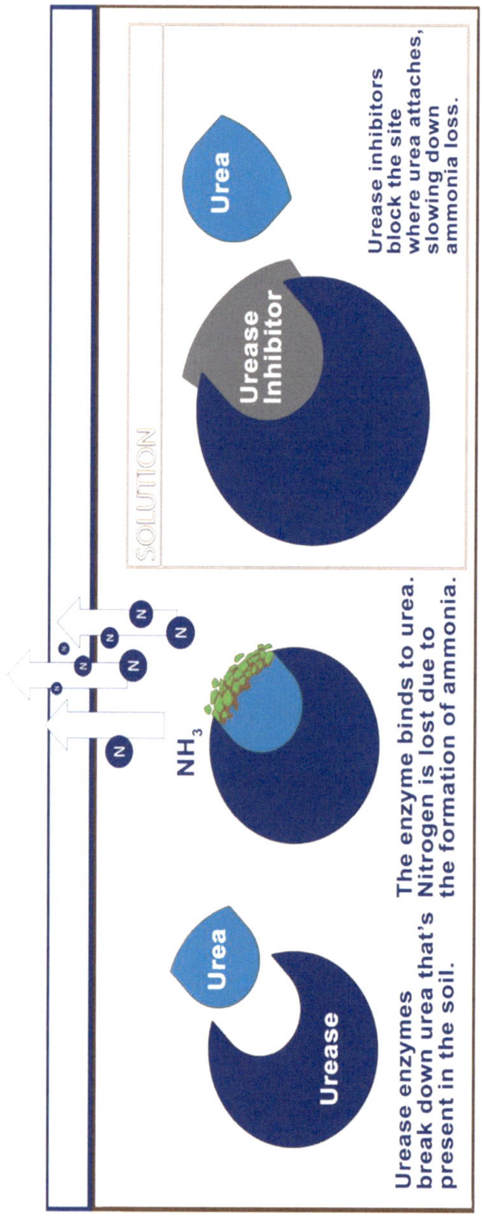

Figure 2 An illustration of how urease inhibitors work. Source: Adapted from: BASF: https://agriculture.basf.com/.

containing urea is increased and adverse environmental impact from fertilizer use is decreased.

Starting in the 1960s, large-scale screening of UIs was carried out by organizations like the National Fertilizer Development Center in Tennessee, USA, and the International Fertilizer Development Center, as well as small-scale efforts at various universities (Kiss and Simihaian, 2002). Metals were long known to inhibit urea hydrolysis and were tested as fertilizer additives; for instance, ethylenediaminetetraacetic acid was used to chelate nickel (Ni^{2+}) metal cofactor in the enzyme active site (Tabatabai, 1977). Shaw (1954) evaluated the action of metals and showed that their potential of urease inhibition followed the order $Ag^{+} \sim Hg^{2+} > Cu^{2+} > Cd^{2+} > Co^{2+} > Ni^{2+} > Zn^{2+} = Sn^{2+} = Mn^{2+} = Pb^{2+}$. These metals inhibit urease by creating a chemical bond with one or more sulphydryl group active sites to produce insoluble sulphites; consequently, the metal with a higher affinity to the enzyme that forms the more insoluble sulphite will be the stronger inhibitor (Shaw, 1954). It is worth noting that the application of heavy metals in soils may cause environmental problems (Briffa et al., 2020).

In the early 1980s, several organic compounds were tested as UIs, e.g. phenyl phosphorodiamdate (PPDA), thiourea (TU) and phosphoryl triamide (PTA) (Mikkelsen and Bock, 1988). In the US, starting in the mid-1990s, hydroquinone and some benzoquinones were known to inhibit urease activity. The one that stood out and most successfully reached the market is *N*-(*n*-butyl) thiophosphoric triamide (NBPT), patented by Allied Chemical Corporation (Chien et al., 2009). Once oxidized in the soil environment (to NBPTO), the triamide competes with urea for the active site of urease, thereby inhibiting its activity. Following the commercial success of NBPT and the large potential market for urea additives, there has been renewed interest in new molecules and formulations of UIs. In the early 2000s, over 14 000 compounds or mixtures of compounds were tested for their effects on soil urease activity, many of which were patented for that purpose (Kiss and Simihaian, 2002; Benini et al., 2004). Although many compounds can inhibit urease, only a handful are non-toxic, effective at low concentrations, chemically stable and compatible with urea-containing fertilizers (Domínguez et al., 2008). Nowadays, the widely used UIs are NBPT, *N*-phenylphosphoric triamide (2-NPT) and cyclohexyl phosphoric triamide (CHPT) (Table 1).

2.2 Nitrification inhibitors

Nitrification is a natural process in soils that converts ammonium to nitrite and then to nitrate (Fig. 1). The first and rate-limiting step of nitrification is ammonia oxidation, which is catalysed by the ammonia monooxygenase (AMO) enzyme. This enzyme is present in ammonia-oxidizing bacteria, ammonia-oxidizing archaea and the recently discovered complete ammonia-oxidizers (comammox) (Daims et al., 2015; van Kessel et al., 2015).

Table 1 EEF types and materials

EEF type	Product name/ abbreviation	Material/compound
UI	Hydroquinone	Hydroquinone
UI	ATS	Ammonium thiosulfate
UI	CTS	Calcium thiosulfate
UI	PPD/PPDA	Phenyl phosphorodiamidate
UI	NBPT	*N*-(*n*-butyl) phosphoric triamide
UI	CHTPT	Cyclohexyl thiophosphoric triamide
UI	CHPT	Cyclohexyl phosphoric triamide
UI	ALS	Ammonium lignosulfonate
UI	NPPT	*N*-(*n*-propyl)-thiophosphoric-triamide
NI	DMPP	3,4-Dimethylpyrazole phosphate
NI	AM	2-Amino-4-chloro-6-methyl pyrimidine
NI	Phenylacetylene	Phenylacetylene
NI	DCD	Dicyandiamide
NI	N-serve/Nitrapyrin	2-Chloro-6-trichloromethyl pyridine
NI	CL-1580	2,4-Diamino-6-trichloromethyl-s-triazine
NI	ATC	4-Amino-l, 2,4-triazole
CRF	PCU	Polymer-coated urea
CRF	PoCU	Polyolefin-coated urea
CRF	SCU	Sulphur-coated urea
CRF	RCU	Resin-coated urea
CRF	Meister	Coated urea with a mixture of polyolefin and silicate mineral

Sources: Fujita and Shoji, 1999; Akiyama et al., 2010; Linquist et al., 2012; Quemada et al., 2013; Decock, 2014; Feng et al., 2016; Gilsanz et al., 2016; Huang et al., 2016; Pan et al., 2016; Qiao et al., 2016; Thapa et al., 2016; Yang et al., 2016; Cai and Akiyama, 2017; Han et al., 2017; Silva et al., 2017; Xia et al., 2017; Li et al., 2018; Ti et al., 2019; Zhang et al., 2019; Gu et al., 2020; Gao et al., 2021; Yang et al., 2021; Yangjin et al., 2021.

Nitrification rate is highest when soils are near field capacity and decreases as moisture content becomes higher or lower. In saturated soils, nitrification nearly stops because of the lack of oxygen (Hu et al., 2015). The rate of nitrification can be controlled by preserving N as NH_4^+, which can be held on soil colloids rather than leached. Nitrate can be leached below the root zone and transferred to surface or sub-surface waters. Under waterlogged conditions, nitrate can be denitrified to N_2O and N_2 by other soil bacteria. N_2 is the most common gas in the atmosphere and is inert, whereas N_2O is a powerful greenhouse gas with a warming potential of 298 times that of CO_2 according to the Intergovernmental Panel on Climate Change (Forster et al., 2007).

NIs are compounds that delay the bacterial oxidation of NH_4^+ to NO_3^- by suppressing the activity of *Nitrosomonas* bacteria in the soil (Trenkel, 2010) (Fig. 3), which slows down the rate of nitrification and decreases associated N_2O emission and NO_3^- leaching. The inhibition of nitrification would reduce NO_3^- availability for denitrification and decrease N_2O and N_2 emissions (Friedl et al., 2017). A recent global meta-analysis indicated that application of NIs could effectively reduce soil denitrification rate and associated N_2 emissions, along with optimal N application rates, the use of ammonium-based fertilizers instead of nitrate-based fertilizers and biochar amendment (Pan et al., 2022).

More than dozens of compounds have been proposed as NIs, either synthetic or biological, but only a few have been evaluated under field conditions for their effectiveness in controlling nitrification in soils, and only several of these have been adopted to some extent worldwide (Subbarao et al., 2006a; Coskun et al., 2017).

As early as in the 1950s, the specific search for nitrification inhibiting substances started, leading to the introduction of nitrapyrin [2-chloro-6 (trichlormethyl) pyridine] to the US market in the 1960s under the trade name N-Serve (Dow Chemical Company, US) (Goring, 1962) and dicyandiamide (DCD) in the European market during a similar period (Zerulla et al., 2001). The fundamental studies conducted since 1960s were pivotal to the rapid progress of EEFs in the succeeding decades (Douglas and Bremner, 1971; Bremner and Douglas, 1971; Hauck, 1980; Bremner, 1985; Bremner, 1986; Bremner and Krogmeier, 1989; McCarty and Bremner, 1989).

The NI 3,4-dimethylpyrazole phosphate (DMPP) was developed by BASF (trademark ENTEC) in 1999. Since then, it has been widely recognized for its low application rate, high efficiency and low toxicity to the environment. As a result, it has been promoted as an advantageous and practical solution for improving fertilizer N management in agriculture (Hofmair, 2000; Zerulla et al., 2001). The commercial products Piadin, a mixture of 1H-1,2,4-triazole and 3-MP as active ingredients (SKW Piesteritz), and Vizura, a mixture of DMPP and 1H-1,2,4-triazole as active ingredients (BASF), were introduced in the early 2000s and 2016, respectively, dominating the German market of synthetic NIs today. Apart from conventional granular urea, NIs can be also applied with urea supergranules, which have been found to be more beneficial to rice growth and yield than prilled urea (Prasad et al., 1984; Malhi and Nyborg, 1985; Sudhakara and Prasad, 1986; Singh and Beauchamp, 1988).

Neem cake was found to possess nitrification inhibition properties (Singh, 2016). A technique for coating urea with neem cake was developed at Indian Agricultural Research Institute (IARI). The neem cake was coated on to urea using a coal-tar kerosene mixture (Prasad et al., 1999). Another technique of coating urea with neem oil micro-emulsion was developed at IARI (Prasad et al.,

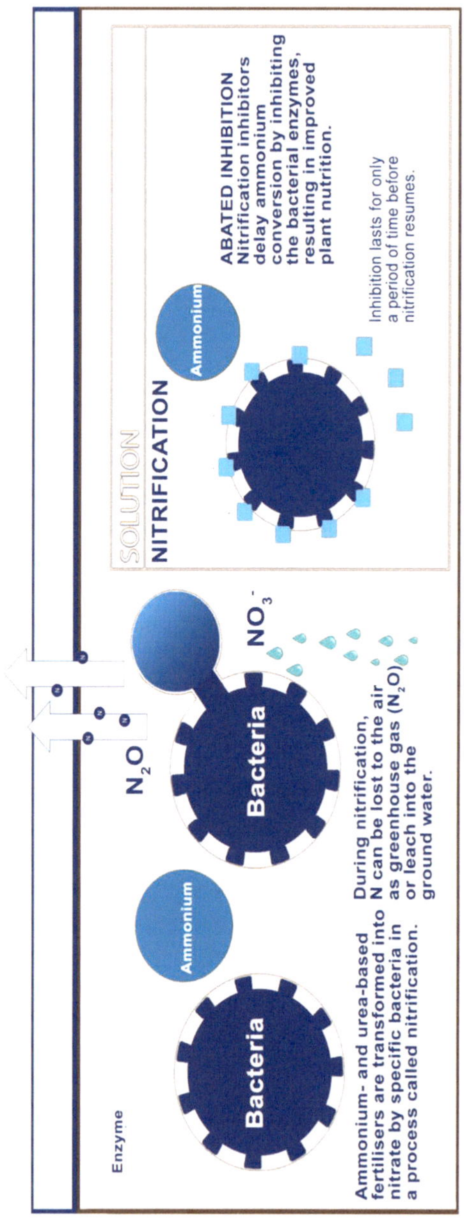

Figure 3 An illustration of how NIs work. Source: Adapted and modified from: BASF, https://agriculture.basf.com/.

1999; Suri et al., 2000). This has been adapted by several companies in India and is sold as neem-coated urea.

With the development of a bioluminescent, recombinant strain of *Nitrosomonas europaea*, new research into NIs from plant root exudates, termed biological nitrification inhibitors (BNIs) has emerged (Subbarao et al., 2006b; 2009; Coskun et al., 2017). Many early studies have presented evidence of BNIs in various ecosystems and tissues of grasses, trees, shrubs, herbs and mosses (Rice and Pancholy, 1974). In cereals, direct isolation of BNIs from root exudates includes methyl 3-(4-hydroxyphenyl) propionate in sorghum (Zakir et al., 2008) and 1,9-decanediol in rice (Sun et al., 2016).

2.3 Controlled-release fertilizer (CRF)

An ideal fertilizer should release nutrients at a rate that matches crop demand throughout its growth stages (Fig. 4). To achieve this, CRFs are encapsulated with a physical barrier or polymeric coating (water-insoluble, semi-permeable or impermeable with pores) which regulates the rate, pattern and duration of the release of N depending on external stimuli such as soil moisture, pH, temperature, biological activity, the composition and thickness of the coating (Fig. 5), thereby reducing Nr losses and increasing NUE (Quemada et al., 2013).

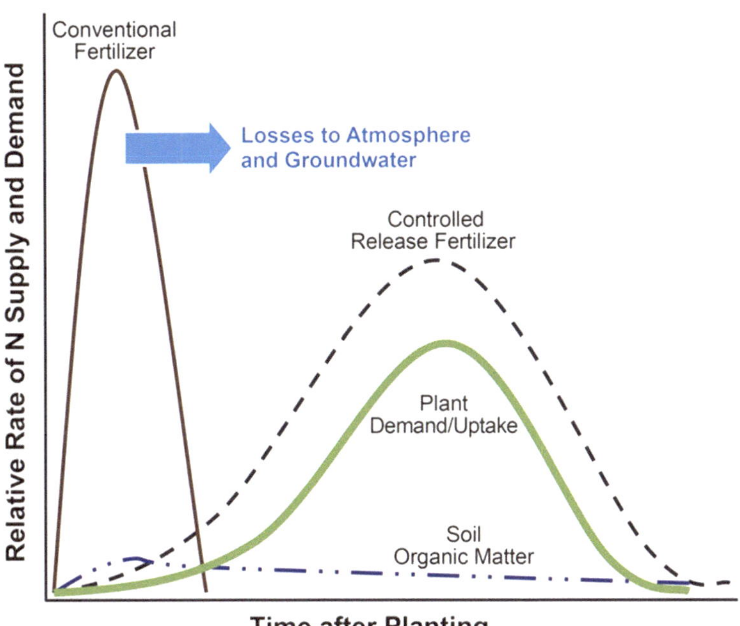

Figure 4 Relative rates of crop demand for N and supply of N from various sources. Source: Adapted from: Chalk et al. (2015).

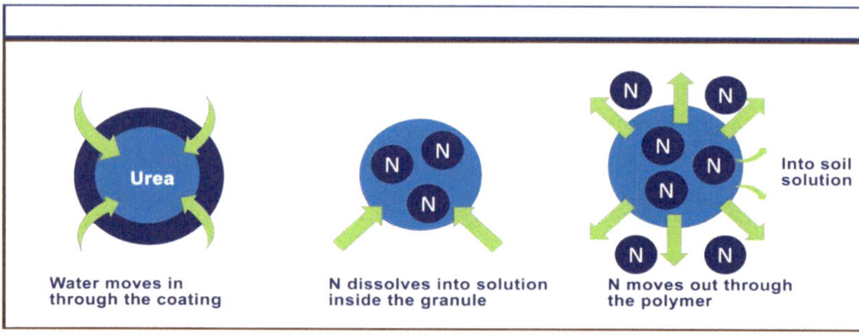

Figure 5 An illustration of how CRFs work. Source: Adapted and modified from: BASF, https://agriculture.basf.com/.

CRFs release nutrients gradually over a period from 20 days to 18 months (Trenkel, 2010). In this regard, CRFs can replace split applications, thus reducing the requirement of multiple field operations and in turn reducing labour and fuel costs (Akiyama et al., 2010). For example, commercial potato producers use three to four applications of N fertilizers for northeast Florida and two applications for southwest Florida. Eliminating extra applications of fertilizer saves the farmer between $12.4 and $17.3 per ha of broadcasting expense (Liu et al., 2014).

Some of the earliest work with sulphur-coated urea (SCU) was conducted by the Tennessee Valley Authority (TVA) in 1961. It was prepared by spraying molten sulphur over granular urea. A wax sealant was then sprayed to seal cracks in the coating, thereby reducing leakage and microbial degradation of the coating (Oertli, 1980; Shaviv, 2000). This is followed by a layer of a conditioner such as attapulgite or talc. The wax sealant may be replaced by an organic polymer layer to produce a polymer-coated SCU.

The concept of encapsulating a soluble fertilizer granule with multiple layers of polymer film to slow or control the availability of nutrients was disclosed in a US patent issued in 1965 to Hansen and assigned to Archer Daniels Midland (Hansen, 1965). The preferred embodiment of this patent involved coatings derived from copolymers of dicyclopentadiene and glyceryl esters of unsaturated fatty acids.

Polyolefin-coated fertilizer (Meister, Chisso and Chisso-Asahi Co., Japan) is processed by adding surfactants to urea and coating the mixture with a thermoplastic resin (Shoji and Gandeza, 1992). Their release rate of urea can be regulated by adjusting the ratio of additives to coating materials. Moreover, the release rate is not influenced by soil factors, except for temperature.

Methylene urea is a type of slow-release fertilizer that can be traced back to 1946 (Yee and Love, 1947). It is created through a manufacturing process where long-chain molecules are created by chemically combining a N source with an aldehyde. Nitrogen release is gradually delayed as these long-chain molecules are decomposed by soil microbes at predictable rates.

However, most conventional coatings made of polymers are not readily biodegradable. This creates a new source of soil pollution as they lead to an undesirable accumulation of plastic residues up to 50 kg/ha/year (Trenkel, 2010). To overcome this environmental obstacle, the research focus is shifting towards biopolymers, which are cheaper than conventional coating materials (Azeem et al., 2020), biodegradable, biocompatible and non-toxic (Cerri et al., 2020). In recent years, different natural organic and inorganic materials have been developed for fertilizers coatings, including latex, clay, gypsum and other coatings. For example, a lac-coated urea was developed by the Indian Lac Research Institute to improve yields in rice-wheat cropping systems (Singh et al., 1999). A new type of chitosan-coated NPK fertilizer, which possessed a three-layer structure, was described by Wu and Liu (2008). Its core was granular water-soluble NPK fertilizers, with the inner coating being chitosan and the outer coating superabsorbent (polyacrylic acid-co-acrylamide) polymer. Gel-based fertilizers work on the principle of absorption of soluble fertilizers within a matrix of polymer gel. They are generally based on polyacrylic acid and its derivates, and are perceived to have an additional function as a water absorbent (Wu and Liu, 2008). In 2022, a commercial biodegradable CRF coating (eqo.x, ICL Specialty Fertilizers) for urea was released, which was designed to meet future European fertilizer regulations.

Several reviews have comprehensively summarized the strengths and weaknesses of a range of natural and synthetic coating materials used in CRFs (Azeem et al., 2014; Dimkpa et al., 2020). Among these materials, polymer-based materials are widely chosen for use as fertilizer coatings because the surface properties, such as the mechanical strength and surface reactivity, are readily modifiable. However, these products come with drawbacks such as high cost, non-biodegradability and complicated synthesis procedures (Trenkel, 2010; Azeem et al., 2014; Dimkpa et al., 2020).

The terms 'controlled-release fertilizer' and 'slow-release fertilizer' are sometimes considered analogous, but they are different (Trenkel, 2010; Shaviv, 2005). The nutrient release pattern of slow-release fertilizer is nearly unpredictable and subject to changes in soil types and climatic conditions. In contrast, the pattern, quantity and time of nutrient release from CRFs can be designed specifically and predicted (Trenkel, 2010). In this chapter, we limit our scope of discussion to CRFs (Table 1).

3 Performance of enhanced efficiency fertilizers

3.1 Effects on biomass production and nitrogen use efficiency

The application of EEFs generally improves crop yields and NUE, based on eight meta-analyses summarized by Kanter and Searchinger (2018). Overall, crop yield can be increased by 7%, 1–9%, 4–10% and 4–6% when applied with

CRF, double inhibitor (UI + NI), UI and NI, respectively (Linquist et al., 2013; Abalos et al., 2014; Feng et al., 2016; Kanter and Searchinger, 2018). NUE can be increased by 8–15%, 5–13% and 12% with the application of double inhibitor (UI + NI), UI and NI, respectively (Linquist et al., 2013; Abalos et al., 2014).

Based on 96 observations, the effect of UIs on yield ranged from −0.8% in sugarcane to +10.2% in wheat (Cantarella et al., 2018). Such variations may be attributed to the difference in the growth cycle of crops and their N response. For example, the yield increase under UI application was higher for cereals and pastures (Cantarella et al., 2018), probably because of their short growth cycles and greater responses to N fertilization (Abalos et al., 2014). On the other hand, excessive N fertilization promotes vegetative growth of cotton and delays maturity, which may reduce yield if deficiency in other nutrients or water occurs at the later growth stages.

The effects of NIs on crop productivity varied among crop types. NIs significantly increase yield and NUE in grassland (14% and 48%, respectively), vegetable systems (5–7% and 48%), irrigated cotton (23% and 46%) (Freney et al., 1993; Chen et al., 1994) and paddy systems (6% and 11%), but not significantly in cereal cropping systems (Li et al., 2018). This may be attributed to the fact that forage and vegetables generally receive higher N applications than cereals, which may increase the capacity of inhibitors in decreasing Nr losses. In addition, the effects of NIs on crop production varied among different soil texture classes (Qiao et al., 2015) and inhibitor types (Yang et al., 2016). For example, Abalos et al. (2014) concluded from a meta-analysis that the overall effect of NIs (DCD) on crop production and NUE in fine-texture soils was lower than in medium or coarse texture soils.

Generally, the application of double inhibitors (UI + NI) significantly increases crop productivity and NUE compared to conventional N fertilizers according to Linquist et al. (2013) and Abalos et al. (2014). However, based on these limited number of studies, the combined and separate use of UI and NI had similar effects on both crop yield and N uptake (Kanter and Searchinger, 2018; Linquist et al., 2013).

Compared to conventional N fertilizers, the effects of CRFs on yield and N uptake are inconsistent (Kanter and Searchinger, 2018). The performance of CRFs on crop yield and NUE is closely related to climatic conditions (e.g. rainfall), soil properties (e.g. soil pH) and crop species (Li et al., 2018). The application of CRFs significantly increased crop yield and NUE in rice-paddy systems (5% and 26%, respectively) (Patil et al., 2010) as well as vegetable production systems (7% and 11%, respectively). However, CRFs had no significant effect on crop yield or NUE of wheat or corn (Li et al., 2018). This may be because dryland systems where wheat and corn are often grown are subject to larger fluctuation of biophysical conditions such as temperature and soil moisture than paddy systems. Such fluctuations might affect N-release patterns from CRFs, and possibly the dynamics of crop growth and N uptake (Cahill et al., 2010; Abalos et al., 2016; Li et al., 2018). A polyolefin-coated fertilizer was shown to have insignificant effects on cotton yield and N uptake in Australia (Chen et al., 2008).

3.2 Effects on reactive nitrogen losses

Lam et al. (2022) summarized 21 meta-analyses on the potential of EEFs to reduce Nr losses from food production systems at both regional and global scales and found that compared to conventional fertilizers, UIs decreased NH_3 emissions by 50–74% (range of the mean reduction). The application of NIs decreased N_2O emissions, Nr leaching and runoff by 27–58%, 17–58% and 45%, respectively. When formulations combining both UIs and NIs were used (double inhibitors), emissions of NH_3 and N_2O decreased by 36–75% and 30–50%, respectively. CRFs can effectively decreased NH_3 emissions, N_2O emissions, N leaching and runoff by 27–83%, 8–77%, 17–92% and 32%, respectively.

UIs inhibit urease activity and slow down urea hydrolysis and subsequently NH_3 volatilization from various agroecosystems (Sommer et al., 2004; Turner et al., 2010; Suter et al., 2013; Pan et al., 2016; Lam et al., 2018a, 2019). For example, compared to urea and S-coated urea, NH_3 emission was reduced by 86% and 37% in 17 days of incubation in soil, respectively (Singh, 2010). Another meta-analysis on 824 observations (Pan et al., 2016) indicated that CHPT resulted in the highest reduction (87%) in NH_3 volatilization, followed by NBPT and phenylphosphorodiamide (PPD), both reducing NH_3 volatilization by about 50%, compared to about 15–30% reduction by ammonium thiosulfate (ATS), PPDA and calcium thiosulfate (CTS). UIs such as ammonium thiosulfate and calcium thiosulfate can inactivate the sulphydryl groups of urease (Saggar et al., 2013) to reduce NH_3 emissions, but their effectiveness depends on edaphic and environmental conditions (Lam et al., 2022).

UIs may reduce N_2O emissions by decreasing the availability of an NH_4^+ substrate for nitrification (Thapa et al., 2016), but their performance varies with crop type, soil texture and management practice. For example, UIs were more effective in decreasing N_2O emissions in corn systems than in wheat systems (36% and 12%, respectively) (Xia et al., 2017). Similarly, N_2O emissions were reduced by an average of 28% with UI applications in coarse-textured soils, by 19% when fertilizers were applied in multiple split doses, and by 32% under field irrigated conditions. However, these results were based on limited studies (Thapa et al., 2016). Previous meta-analyses show that the overall effects of UIs on N_2O emissions were inconsistent (Akiyama et al., 2010; Thapa et al., 2016; Xia et al., 2017). This was probably because hydrolysis of urea does not directly affect N_2O emissions and the substrate for nitrification is affected by plant N uptake and soil N immobilization.

NIs have been widely studied for mitigating N_2O emissions from agricultural soils. On average, NIs significantly decreased N_2O emissions by 31–48% compared with fertilizers (Akiyama et al., 2010; Qiao et al., 2015; Thapa et al., 2016). The effects of NIs on N_2O emissions varied with land-use types and cropping systems; for example, NIs was found the most effective in grassland (−54%) followed by upland fields (−34%) and paddy fields (−30%) (Akiyama et al., 2010). DCD, DMPP and nitrapyrin have been successfully marketed as NIs (Frye, 2005; Subbarao et al., 2006a). Nitrapyrin has been used extensively in North America, where it is injected directly into the soil along with anhydrous NH_3 to decrease N_2O emission. DMPP is reported to be highly specific in inhibiting nitrification and is effective at a lower application rate than other NIs (Dittert et al., 2001). In addition, DMPP has about the same relative mobility as NH_4^+, enhancing its ability to inhibit nitrification effectively (Pasda et al., 2001). Also, NIs differ in water solubility and volatility, which influence their effectiveness (Zerulla et al., 2001; Cantarella et al., 2018). It should be noted that environmental factors alter the effectiveness of NIs (Lam et al., 2022).

However, the use of NIs prolongs the retention of NH_4^+ in soil, which could potentially increase NH_3 emission, the redeposition of which would subsequently result in indirect N_2O emission (Kim et al., 2012; Soares et al., 2012; Zaman and Nguyen, 2012). As reviewed by Lam et al. (2017), NIs decreased direct N_2O emission by 8–57%, but generally increased NH_3 emission by 3–65%. This demonstrated that the beneficial effect of NIs in decreasing direct N_2O emission can be undermined or even outweighed by an increase in NH_3 volatilization (Lam et al., 2017, 2018b). This 'double-edged sword' effect of using NIs suggested that appropriate NH_3 mitigation strategies should be taken where NIs are being used to decrease direct N_2O emissions, to enable effective and viable climate change mitigation.

Pan et al. (2016) conducted a meta-analysis to comprehensively analyse the effects of CRFs on NH_3 volatilization. The analysis revealed that CRFs reduced NH_3 volatilization by 68% on average. CRFs such as thermoplastic resin-coated urea, sulphur-coated urea and polyfin-coated urea significantly reduced NH_3 volatilization by 83%, 78% and 69%, respectively (Pan et al., 2016). The effects of CRFs on NH_3 volatilization vary with cropping systems. For example, CRF appeared the most effective in paddy systems, reducing NH_3 losses by 83% on average (Li et al., 2018). Polymer-coated urea fertilizers reduced NH_3 volatilization by 80% in a Tenosol (Arenosol, FAO classification) soil grown with perennial ryegrass (Lam et al., 2019). Sulphur-coated fertilizers can also reduce NH_3 volatilization by 61–69% in a yellow-brown soil planted with cabbage (Shan et al., 2015). Overall, CRFs significantly decreased N_2O emissions by 14–58%, compared to conventional fertilizers (Akiyama et al., 2010). However, the effects of CRFs on N_2O mitigation vary greatly with land-use and soil types (Lam et al., 2022). Feng et al. (2016) reported that N_2O emission was significantly decreased by 18% and 44% by CRFs in medium- and fine-textured soils, respectively, but was unaffected in a coarse-textured soil. This may be attributed to a higher cation exchange capacity, and subsequently NH_4^+ adsorption in a clayey soil, compared with a sandy soil (Jarecki et al., 2008).

4 Development of new generation enhanced efficiency fertilizers

4.1 Insights from meta-analyses worldwide

Extensive research has been conducted to assess the potential of EEFs in decreasing Nr losses through various pathways. The results of these research efforts have been quantitatively synthesized in 21 national and global meta-analyses across various agricultural systems including cropland, grassland, rice paddy and upland, and summarized in Lam et al. (2022). Based on these meta-analyses, although EEFs are generally effective in reducing Nr losses through targeted pathways, there is still plenty of room for improvement in EEFs performance. First, the mean reduction in Nr losses by using EEFs is mostly <50% (seldom reaching >70%). In contrast, some individual studies reported >90% reduction of Nr losses by using EEFs (Grant et al., 1996; Hadi et al., 2008; Smith et al., 2008). This difference in percent reduction suggests that the full potential of these EEFs has not been realized under certain circumstances. Indeed, the wide range of responses of a particular coating or inhibitor suggests that its performance is not universal and is less effective under certain conditions. Second, the average effect of a certain EEF on a certain Nr loss pathway varies with meta-analyses. While this variation could be caused by the inclusion of different studies in these meta-analyses, more importantly the targeted ecosystems and geographical locations were not identical. This highlights the need of designing inhibitors and coatings tailored for specific edaphic, environmental and agricultural settings. However, currently there are only three commercially available UIs (dominated by NBPT, Cantarella et al., 2018) and a handful for NIs (DMPP, DCD and nitrapyrin, Beeckman et al., 2018). More options are available for controlled-release coatings but they are expensive (Chen et al., 2018), which has hindered their wide adoption especially in developing countries (Mujerij et al., 2012). Third, the mechanisms underlying why EEFs do not work well are poorly understood, hence impairing the advancement of EEFs.

4.2 Research needs: New generation of enhanced efficiency fertilizers

The search for inhibitors and controlled-release coatings started a few decades ago (see Section 2). We need to be proactive in designing and engineering next generation of EEFs that target-specific edaphic, environmental and agricultural settings to achieve sustainable agricultural intensification under growing population pressure. To advance fertilizer development and innovation, it is not sufficient to rely solely on expertise in agronomy and soil science; concerted research efforts from multi-disciplinary experts from e.g. synthetic chemistry, chemical engineering plant physiology, and nanotechnology are also required to revolutionize fertilizer development.

Innovative ideas on the new generation of EEFs have been proposed (Lam et al., 2022). Instead of massive random screening of inhibitors or modifying fertilizer formulations using the existing compounds, new molecules or compounds should be designed that target the known or anticipated issues hindering the effectiveness of the inhibitors. Chemical and functional studies will be required to underpin efficient inhibitor design for high stability and longevity (Lam et al., 2022). For example, NBPT, the most widely used UI, readily undergoes hydrolytic degradation, resulting in a rapid loss of inhibitory activity (Dimkpa et al., 2020, Cantarella et al., 2018). In this regard, *N,N′*-disubstituted acylthioureas have been shown as potential UIs that can cope with increased temperature (Brito et al., 2015), suggesting that their inhibitory performance could be improved by changing substituents. Likewise, as a widely used NI, DMPP undergoes degradation in soils, resulting in side-chain oxidation and dimerization (Sidhu et al., 2021). Instead of having one NI for all soil conditions, it has been proposed that readily accessible 1,2,3-triazoles approach allow the tuning of the inhibitor properties, including polarity, hydrophilicity, lipophilicity, to target various soil pH and soil temperatures (Taggert et al., 2021). For instance, a study has recently demonstrated that 1,2,3-triazoles with short non-polar substituents outperform DMPP in neutral soils at 35°C (Taggert et al., 2021).

As for controlled-release coatings, polysaccharides and polyphenols are becoming more popular as desirable candidates as they are biodegradable, relatively inexpensive and can offer tunable porosities and surface functional groups that help control the release of the encapsulated fertilizers (Lam et al., 2022). Specifically, urea granules have recently been encapsulated within liquid marbles stabilized with various functional monomers, creating a polymer film (Schmücker et al., 2018) that offers advantages over conventional controlled-release coatings, such as its tunable surface chemistry and ease of manufacture. The technology allows control over water permeation rates into the capsule via manipulation of the pore size of the polymer film coating. A recent study showed that metal-phenolic networks (MPNs) can provide a physical barrier against water, thereby controlling the dissolution of urea and its release into soil (Mazaheri et al., 2022). This MPN fabrication method is a new chapter in creating inexpensive, environmentally friendly materials in CRFs. These supramolecular coordination structures consist of metal ions and naturally occurring polyphenols that self-assemble as films or coatings under suitable synthesis conditions (Filippidi et al., 2017; Yun et al., 2018). The metal ions have the additional benefit of providing micronutrients to the soil-plant systems to address issues of micronutrient deficiency. The diverse and desirable properties of MPN technologies enable the release of stimuli-responsive nutrients targeting a range of soil types, seasonal moistures and temperatures.

Plant-oriented approaches to designing a controlled-release coating have also been proposed. The plant-microbe interaction is an intricate process that involves underground signalling and communication, followed by modifications in the plant's biological, morphological and biochemical components. Some proteins, lipids, small RNA's and metabolites that pass between plants and microbes are the signalling molecules to facilitate the 'dialogue' between plants and microbes (Welti et al., 2007; Leach et al., 2017). Research shows that plants can influence how fungi, bacteria and other microbes behave by sending these chemical signals, especially when under stresses (Trivedi et al., 2020). These signalling molecules can be incorporated into the coating of fertilizers and, upon release, would attract beneficial microbes to improve plants' nutrient acquisition and tolerance to environmental stresses (Lam et al., 2022). Biosensors that respond to these signalling molecules may also be included in the fertilizer coatings such that nutrients are released via the coatings when the sensors detect these stress molecules in soil (Monreal et al., 2016; Dimkpa et al., 2020; Lam et al., 2022).

Nanotechnology, which involves the use of nanoscale materials, has been utilized to enhance the performance of N fertilizers (Dimkpa and Dindraban, 2017). Various nanomaterials, such as carbon-based, mineral-based and metal and metal oxide nanoparticles, have demonstrated the ability to effectively deliver nutrients through roots, leaves and seed coatings (Kottegoda et al., 2017; Gao et al., 2018; Lowry et al., 2019). These nanomaterials can serve as the nutrient itself, or as a platform to deliver the nutrients to desired locations in crops (Mastronardi et al., 2017). Nano urea has been reported to enhance crop productivity and reduce N leaching and runoff (Manjunatha et al., 2016; Duhan et al., 2017; Lowry et al., 2019). However, there are concerns regarding the high cost, unsubstantiated claims of the small amount required (ignorance of N mass balance), and potential unintended environmental and ecological consequences of using nanomaterials. The potential toxicity of nanomaterials on plants, microorganisms, animals and human need to be carefully evaluated (Huang et al., 2015).

5 Enhanced efficiency fertilizer adoption: societal costs and benefits

Farmers traditionally consider the cost of fertilizer inputs and the benefits they expect to receive when making decisions about using fertilizers to grow crops and pastures (Dillon and Anderson, 2012). The adoption of currently available EEFs by farmers is limited due to perceived lack of economic viability. Prices for fertilizers stabilized with UIs and NIs could be 20–30% higher and for CRFs multiple times more than conventional fertilizers (Trenkel, 2010).

However, adverse impacts of Nr losses, referred to as externalities, which cause unintended loss of welfare (societal cost) cannot be ignored. The three major categories for societal cost are (i) loss of life years and human health, (ii) loss of biodiversity and ecosystem services and (iii) climate change (Brink et al., 2011). These social impacts are caused by multiple factors and are highly dependent on resilience and resistance of humans, ecosystems and climate. In the absence of a market for these societal costs, expressing societal cost in economic (monetary) terms is based on 'willingness to pay' (WTP) of people to prevent these effects, rather than following a standard economic procedure (Tully and Winer, 2014). Using various sources of WTP data, van Grinsven et al. (2013) estimated that the costs of damage of Nr in the EU27 in 2008 to the environment, resources, human health, ecosystems, climate and society (societal costs) ranged between €75 and billion per year €485 billion per year, with around €35–230 billion per year related to agriculture (van Grinsven et al., 2013).

The quantity of Nr losses associated with fertilizer application was obtained by scaling N_2O emission, NH_3 volatilization and N leaching factors for direct and indirect Nr losses with the actual levels of Nr inputs (synthetic N fertilizer, manure deposited on the fields and used as organic fertilizer, cultivated biological N fixation, atmospheric N deposition and irrigation) for each member state in the EU27 (IPCC, 2006; Lesschen et al., 2011; van Grinsven et al., 2013; FAOSTAT, 2021). A share of around 44% NH_3 and NO_x emissions, 57% N_2O emission and N leaching and runoff is related to fertilizer application. The annual societal costs of pollution by fertilizer N therefore ranges between €20 billion per year and €120 billion per year. The total agricultural use of fertilizer N in the EU27 in 2008 was estimated to be around 10 Tg per year, suggesting that the societal costs per kg N fertilizer applied range between €2 and €12, which is much higher than the cost of N fertilizer itself (€0.7–0.9 per kg N, based on the average fertilizer prices of urea, calcium ammonium nitrate, urea ammonium nitrate and anhydrous ammonia in 2006–10) (Silva, 2011) and the direct economic values of agricultural products at €0.6–3 per kg N (using N response curves and world market prices of winter wheat, milk and oilseed rape) (Brink et al., 2011; van Grinsven et al., 2013).

Using the same methodology, total annual health and environmental damages of anthropogenic Nr use (agricultural and industrial) was estimated to be US$210 billion in the US in the early 2000s (range: US$81–$441 billion per year) (Sobota et al., 2015). Another approach to valuing societal costs of Nr pollution is to use the cost of repairing the damage that is caused. The US spent US$293 million in 1985–2019 to remediate agricultural non-point source pollution in the Great Lakes Basin ecosystem (Hartig et al., 2020). Australia has spent AU$950 million since 2013 in an attempt to rectify the deteriorating water quality of the Great Barrier Reef associated with agricultural fertilizer use (Kroon et al., 2016). A recent study estimated that, in China where croplands contribute to more than one-third of Nr discharge into water sources, around US$100 billion in infrastructure would be needed to restore safe, sustainable N levels in bodies of water (Yu et al., 2019).

Identifying and estimating the magnitude of the societal cost of extra units of Nr losses from the agricultural production system highlights the potential gains from reducing the losses of fertilizer N from the use of EEFs. Based on (i) the findings of 21 meta-analyses summarized by Lam et al. (2022) on the impact of EEFs on Nr losses from food production systems, and (ii) the societal costs of Nr losses in the EU27 (van Grinsven et al., 2013), fully replacing conventional fertilizers by UIs, NIs, double inhibitors and CRFs would reduce the societal cost of pollution due to fertilizer N use by €0−45, €(-5)−85, €5−30 and €5−80 billion per year, respectively. Although the cost of EEFs (€0.8–3 per kg N) are generally higher than the conventional fertilizer (€0.7–0.9 per kg N) (Silva, 2011; Carson and Ozores-Hampton, 2014), EEFs are more cost-effective considering the increased direct economic values of agricultural products (estimated at €0.7−4 per kg N) and the decreased societal costs (€1.5−3 per kg N applied). The benefits would be higher considering the reduction in N input when EEFs are used.

An integrated assessment of the societal costs and benefits of using EEFs in replacement of conventional fertilizers is critical to wide adoption of these environmentally friendly fertilizer products. Specifically, a cost-benefit analysis (CBA) can be carried out at varying levels of complexity, scale, comprehensiveness and completeness, from the perspective of an investment decision of an individual, accounting for the private costs and revenues only; and from the perspective of the society as a whole taking into account the impact on social welfare. A comprehensive economic valuation provides valuable insights into the trade-offs between the positive and negative effects of human interventions on N cycles and the adoption of new fertilizer technologies. This type of integrated CBA on a regional, national or global scale is lacking, but is urgently needed.

Recently, researchers have identified a package of the most effective on-farm field mitigation measures to abate N pollution and estimated the costs and benefits of implementing these measures (Gu et al., 2023). The use of EEFs is one of these measures and is regarded as a typical Tier-1 measure, a measure with low technical thresholds, high mitigation efficacy, low implementation cost and high acceptance by farmers (Gu et al., 2023). Governments can incentivize the adoption of EEFs by subsidizing them at the same or lower price than conventional fertilizers. This approach can encourage more farmers to use EEFs, provided that they can achieve similar or greater yields with the same or lower input costs. Innovative policies such as a Nitrogen Credit System, which acknowledges the responsibilities and limitations of the multiple parties along the food chain, including farmers, suppliers, processors, retailers, consumers and governments (Gu et al., 2021), could be implemented to incentivize and subsidize the adoption of certified environmentally friendly practices to mitigate N pollution.

6 Conclusion and future trends in research

The use of N fertilizers to meet growing food demand is inevitable, but the Nr losses from fertilizer use can be greatly decreased. EEFs were developed a few decades ago, with a handful of products being commercially available. These fertilizer products show promise in decreasing Nr losses, improving NUE and sustaining crop production, but their inconsistent performance has not been well understood or rectified. The traditional, non-systematic approach hampers breakthroughs in the design and development of EEFs because it does not address the specific limitations of the existing products. An entirely new multi-disciplinary approach that couples agronomy and soil science with synthetic chemistry, chemical engineering, nanotechnology, plant biochemistry and economics is required for creating more efficient coatings and inhibitors – the next-generation EEFs. The societal benefits of using these EEFs to reduce environmental damage should be recognized and incorporated into fertilizer decision and policy making. Incentives, such as subsidies and tax credits, should be provided to encourage wider adoption of EEFs. These fertilizers are crucial to addressing the triple challenges of food security, environmental degradation and climate change, as well as for improving the efficiency and profitability of food production.

7 Where to look for further information

The following articles provide a good overview of the subject:

- Dimkpa, C.O., Fugice, J., Singh, U. and Lewis, T. D. 2020. Development of fertilizers for enhanced nitrogen use efficiency–Trends and perspectives. *Science of the Total Environment* 731, 139113.
- Gu, B., Zhang, X., Lam, S. K., Yu, Y., van Grinsven, H. J. M., Zhang, S., Wang, X., Bodirsky, B. L., Wang, S., Duan, J., Ren, C., Bouwman, L., de Vries, W., Xu, J., Sutton, M. A. and Chen, D. 2023. Cost-effective mitigation of nitrogen pollution from global croplands. *Nature* 613, 77–84.

- Kanter, D. R. and Searchinger, T. D. 2018. A technology-forcing approach to reduce nitrogen pollution. *Nature Sustainability* 1, 544–552.
- Lam, S. K., Wille, U., Hu, H. W., Caruso, F., Mumford, K., Liang, X., Pan, B., Malcolm, B., Roessner, U., Suter, H. and Stevens, G. 2022. Next-generation enhanced-efficiency fertilizers for sustained food security. *Nature Food* 3, 575–580.
- Singh, B. 2016. Agronomic benefits of neem coated urea–Ā review. *International Fertilizer Association Review Papers*. Paris: International Fertilizer Association.
- Trenkel, M. E. 2010. *Slow-and Controlled-Release and Stabilized Fertilizers: An Option for Enhancing Nutrient Use Efficiency in Agriculture*. International Fertilizer Industry Association (IFA).
- IFA. 2022. *Reducing Emissions from Fertilizer Use Report*. International Fertilizer Industry Association (IFA). https://www.fertilizer.org/resource/reducing-emissions-from-fertilizer-use-report/.

Key research in this area can be found at the following organisations:

- International Fertilizer Association. https://www.fertilizer.org/.
- The International Fertilizer Development Center. https://ifdc.org/.
- ARC Research Hub for Smart Fertilisers at The University of Melbourne. https://smartfertiliserhub.org.au/.
- Efficient Fertilizer Consortium. https://foundationfar.org/news/ffar-launches-efficient-fertilizer-consortium/.
- United States Department of Agriculture. https://www.usda.gov/.

8 References

Abalos, D., Jeffery, S., Drury, C. F. and Wagner-Riddle, C. 2016. Improving fertilizer management in the US and Canada for N_2O mitigation: understanding potential positive and negative side-effects on corn yields. *Agriculture, Ecosystems and Environment* 221, 214–221.

Abalos, D., Jeffery, S., Sanz-Cobena, A., Guardia, G. and Vallejo, A. 2014. Meta-analysis of the effect of urease and nitrification inhibitors on crop productivity and nitrogen use efficiency. *Agriculture, Ecosystems and Environment* 189, 136–144.

Akiyama, H., Yan, X. and Yagi, K. 2010. Evaluation of effectiveness of enhanced-efficiency fertilizers as mitigation options for N_2O and NO emissions from agricultural soils: meta-analysis. *Global Change Biology* 16(6), 1837–1846.

Azeem, B., KuShaari, K., Naqvi, M., Kok Keong, L., Almesfer, M. K., Al-Qodah, Z., Naqvi, S. R. and Elboughdiri, N. 2020. Production and characterization of controlled release urea using biopolymer and geopolymer as coating materials. *Polymers* 12(2), 400.

Azeem, B., KuShaari, K., Man, Z. B., Basit, A. and Thanh, T. H. 2014. Review on materials & methods to produce controlled release coated urea fertilizer. *Journal of Controlled Release* 181, 11–21.

Beeckman, F., Motte, H. and Beeckman, T. 2018. Nitrification in agricultural soils: impact, actors and mitigation. *Current Opinion in Biotechnology* 50, 166–173.

Benini, S., Rypniewski, W. R., Wilson, K. S., Mangani, S. and Ciurli, S. 2004. Molecular details of urease inhibition by boric acid: insights into the catalytic mechanism. *Journal of the American Chemical Society* 126(12), 3714–3715.

Bouwman, A. F., Boumans, L. J. M. and Batjes, N. H. 2002. Emissions of N_2O and NO from fertilized fields: summary of available measurement data. *Global Biogeochemical Cycles* 16(4), 1058.

Bremner, J. M. 1985. Iowa State University Research Foundation ISURF. Method of inhibiting nitrification of ammonium nitrogen. U.S. Patent 4,552,581.

Bremner, J. M. 1986. Inhibition of nitrogen transformations in soil. In *Proceedings of IV International Symposium on Microbial Ecology* (Megusar, F. and Gantar, M., Eds), pp. 337–342.

Bremner, J. M. and Douglas, L. A. 1971. Inhibition of urease activity in soils. *Soil Biology and Biochemistry* 3(4), 297–307.

Bremner, J. M. and Krogmeier, M. J. 1989. Effects of nitrification inhibitors on germination of various seeds in soil. *Biology and Fertility of Soils* 8(4), 369–372.

Briffa, J., Sinagra, E. and Blundell, R. 2020. Heavy metal pollution in the environment and their toxicological effects on humans. *Heliyon* 6(9), e04691.

Brink, C., van Grinsven, H. J. M., Jacobsen, B. H. and Klimont, Z. 2011. Costs and benefits of nitrogen in the environment. In *European Nitrogen Assessment*. Cambridge: Cambridge University Press, pp. 612. http://www.nine-esf.org/sites/nineesf.org/files/ena_doc/ENA_pdfs/ENA_c22.pdf and Supplementary material http://www.nine-esf.org/sites/nine-esf.org/files/ena_doc/ ENA_supp/ENA_supp_c22.pdf.

Brito, T. O., Souza, A. X., Mota, Y. C. C., Morais, V. S. S., de Souza, L. T., de Fátima, Â., Macedo, F. and Modolo, L. V. 2015. Design, syntheses and evaluation of benzoylthioureas as urease inhibitors of agricultural interest. *RSC Advances* 5(55), 44507–44515.

Cahill, S., Osmond, D., Weisz, R. and Heiniger, R. 2010. Evaluation of alternative nitrogen fertilizers for corn and winter wheat production. *Agronomy Journal* 102(4), 1226–1236.

Cai, Y. and Akiyama, H. 2017. Effects of inhibitors and biochar on nitrous oxide emissions, nitrate leaching, and plant nitrogen uptake from urine patches of grazing animals on grasslands: a meta-analysis. *Soil Science and Plant Nutrition* 63(4), 405–414.

Cantarella, H., Otto, R., Soares, J. R. and de Brito Silva, A. G. B. 2018. Agronomic efficiency of NBPT as a urease inhibitor: a review. *Journal of Advanced Research* 13, 19–27.

Carson, L. and Ozores-Hampton, M. 2014. Description of enhanced-efficiency fertilizers for use in vegetable production. https://edis.ifas.ufl.edu/publication/HS1247. Accessed on 5 April 2021.

Cerri, B. C., Borelli, L. M., Stelutti, I. M., Soares, M. R. and da Silva, M. A. 2020. Evaluation of new environmental friendly particulate soil fertilizers based on agroindustry wastes biopolymers and sugarcane vinasse. *Waste Management* 108, 144–153.

Chalk, P. M., Craswell, E. T., Polidoro, J. C. and Chen, D. 2015. Fate and efficiency of ^{15}N-labelled slow-and controlled-release fertilizers. *Nutrient Cycling in Agroecosystems* 102(2), 167–178.

Chen, D. L., Freney, J. R., Mosier, A. R. and Chalk, P. M. 1994. Reducing denitrification loss with nitrification inhibitors following presowing applications of urea to a cottonfield. *Australian Journal of Experimental Agriculture* 34(1), 75–83.

Chen, D., Freney, J. R., Rochester, I., Constable, G. A., Mosier, A. R. and Chalk, P. M. 2008. Evaluation of a polyolefin coated urea (Meister) as a fertilizer for irrigated cotton. *Nutrient Cycling in Agroecosystems* 81(3), 245–254.

Chen, J., Lü, S., Zhang, Z., Zhao, X., Li, X., Ning, P. and Liu, M. 2018. Environmentally friendly fertilizers: a review of materials used and their effects on the environment. *Science of the Total Environment* 613–614, 829–839.

Chien, S. H., Prochnow, L. I. and Cantarella, H. 2009. Recent developments of fertilizer production and use to improve nutrient efficiency and minimize environmental impacts. *Advances in Agronomy* 102, 267–322.

Coskun, D., Britto, D. T., Shi, W. and Kronzucker, H. J. 2017. Nitrogen transformations in modern agriculture and the role of biological nitrification inhibition. *Nature Plants* 3(6), 17074.

Daims, H., Lebedeva, E. V., Pjevac, P., Han, P., Herbold, C., Albertsen, M., Jehmlich, N., Palatinszky, M., Vierheilig, J., Bulaev, A., Kirkegaard, R. H., von Bergen, M., Rattei, T., Bendinger, B., Nielsen, P. H. and Wagner, M. 2015. Complete nitrification by Nitrospira bacteria. *Nature* 528(7583), 504–509.

Davidson, E. A. 2009. The contribution of manure and fertilizer nitrogen to atmospheric nitrous oxide since 1860. *Nature Geoscience* 2(9), 659–662.

Decock, C. 2014. Mitigating nitrous oxide emissions from corn cropping systems in the Midwestern US: potential and data gaps. *Environmental Science and Technology* 48(8), 4247–4256.

Dillon, J. L. and Anderson, J. R. 2012. *The Analysis of Response in Crop and Livestock Production*. Oxford: Pergamon Press.

Dimkpa, C. O. and Bindraban, P. S. 2017. Nanofertilizers: new products for the industry? *Journal of Agricultural and Food Chemistry* 66(26), 6462–6473.

Dimkpa, C. O., Fugice, J., Singh, U. and Lewis, T. D. 2020. Development of fertilizers for enhanced nitrogen use efficiency-Trends and perspectives. *Science of the Total Environment* 731, 139113.

Dittert, K., Bol, R., King, R., Chadwick, D. and Hatch, D. 2001. Use of a novel nitrification inhibitor to reduce nitrous oxide emission from ^{15}N-labelled dairy slurry injected into soil. *Rapid Communications in Mass Spectrometry* 15(15), 1291–1296.

Domínguez, M. J., Sanmartín, C., Font, M., Palop, J. A., San Francisco, S., Urrutia, O., Houdusse, F. and García-Mina, J. M. 2008. Design, synthesis, and biological evaluation of phosphoramide derivatives as urease inhibitors. *Journal of Agricultural and Food Chemistry* 56(10), 3721–3731.

Douglas, L. A. and Bremner, J. M. 1971. A rapid method of evaluating different compounds as inhibitors of urease activity in soils. *Soil Biology and Biochemistry* 3(4), 309–315.

Duhan, J. S., Kumar, R., Kumar, N., Kaur, P., Nehra, K. and Duhan, S. 2017. Nanotechnology: the new perspective in precision agriculture. *Biotechnology Reports* 15, 11–23.

Erisman, J. W., Galloway, J. N., Seitzinger, S., Bleeker, A., Dise, N. B., Petrescu, A. R., Leach, A. M. and de Vries, W. 2013. Consequences of human modification of the global nitrogen cycle. *Philosophical Transactions of the Royal Society of London Series B* 368(1621), 1–9.

Ernst, J. W. and Massey, H. F. 1960. The effects of several factors on volatilization of ammonia formed from urea in the soil. *Soil Science Society of America Journal* 24(2), 87–90.

FAOSTAT 2021. *Food and Agriculture Organization of the United Nations Statistical Databases, Inputs-Fertilizers by Nutrient Sheet*. https://www.fao.org/faostat/en/#data/RFN.

Feng, J., Li, F., Deng, A., Feng, X., Fang, F. and Zhang, W. 2016. Integrated assessment of the impact of enhanced-efficiency nitrogen fertilizer on N_2O emission and crop yield. *Agriculture, Ecosystems and Environment* 231, 218–228.

Filippidi, E., Cristiani, T. R., Eisenbach, C. D., Waite, J. H., Israelachvili, J. N., Ahn, B. K. and Valentine, M. T. 2017. Toughening elastomers using mussel-inspired iron-catechol complexes. *Science* 358(6362), 502–505.

Forster, P., Ramaswamy, V., Artaxo, P., Berntsen, T., Betts, R., Fahey, D. W., Haywood, J., Lean, J., Lowe, D. C., Myhre, G., Nganga, J., Prinn, R., Raga, G., Schulz, M. and Van Dorland, R. 2007. Changes in atmospheric constituents and in radiative forcing. In *Climate Change 2007: The Physical Science Basis. Contribution of Working Group I to the Fourth Assessment Report of the Intergovernmental Panel on Climate Change* [Solomon, S., Qin, D., Manning, M., Chen, Z., Marquis, M., Averyt, K. B., Tignor, M. and Miller, H. L. (Eds)]. Cambridge and New York, NY: Cambridge University Press.

Fowler, D., Coyle, M., Skiba, U., Sutton, M. A., Cape, J. N., Reis, S., Sheppard, L. J., Jenkins, A., Grizzetti, B., Galloway, J. N. and Vitousek, P. 2013. The global nitrogen cycle in the twenty-first century. *Philosophical Transactions of the Royal Society of London Series B* 368(1621), 20130164.

Freney, J. R., Chen, D. L., Mosier, A. R., Rochester, I. J., Constable, G. A. and Chalk, P. M. 1993. Use of nitrification inhibitors to increase fertilizer nitrogen recovery and lint yield in irrigated cotton. *Fertilizer Research* 34(1), 37–44.

Friedl, J., Scheer, C., Rowlings, D. W., Mumford, M. T. and Grace, P. R. 2017. The nitrification inhibitor DMPP (3, 4-dimethylpyrazole phosphate) reduces N_2 emissions from intensively managed pastures in subtropical Australia. *Soil Biology and Biochemistry* 108, 55–64.

Frye, W. 2005. Nitrification inhibition for nitrogen efficiency and environment protection. In IFA International Workshop on Enhanced-Efficiency Fertilizers, pp. 28–30.

Fujita, T. and Shoji, S. 1999. Kinds and properties of Meister fertilizers. In *Di Dalam Meister Controlled Release Fertilizer-Properties and Utilization*. Sendai: Konno Printing Company Ltd.

Galloway, J. N., Aber, J. D., Erisman, J. W., Seitzinger, S. P., Howarth, R. W., Cowling, E. B. and Cosby, B. J. 2003. The nitrogen cascade. *BioScience* 53(4), 341–356.

Galloway, J. N., Townsend, A. R., Erisman, J. W., Bekunda, M., Cai, Z., Freney, J. R., Martinelli, L. A., Seitzinger, S. P. and Sutton, M. A. 2008. Transformation of the nitrogen cycle: recent trends, questions, and potential solutions. *Science* 320(5878), 889–892.

Gao, J., Luo, J., Lindsey, S., Shi, Y., Sun, Z., Wei, Z. and Wang, L. 2021. Benefits and risks for the environment and crop production with application of nitrification inhibitors in China. *Journal of Soil Science and Plant Nutrition* 21(1), 497–512.

Gao, X., Avellan, A., Laughton, S., Vaidya, R., Rodrigues, S. M., Casman, E. A. and Lowry, G. V. 2018. CuO nanoparticle dissolution and toxicity to wheat (Triticum aestivum) in rhizosphere soil. *Environmental Science and Technology* 52(5), 2888–2897.

Gilsanz, C., Báez, D., Misselbrook, T. H., Dhanoa, M. S. and Cárdenas, L. M. 2016. Development of emission factors and efficiency of two nitrification inhibitors, DCD and DMPP. *Agriculture, Ecosystems and Environment* 216, 1–8.

Goring, C. A. I. 1962. Control of nitrification by 2-chloro-6-(trichloro-methyl) pyridine. *Soil Science* 93(3), 211–218.

Grant, C. A., Brown, K. R., Bailey, L. D. and Jia, S. 1996. Volatile losses of NH_3 from surface-applied urea and urea ammonium nitrate with and without the urease inhibitors NBPT or ammonium thiosulphate. *Canadian Journal of Soil Science* 76(3), 417–419.

Gu, B., van Grinsven, H. J. M., Lam, S. K., Oenema, O., Sutton, M. A., Mosier, A. and Chen, D. 2021. A credit system to solve agricultural nitrogen pollution. *Innovation* 2(1), 100079.

Gu, B., Zhang, X., Lam, S. K., Yu, Y., van Grinsven, H. J. M., Zhang, S., Wang, X., Bodirsky, B. L., Wang, S., Duan, J., Ren, C., Bouwman, L., de Vries, W., Xu, J., Sutton, M. A. and Chen, D. 2023. Cost-effective mitigation of nitrogen pollution from global croplands. *Nature* 613(7942), 77–84.

Gu, J., Wu, Y., Tian, Z. and Xu, H. 2020. Nitrogen use efficiency, crop water productivity and nitrous oxide emissions from Chinese greenhouse vegetables: a meta-analysis. *Science of the Total Environment* 743, 140696.

Hadi, A., Jumadi, O., Inubushi, K. and Yagi, K. 2008. Mitigation options for N2O emission from a corn field in Kalimantan, Indonesia. *Soil Science and Plant Nutrition* 54(4), 644-649.

Han, Z., Walter, M. T. and Drinkwater, L. E. 2017. N_2O emissions from grain cropping systems: a meta-analysis of the impacts of fertilizer-based and ecologically-based nutrient management strategies. *Nutrient Cycling in Agroecosystems* 107(3), 335-355.

Hansen, L. I. 1965. Granular fertilizer having a plurality of coatings and the process of making. U.S. Patent No. 3,223,518.

Hartig, J. H., Krantzberg, G. and Alsip, P. 2020. Thirty-five years of restoring Great Lakes Areas of Concern: gradual progress, hopeful future. *Journal of Great Lakes Research* 46(3), 429–442.

Hauck, R. D. 1980. Mode of action of nitrification inhibitors. In Stelly, M. et al. (Eds) *Nitrification Inhibitors–Potentials and Limitations*. ASA Special Publication, 38, pp. 19–32. Madison: American Society of Agronomy.

Hofmair, W. 2000. DMPP-ein neuer Nitrifikationsinhibitor (Wirkstoff-Wirksamkeit-Einsatzgebiet). *Arbeitsgemeinschaft landwirtschaftlicher Versuchsanstalten, Jahrestagung* 24.

Hu, H. W., Chen, D. and He, J. Z. 2015. Microbial regulation of terrestrial nitrous oxide formation: understanding the biological pathways for prediction of emission rates. *FEMS Microbiology Reviews* 39(5), 729–749.

Huang, S., Lv, W., Bloszies, S., Shi, Q., Pan, X. and Zeng, Y. 2016. Effects of fertilizer management practices on yield-scaled ammonia emissions from croplands in China: a meta-analysis. *Field Crops Research* 192, 118–125.

Huang, S., Wang, L., Liu, L., Hou, Y. and Li, L. 2015. Nanotechnology in agriculture, livestock, and aquaculture in China. A review. *Agronomy for Sustainable Development* 35(2), 369–400.

IFASTAT 2023. *Fertilizer Consumption*. IFASTAT, International Fertilizer Association (IFA). Accessed on 1 May 2023.

IPCC 2006. *IPCC Guidelines for National Greenhouse Gas Inventories* Vol. 4 [Eggleston, H. S., Buendia, L., Miwa, K., Ngara, T. and Tanabe, K. (Eds)]. Kanagawa: Institute for Global Environmental Strategies.

Jarecki, M. K., Parkin, T. B., Chan, A. S., Hatfield, J. L. and Jones, R. 2008. Greenhouse gas emissions from two soils receiving nitrogen fertilizer and swine manure slurry. *Journal of Environmental Quality* 37(4), 1432–1438.

Kanter, D. R. and Searchinger, T. D. 2018. A technology-forcing approach to reduce nitrogen pollution. *Nature Sustainability* 1(10), 544–552.

Kim, D. G., Saggar, S. and Roudier, P. 2012. The effect of nitrification inhibitors on soil ammonia emissions in nitrogen managed soils: a meta-analysis. *Nutrient Cycling in Agroecosystems* 93(1), 51–64.

Kiss, S. and Simihaian, M. 2002. *Improving Efficiency of Urea Fertilizers by Inhibition of Soil Urease Activity*. Dordrecht: Springer Science & Business Media.

Kissel, D. E., Cabrera, M. L. and Ferguson, R. B. 1988. Reactions of ammonia and urea hydrolysis products with soil. *Soil Science Society of America Journal* 52(6), 1793–1796.

Kottegoda, N., Sandaruwan, C., Priyadarshana, G., Siriwardhana, A., Rathnayake, U. A., Berugoda Arachchige, D. M., Kumarasinghe, A. R., Dahanayake, D., Karunaratne, V. and Amaratunga, G. A. 2017. Urea-hydroxyapatite nanohybrids for slow release of nitrogen. *ACS Nano* 11(2), 1214–1221.

Kroon, F. J., Thorburn, P., Schaffelke, B. and Whitten, S. 2016. Towards protecting the Great Barrier Reef from land-based pollution. *Global Change Biology* 22(6), 1985–2002.

Lam, S. K., Suter, H., Mosier, A. R. and Chen, D. 2017. Using nitrification inhibitors to mitigate agricultural N_2O emission: a double-edged sword? *Global Change Biology* 23(2), 485–489.

Lam, S. K., Suter, H., Bai, M., Walker, C., Davies, R., Mosier, A. R. and Chen, D. 2018a. Using urease and nitrification inhibitors to decrease ammonia and nitrous oxide emissions and improve productivity in a subtropical pasture. *Science of the Total Environment* 644, 1531–1535.

Lam, S. K., Suter, H., Davies, R., Bai, M., Mosier, A. R., Sun, J. and Chen, D. 2018b. Direct and indirect greenhouse gas emissions from two intensive vegetable farms applied with a nitrification inhibitor. *Soil Biology and Biochemistry* 116, 48–51.

Lam, S. K., Suter, H., Bai, M., Walker, C., Mosier, A. R., van Grinsven, H. and Chen, D. 2019. Decreasing ammonia loss from an Australian pasture with the use of enhanced efficiency fertilizers. *Agriculture, Ecosystems and Environment* 283, 106553.

Lam, S. K., Wille, U., Hu, H. W., Caruso, F., Mumford, K., Liang, X., Pan, B., Malcolm, B., Roessner, U., Suter, H., Stevens, G., Walker, C., Tang, C., He, J. Z. and Chen, D. 2022. Next-generation enhanced-efficiency fertilizers for sustained food security. *Nature Food* 3(8), 575–580.

Leach, J. E., Triplett, L. R., Argueso, C. T. and Trivedi, P. 2017. Communication in the phytobiome. *Cell* 169(4), 587–596.

Lesschen, J. P., van den Berg, M., Westhoek, H. J., Witzke, H. P. and Oenema, O. 2011. Greenhouse gas emission profiles of European livestock sectors. *Animal Feed Science and Technology* 166–167, 16–28.

Li, T., Zhang, W., Yin, J., Chadwick, D., Norse, D., Lu, Y., Liu, X., Chen, X., Zhang, F., Powlson, D. and Dou, Z. 2018. Enhanced-efficiency fertilizers are not a panacea for resolving the nitrogen problem. *Global Change Biology* 24(2), e511–e521.

Lin, B. L., Sakoda, A., Shibasaki, R. and Suzuki, M. 2001. A modelling approach to global nitrate leaching caused by anthropogenic fertilisation. *Water Research* 35(8), 1961–1968.

Linquist, B. A., Adviento-Borbe, M. A., Pittelkow, C. M., van Kessel, C. and van Groenigen, K. J. 2012. Fertilizer management practices and greenhouse gas emissions from rice systems: a quantitative review and analysis. *Field Crops Research* 135, 10–21.

Linquist, B. A., Liu, L., van Kessel, C. and van Groenigen, K. J. 2013. Enhanced efficiency nitrogen fertilizers for rice systems: meta-analysis of yield and nitrogen uptake. *Field Crops Research* 154, 246–254.

Liu, G., Zotarelli, L., Li, Y., Dinkins, D., Wang, Q. and Ozores-Hampton, M. 2014. *Controlled-Release and Slow-Release Fertilizers as Nutrient Management Tools*. US Department of Agriculture, UF/IFAS Extension Service, University of Florida, IFAS.

Lowry, G. V., Avellan, A. and Gilbertson, L. M. 2019. Opportunities and challenges for nanotechnology in the agri-tech revolution. *Nature Nanotechnology* 14(6), 517–522.

Malhi, S. S. and Nyborg, M. 1985. Methods of placement for increasing the efficiency of N fertilizers applied in the fall. *Agronomy Journal* 77(1), 27–32.

Manjunatha, S. B., Biradar, D. P. and Aladakatti, Y. R. 2016. Nanotechnology and its applications in agriculture: a review. *Journal of Farm Science* 29(1), 1–13.

Mastronardi, E., Monreal, C. and DeRosa, M. C. 2017. Personalized medicine for crops? Opportunities for the application of molecular recognition in agriculture. *Journal of Agricultural and Food Chemistry* 66(26), 6457–6461.

Mazaheri, O., Alivand, M. S., Zavabeti, A., Spoljaric, S., Pan, S., Chen, D., Caruso, F., Suter, H. C. and Mumford, K. A. 2022. Assembly of metal-phenolic networks on water-soluble substrates in nonaqueous media. *Advanced Functional Materials* 32(24), 2111942.

McCarty, G. W. and Bremner, J. M. 1989. Inhibition of nitrification in soil by heterocyclic nitrogen compounds. *Biology and Fertility of Soils* 8(3), 204–211.

Mikkelsen, R. L. and Bock, B. R. 1988. Urea fertilizers. Bull. Y-206. National Fertilizer Development Center, Tennessee Valley Authority, Muscle Shoals, Alabama, Bulletin, p. 175.

Monreal, C. M., DeRosa, M., Mallubhotla, S. C., Bindraban, P. S. and Dimkpa, C. 2016. Nanotechnologies for increasing the crop use efficiency of fertilizer-micronutrients. *Biology and Fertility of Soils* 52(3), 423–437.

Mujeri, M. K., Shahana, S., Chowdhury, T. T. and Haider, K. T. 2012. Improving the effectiveness, efficiency and sustainability of fertilizer use in South Asia. South Asia: Global Development Network.http://www.gdn.int/admin/uploads/editor/files/SA_3_ResearchPaper_Fertilizer_Efficiency.Pdf.Accessed on 4 May 2015.

Oertli, J. J. 1980. Controlled-release fertilizers. *Fertilizer Research* 1(2), 103–123.

Pan, B., Lam, S. K., Mosier, A., Luo, Y. and Chen, D. 2016. Ammonia volatilization from synthetic fertilizers and its mitigation strategies: a global synthesis. *Agriculture, Ecosystems and Environment* 232, 283–289.

Pan, B., Xia, L., Lam, S. K., Wang, E., Zhang, Y., Mosier, A. and Chen, D. 2022. A global synthesis of soil denitrification: driving factors and mitigation strategies. *Agriculture, Ecosystems and Environment* 327, 107850.

Pasda, G., Hahndel, R. and Zerulla, W. 2001. Effect of fertilizers with the new nitrification inhibitor DMPP (3, 4-dimethylpyrazole phosphate) on yield and quality of agricultural and horticultural crops. *Biology and Fertility of Soils* 34(2), 85–97.

Patil, M. D., Das, B. S., Barak, E., Bhadoria, P. B. and Polak, A. 2010. Performance of polymer-coated urea in transplanted rice: effect of mixing ratio and water input on nitrogen use efficiency. *Paddy and Water Environment* 8, 189–198.

Prasad, R., Singh, S. and De, R. 1984. Effect of time and method of application on the relative efficiency of prilled urea and urea supergranules for rice. *Journal of Agricultural Science* 103(3), 539–542.

Prasad, R., Singh, S., Saxena, V. S. and Devkumar, C. 1999. Coating of prilled urea with neem (*Azadirachta indica* Juss) oil for efficient nitrogen use in rice. *Naturwissenschaften* 86(11), 538–539.

Qiao, C., Liu, L., Hu, S., Compton, J. E., Greaver, T. L. and Li, Q. 2015. How inhibiting nitrification affects nitrogen cycle and reduces environmental impacts of anthropogenic nitrogen input. *Global Change Biology* 21(3), 1249–1257.

Quemada, M., Baranski, M., Nobel-de Lange, M. N. J., Vallejo, A. and Cooper, J. M. 2013. Meta-analysis of strategies to control nitrate leaching in irrigated agricultural systems and their effects on crop yield. *Agriculture, Ecosystems and Environment* 174, 1–10.

Rice, E. L. and Pancholy, S. K. 1974. Inhibition of nitrification by climax ecosystems. III. Inhibitors other than tannins. *American Journal of Botany* 61(10), 1095–1103.

Saggar, S., Singh, J., Giltrap, D. L., Zaman, M., Luo, J., Rollo, M., Kim, D. G., Rys, G. and Van der Weerden, T. J. 2013. Quantification of reductions in ammonia emissions from fertiliser urea and animal urine in grazed pastures with urease inhibitors for agriculture inventory: New Zealand as a case study. *Science of the Total Environment* 465, 136–146.

Schmücker, C., Stevens, G. W. and Mumford, K. A. 2018. Liquid marble formation and solvent vapor treatment of the biodegradable polymers polylactic acid and polycaprolactone. *Journal of Colloid and Interface Science* 514, 349–356.

Shan, L., He, Y., Chen, J., Huang, Q. and Wang, H. 2015. Ammonia volatilization from a Chinese cabbage field under different nitrogen treatments in the Taihu Lake Basin, China. *Journal of Environmental Sciences* 38, 14–23.

Shaviv, A. 2000. Advances in controlled-release fertilizers. *Advances in Agronomy* 71, 1–49.

Shaviv, A. 2005. Environmental friendly nitrogen fertilization. *Science in China Series C Life Sciences* 48(S2), 937–947.

Shaw, W. H. R. 1954. The inhibition of urease by various metal ions. *Journal of the American Chemical Society* 76(8), 2160–2163.

Shoji, S. and Gandeza, A. T. 1992. *Controlled Release Fertilizers with Polyolefin Resin Coating*. Sendai: Konno Printing Co.

Sidhu, P. K., Taggert, B. I., Chen, D. and Wille, U. 2021. Degradation of the nitrification inhibitor 3,4-dimethylpyrazole phosphate (DMPP) in soils: indication of chemical pathways. *ACS Agricultural Science and Technology* 1(5), 540–549.

Silva, G. 2011. Fertilizer prices on the rise. Michigan State University Extension, February 22, 2011. Available in: https://www.canr.msu.edu/news/fertilizer_prices_on_the_rise. Accessed on September 5, 2022.

Silva, A. G. B., Sequeira, C. H., Sermarini, R. A. and Otto, R. 2017. Urease inhibitor NBPT on ammonia volatilization and crop productivity: a meta-analysis. *Agronomy Journal* 109(1), 1–13.

Singh, B. 2016. Agronomic benefits of neem coated urea–a review. *International Fertilizer Association Review Papers*. Paris: International Fertilizer Association.

Singh, M. K., Verma, U. N., Pal, S. K. and Thakur, R. 1999. Efficiency of slow-release N-fertilizers from rice and their residual effect on wheat. *Indian Journal of Agronomy* 44, 216-218.

Singh, U. 2010. Nutrient efficient fertilizers and practices relating to food security and sustainability. TFI/FIRT Fertilizer Outlook and Technology Conference, November 16–18, Savannah, GA.

Singh, Y. and Beauchamp, E. G. 1988. Response of winter wheat to fall-applied large urea granules with dicyandiamide. *Canadian Journal of Soil Science* 68(1), 133–142.

Smith, L. C., deKlein, C. A. M. and Catto, W. D. 2008. Effect of dicyandiamide applied in a granular form on nitrous oxide emissions from a grazed dairy pasture in Southland, New Zealand. *New Zealand Journal of Agricultural Research* 51(4), 387–396.

Soares, J. R., Cantarella, H. and de Campos Menegale, M. L. 2012. Ammonia volatilization losses from surface-applied urea with urease and nitrification inhibitors. *Soil Biology and Biochemistry* 52, 82–89.

Sobota, D. J., Compton, J. E., McCrackin, M. L. and Singh, S. 2015. Cost of reactive nitrogen release from human activities to the environment in the United States. *Environmental Research Letters* 10(2), 025006.

Sommer, S. G., Schjoerring, J. K. and Denmead, O. T. 2004. Ammonia emission from mineral fertilizers and fertilized crops. *Advances in Agronomy* 82, 557–622.

Subbarao, G. V., Ito, O., Sahrawat, K. L., Berry, W. L., Nakahara, K., Ishikawa, T., Watanabe, T., Suenaga, K., Rondon, M. and Rao, I. M. 2006a. Scope and strategies for regulation of nitrification in agricultural systems-challenges and opportunities. *Critical Reviews in Plant Sciences* 25(4), 303–335.

Subbarao, G. V., Ishikawa, T., Ito, O., Nakahara, K., Wang, H. Y. and Berry, W. L. 2006b. A bioluminescence assay to detect nitrification inhibitors released from plant roots: a case study with *Brachiaria humidicola*. *Plant and Soil* 288(1–2), 101–112.

Subbarao, G. V., Nakahara, K., Hurtado, M. P., Ono, H., Moreta, D. E., Salcedo, A. F., Yoshihashi, A. T., Ishikawa, T., Ishitani, M., Ohnishi-Kameyama, M., Yoshida, M., Rondon, M., Rao, I. M., Lascano, C. E., Berry, W. L. and Ito, O. 2009. Evidence for biological nitrification inhibition in *Brachiaria* pastures. *Proceedings of the National Academy of Sciences of the United States of America* 106(41), 17302-17307.

Sudhakara, K. and Prasad, R. 1986. Relative efficiency of prilled urea, urea supergranules (USG) and USG coated with neem cake or DCD for direct-seeded rice. *Journal of Agricultural Science, Cambridge* 106, 185–190.E.

Sun, L., Lu, Y., Yu, F., Kronzucker, H. J. and Shi, W. 2016. Biological nitrification inhibition by rice root exudates and its relationship with nitrogen-use efficiency. *New Phytologist* 212(3), 646–656.

Suri, I. K., Mathews, S. and Saxena, V. S. 2000. Research note. Coating of prilled urea with neem trial at KRIBHCO's Hazira plant. *Fertiliser News* 45(12), 71–72.

Suter, H., Sultana, H., Turner, D., Davies, R., Walker, C. and Chen, D. 2013. Influence of urea fertiliser formulation, urease inhibitor and season on ammonia loss from ryegrass. *Nutrient Cycling in Agroecosystems* 95(2), 175–185.

Sutton, M. A., Bleeker, A., Howard, C. M., Bekunda, M., Grizzetti, B., de Vries, W., van Grinsven, H. J. M., Abrol, Y. P., Adhya, T. K., Billen, G., Davidson, E. A., Datta, A., Diaz, R., Erisman, J. W., Liu, X. J., Oenema, O., Palm, C., Raghuram, N., Reis, S., Scholz, R. W., Sims, T., Westhoek, H. and Zhang, F. S. 2013. *Our Nutrient World: The Challenge to Produce More Food and Energy with Less Pollution*. Global Overview of Nutrient Management. Edinburgh: Centre for Ecology and Hydrology.

Tabatabai, M. A. 1977. Effects of trace elements on urease activity in soils. *Soil Biology and Biochemistry* 9(1), 9–13.

Taggert, B. I., Walker, C., Chen, D. and Wille, U. 2021. Substituted 1, 2, 3-triazoles: a new class of nitrification inhibitors. *Scientific Reports* 11(1), 14980.

Terman, G. L. 1980. Volatilization losses of nitrogen as ammonia from surface-applied fertilizers, organic amendments, and crop residues. *Advances in Agronomy* 31, 189–223.

Thapa, R., Chatterjee, A., Awale, R., McGranahan, D. A. and Daigh, A. 2016. Effect of enhanced efficiency fertilizers on nitrous oxide emissions and crop yields: a meta-analysis. *Soil Science Society of America Journal* 80(5), 1121–1134.

Ti, C., Xia, L., Chang, S. X. and Yan, X. 2019. Potential for mitigating global agricultural ammonia emission: a meta-analysis. *Environmental Pollution* 245, 141–148.

Tian, H., Bian, Z., Shi, H., Qin, X., Pan, N., Lu, C., Pan, S., Tubiello, F. N., Chang, J., Conchedda, G., Liu, J., Mueller, N., Nishina, K., Xu, R., Yang, J., You, L. and Zhang, B. 2022. History of anthropogenic nitrogen inputs (HaNi) to the terrestrial biosphere: a 5 arcmin resolution annual dataset from 1860 to 2019. *Earth System Science Data* 14(10), 4551–4568.

Townsend, A. R., Howarth, R. W., Bazzaz, F. A., Booth, M. S., Cleveland, C. C., Collinge, S. K., Dobson, A. P., Epstein, P. R., Holland, E. A., Keeney, D. R., Mallin, M. A., Rogers, C. A., Wayne, P. and Wolfe, A. H. 2003. Human health effects of a changing global nitrogen cycle. *Frontiers in Ecology and the Environment* 1(5), 240–246.

Trenkel, M. E. 2010. *Slow-and Controlled-Release and Stabilized Fertilizers: An Option for Enhancing Nutrient Use Efficiency in Agriculture*. International Fertilizer Industry Association (IFA).

Trivedi, P., Leach, J. E., Tringe, S. G., Sa, T. and Singh, B. K. 2020. Plant-microbiome interactions: from community assembly to plant health. *Nature Reviews. Microbiology* 18(11), 607–621.

Tully, S. M. and Winer, R. S. 2014. The role of the beneficiary in willingness to pay for socially responsible products: A meta-analysis. *Journal of Retailing* 90(2), 255–274.

Turner, D. A., Edis, R. B., Chen, D., Freney, J. R., Denmead, O. T. and Christie, R. 2010. Determination and mitigation of ammonia loss from urea applied to winter wheat with N-(n-butyl) thiophosphorictriamide. *Agriculture, Ecosystems and Environment* 137(3–4), 261–266.

Van Grinsven, H. J., Holland, M., Jacobsen, B. H., Klimont, Z., Sutton, M. A. and Jaap Willems, W. 2013. Costs and benefits of nitrogen for Europe and implications for mitigation. *Environmental Science and Technology* 47(8), 3571–3579.

Van Kessel, M. A., Speth, D. R., Albertsen, M., Nielsen, P. H., Op den Camp, H. J., Kartal, B., Jetten, M. S. and Lücker, S. 2015. Complete nitrification by a single microorganism. *Nature* 528(7583), 555-559.

Wang, C., Houlton, B. Z., Dai, W. and Bai, E. 2017. Growth in the global N_2 sink attributed to N fertilizer inputs over 1860 to 2000. *Science of the Total Environment* 574, 1044–1053.

Wei, Z., Hoffland, E., Zhuang, M., Hellegers, P. and Cui, Z. 2021. Organic inputs to reduce nitrogen export via leaching and runoff: a global meta-analysis. *Environmental Pollution* 291, 118176.

Welti, R., Shah, J., Li, W., Li, M., Chen, J., Burke, J. J., Fauconnier, M. L., Chapman, K., Chye, M. L. and Wang, X. 2007. Plant lipidomics: discerning biological function by profiling plant complex lipids using mass spectrometry. *Frontiers in Bioscience: A Journal and Virtual Library* 12, 2494–2506.

White, R. E., Cai, G., Chen, D., Fan, X. H., Pacholski, A., Zhu, Z. L. and Ding, H. 2002. Gaseous nitrogen losses from urea applied to maize on a calcareous fluvo-aquic soil in the North China Plain. *Soil Research* 40(5), 737–748.

Wu, L. and Liu, M. 2008. Preparation and properties of chitosan-coated NPK compound fertilizer with controlled-release and water-retention. *Carbohydrate Polymers* 72(2), 240–247.

Xia, L., Lam, S. K., Chen, D., Wang, J., Tang, Q. and Yan, X. 2017. Can knowledge-based N management produce more staple grain with lower greenhouse gas emission and reactive nitrogen pollution? A meta-analysis. *Global Change Biology* 23(5), 1917–1925.

Xia, L., Lam, S. K., Wolf, B., Kiese, R., Chen, D. and Butterbach-Bahl, K. 2018. Trade-offs between soil carbon sequestration and reactive nitrogen losses under straw return in global agroecosystems. *Global Change Biology* 24(12), 5919–5932.

Yang, M., Fang, Y., Sun, D. and Shi, Y. 2016. Efficiency of two nitrification inhibitors (dicyandiamide and 3, 4-dimethypyrazole phosphate) on soil nitrogen transformations and plant productivity: a meta-analysis. *Scientific Reports* 6(1), 22075.

Yang, M., Zhu, X., Bai, Y., Sun, D., Zou, H., Fang, Y. and Zhang, Y. 2021. Coated controlled-release urea creates a win-win scenario for producing more staple grains and resolving N loss dilemma worldwide. *Journal of Cleaner Production* 288, 125660.

Yangjin, D., Wu, X., Bai, H. and Gu, J. 2021. A meta-analysis of management practices for simultaneously mitigating N_2O and NO emissions from agricultural soils. *Soil and Tillage Research* 213, 105142.

Yee, J. Y. and Love, K. S. 1947. Nitrification of urea-formaldehyde reaction products. *Soil Science Society of America Journal* 11(C), 389–392.

Yu, C., Huang, X., Chen, H., Godfray, H. C. J., Wright, J. S., Hall, J. W., Gong, P., Ni, S., Qiao, S., Huang, G., Xiao, Y., Zhang, J., Feng, Z., Ju, X., Ciais, P., Chr, N. Stenseth, Hessen, D. O., Sun, Z., Yu, L., Cai, W., Fu, H., Huang, X., Zhang, C., Liu, H. and Taylor, J. 2019. Managing nitrogen to restore water quality in China. *Nature* 567, 516–520.

Yun, G., Besford, Q. A., Johnston, S. T., Richardson, J. J., Pan, S., Biviano, M. and Caruso, F. 2018. Self-assembly of Nano-to macroscopic metal-phenolic materials. *Chemistry of Materials* 30(16), 5750–5758.

Zakir, H. A. K. M., Subbarao, G. V., Pearse, S. J., Gopalakrishnan, S., Ito, O., Ishikawa, T., Kawano, N., Nakahara, K., Yoshihashi, T., Ono, H. and Yoshida, M. 2008. Detection, isolation and characterization of a root-exuded compound, methyl 3-(4-hydroxyphenyl) propionate, responsible for biological nitrification inhibition by sorghum (Sorghum bicolor). *New Phytologist* 180(2), 442–451.

Zaman, M. and Nguyen, M. L. 2012. How application timings of urease and nitrification inhibitors affect N losses from urine patches in pastoral system. *Agriculture, Ecosystems and Environment* 156, 37–48.

Zerulla, W., Barth, T., Dressel, J., Erhardt, K., Horchler von Locquenghien, K., Pasda, G., Rädle, M. and Wissemeier, A. 2001. 3, 4-Dimethylpyrazole phosphate (DMPP)–a new nitrification inhibitor for agriculture and horticulture. *Biology and Fertility of Soils* 34(2), 79–84.

Zhang, W., Liang, Z., He, X., Wang, X., Shi, X., Zou, C. and Chen, X. 2019. The effects of controlled release urea on maize productivity and reactive nitrogen losses: a meta-analysis. *Environmental Pollution* 246, 559–565.

Zhou, M. and Butterbach-Bahl, K. 2014. Assessment of nitrate leaching loss on a yield-scaled basis from maize and wheat cropping systems. *Plant and Soil* 374(1–2), 977–991.

Chapter 8

Banding nitrogen fertilisers and the implications for enhanced efficiency fertiliser technology

Chelsea K. Janke[1], The University of Queensland, Australia; Cristina Martinez[1,2], The University of Queensland and The Grains Research and Development Corporation, Australia; and Yash Dang[1] and Michael J. Bell[1,3], The University of Queensland, Australia

1 Introduction

Globally recognised fertiliser best management practices (BMPs) incorporate four fundamental considerations. These include the right placement of fertiliser, the right timing of application, the right rate and the right type of fertiliser product, which combine to form the 4R Nutrient Stewardship Framework (Fixen, 2020). The aim of fertiliser BMPs is to optimise the use of applied nutrients by maximising crop uptake and minimising nutrient losses, thereby addressing economic, environmental and social goals (International Fertilizer Association, 2009). With a growing focus on environmental, social and governance frameworks across sectors, optimising nutrient use efficiency is more relevant than ever. The following chapter discusses two aspects of the

[1] School of Agriculture and Food Sustainability, The University of Queensland, Australia.
[2] The Grains and Research Development Corporation, Australia.
[3] Centre for Crop Science, Queensland Alliance for Agriculture and Food Innovation, The University of Queensland, Australia.

http://dx.doi.org/10.19103/AS.2024.0135.12

4R Nutrient Stewardship Framework which are the subject of renewed interest viz. optimising fertiliser placement and the use of enhanced efficiency fertilisers (EEFs).

1.1 Methods of fertiliser application – a focus on banding

The right placement of nitrogen (N) fertilisers is a key component of the 4R Nutrient Stewardship Framework, with a wide range of application methods available (outlined in Table 1). Briefly, these include broadcasting, zonal placement, incorporation and banded application for granular and solid fertilisers. Liquid fertiliser application strategies include starter solution, foliar application, fertigation, sub-surface injection and aerial application (Table 1). Despite the wide range of N fertiliser application methods, the placement method employed is typically based on physical, financial or logistical constraints and the economic value of the commodity. While recent increases in economic and environmental drivers for improving N use efficiency (NUE) are

Table 1 Overview of fertiliser application methods

	Method	Description
Granular and solid fertilisers	Broadcasting	Uniform spread of fertiliser over the surface of the field
	Zonal placement	Surface placement of solid fertilisers in a specific zone, usually with reference to the location of the seed or crop roots
	Incorporation	Usually following broadcast or zone placement strategies and involves the shallow mixing or covering of fertiliser with soil
	Banding	The sub-surface application of solid or granular fertilisers in a continuous band, which is covered with soil. Band placement is usually based on proximity to the crop root zone
Liquid fertilisers	Starter solution	A nutrient solution is applied at seeding or with young plants as they are transplanted into fields
	Foliar application	A nutrient solution is sprayed on the foliage of growing plants
	Fertigation	Water soluble nutrients are applied uniformly across the field with water through irrigation infrastructure
	Injection	Liquid fertilisers are applied in concentrated strips on the soil surface or in furrows, either under pressure or with no pressure. In some instances, injection will involve sub-surface placement, similar to banding
	Aerial application	Fertiliser solution or granules applied uniformly to the crop when ground application is not practical

resulting in a trend of increasing adoption of sub-surface N fertiliser banding, this trend is not always consistent. For example, there are moves to transition from traditional sub-surface banding to broadcasting fertilisers occurring in cropping systems which are characterised by large areas of land (e.g. broadacre cropping in the cereal belt of NE Australia) in response to logistical considerations, as the traditional sub-surface banding of fertilisers across many thousands of hectares is relatively slow and expensive. In contrast, the higher value of many horticultural commodities can justify the additional time and labour costs needed to adopt precision fertiliser placement that involves tillage and sub-surface banding.

Banding of N fertilisers below the soil surface can improve NUE relative to other fertiliser application methods by two key mechanisms. Firstly, banded applications can be placed closer to the root zone, maximising the opportunity for crop uptake. This mechanism is especially important in agricultural systems where surface application and/or minimum tillage practices limit the opportunities for broadcast fertiliser-N to be incorporated into soil, potentially causing N to be stranded on the soil surface (e.g. dryland cropping; Armstrong et al., 2015). Secondly, sub-surface banded applications minimise the opportunity for N losses (Chen et al., 2016; Sandral et al., 2017). The N fertiliser placed on the soil surface is particularly vulnerable to gaseous loss through the process of volatilisation and aqueous loss via overland flow (removal of N in water travelling over soil surfaces), both of which can result in significant N losses and have detrimental impacts on the natural environment (Hatfield and Follett, 2008). Sub-surface banding can mitigate some of these losses by (1) the re-capture of any gaseous ammonia (NH_3) generated by the soil solution above the fertiliser band, preventing loss from the soil (Rochette et al., 2013b) and (2) physically moving the fertiliser-N from overland flow pathways.

1.2 Enhanced efficiency fertilisers

The right type of fertiliser product is a fundamental element of fertiliser BMPs, with EEFs attracting renewed interest as a tool for mitigating environmental losses and improving NUE. The EEFs are defined as fertiliser products which facilitate improved uptake of nutrients by plants and reduce the environmental loss potential when compared to an appropriate reference product (Association of American Plant Food Control Officials, 2013). Globally, urea is considered the 'standard' fertiliser-N product; the discussion on EEFs in this chapter, therefore, focuses on technologies which are applied to urea-based fertilisers.

The EEFs primarily use physical or chemical approaches to regulate the supply and availability of N to plants, with the aim of reducing the risk of N loss

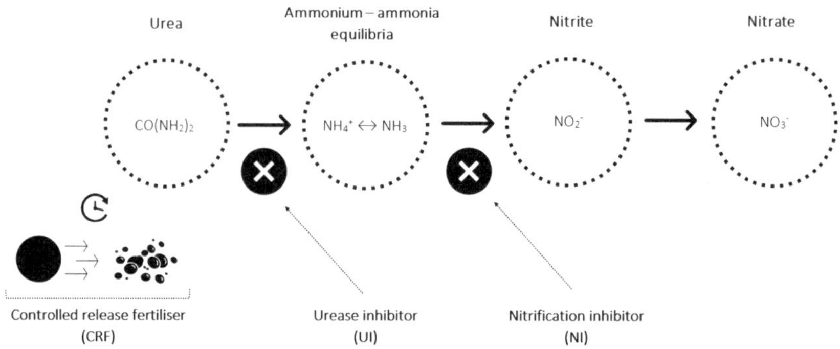

Figure 1 The primary modes of action for the most widely used urea-based enhanced efficiency fertilisers, including: the controlled release of urea (CRF), the inhibitor of urease activity (UI) and the inhibition of the first step in the nitrification process (NI).

by improving the synchrony between N availability and crop demand (Fig. 1). Globally, the most widely used EEF technologies include:

- Controlled N release achieved by using physical barriers between fertiliser granules and soil solution (e.g. coatings); and
- Chemical stabilisation of N through co-application of chemicals which inhibit key steps of N transformations in soil (e.g. urease inhibitors [UIs] and nitrification inhibitors [NIs]).

Selection of the most appropriate EEF product will vary depending on the local loss pathways and patterns of crop N demand. For example, a UI may be most effective in situations where N is surface applied and susceptible to volatilisation losses in the weeks immediately following fertilisation. A NI may be most effective in scenarios where the risk of N losses to leaching or denitrification is high. Alternatively, a controlled-release fertiliser (CRF) with a relatively long release profile (>80 days) may be more effective in crops which have an extended period of N uptake relative to shorter season crops. Understanding the N demand of the crop, the dominant N loss pathways in a field and the seasonal rainfall forecast are all important for selecting an EEF that will improve NUE.

1.3 Banding enhanced efficiency fertilisers

While both the fertiliser application method (right placement) and fertiliser technology (right product) are crucial to improving on-farm NUE, the interaction of these two management strategies is often overlooked. These interactions may negatively influence the effectiveness of EEF products and limit adoption, with EEFs considered ineffective when the mode of application may have been the factor that impeded EEF efficacy.

A vast body of literature reports on the outcomes of EEF use in broadcast and/or incorporated fertiliser applications (Alonso-Ayuso et al., 2016; Barker and Sawyer, 2017; Engel et al., 2015; Rowlings et al., 2016). Recent studies have highlighted the impacts of banding on the mode of action of EEF technologies (Janke et al., 2020; Janke et al., 2021; Janke et al., 2022; Martinez et al., 2021) and suggested that differing application methods, specifically banded *cf.* broadcast/mixed (dispersed), may help explain why reported field outcomes of EEFs are often variable (Alonso-Ayuso et al., 2016; Maharjan et al., 2017; Sanz-Cobena et al., 2012; Wang et al., 2016b). Given the literature coverage on dispersed EEF applications, this chapter focuses on the impact of banded fertiliser application on EEF efficacy. Firstly, the key reactions within urea-based fertiliser bands are summarised (Section 2) to provide context for subsequent sections, where the impact of banding (Section 3) and key influencing factors on CRFs and stabilised fertilisers (Section 4) are discussed. A case study is presented in Section 5, while Section 6 summarises the state of the art on EEF banding and outlines future trends and knowledge gaps within the 4R Nutrient Stewardship Framework.

2 Nitrogen dynamics within bands of urea

The reactions and transformations of N within and surrounding urea fertiliser bands vary from those surrounding urea granules which are dispersed in soil. This is an outcome of concentrating the placement of urea granules, increasing the localised concentration of urea-N and the by-products of subsequent reactions. The extent to which banding impacts N dynamics is dependent on the in-band concentration (which is a factor of the application rate and band spacing), a range of soil characteristics and responses to key environmental variables. The reactions of N in urea bands and the influencing factors are discussed in this section to provide a fundamental understanding of the 'banding effect', which is central to the discussion on banded EEF efficacy in subsequent sections.

2.1 Initial reactions of urea fertiliser in soil

Urea fertiliser typically contains 46% N and is rapidly hydrolysed in soil solution by an extracellular enzyme, urease (urea amidohydrolase). In soil, this enzyme primarily exists both intracellularly and extracellularly on soil clay and humic phases (Bakshi and Varma, 2010). Urease catalyses reactions between urea [$CO(NH_2)_2$] and water (H_2O) to form ammonium ($2NH_4^+$) and bicarbonate (HCO_3^-) ions (Reaction 1).

$$CO(NH_2)_2 + 2H_2O + H^+ \rightarrow 2NH_4^+ + HCO_3^- \qquad \text{Reaction 1}$$

Bicarbonate ions may further react with hydrogen ions (protons; H^+) in soil, producing carbon dioxide (CO_2; Reaction 2).

$$HCO_3^- + H^+ \leftrightarrow CO_2 + H_2O \qquad \text{Reaction 2}$$

Both reactions (1 and 2) consume H^+ which is present in the soil, as shown in Reaction 3. This has the effect of raising the pH of the immediate soil solution, although this effect is usually transient around individual urea granules. Some NH_4^+ ions generated during urea hydrolysis can be adsorbed onto the negatively charged surfaces of clays or humic fractions in soil organic matter (SOM) by displacing other cations or H^+ ions, which enter the soil solution (Reaction 3).

$$H^+\text{-soil} + NH_4^+ \leftrightarrow NH_4^+\text{-soil} + H^+ \qquad \text{Reaction 3}$$

Ions of NH_4^+ which do not interact with soil exchange surfaces but remain in solution may be converted to aqueous ammonia (NH_3), which exists in equilibrium with NH_4^+ as a function of the concentration of each of the N species, pH (H^+ concentration) and the pKa value of the reaction at a given temperature (Emerson et al., 1975; Erickson, 1985; Reaction 4).

$$NH_4^+ \leftrightarrow NH_3 + H^+ \qquad \text{Reaction 4}$$

2.2 Impact of banding on urea-nitrogen reactions

When urea fertiliser is applied in sub-surface bands, a significant portion of Reactions 1 and 2 (i.e. hydrolysis of urea and bicarbonate reaction with H^+) can occur in a very localised zone. This is especially the case in heavier textured soils (e.g. clays), where the consumption of H^+ occurs from a relatively small volume of soil and there is significant alkalinisation in the soil within and immediately surrounding a urea fertiliser band (Andrews et al., 1984; Bloom and Skylberg, 2012). Conversely, in lighter textured soils (e.g. sandy soils), these reactions occur across relatively larger volumes of soil and changes in the soil pH may be less extreme (Janke et al., 2021). This urea band effect has been demonstrated in various field studies (Hauck and Stephenson, 1965; Pang et al., 1973; Rochette et al., 2013a), with the local pH around a hydrolysing urea band increasing to as high as pH 9.3 in some soils.

The coincidence of alkaline conditions and high concentrations of NH_4^+ in urea fertiliser bands has the effect of shifting the equilibrium reaction ($NH_4^+ \leftrightarrow NH_3$) to the right, forming high concentrations of aqueous NH_3. Indeed, the formation of phytotoxic NH_3 concentrations has been observed around urea bands (Creamer and Fox, 1980; Hauck and Stephenson, 1965; Passioura and Wetselaar, 1972). This can have adverse implications for crop access to applied N and may also kill seedlings which are close to the fertiliser band. Furthermore, aqueous NH_3 is also in equilibrium with gaseous NH_3. Thus, where concentrations of aqueous NH_3 are high, there is a concomitant increase in volatile NH_3. When this occurs in sub-surface bands, gaseous NH_3 formed within the fertiliser band may re-dissolve in soil solution above the band where aqueous NH_3 concentrations are relatively much lower. In this soil solution, the pH and NH_4^+ concentrations are also likely to be much lower, so re-dissolved NH_3 may shift back to NH_4^+. A final notable effect that occurs around urea bands is an increase in the local electrical conductivity (EC), driven by significant increases in the salt concentration following fertiliser addition. Several studies have observed this effect around urea bands in both laboratory studies (Janke et al., 2021; Janke et al., 2022) and in the field (Gao et al., 2015; Janke et al., 2020).

The significant modification of soil chemistry from these combined effects can have an inhibitory effect on the function of many soil microbes (Hauck and Stephenson, 1965), which are the main drivers of N transformations in the soil. In particular, these chemical shifts can inhibit the activity of the nitrifying community, which alters the N transformation processes that would usually occur around a urea granule. For example, N held in dispersed urea granules would usually undergo relatively rapid hydrolysis to NH_4^+, followed by subsequent nitrification to NO_3^-, with the entire process taking place over a period of 2–4 days in most soils (Antil et al., 1992; Christianson et al., 1993). In a urea fertiliser band, the change in osmotic pressure (Ying et al., 2017) following fertiliser application and the significant increases in pH and aqueous NH_3 (Bae et al., 2001; Claros et al., 2010) inhibit the nitrification process such that most of the applied N is maintained as NH_4^+ for much longer than dispersed applications.

2.3 Factors which influence the 'band effect'

The volume of soil immediately influenced by a saturated fertiliser solution and the ensuing reactions has been termed the 'fertosphere' (Janke et al., 2019; Meyer et al., 2020). The size (i.e. how far from the band the effects are observed) and persistence of biochemical changes in the fertosphere is a function of (1) the in-band concentration, (2) interactions with key soil characteristics and (3) the influence of environmental factors such as soil water content and temperature, which drive N distribution and transformation rates.

2.3.1 In-band concentration

The in-band urea concentration influences the extent and duration of chemical changes within the fertosphere. Firstly, the addition of the salt-based fertiliser directly increases the EC, with higher in-band concentrations corresponding to greater increases in EC. Secondly, the larger the amount of urea which undergoes hydrolysis in soil solution (i.e. the higher the in-band concentration), the greater the consumption of H^+ (i.e. Reactions 1 and 2), and thus the larger the localised changes in pH. Likewise, increases in the amount of urea hydrolysed produce greater concentrations of NH_4^+. Consequently, in combination with pH increases, the formation of aqueous NH_3 also increases with increasing in-band concentration. The greater the magnitude of these changes, the stronger the inhibitory effect on nitrification and the longer N is preserved as NH_4^+ within the fertosphere (Fig. 2; Janke et al., 2020). Where in-band concentrations are high (i.e. high application rates or wide band spacings, or both), urea may even exceed the immediate processing capacity of urease in the relatively small volume of soil in which fertiliser is applied (Bhat et al., 2011; Singh and Nye, 1984). This can result in the preservation of some of the applied N as urea for periods much longer than the typical 2–4 days for complete hydrolysis of urea applied to soil. Indeed, when in-band concentrations are high, urea bands

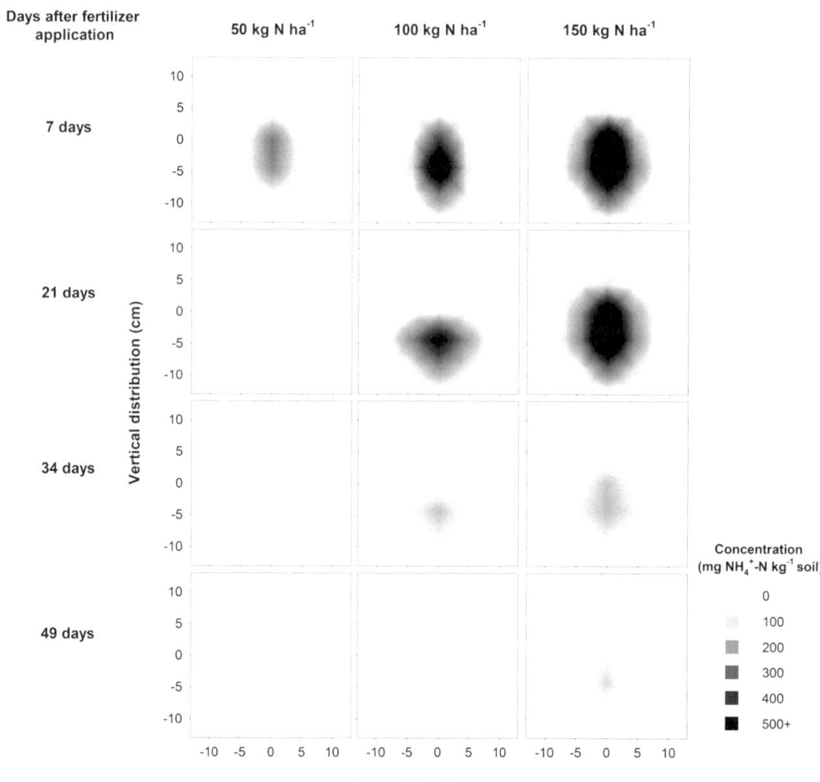

Figure 2 Lateral and vertical distribution of NH_4^+-N from urea bands applied at increasing rates of N application

have been described as behaving 'like a slow-release fertiliser' due to the preservation of N in relatively stable N forms and slowed conversion to NO_3^- (Martinez et al., 2022).

2.3.2 Soil characteristics

Properties influencing the distribution of N in soil appear to be the dominant soil factors which impact the magnitude and to a lesser extent the persistence, of chemical modifications within the fertosphere. In static systems (i.e. negligible mass flow), where diffusion is the primary process of N dispersal, the impedance of the soil is the main factor controlling N redistribution. Impedance is primarily a function of soil texture and water content, with coarser textured soils and higher volumetric water contents creating a less tortuous pathway (tortuosity is a term describing the complexity of pore geometry for fluid and solute movement) that facilitates greater diffusion of N (Nye, 1980). The distribution of N is especially important for the movement of urea-N and aqueous NH_3 (arising from high concentrations of NH_4^+ and localised pH increases) in the early stages of urea hydrolysis, as the distribution of these N molecules determines the size (volume of soil impacted by chemical changes) and intensity (magnitude of chemical changes) in the fertosphere. For example, in the absence of mass flow, soils with low impedance may enable urea-N and aqueous NH_3 to readily diffuse from the fertiliser band down concentration gradients (i.e. from the concentrated fertiliser band to the bulk soil), dispersing the effects of the ureolytic reactions through larger soil volumes. In contrast, when N diffusion is restricted, all the reactions and subsequent impacts on soil chemistry occur in a smaller volume of soil, increasing the magnitude of chemical modification. It is important to note that substantial N distribution in soil profiles usually occurs in response to mass flow. This is primarily regulated by environmental conditions (see Section 2.3.3) although the extent and rate with which the water will permeate through soil are also moderated by soil characteristics. For example, clay and SOM are small, charged particles which slow water movement (due to the tortuosity associated with small particles) and bind water molecules (by physical incorporation into layered clay particles and by chemical binding of the partially positive side of water molecules to the negatively charged sites of organic and clay compounds). In contrast, sand particles are much larger and uncharged, resulting in relatively low tortuosity for water movement and a poor ability to retain water molecules. Mass flow, as an environmental variable, is discussed in more detail in the subsequent section.

Soil characteristics which are responsible for buffering some of the chemical changes that occur in soil following urea application may also influence banded N dynamics. Clay and SOM content have both been linked to the capacity of soils to buffer changes in pH (Jansen van Rensburg et al., 2009; Nelson and

Su, 2010). These soil constituents have a large number of surface charge sites which can adsorb the H^+ ions which are released into the soil solution to maintain surface-solution H^+ equilibrium during urea hydrolysis (Bloom and Skylberg, 2012), resulting in rapid buffering of the pH increases. However, in the case of clay, soil impedance is also increased as the small particles increase the impedance within the soil. Combined with a higher charge density, soils with a high clay content may both constrain and buffer chemical changes which occur within the fertosphere. In contrast, soils with lower clay and SOM content (i.e. lower impedance) exhibit wider diffusion of N out from the urea fertiliser band, but the inherently low capacity to buffer chemical changes may result in a wider soil volume which is subject to significant chemical changes (Janke et al., 2021).

Soil biological communities also influence the conditions within the fertosphere around urea fertiliser bands. Soil microbes and their associated enzymes mediate the rate of N transformations, indirectly influencing the distribution of N by regulating the forms and concentration of N that is present. The urease enzyme is responsible for the initial hydrolysis of urea-N, forming NH_4^+ and increasing the local pH. Urease exists both intracellularly and extracellularly and is positively correlated with clay and humic phases in soil (Bakshi and Varma, 2010; Zantua and Bremner, 1977). The potential activity of urease, therefore, varies between soils, with those containing higher amounts of urease having a greater capacity to hydrolyse urea and facilitate quicker and larger changes to fertosphere chemistry (Janke et al., 2019). Urease is, therefore, a key soil constituent which drives the development of fertosphere conditions. The microbially mediated nitrification process also influences fertosphere conditions by modifying NH_4^+ concentrations and producing an acidifying effect over time (Bolan et al., 1991). Soil nitrification potential differs between soils in response to a range of variables, with the soil pH and physical composition (e.g. proportions of sand, silt, clay, SOM, etc.) being the dominant soil-dependant factors (Sahrawat, 2008). However, the influence of soil nitrification potential on fertosphere conditions is secondary, in that factors influencing nitrification may only impact N dynamics once the inhibitory effects associated with urea hydrolysis have dissipated.

2.3.3 Environmental variables

The dominant environmental factors which influence the formation and persistence of the fertosphere around urea bands are (1) rainfall and irrigation and (2) temperature.

The nature of water application to soil (i.e. volume, duration and frequency), in combination with soil physical properties (i.e. tortuosity, chemical interactions; see Section 2.3.2), determine the rate of water movement through the soil profile. Water moving through fertiliser bands can aid in the dissipation of fertosphere effects by enhancing the dispersal of N via mass flow – the transport of dissolved nutrients in the water. This process is typically characterised by a nonlinear penetrating front (termed hydrodynamic dispersion), due to the irregular pore space and dimensions that result in variable velocities of water permeating through soil (Novak and Hlavacikova, 2019). The effects of mass flow and hydrodynamic dispersion are particularly obvious in the distribution of urea-N and aqueous NH_3 in soils with low impedance, where the process of solute and water transport dilutes and disperses the undesirable chemical conditions associated with urea bands.

In addition to water movement in soil profiles and its impact on N distribution, the general water content (i.e. the amount of water contained in pore spaces) and water potential (i.e. the availability of water) in soils will influence the fertosphere environment by regulating N transformation rates. Soil water status is an important regulator of microbial activity, where a water potential of −0.01 MPa is generally considered to be optimal for most soil microorganisms (Gupta et al., 2008). When water potential is low, microbial activity is decreased, causing a slowing of many of the microbially mediated processes, including N cycling. Depending on when dry conditions around a fertiliser band occur, this may result in an extended period of modified fertosphere conditions (e.g. where urea hydrolysis has increased the pH and $NH_4^+ \leftrightarrow$ aqueous NH_3 equilibria, but drying conditions subsequently limit nitrification) or delayed formation of the fertosphere (e.g. where urea is applied to dry soil and hydrolysis is limited until soil water status improves; Martinez et al., 2021). Likewise, excessive soil water will also inhibit N cycling but may also enhance the activity of denitrifying microbes and increase emissions of nitrous oxide (N_2O), a potent greenhouse gas, and dinitrogen (N_2). There is evidence to suggest the fertosphere environment created by banding may exacerbate N_2O emissions relative to broadcast applications under wet conditions (Cavigelli et al., 2012). Although the exact mechanism for this is not clear, it may be an outcome of nitrite (NO_2^-) accumulation in the fertosphere (e.g. Janke et al., 2019) which can react to form N_2O via biological or chemical pathways (Venterea, 2007). However, if N fertiliser bands are placed deep (i.e. ≥0.2 m) in the soil profile, fertiliser-derived N_2O emissions may be considerably reduced (Rychel et al., 2020; Wu et al., 2021).

Like soil water status, the temperature of the soil also influences the formation and conditions of the fertosphere by regulating the rate of microbially driven N reactions. Although optimal temperatures for microbes vary considerably, higher temperatures generally have a stimulating effect and cooler temperatures decrease microbial activity. Accordingly, warmer temperatures enhance ureolytic reactions (Carmona et al., 1990), resulting in the more rapid and intense formation of the fertosphere, whereas cooler temperatures slow the development of chemical changes which occur in the fertosphere (Creamer and Fox, 1980).

2.4 Implications for nitrogen use efficiency and nitrogen losses

Fertiliser banding is generally recognised as a strategy for enhancing NUE. This is primarily due to the proximal placement of N to the root zone, which maximises the opportunity for the crop to intercept and utilise the applied N (Sanchez et al., 1990; Sandral et al., 2017). However, since the development of fertosphere conditions can create an inhibitory effect on nitrification and result in NH_4^+ accumulation, phytotoxic conditions often develop in and around urea fertiliser bands (Court et al., 1964a; Court et al., 1964b; Passioura and Wetselaar, 1972). This may be a direct result of the excessive influx of NH_4^+ into plant cells (Goyal and Huffaker, 1984) or an indirect effect of shifting the $NH_4^+ \leftrightarrow NH_3$ equilibria in favour of NH_3, resulting in aqueous NH_3 increasing to toxic concentrations. In some instances, incomplete nitrification may occur, resulting in an accumulation of NO_2^- which is also toxic to many plants under differing conditions. Other effects of the fertosphere may also have an impact on the toxicity of the band zone to plant roots. For example, the highly saline solution of the fertosphere may disrupt root growth and/or induce ion cytotoxicity (build-up of ions in plant tissue cells) ultimately causing death (Isayenkov and Maathuis, 2019). Likewise, the alkaline conditions of the fertosphere may induce other nutrient deficiencies (e.g. phosphorus, iron, manganese, zinc and boron), in addition to creating general and aluminium-induced phytotoxicity (Brautigan et al., 2012). Combined, these conditions create an environment that can negatively impact plant growth, potentially negating the benefits of increased N interception. However, as discussed in prior sections, the development and persistence of these fertosphere conditions depend on a range of factors. Consequently, the impact of the fertosphere on N access and utilisation can vary from significant to minimal.

The secondary outcome of sub-surface fertiliser banding is the potential to decrease N losses, most notably via reductions in the emission of some gaseous forms and movement in overland flow. Whilst mitigation of overland flow losses is a direct result of avoiding N placement in the area where these loss events take place (i.e. the soil surface), reductions of N emissions by sub-surface fertiliser bands are somewhat more complex. Banding N fertilisers can mitigate NH_3 volatilisation losses by up to 7% for each additional centimetre of covering soil (Rochette et al., 2013b), although this effect will be dependent on how deep the band is placed and how well the band is covered. The soil covering the band captures gaseous NH_3 as it diffuses, allowing a re-dissolution back into the solution. There is also some evidence for significant reductions in N_2O emissions to levels similar to unfertilised conditions when N fertiliser bands are incorporated deeper (0.2 m) into the soil profile (Rychel et al., 2020). It is hypothesised that shallow-placed fertiliser bands (i.e. ≤0.1 m) may be more susceptible to higher N_2O emissions due to being exposed to higher levels of microbial activity and potentially greater reductions in O_2 under heavy rainfall or irrigation events when compared to surface applications or deep band placement (Burger and Venterea, 2011; Rychel et al., 2020; Smith et al., 2012), although there has been little investigation into these processes. Irrespective of band depth, the accumulation of NO_2^- which occurs in concentrated urea fertiliser bands due to incomplete nitrification (Janke et al., 2019), may form a key substrate for N_2O production (Venterea, 2007) and is not usually present in high concentrations when fertiliser is broadcast and/or dispersed in soil.

Overall, banding N fertilisers can mitigate significant N losses associated with broadcast and incorporated fertiliser applications, although bands must be placed to a sufficient depth and have adequate soil coverage. Further, reducing N losses by applying fertiliser in sub-surface bands does not always translate to improved crop NUE; a range of soil, environmental and crop factors regulate the formation of fertosphere conditions, distribution of N from the band and the accessibility of N by roots.

3 Controlled-release fertilisers: nitrogen dynamics and implications

Controlled-release fertilisers delay the rate of nutrient release into soil, with the aim to improve the synchronisation of N release with plant uptake and increase fertiliser N recovery (Trenkel, 2010). This is typically achieved by coating a water-soluble nutrient core (i.e. urea) with semi-permeable substances including sulphur, polymers and biodegradable plant oils (Azeem et al., 2014; Chen et al., 2008). Soil water moves into the CRF granules, dissolving the soluble nutrient core and increasing osmotic pressure within the granule. Depending on the structural integrity of the coating, the granule may (1) undergo a 'failure mechanism', whereby the coating ruptures and nutrients within the granule are released at once or (2) swell and release nutrients gradually through the now-porous coat along a concentration gradient from the granule core into the surrounding soil. This process is referred to as a 'multi-stage diffusion' model which generates a sigmoidal pattern of nutrient release that is typically the closest match to the temporal pattern of plant nutrient uptake (Shaviv, 2001). However, the synchrony of N availability with crop demand may be hindered by a suboptimal composition of the coating, while variation in soil water matric potential and temperature may alter the nutrient release pattern of CRFs (Azeem et al., 2014; Verburg et al., 2021). Understanding the relationship between coating type and soil matric potential is especially important, as low coating permeability can delay water adsorption and low matric potentials can reduce subsequent N release (Verburg et al., 2021). Soil temperature will influence N dynamics whereby increasing temperatures may increase the water vapour pressure between the CRF and surrounding soil and/or increase the permeability of the granule coating to water (Gandeza et al., 1991), with both proposed mechanisms increasing the rate of water adsorption by CRFs and initiating N release sooner. The selection of CRFs therefore needs to consider the response of the coating composition to the expected soil water matric potentials and temperatures to optimise N release dynamics with crop uptake patterns.

Historically, a wide range of products have been used to coat N-based granules, with a recent and extensive review of these materials covered by Lawrencia et al. (2021). However, the following sections focus on the types of coatings that are most prevalent (synthetic polymers) and those likely to provide sustainable solutions for future fertiliser use (biodegradable).

3.1 Synthetic polymer coatings

Synthetic polymer-coated products are the most common CRFs on the market and are typically composed of thermoplastic resins. However, they are only slowly biodegradable and thus are a potential environmental pollutant (Azeem et al., 2014; Trenkel, 2010). Polymer-coated CRFs have been shown to be largely unaffected by soil pH, salinity, texture, cation exchange capacity or microbial activity, and therefore N release patterns are relatively predictable (Trenkel, 1997). The rate of nutrient release is principally governed by the thickness and permeability of the granule coating, soil temperature and soil water matric potential (Azeem et al., 2014; Verburg et al., 2021). Relative to more fragile coatings (e.g. sulphur), polymer-coated fertilisers are more likely to consistently follow the diffusion mechanism of release and better synchronise nutrient release with plant N uptake patterns (Azeem et al., 2014; Naz and Sulaiman, 2016).

Applications of synthetic polymer-coated urea (PCU) have been shown to result in continuous N supply under field conditions, leading to increased yields and N recovery relative to standard urea in cotton (Geng et al., 2016) and maize (Gagnon et al., 2012; Noellsch et al., 2009). In these studies, PCU was broadcast and incorporated into the soil by tillage. However, other studies that investigated PCU under banded application have found PCU to perform similarly (Martinez et al., 2021) or worse (Dang et al., 2021; Martinez et al., 2022) than standard urea at an equivalent rate of application for irrigated maize. In these studies, N release seemed to be poorly matched to maize demand, despite a slow and controlled release of N from granules in adjacent microplots without plants. Banding of PCU appears to delay the release of N from CRFs by lowering the concentration gradient between granules in close proximity and the soil solution, thereby slowing the diffusion of N from granules (Janke et al., 2022; Fig. 3). This may have the effect of restricting the supply of N, especially during the early phases of crop growth; crops with early demand for N (i.e. short-season crops) may not be able to access sufficient N from bands of CRFs during this period. This would be accentuated when the soil moisture content in the fertilised layer fluctuates in response to plant uptake, evaporation and sporadic rainfall events. There is some evidence to suggest that long-season crops may better utilise N released from banded CRF products (e.g. sugarcane; Dowie et al., 2019; Wang et al., 2016a), although whether this is due to the

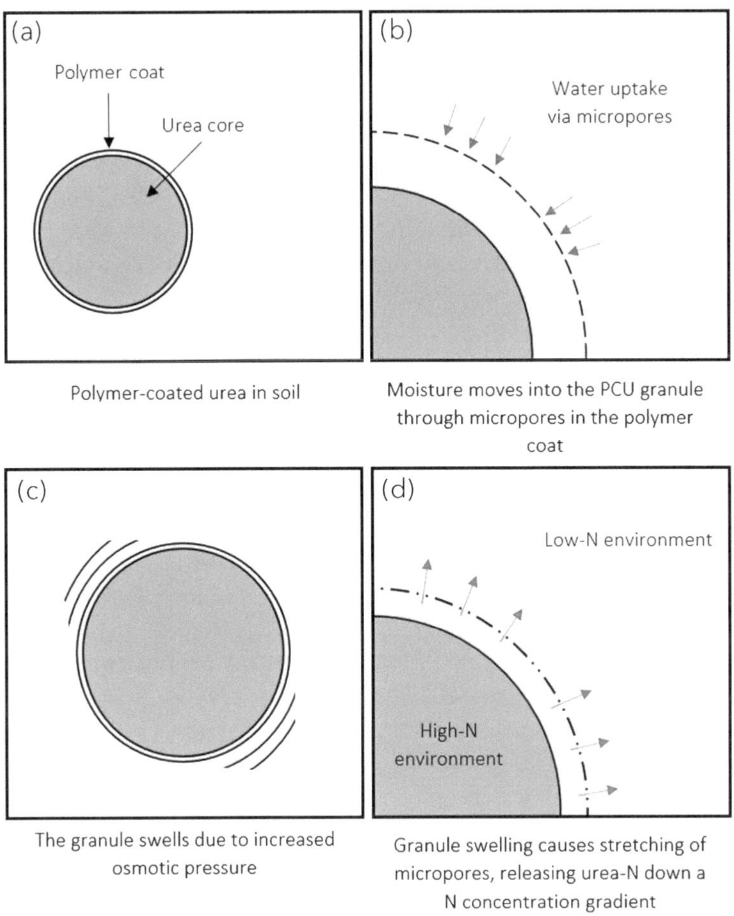

Figure 3 Mechanics of soil water imbibement into polymer-coated nitrogen fertiliser granules and subsequent release of nitrogen to soil solution

extended period of crop N demand or the wetter climatic conditions in which these studies were conducted has not been determined.

In addition to the sometimes-poor temporal synchrony, soil characteristics can also influence the efficacy of banded CRFs. The N released from banded PCUs in heavy textured soils remains in a confined volume of the soil close to the zone of fertiliser placement, which is typically in the 0–10 cm depth of the soil profile (Dang et al., 2021; Martinez et al., 2021). Under rainfed conditions, the limited dispersal of N from CRF bands can make this N periodically inaccessible to crops, as roots move into deeper soil layers seeking water during drier periods. However, in coarser textured soil and in situations where N leaching represents a major loss pathway, the limited release and distribution of N from banded CRFs may deliver positive outcomes in terms of greater in-crop NUE and N retention (Connellan et al., 2021; Dowie et al., 2019). Thus, the efficacy of CRF products to deliver NUE benefits is highly dependent on local soil characteristics and moisture availability, which dictate the dominant N loss pathways (i.e. soil properties, climatic conditions) and the ability of a crop to access N over an extended period.

In some situations, concentrated bands of PCU in surface soil layers may also be a poor option for the mitigation of N_2O emissions when compared to other EEF technologies (e.g. NIs; Dang et al., 2021; Martinez et al., 2021). This is a consequence of the characteristic slow release of N from PCUs which results in a sustained supply of rapidly nitrified N, particularly later in the season when the majority of crop N uptake may be largely complete. This late-season mineral N supply is therefore vulnerable to loss. Such outcomes have been demonstrated in many studies observing CRF dynamics (e.g. Besen et al., 2021; Dang et al., 2021; Graham et al., 2018; Soares et al., 2015), with CRFs shown to be relatively ineffective at reducing N_2O emissions compared to products that employ NIs. A particular concern is that banded PCU may result in an accumulation of NO_2^- (Janke et al., 2019; Maharjan and Venterea, 2013) which can exacerbate N_2O emissions compared to broadcast applications (Halvorson and Grosso, 2013). However, in instances where denitrifying events occur shortly after fertilisation, PCU in either banded or broadcast applications may protect against the immediate spikes in N_2O losses (Scheer et al., 2016). Other factors such as (1) tillage (Halvorson and Grosso, 2013); (2) climate and water regime (i.e. rainfed vs irrigated conditions; Dell et al., 2014; or summer vs winter crop seasons; Besen et al., 2021) and (3) soil type (e.g. alkaline vs acid soils; Thapa et al., 2016) are also known to influence the efficacy of CRFs for mitigating N_2O emissions.

The efficacy of PCU bands under leaching conditions is also variable. Whilst the coating and slow-release mechanism may protect PCU-N from singular episodes of rainfall or irrigation *cf.* urea, this may not be the case under conditions of prolonged rainfall events. Studies indicate that these extended wet periods can result in NO_3^- leaching rates similar to, or greater than, urea (DeBruin et al., 2021; Martinez et al., 2021; Xie et al., 2020), although the patterns of N leaching in the soil profile are likely to vary between fertiliser types. The N released from PCU may be particularly vulnerable to leaching losses, as lower concentrations of urea in soil solution mean that the characteristic increases in pH, aqueous NH_3 and EC around urea bands do not occur around PCU bands (Janke et al., 2022). The absence of these conditions facilitates the rapid conversion of released N to NO_3^-, increasing the potential for leaching losses in the right conditions.

Both spatial and temporal asynchrony of N supply from banded CRFs relative to plant demand is, therefore, possible, especially for shorter-season crops and under conditions with intermittent water supply. Effective utilisation of N from banded CRFs requires a comprehensive knowledge of the factors influencing N release from these products, with the current understanding of CRF behaviour suggesting their use in banded applications may only be suitable as a basal supply of N in long-season crops, particularly in coarse-textured soils in high rainfall environments, and/or where conditions are expected to favour N losses shortly after fertiliser application.

3.2 Biodegradable coatings

Whilst polymer-coated fertilisers are the dominant CRF products used in commercial settings, the persistence of the polymer 'shells' in soil (*in-situ* in fields) or in off-site receiving water bodies through runoff is of concern (Lian et al., 2021). This has led to the development of biodegradability criteria for CRFs in some fertiliser regulatory legislations (International Fertilizer Association, 2019), which will likely be more widely adopted in the coming years. In response, fertiliser companies are developing new biodegradable coatings (e.g. plant oil-coated urea, POCU) that offer an opportunity to minimise environmental pollution whilst retaining the desirable attributes of a functional and predictable coated product. Several plant-derived materials including lignin, modified starches, cellulose and plant-oil-based polymers have been suggested as potentially suitable fertiliser coatings (Azeem et al., 2014). As biodegradable substances, they are susceptible to microbial decomposition and therefore do not accumulate in the soil. Furthermore, these materials are generally less expensive than conventional plastics and are often waste products from various processes.

The performance of biodegradable products relative to polymer coatings has been mostly investigated under controlled laboratory conditions. For example, Feng et al. (2019) successfully demonstrated the gradual release of N from urea coated with polyurethane derived from soybean oil into water incubated at 25°C. Bortoletto-Santos et al. (2020) modified the coating thickness of various biodegradable polyurethanes and monitored the release of urea-N under controlled conditions, while Boberski et al. (2022) recently evaluated hemp oil-coated ammonium nitrate granules, and specifically the effects of the number of oil layers on the release of N into water. While field studies of biodegradable CRFs are not widely documented, a recent series of field-based studies by Martinez et al. (2021) evaluated the performance of POCU relative to PCU and granular urea applied to irrigated maize. Their results indicated similar release and uptake dynamics between the two CRFs that were targeting a 90-day release period. However, there was evidence of coating failure in a proportion of the POCU granules, with early spikes in urea hydrolysis suggesting a lack of structural integrity in withstanding osmotic pressure increases as water moved into granules. Aside from the coating material composition, coating thickness and granule radius are critical factors for coating integrity and may be important factors in optimising the performance of biodegradable CRFs (Hauck, 1985; Martinez et al., 2021; Shaviv et al., 2003). With increasing social and legislative pressures, the development and adoption of biodegradable CRFs are increasing, and these types of CRFs are likely to dominate future CRF markets.

4 Chemically stabilised fertilisers: nitrogen dynamics and implications

The rate of N transformation in the soil can be slowed or inhibited by the application of certain compounds that regulate the formation of N species that are vulnerable to loss processes (Trenkel, 2010). When applied to urea granules, these products are broadly referred to as 'stabilised' fertilisers because they slow the formation of more reactive N forms, potentially increasing the window of opportunity for plant uptake. There are two key processes that are targeted by stabilised fertilisers, viz. the hydrolysis of urea to ammonium carbonate ($[NH_4]CO_3$) by UIs (Reaction 1) and the oxidisation of NH_3 to NO_2^- (i.e. the first step of nitrification) by NIs.

4.1 Urease inhibitors

The UIs restrict the activity of the soil enzyme urease, responsible for the hydrolysis of urea through a range of competitive mechanisms specific to the inhibitory molecule. A wide range of chemicals have been tested for inhibition of urease activity (Byrnes and Freney, 1995), with the most suitable compounds for soil application being the phosphoryl di- and tri-amides, which are structural analogues of urea. These typically either chelate nickel, an essential component for urease activity, or interact with a key functional group in the urease molecule, precluding urea hydrolysis (Medina and Radel, 1988).

N-(n-butyl) thiophosphoric triamide (NBPT) is the most widely used UI in the fertiliser industry. Once applied to soil, NBPT will convert to its oxygen analogue, N-butyl phosphoric triamide (NBPTO) (Creason et al., 1990). In this form, the compound will competitively exclude urea molecules from urease by forming a hydrogen bond which acts as a tridentate ligand to bind the two nickel atoms and residual carbamate oxygen at the active site of urease (Manunza et al., 1999). Despite widespread adoption, the effectiveness of NBPT decreases rapidly with time, with various studies indicating no significant differences to untreated soil after *ca.* 5–15 days following broadcast or incorporated applications (Cantarella et al., 2018; Sanz-Cobena et al., 2012; Singh et al., 2013). Several other factors which exacerbate the loss of UI efficacy include (1) soil moisture contents above *ca.* 60% water-filled pore space (Sanz-Cobena et al., 2016), where the increasingly oxygen-limited environment reduces the conversion of NBPT to NBPTO; (2) high soil temperatures which enhance enzymatic activity (Bakshi and Varma, 2010) to a level that NBPTO degradation outpaces the rate of NBPTO formation (Carmona et al., 1990) and (3) soils of acidic to slightly alkaline pH, where NBPT and NBPTO degrade more rapidly than in strongly alkaline conditions (Engel et al., 2015).

The UIs are primarily used to prevent NH_3 volatilisation losses, since they limit the formation of NH_4^+. Consequently, UIs tend to deliver greater benefits when urea is surface applied in situations where volatilisation losses are high (Chen et al., 2016; Mahler, 2001). When urea is sub-surface banded, volatilisation losses are restricted by soil above the fertiliser band, and the conditions observed in urea bands may also create a transient inhibitory effect on urea hydrolysis, limiting the relative efficacy of UIs. In fact, UIs may be most effective at short distances away from the fertiliser band (e.g. 1–5 cm from the fertiliser band) where some urea has diffused beyond the fertosphere conditions that may be inhibiting urease activity. Soil characteristics and moisture dynamics that effectively determine urea dispersion out from concentrated bands will determine the effectiveness of such a targeted application strategy (Janke et al., 2021). In general, studies which have examined the efficacy of banded UI products have concluded that UIs have limited beneficial impact in banded applications compared to untreated bands (Janke et al., 2020), with sub-surface bands performing similarly to broadcast/dispersed applications of UI-treated urea in terms of reduced NH_3 volatilisation (Cantarella et al., 2018; Singh et al., 2013). Some benefit to banding UIs may occur where N is preserved as urea up until an early rainfall or irrigation event, allowing dispersal of urea through larger volumes of soil and/or deeper into the soil profile, prior to hydrolysis (Janke et al., 2020). This creates a larger N-enriched volume of soil which may be more effectively exploited by plants and may be especially beneficial in situations of poor N fertility in deeper soil layers.

4.2 Nitrification inhibitors

The NIs block the oxidisation of NH_3 by inhibiting the ammonia mono-oxygenase (AMO) and hydroxylamine (HAO) oxidoreductase enzymatic pathways. Typically, AMO is the target of NI compounds with a range of mechanisms by which inhibitory compounds act to impede this enzymatic process (Hyman et al., 1990; Hyman and Wood, 1985; Juliette et al., 1993; McCarty and Bremmer, 1986). By nature, mono-oxygenase enzymes - including AMO - can utilise a wide variety of substrates (Gunsalus et al., 1975); there are over 40 known compounds upon which this class of enzymes can act and which consequently may competitively inhibit NH_3 oxidation (McCarty, 1999). A more extensive review of the differing types and modes of action of Nis is covered in Janke (2020).

Despite the large range of compounds that are inhibitory to nitrification, relatively few compounds are utilised in the production of NI-stabilised fertiliser products. While this is largely a result of inhibitory compounds demonstrating toxic effects to plants and high production costs (e.g. synthesis of inhibitor and application to N fertilisers), the rapid degradation of NIs under field conditions can result in product instability, and so render widespread use even more problematic. In the field, the efficacy of NIs is greatly moderated by a number of environmental factors including: (1) temperature, where increases lead to more rapid degradation of the NI (Chen et al., 2015; Kelliher et al., 2008); (2) soil type, where differing combinations of soil properties will influence NI efficacy (Shi et al., 2016; Wu et al., 2007); (3) SOM, where increasing SOM content increases sorption of the NI, reducing its zone of inhibition and effectiveness (Goring, 1962; Hendrickson and Keeney, 1979; Lewis and Stefanson, 1975; Shi et al., 2016); (4) moisture, where increasing moisture limits the aerobic nitrification pathway and thus the ability of the NI to deliver benefits (Menendez et al., 2012) and (5) the aerobic state of the soil, where the NI degradation rate is higher under aerobic conditions (Balaine et al., 2015). The relationship of soil pH to NI effectiveness is still unclear (Hendrickson et al., 1978; Shi et al., 2016).

In banded applications, NI products prolong the inhibitory effects observed in urea bands (Janke et al., 2020). The duration and efficacy of this effect are moderated by soil characteristics and soil water dynamics, especially as they influence the co-localisation of the inhibitor and $NH_4^+ \leftrightarrow NH_3$ distribution. The NIs are unlikely to disperse far from the fertiliser band (Azam et al., 2001; Janke et al., 2021), and this is particularly the case in soils with characteristics which limit diffusive flow (e.g. high clay, SOM) and with low water content (Barth et al., 2001; Marsden et al., 2016). In the case of NIs that are positively charged (e.g. 3, 4-dimethylpyrazole phosphate), the inhibitory molecule can interact with the soil cation exchange, further reducing distribution and efficacy. While NH_4^+-N is relatively immobile in soil, under the conditions created by maintaining

NH_4^+ with NI application (i.e. high pH, and especially high NH_4^+ concentrations), there is significant formation of aqueous NH_3, which is considerably more mobile. In NI bands, much of the applied N can therefore diffuse or leach as aqueous NH_3, while the inhibitor will remain relatively close to the zone of application. This creates a spatial dislocation of inhibitor and substrate, therefore limiting the efficacy of the NI. This scenario is most likely in situations where there is considerable mass flow through soil and/or where soil characteristics permit rapid redistribution of N from fertiliser bands. However, where mass flow and N distribution are more limited, there is evidence that banded NIs can be highly effective (Janke et al., 2020). In fact, under these conditions, the inhibitory period that occurs under banded application may be significantly extended, compared to the widely reported 14- to 30-day durations of effectiveness when NI products are broadcast and/or incorporated (Chen et al., 2010; Guardia et al., 2018; Irigoyen et al., 2003). This is possibly an outcome of the fertosphere conditions temporarily protecting the inhibitor from microbial degradation and/or the localised and high concentration of the banded inhibitory compound having a protective effect by both physically and chemically limiting microbial access. These hypotheses have not been demonstrated and require further research. Further, while this extended inhibitory effect may be beneficial in some environments (e.g. prolonged wet conditions where denitrification losses may be high), it may also result in a limited redistribution of N through the root zone, thus potentially limiting accessibility by crops. This may be especially problematic in dryland cropping systems where N can become stranded in dry topsoil layers.

5 Case study: plant uptake and soil nitrogen dynamics from banded enhanced efficiency fertilisers

5.1 Field experiment design

We examined N responses in irrigated maize on a Vertisol (classified according to the International Union of Soil Sciences Working Group, 2015) from banded urea, NI, UI and PCU products applied at rates of 125 kg N ha^{-1} (sub-optimal relative to crop N demand) and 250 kg N ha^{-1} (optimal or supra-optimal). To confirm these rates were sub/supra optimal for the maize crop, six rates of urea (0, 62.5, 125, 175, 250 and 300 kg N ha^{-1}) were tested to develop a reference N response curve, against which EEFs were benchmarked. Treatments were replicated four times in a randomised block design. Fertilisers were band-applied at a soil depth of 0.1 m with a spacing of 0.75 m, immediately prior to sowing of the maize crop. A maize hybrid (PAC 606IT) was planted at a rate of 70 000 plants ha^{-1} at a depth of 0.06 m, adjacent to fertiliser bands.

Soil samples from urea and EEF treatments applied at 250 kg N ha^{-1} were used to quantify the mineral-N remaining in the soil after the harvest of the maize crop. The lateral distance of sampling positions in relation to the fertiliser

N-band was kept consistent across all treatments. Soil samples were taken on (−0.025–0.025 m), near (0.05–0.1 m) and in the mid-point between (0.35–0.40 m) pairs of N-bands. Samples were dried at 40°C, ground and analysed for concentrations of NO_3^--N and NH_4^+-N. The total mass of NO_3^- - N and NO_3^+-N was calculated by multiplying concentrations with bulk density and the depth of the soil layer. More details and full results are outlined in Dang et al. (2021) and Martinez et al. (2022).

5.2 Results and discussion

Maize grain yield and crop N uptake were highly responsive to N fertilisation (Fig. 4). The grain yield of maize increased significantly with increasing rates of urea up to 175 kg N ha^{-1} (Fig. 4a), with a trend for only very small increases at higher rates. Maximum potential yield was 11.9 t ha^{-1} and a 90% relative yield of 11.2 t ha^{-1} was recorded at 175 kg N ha^{-1}. Similarly, crop N accumulation in biomass continued to increase significantly up to rates of 175–250 kg N ha^{-1} in urea treatments (Fig. 4a).

In situations such as this, where there was a strong crop demand for N, positive benefits from EEF use would be expected in conditions that favoured fertiliser N loss. Since the use of sub-surface fertiliser banding is known to reduce NH_3 volatilisation relative to broadcast surface applications (Liu et al., 2019; Steusloff et al., 2019) and the Vertisol field site was characterised by slow internal drainage rates, the most likely N loss pathway in this situation was denitrification (Migliorati et al., 2016). The seasonal conditions that would have favoured denitrification losses (periods of waterlogging during early growth, when crop N acquisition was low) did occur in the first 40–50 days after sowing and resulted in significant spikes in N_2O emissions in the treatments receiving urea alone, and to a lesser extent, in the PCU (see Fig. 2 in Dang et al., 2021). However, despite these observations, banded EEFs (and particularly banded

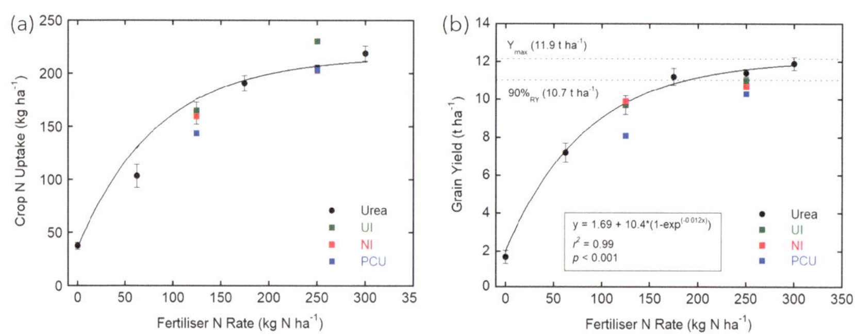

Figure 4 Maize (a) crop N uptake at 103 days after sowing (between milk and physiological maturity stage) and (b) grain yield at two rates of N application for urea and EEFs. Data partially reproduced from Martinez et al. (2022)

NI-urea) did not lead to improved outcomes in either crop N uptake (Fig. 4a) or grain yield (Fig. 4b) relative to urea applied at the same rate. The reasons for this are not clear but could be related to either conditions that were not conducive to generation of large N losses (presumably as N_2) or to restricted N distribution in the EEF products that limited crop acquisition of fertiliser N.

The results from the PCU products were potentially a good example of asynchrony in the rate of N release and/or small volumes of N enrichment with a banded CRF product. Yields were significantly lower at both the sub-optimal (8.1 t ha^{-1}) and supra-optimal (10.3 t ha^{-1}) application rates compared to urea (9.7 and 11.4 t ha^{-1}, respectively). Likewise, crop N uptake in the PCU treatments was consistently lower, although this was not significant at either fertiliser application rate (Fig. 4a). Evidence for this delayed N release was supported in a study by Janke et al. (2020) at the same site that demonstrated a slowed release of N over 71 days for banded PCU, with a smaller zone of N-enriched soil compared to other N fertilisers (urea, NI and UI). This smaller volume of N enrichment, combined with delayed N release from the PCU granules, likely limited plant access to fertiliser N early in the growing season - despite the proximity of the band placement to the seeding row (Martinez et al., 2021).

The amount and distribution of NH_4^+-N in the root zone at the end of the maize growing season for the 250 kg N ha^{-1} rate (to a depth of 0.9 m) was similar for urea and all EEF products (Fig. 5a), although there were suggestions of some differences in NO_3^--N - both in quantities (Fig. 5a) and distribution (Fig. 5b). The most residual NO_3^--N was found in the UI treatment, especially in deeper profile layers (Fig. 5b). This trend for deeper movement of N from UI-treated urea would be consistent with longer preservation of N as urea that was redistributed deeper into the profile with early season rainfall and irrigation. There was little difference between the NI or PCU products and urea in terms of the amount of residual N, but there was further evidence of these

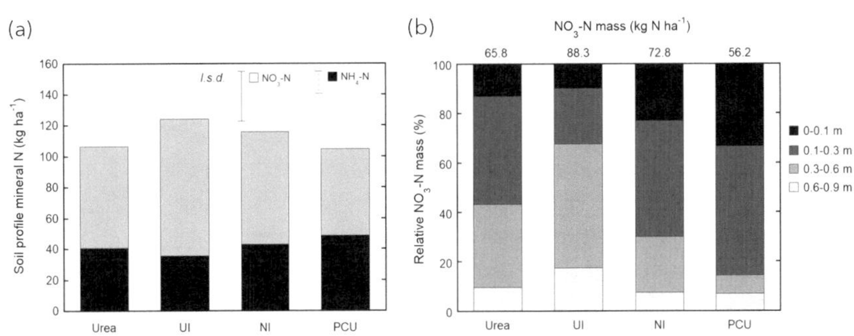

Figure 5 Residual (a) soil profile mineral N (kg ha^{-1}) for urea and EEFs applied at 250 kg N ha^{-1}, with vertical bars representing a *l.s.d.* at 5% significance for NO_3^--N (solid) and NH_4^+-N (dashed) and (b) the relative No_3N mass (%) depth distribution for banded fertiliser treatments. Data reproduced from Dang et al. (2021)

EEF products slowing the generation of NO_3^--N and affecting N distribution within the soil profile (Fig. 5b). Urea and NI-urea treatments retained 55% and 70%, respectively, of the total profile NO_3^--N residual from fertiliser in the top 0-0.30 m of the soil profile. This is consistent with the concentrated urea bands creating an inhibitory effect on nitrification, resulting in similar N dynamics to banded NI urea (Janke et al., 2020). In contrast, 85% of total NO_3^--N was found in the 0–0.30 m profile layer in the banded PCU treatments, which is consistent with delayed N release and redistribution within the soil profile, and potentially restricted root access to fertiliser N.

This study is a good example of the challenges facing the development of more effective fertiliser N practices using EEF technologies. Whilst differences in N dynamics around urea and EEF bands can be demonstrated, and reductions in N_2O emissions achieved (at least over the duration of a crop growing season), this does not necessarily result in improved fertiliser N recovery and use by crops. Variable rates of N release and changed volumes of soil N enrichment may negate any benefits from a reduction in N losses unless the loss environment is particularly severe, with banded applications accentuating these effects. Given the additional cost associated with EEF technologies, it is these high-loss situations where benefits from EEF use are most likely to occur, but further research on the impact of banding EEF products in these conditions is needed.

6 Conclusion and future trends

Sub-surface banding of urea-based fertilisers, particularly at higher in-band concentrations typical of row crops grown at wider row spacings, significantly modifies soil biochemistry and, consequently, the N cycling dynamics of applied urea fertiliser. These changes can limit nitrification rates and prolong the presence of high NH_4^+ concentrations in a high pH environment that favours the formation of aqueous NH_3. In effect, these changes may limit root access to applied N and potentially reduce the risks of leaching and denitrification for varying durations. The extent of this is dependent on soil characteristics that relate to diffusion and pH buffering and soil water dynamics, all of which may increase the fertosphere volume and dilute the fertosphere conditions. Consequently, the band zone creates an environment where the rate of N transformations is already constrained to varying extents, and these conditions will impact the efficacy and longevity of urease and NIs.

In particular, UIs appear to be of limited benefit in banded applications, where NH_3 volatilisation losses (the primary loss pathway targeted by UIs) are already reduced in sub-surface urea bands by the physical protection offered by covering the soil. Further, where N is maintained as urea, mass flow may leach uncharged urea molecules away from the relatively immobile UI compounds during rainfall or irrigation events, rendering the UI ineffective. Interestingly, the use of a UI to facilitate deeper leaching of urea-N may be a potential solution for increasing the volume of fertiliser N enrichment in heavier textured soils where the co-location of water and N in the soil profile is a critical success factor for crop performance.

The NIs may be effective if they persist long enough to be present when the conditions in and around urea bands became more favourable for nitrification. Soil characteristics which influence solute distribution are key factors regulating the efficacy of this technology in banded applications, where there may be asynchrony of the substrate-N and the inhibitor in soils which facilitate N diffusion (i.e. as aqueous NH_3) but restrict movement of the, often charged, inhibitory molecules. The duration of inhibition is influenced by the in-band concentrations of both the applied urea and the NI, with higher concentrations of both facilitating longer periods of N preserved as NH_4^+. This is an important consideration for short-season crops where N uptake in the first half of the growing season is critical; maintaining inhibitory conditions in and around fertiliser bands may limit the volume of N enrichment and restrict access by plant roots. These effects can be particularly problematic in rainfed agriculture dependent on access to water stored in deeper profile layers.

Compared to soluble forms of urea, CRF products may protect N from leaching and denitrification losses in isolated instances of high rainfall soon after fertiliser application. However, substrate-limited rates of ureolytic activity limit the formation of conditions inhibitory to nitrification, meaning that released N is rapidly nitrified and vulnerable to loss over extended periods of time. Further, banding CRF products may promote even slower N release by interfering with the development of strong granule-soil solution concentration gradients, which are necessary for the diffusive release of N in these fertilisers. These effects can limit the volume of the N-enriched zone around CRF bands, potentially reducing the accessibility by crop roots. Banded CRFs appear best suited for (1) mitigating N losses which are expected shortly after fertiliser application and (2) delivering a low, basal supply of N to plants that have regularly moist root zones in proximity to the fertiliser band.

The current state-of-the-art suggests that there are a number of situations where EEF use is not delivering the expected NUE benefits, with banded applications especially problematic. Whilst this may, in part, be related to factors such as soil type, seasonal conditions and crop requirements (e.g. crops with wide row spacings, high N requirements, on heavier textured soils or grown under conditions of fluctuating moisture availability), there are important knowledge gaps that need to be addressed to optimise NUE benefits from EEF technologies in different situations. These are listed below and highlight the future direction of banded EEF research.

- Precise quantification of N dynamics and uptake from banded EEFs using isotopically labelled fertiliser products. This is necessary for holistic evaluations which capture (1) soil release dynamics, (2) subsequent N cycling processes, (3) agronomic performance (N uptake and recovery, yield, grain N) and (4) identification and measurement of N loss pathways.
- Determining the mechanisms driving the reduction in inhibitor functionality that occur in banded applications of stabilised N products. Whilst the relatively short 'effective' period for NIs and UIs is well documented, the reason for the loss-of-function is relatively unexplored. There is some evidence to suggest that the conditions of concentrated fertiliser bands may either delay or increase the rates of degradation of inhibitory molecules. Further, the influence of mass flow on inhibitor distribution is especially relevant when N fertiliser is banded, particularly in soils where dislocation of substrate-N and inhibitor is highly likely. The movement of inhibitors with the mass flow is poorly studied due to the complexity and high analytical costs associated with extraction from soil and analysis of inhibitory compounds. However, methods that enable the collection of soil solutions (Janke et al., 2021; Oudyn and Janke, 2018) or stable isotope labels (Marsden et al., 2016) could be further explored to address inhibitor distribution questions. Understanding the distribution dynamics of inhibitors and N-substrates in response to mass flow in different soil types is pertinent to understanding the soil and soil–water constraints relating to banded inhibitor efficacy.
- Most studies of EEFs focus on soil N dynamics, losses and agronomic responses but not on the effects of banded EEFs on microbial communities and their functions. This understanding is important, given that these communities are the key mediators of soil N cycling and availability, but direct measurement of the impact of differing EEF formulations on soil health and functionality is also becoming increasingly important. Such studies are particularly relevant for the development of the next generation of EEF formulations that contain 'biostimulants' and other biological agents which act on N cycling processes. In particular, research is necessary to understand the interactions between differing N fertosphere volumes and the effectiveness of any associated biostimulants/bio-modifers, with particular importance for banded applications.
- Broader economic evaluation of EEF application in broadacre cropping systems. The EEFs are always more expensive when compared to standard urea, due to their value-add technologies. Whilst there is extensive research on the agronomic efficacy of EEFs, the associated economic analysis of product performance has largely been constrained to cost-benefit analyses based on productivity alone. More detailed analyses that include an assessment of potential environmental credits due to mitigated

N losses are required to provide a more complete assessment of the economic value that can be derived from these technologies.

7 Where to look for further information

The International Fertiliser Association (IFA) is the only global fertiliser association and their online website (www.fertilizer.org) provides access to several resources (publications, podcasts, webinars) on topics relevant to plant nutrition.

8 References

Alonso-Ayuso, M., Grabriel, J. L. and Quemada, M. (2016). Nitrogen use efficiency and residual effect of fertilizers with nitrification inhibitors. *European Journal of Agronomy* 80, 1-8.

Andrews, R. K., Blakeley, R. L. and Serner, B. (1984). Urea and urease. *Advances in Inorganic Biochemistry* 6, 245-283.

Antil, R. S., Gangwar, M. S. and Kumar, V. (1992). Transformation and movement of urea in soil as influenced by water application rate, moisture management regime, and initial moisture content. *Arid Soil Research and Rehabilitation* 6(4), 319-325.

Armstrong, R. D., Dunsford, K., Mclaughlin, M. J., Mcbeath, T., Mason, S. and Dunbabin, V. M. (2015). Phosphorus and nitrogen fertiliser use efficiency of wheat seedlings grown in soil from contrasting tillage systems. *Plant and Soil* 396(1-2), 297-309. https://doi:10.1007/s11104-015-2586-2.

Association of American Plant Food Control Officials. (2013). *Official Publication Number 65. Official Publication.* Association of American Plant Food Control Officials.

Azam, F., Benckiser, G., Muller, C. and Ottow, J. (2001). Release, movement and recovery of 3,4-dimethylpyrazole phosphate (DMPP), ammonium, and nitrate from stabilized nitrogen fertilizer granules in a silty clay soil under laboratory conditions. *Biology and Fertility of Soils* 34(2), 118-125. https://doi.org/10.1007/s003740100384.

Azeem, B., Kushaari, K., Man, Z. B., Basit, A. and Thanh, T. H. (2014). Review on materials and method to produce controlled release coated urea fertilizer. *Journal of Controlled Release* 181, 11-21. https://doi.org/10.1016/j.jconrel.2014.02.020.

Bae, W., Baek, S., Chung, J. and Lee, Y. (2001). Optimal operation factors for nitrite accumulation in batch reactors. *Biodegradation* 12(5), 359-366. https://doi-org.ezproxy.library.uq.edu.au/10.1023/A:1014308229656.

Bakshi, M. and Varma, A. (2010). Soil enzyme: The state-of-art. In: Shukla, G. and Varma, A. (Eds.), *Soil Enzymology*. Berlin Heidelberg: Springer, pp. 1-23.

Balaine, N., Clough, T. J., Kellier, F. M. and Koten, C. V. (2015). Soil aeration affects the degradation rate of the nitrifiation inhibitor dicyandiamide. *Soil Research* 53(2), 137-143.

Barker, D. and Sawyer, J. (2017). Evaluation of nitrogen fertilizer additives for enhanced efficiency in corn on Iowa soils. *Crop, Forage and Turfgrass Management* 3(1), 1-6.

Barth, G., Tucher, Sv. and Schmidhalter, U. (2001). Influence of soil parameters on the effect of 3,4-dimethylpyrazole-phosphate as a nitrification inhibitor. *Biology and Fertility of Soils* 34(2), 98-102.

Besen, M. R., Riberio, R. H., Bayer, C. and Piva, J. T. (2021). Nitrous oxide emissions from enhanced-efficiency nitrogen fertilizers applied to annual crops in a subtropical ecosystem. *Nutrient Cycling in Agroecosystems* 121(1), 85–98.

Bhat, M., Murthy, D. and Saidutta, M. (2011). Urea hydrolysis in saturated loam soil. *ARPN Journal of Agricultural and Biological Science* 6, 60–63.

Bloom, P. and Skylberg, U. (2012). Soil pH and pH buffering. In: Huang, P., Li, Y. and Sumner, M. (Eds.), *Handbook of Soil Sciences: Properties and Processes*. Boca Raton: CRC Press.

Boberski, P., Torchala, K., Studnik, H., Wojcik, J., Glowka, M. and Kuznik, N. (2022). Biobased materials for the development of biodegradable slow-release fertilizers. *Coatings* 12(8), 1215. https://doi.org/10.3390/coatings12081215.

Bolan, N. S., Hedley, M. J. and White, R. E. (1991). Processes of soil acidification during nitrogen cycling with emphasis on legume based pastures. *Plant and Soil* 134(1), 53–63.

Bortoletto-Santos, R., Guimaraes, G. G. F., Junior, V. R., Cruz, D. F. D., Polito, W. L. and Ribeiro, C. (2020). Biodegradable oil-based polymeric coatings on urea fertilizer: N release kinetic transformations of urea in soil. *Scientia Agricola* 77(1).

Brautigan, D. J., Rengasamy, P. and Chittleborough, D. J. (2012). Aluminium speciation and phytotoxicity in alkaline soils. *Plant and Soil* 360(1–2), 187–196. https://doi-org.ezproxy.library.uq.edu.au/10.1007/s11104-012-1232-5.

Burger, M. and Venterea, R. T. (2011). Effects of nitrogen fertilizer types on nitrous oxide emissions. In: Guo, L., Gunasekara, A. S. and Mcconnell, L. (Eds.), *Understanding Greenhouse Gas Emissions From Agricultural Management*. Washington, DC: American Chemical Society. pp. 179–202.

Byrnes, B. H. and Freney, J. R. (1995). Recent developments of the use of urease inhibitors in the tropics. *Fertilizer Research* 42(1–3), 251–259. https://doi.org/10.1007/BF00750519.

Cantarella, H., Otto, R., Soares, J. R. and Silva, A. G. B. (2018). Agronomic efficiency of NBPT as a urease inhibitor: A review. *Journal of Advanced Research* 13, 19–27.

Carmona, G., Christianson, C. B. and Byrnes, B. H. (1990). Temperature and low concentration effects of the urease inihibitor N-(n-butyl)thiophosphoric triamide (nBTPT) on ammonia volatilization from urea. *Soil Biology and Biochemistry* 22(7), 933–937.

Cavigelli, M. A., Grosso, S. J. D., Liebig, M. A., Snyder, C. S., Fixen, P. E., Venterea, R. T., Leytem, A. B., McLain, J. E. and Watts, D. B. (2012). US agricultural nitrous oxide emissions: Context, status, and trends. *Frontiers in Ecology and the Environment* 10(10), 537–546. https://doi.org/10.1890/120054.

Chen, D., Suter, H. C., Islam, A. and Edis, R. (2010). Influence of nitrification inhibitors on nitrification and nitrous oxide (N_2O) emission from a clay loam soil fertilized with urea. *Soil Biology and Biochemistry* 42(4), 660–664.

Chen, D., Suter, H., Islam, A., Edis, R., Freney, J. R. and Walker, C. N. (2008). Prospects of improving efficiency of fertiliser nitrogen in Australian agriculture: A review of enhanced efficiency fertilisers. *Soil Research* 46(4), 289–301. https://doi.org/10.1071/SR07197.

Chen, Q., Qi, L., Bi, Q., Dai, P., Sun, D., Sun, C., Liu, W., Lu, L., Ni, W. and Lin, X. (2015). Comparative effects of 3,4-dimethylpyrazole phosphate (DMPP) and dicyandiamide (DCD) on ammonia-oxidizing bacteria and archaea in vegetable soil. *Applied*

Microbiology and Biotechnology 99(1), 477–487. https://doi.org/10.1007/s00253-014-6026-7.

Chen, Z., Wang, H., Liu, X., Liu, Y., Gao, S. and Zhou, J. (2016). The effect of N fertilizer placement on the fate of Urea-^{15}N and yield of winter wheat in Southeast China. *PLoS ONE* 11(4), e0153701.

Christianson, C. B., Baethgen, W. E., Carmona, G. and Howard, R. G. (1993). Microsite reactions of urea-nBPT fertilizer on the soil surface. *Soil Biology and Biochemistry* 25(8), 1107–1117.

Claros, J., Jimenez, E., Borras, L., Aguado, D., Seco, A., Ferrer, J. and Serralta, J. (2010). Short-term effect of ammonia concentration and salinity on activity of ammonia oxidizing bacteria. *Water Science and Technology* 61(12), 3008–3016.

Connellan, J., Thompson, M., Salter, B., Panitz, J. and Olayemi, M. (2021). *Cane Farmer Trials of Enhanced Efficiency Fertiliser in the Catchments of the Great Barrier Reef*. Queensland, Australia: Sugar Research Australia.

Court, M. N., Stephen, R. C. and Waid, J. S. (1964a). Toxicity as a cause of the inefficiency of urea as a fertilizer. II. Experimental. *Journal of Soil Science* 15(1), 49–65.

Court, M. N., Stephen, R. C. and Waid, J. S. (1964b). Toxicity as a cause of the inefficiency of urea as a fertilizers. I. Review. *Journal of Soil Science* 15(1), 42–48.

Creamer, F. L. and Fox, R. H. (1980). The toxicity of banded urea or diammonium phosphate to corn as influenced by soil temperature, moisture, and pH. *Soil Science Society of America Journal* 44(2), 296–300. https://doi-org.ezproxy.library.uq.edu.au/10.2136/sssaj1980.03615995004400020019x.

Creason, G. L., Schmitt, M. R., Douglass, E. A. and Henderickson, L. L. (1990). Urease inhibtory activity associated with N-(n-butyl)thiophosphoric triamide is due to formation of its oxon analog. *Soil Biology and Biochemistry* 22(2), 209–211. https://doi.org/10.1016/0038-0717(90)90088-H.

Dang, Y. P., Martinez, C., Smith, D., Rowlings, D., Grace, P. and Bell, M. (2021). Maize production and nitrous oxide emissions from enhanced efficiency nitrogen fertilizers. *Nutrient Cycling in Agroecosystems* 121(2–3), 191–208. https://doi.org/10.1007/s10705-021-10171-4.

Debruin, J. L., Gorowara, R. L., Schussler, J., Pape, R., Grafton, M., Liu, L., Macchia, J., Kendra, S., Zhang, J. and Burch, R. (2021). Evaluating polymer-coated fertilizer prototypes designed for planting along with maize seed. *Agronomy Journal* 113(2), 1619–1639. https://doi.org/10.1002/agj2.20559.

Dell, C. J., Han, K., Bryant, R. B. and Schmidt, J. P. (2014). Nitrous oxide emissions with enhanced efficiency nitrogen fertilizers in a rainfed system. *Agronomy Journal* 106(2), 723–731. https://doi.org/10.2134/agronj2013.0108.

Dowie, J., Thompson, M. and Anderson, A. (2019). A three-year assessment of controlled-release and nitrification-inhibiting fertilisers in the Burdekin. *Proceedings of the Australian Society of Sugar Cane Technologists* 41, 547–557.

Emerson, K., Russo, R. C., Lund, R. E. and Thurston, R. V. (1975). Aqueous ammonia equilibrium calculations: Effect of pH and temperature. *Journal of the Fisheries Research Board of Canada* 32(12), 2379–2383. https://doi-org.ezproxy.library.uq.edu.au/10.1139/f75-274.

Engel, R. E., Towey, B. D. and Gravens, E. (2015). Degradation of the urease inhibitor NBPT as affected by soil pH. *Soil Science Society of America Journal* 79(6), 1674–1683.

Erickson, R. J. (1985). An evaluation of mathematical models for the effects of pH and temperature on ammonia toxicity to aquatic organisms. *Water Research* 19(8), 1047-1058. https://doi.org/10.1016/0043-1354(85)90375-6.

Feng, G. D., Ma, Y., Zhang, M., Jia, Py., Hu, Lh., Zhou, Y. H. and Zhou, Yh. (2019). Polyurethane-coated urea using fully vegetable oil-based polyols: Design, nutrient release and degradation. *Progress in Organic Coatings* 133, 267-275.

Fixen, P. E. (2020). A brief account of the genesis of 4R nutrient stewarship. *Agronomy Journal* 112(5), 4511-4518. https://doi.org/10.1002/agj2.20315.

Gagnon, B., Ziadi, N. and Grant, C. (2012). Urea fertilizer forms affect grain corn yield and nitrogen use efficiency. *Canadian Journal of Soil Science* 92(2), 341-351. https://doi.org/10.4141/cjss2011-074.

Gandeza, A. T., Shoji, S. and Yamada, I. (1991). Simulation of crop response to polyolefin-coated urea: I. Field dissolution. *Soil Science Society of America Journal* 55(5), 1462-1467. https://doi.org/10.2136/sssaj1991.03615995005500050044x.

Gao, X., Asgedom, H., Tenuta, M. and Flaten, D. N. (2015). Enhaced efficiency urea sources and placement effect on nitrous oxide emissions. *Agronomy Journal* 107(1), 265-277. https://doi.org/10.2134/agronj14.0213.

Geng, J., Ma, Q., Chen, J., Zhang, M., Li, C., Yang, Y., Yang, X., Zhang, W. and Liu, Z. (2016). Effects of polymer coated urea and sulfur fertilization on yield, nitrogen use efficiency and leaf senescence of cotton. *Field Crops Research* 187, 87-95.

Goring, C. A. I. (1962). Control of nitrification by 2-chloro-6-(trichloromethyl)-pyridine. *Soil Science* 93(3), 211-218.

Goyal, S. S. and Huffaker, R. C. (1984). Nitrogen toxicity in plants. In: Hauck, R. D. (Ed.), *Nitrogen in Crop Production*. Madison, Wisconsin, USA: American Society of Agronomy, Crop Science Society of America, Soil Science Society of America. pp. 97-118.

Graham, R. F., Greer, K. D., Villamil, M. B., Nafziger, E. D. and Pittelkow, C. M. (2018). Enhanced-efficiency fertilizer impacts on yield-scaled nitrous oxide emissions in maize. *Soil Science Society of America Journal* 82(6), 1469-1481. https://doi.org/10.2136/sssaj2018.05.0196.

Guardia, G., Marsden, K. A., Vellejo, A., Jones, D. L. and Chadwick, D. R. (2018). Determining the influence of environment and edaphic factors on the fate of the nitrification inhibitors DCD and DMPP in soil. *Science of the Total Environment* 624, 1202-1212. https://doi.org/10.1016/j.scitotenv.2017.12.250.

Gunsalus, I. C., Pederson, T. C. and Sligar, S. G. (1975). Oxygenase-catalyzed biological hydroxylations. *Annual Review of Biochemistry* 44, 377-407.

Gupta, R. K., Abrol, I. P., Finkl, C. W., Kirkham, M. B., Arbestain, M. C., Macías, F., Chesworth, W., Germida, J. J., Loeppert, R. H., Cook, M. G., Schwab, G. O., Konstankiewicz, K., Pytka, J., Oertli, J. J., Singer, A., Edmonds, W. J. and Feng, Y. (2008). Soil microbiology. In: Chesworth, W. (Ed.), *Encyclopedia of Soil Science*. Dordrecht: Springer, pp. 673-678.

Halvorson, A. D. and Grosso, S. J. D. (2013). Nitrogen placement and source effects on nitrous oxide emissions and yields of irrigated corn. *Journal of Environmental Quality* 42(2), 312-322. https://doi.org/10.2134/jeq2012.0315.

Hatfield, J. L. and Follett, R. F. (2008). *Nitrogen in the Environment: Sources, Problems, and Management*. San Diego: Elsevier Academic Press.

Hauck, R. (1985). Slow release and bio-inhibitor-amended nitrogen fertilizers. In: Engelstad, O. (Ed.), *Fertiliser Technology and Use*. Madison, WI: SSSA.

Hauck, R. D. and Stephenson, H. F. (1965). Nitrogen sources, nitrification of nitrogen fertilizers. Effect of nitrogen source, size and pH of granule, and concentration. *Journal of Agricultural and Food Chemistry* 13(6), 486–492.

Hendrickson, L. L. and Keeney, D. R. (1979). A Bioassay to determine the effect of organic matter and pH on the effectiveness of nitrapyrin (N-serve) as a nitrification inhibitor. *Soil Biology and Biochemistry* 11(1), 51–55.

Hendrickson, L. L., Walsh, L. M. and Kenney, D. R. (1978). Effectiveness of nitrapyrin in controlling nitrification of fall and spring-applied anhydrous ammonia 1. *Agronomy Journal* 70(5), 704–708.

Hyman, M. R., Kim, C. Y. and Arp, D. J. (1990). Inhibition of ammonia monooxygenase in *Nitrosomonas europaea* by carbon disulfide. *Journal of Bacteriology* 172(9), 4775–4782.

Hyman, M. R. and Wood, P. M. (1985). Interaction of ammonia monooxygenase from *Nitrosomonas europaea* by acetylene. *Biochemical Journal* 227(3), 719–725.

International Fertilizer Association. (2009). *The Global "4R" Nutrient Stewardship Framework. IFA Task Force on Fertilizer Best Management Practices.* Paris, France: International Fertilizer Industry Association (IFA).

International Union of Soil Sciences Working Group. (2015). World reference base for soil resources 2014. *World Soil Resources Reports*. Rome.

Irigoyen, I., Muro, J., Azpilikueta, M., Aparicio-Tejo, P. and Lamsfus, C. (2003). Ammonium oxidation kinetics in the presence of nitrification inhibitors DCD and DMPP at various temperatures. *Soil Research* 41(6), 1177–1183.

Isayenkov, S. V. and Maathuis, F. J. M. (2019). Plant salinity stress: Many unanswered questions remain. *Frontiers in Plant Science* 10, 80. https://doi.org/10.3389/fpls.2019.00080.

Janke, C., Fujinuma, R., Moody, P. and Bell, M. (2021). Biochemical changes and distribution of nitrogen from bands of stabilised N-fertilizers in contrasting soils. *Geoderma* 382, 114770. https://doi.org/10.1016/j.geoderma.2020.114770.

Janke, C. K. (2020). The impact of banding urea and enhanced efficiency fertilizers on nitrogen transformations and the implications for nitrogen use efficiency in high-risk environments. Doctor of Philosophy, The University of Queensland.

Janke, C. K., Fujinuma, R., Moody, P. and Bell, M. J. (2019). Biochemical effects of banding limit the benefits of nitrification inhibition and controlled release technology in the fertosphere of high N-input systems. *Soil Research* 57(1), 28–40. https://doi.org.ezproxy.library.uq.edu.au/10.1071/SR18211.

Janke, C. K., Moody, P. and Bell, M. J. (2020). Three-dimensional dynamics of nitrogen from banded enhanced efficiency fertilizers. *Nutrient Cycling in Agroecosystems* 118(3), 227–247. https://doi.org/10.1007/s10705-020-10095-5.

Janke, C. K., Moody, P., Fujinuma, R. and Bell, M. (2022). The impact of banding polymer-coated urea on nitrogen availability and distribution in contrasting soils. *Journal of Soil Science and Plant Nutrition* 22(3), 3081–3095. https://doi.org/10.1007/s42729-022-00869-x.

Jansen van Rensburg, H. G., Claassens, A. S. and Beukes, D. J. (2009). Relationships between soil buffer capacity and selected soil properties in a resource-poor farming area in the Mpumalanga Province of South Africa. *South African Journal of Plant and Soil* 26(4), 237–243. https://doi.org/10.1080/02571862.2009.10639961.

Juliette, L. Y., Hyman, M. R. and Arp, D. J. (1993). Inhibition of ammonia oxidation in *Nitrosomonas europaea* by sulphur compounds: Thioethers are oxidized to sulfoxides by ammonia monooxygenase. *Applied and Environmental Microbiology* 59(11), 3718-3727. https://doi.org/10.1128/aem.59.11.3718-3727.1993.

Kelliher, F. M., Clough, T. J., Clark, H., Rhys, G. and Sedcole, J. R. (2008). The temperature dependence of dicyandiamide (DCD) degradation in soils: A data synthesis. *Soil Biology and Biochemistry* 40(7), 1878–1882.

Lawrencia, D., Wong, S. K., Low, D. Y. S., Goh, B. H., Goh, J. K., Ruktanonchai, U. R., Soottitantawat, A., Lee, L. H. and Tang, S. Y. (2021). Controlled release fertilizers: A review on coating materials and mechanisms of release. *Plants* 10(2), 238. https://doi.org/10.3390/plants10020238.

Lewis, D. C. and Stefanson, R. C. (1975). Effect of N-serve on nitrogen transformations and wheat yields in some Australian soils. *Soil Science* 119(4), 273–279.

Lian, J., Liu, W., Meng, L., Wu, J., Zeb, A., Cheng, L., Lian, Y. and Sun, H. (2021). Effects of microplastics derived from polymer-coated fertilizer on maize growth, rhizosphere, and soil properties. *Journal of Cleaner Production* 318, 128571. https://doi.org/10.1016/j.jclepro.2021.128571.

Liu, S., Wang, X., Yin, X., Savoy, H. J., Mcclure, A. and Essington, M. E. (2019). Ammonia volatilization loss and corn nitrogen nutrition and productivity with efficiency enhanced UAN and urea under No-tillage. *Scientific Reports* 9(1), 6610. https://doi.org/10.1038/s41598-019-42912-5.

Maharjan, B., Ferguson, R. B. and Slater, G. P. (2017). Irrigated corn productivity as influenced by nitrogen source, rate, and climatic conditions. *Agronomy Journal* 109(6), 2957-2965. https://doi-org.ezproxy.library.uq.edu.au/10.2134/agronj2017.04.0209.

Maharjan, B. and Venterea, R. T. (2013). Nitrite intensity explains N management effects on N_2O emissions in maize. *Soil Biology and Biochemistry* 66, 229–238. https://doi.org/10.1016/j.soilbio.2013.07.015.

Mahler, R. (2001). Fertilizer placement. In: *UI Extension Publications and Multimedia Catalog, Fertilizers and Soils, General*. University of Idaho Extension.

Manunza, B., Deiana, S., Pintore, M. and Gessa, C. (1999). The binding mechanis of urea, hydroxamic acid and N-(N-butyl)-phosphoric triamide to the urease active site. A comparative molecular dynamics study. *Soil Biology and Biochemistry* 31(5), 789–796.

Marsden, K. A., Marin-Martinez, A. J., Vallejo, A., Hill, P. W., Jones, D. L. and Chadwick, D. R. (2016). The mobility of nitrification inhibitors under simulated ruminant urine deposition and rainfall: A comparison between DCD and DMPP. *Biology and Fertility of Soils* 52(4), 491–503.

Martinez, C., Clarke, D., Dang, Y. P., Janke, C. and Bell, M. J. (2021). Integrated field assessment of nitrogen release dynamics and crop recovery of band-applied controlled-release fertilisers. *Plant and Soil* 466(1–2), 257–273. https://doi.org/10.1007/s11104-021-05043-3.

Martinez, C., Janke, C., Dang, Y. and Bell, M. (2022). *Nitrogen Release Dynamics of Enhanced Efficiency Fertilisers (EEFs): Placement, Soil Factors and Plant Uptake*. GRDC Update Papers.

McCarty, G. W. (1999). Modes of action of nitrification inhibitors. *Biology and Fertility of Soils* 29(1), 1–9. https://doi.org/10.1007/s003740050518.

McCarty, G. W. and Bremmer, J. M. (1986). Inhibition of nitrification in soil by acetylenic compounds. *Soil Science Society of America Journal* 50(5), 1198-1201. https://doi.org/10.2136/sssaj1986.03615995005000050021x.

Medina, R. and Radel, R. (1988). Mechanisms of urease inhibition. In: Brock, B. and Kissel, D. (Eds.), *Ammonia Volatilization From Urea Fertilizers*. Alabama: National Fertilizer Development Centre.

Menendez, S., Barrena, I., Setien, I., Gonzalez-Murua, C. and Estavillo, J. M. (2012). Efficiency of nitrification inhibitor DMPP to reduce nitrous oxide emissions under different temperature and moisture conditions. *Soil Biology and Biochemistry* 53, 82–89.

Meyer, G., Bell, M. J., Doolette, C. L., Brunetti, G., Zhang, Y., Lombi, E. and Kopittke, P. M. (2020). Plant-available phosphorus in highly concentrated fertilizer bands: Effects of soil type, phosphorus form, and coapplied potassium. *Journal of Agricultural and Food Chemistry* 68(29), 7571–7580. https://doi.org/10.1021/acs.jafc.0c01287.

Migliorati, M. D. A., Bell, M., Lester, D., Rowlings, D. W., Scheer, C., de Rosa, D. and Grace, P. R. (2016). Comparison of grain yields and N_2O emission on oxisol and vertisol soils in response to fertiliser N applied as urea or urea coated with the nitrification inhibitor 3, 4-dimethylpyrazole phosphate. *Soil Research* 54(5), 552–564. http://doi.org/10.1071/SR15336.

Naz, M. Y. and Sulaiman, S. A. (2016). Slow release coating remedy for nitrogen loss from conventional urea: A review. *Journal of Controlled Release* 225, 109–120. https://doi.org/10.1016/j.jconrel.2016.01.037.

Nelson, P. N. and Su, N. (2010). Soil pH buffering capacity: A descriptive function and its application to some acidic tropical soils. *Soil Research* 48(3), 201–207. https://doi.org/10.1071/SR09150.

Noellsch, A. J., Motavalli, P. P., Nelson, K. A. and Kitchen, N. R. (2009). Corn response to conventional and slow-release nitrogen fertilizers across a claypan landscape. *Agronomy Journal* 101(3), 607–614. https://doi.org/10.2134/agronj2008.0067x.

Novak, V. and Hlavacikova, H. (2019). *Applied Soil Hydrology*. Cham: Springer.

Nye, P. H. (1980). Diffusion of ions and uncharged solutes in soil and soil clays. *Advances in Agronomy* 31, 225–272.

Oudyn, F. W. and Janke, C. K. (2018). Using cotton wicks as an alternative to soil extracts to monitor the movement of urea and chemical inhibitors in fertiliser band application experiments. National Soils Conference. Canberra, Australia.

Pang, P. C., Hedlin, R. A. and Cho, C. M. (1973). Transformation and movement of band-applied urea, ammonium sulfate, and ammonium hydroxide during incubation in several Manitoba soils. *Canadian Journal of Soil Science* 53(3), 331-341.

Passioura, J. B. and Wetselaar, R. (1972). Consequences of banding nitrogen fertilizers in soil. II. Effects on the growth of wheat roots. *Plant and Soil* 36(1–3), 461–473.

Rochette, P., Angers, D. A., Chantigny, M. H., Gasser, M., MacDonald, J. D., Pelster, D. E. and Bertrand, N. (2013a). NH_3 volatilization, soil NH_4^+ concentration and soil pH following subsurface banding of urea at increasing rates. *Canadian Journal of Soil Science* 93(2), 261-268.

Rochette, P., Angers, D. A., Chantigny, M. H., Gasser, M. O., MacDonald, J. D., Pelster, D. E. and Bertrand, N. (2013b). Ammonia volatilization and nitrogen retention: How deep to incorporate urea? *Journal of Environmental Quality* 42(6), 1635-1642.

Rowlings, D. W., Scheer, C., Liu, S. and Grace, P. R. (2016). Annual nitrogen dynamics and urea fertilizer recoveries from a dairy pasture using ^{15}N; effect of nitrification inhibitor DMPP and reduced application rates. *Agriculture, Ecosystems and Environment* 216, 216–225. https://doi.org/10.1016/j.agee.2015.09.025.

Rychel, K., Meurer, K. H. E., Borjesson, G., Strömgren, M., Getahun, G. T., Kirchmann, H. and Kätterer, T. (2020). Deep N fertilizer placement mitigated N_2O emissions in a

Swedish field trial with cereals. *Nutrient Cycling in Agroecosystems* 118(2), 133–148. https://doi.org/10.1007/s10705-020-10089-3.

Sahrawat, K. L. (2008). Factors affecting nitrification in soils. *Communications in Soil Science and Plant Analysis* 39(9–10), 1436–1446. https://doi.org/10.1080/00103620802004235.

Sanchez, C. A., Swanson, S. and Porter, P. S. (1990). Banding P to improve fertilizer use efficiency of lettuce. *Journal of the American Society for Horticultural Science* 115(4), 581–584.

Sandral, G., Tavakkoli, E., Harris, F., Koetz, E. and Angus, J. Improving nitrogen fertiliser use efficiency in wheat using mid-row banding. In: *Proceedings of the 18th Australian Society of Agronomy Conference*, Ballarat, 2017.

Sanz-Cobena, A., Abalos, D., Meijide, A., Sanchez-Martin, L. and Vallejo, A. (2016). Soil moisture determine the effectiveness of two urease inhibitors to decrease N_2O emission. *Mitigation and Adaptation Strategies for Global Change* 21, 1131–1144.

Sanz-Cobena, A., Sanchez-Martin, L., Garcia-Torres, L. and Vallejo, A. (2012). Gaseous emission of N_2O and NO and NO_3- leaching from urea applied with urease and nitrification inhibitors to a maize (*Zea mays*) crop. *Agriculture, Ecosystems and Environment* 149, 64–73.

Scheer, C., Rowlings, D. W., Migliorati, M. D. A., Lester, D. W., Bell, M. J. and Grace, P. R. (2016). Effect of enhanced efficiency fertilisers on nitrouse oxide emission in a sub-tropical cereal cropping systems. *Soil Research* 54(5), 544–551. https://doi.org/10.1071/SR15332.

Shaviv, A. (2001). Advances in controlled-release fertilizers. *Advances in Agronomy* 71, 1–49. https://doi.org/10.1016/S0065-2113(01)71011-5.

Shaviv, A., Raban, S. and Zaidel, E. (2003). Modeling controlled nutrient release from polymer coated fertilizers: Diffusion release from single granules. *Environmental Science and Technology* 37(10), 2251–2256. https://doi-org.ezproxy.library.uq.edu.au/10.1021/es011462v.

Shi, X., Hu, H. W., Muller, C., He, J. Z., Chen, D. and Suter, H. C. (2016). Effects of the nitrification inhibitor 3,4-dimethylpyrazole phosphate on nitrification and nitrifiers in two contrasting agricultural soils. *Applied and Environmental Microbiology* 82(17), 5236–5248. https://doi.org/10.1128/AEM.01031-16.

Singh, J., Kunhikrishnan, A., Bolan, N. S. and Saggar, S. (2013). Impact of urease inhibitor on ammonia and nitrous oxide emissions from temperate pasture soil cores receiving urea fertilizer and cattle urine. *Science of the Total Environment* 465, 56–63.

Singh, R. and Nye, P. H. (1984). The effect of soil pH and high urea concentration on urease activity in soil. *Journal of Soil Science* 35(4), 519–527.

Smith, K., Watts, D., Way, T., Torbert, H. and Prior, S. (2012). Impact of tillage and fertilizer application method on gas emissions in a corn cropping system. *Pedosphere* 22(5), 604–615.

Soares, J. R., Cantarella, H., Vargas, V. P., Carmo, J. B., Martins, A. A., Sousa, R. M. and Andrade, C. A. (2015). Enhanced-efficiency fertilizers in nitrous oxide emissions from urea applied to sugarcane. *Journal of Environmental Quality* 44(2), 423–430. https://doi.org/10.2134/jeq2014.02.0096.

Steusloff, T. W., Nelson, K. A., Motavalli, P. P. and Singh, G. (2019). Fertilizer placement affects corn and nitrogen use efficiency in a claypan soil. *Agronomy Journal* 111(5), 2512–2522. https://doi.org/10.2134/agronj2019.02.0108.

Thapa, R., Chatterjee, A., Awale, R., Mcgranahan, D. A. and Daigh, A. (2016). Effect of enhanced efficiency fertilizers on nitrous oxide emission and crop yields: A meta-analysis. *Soil Science Society of America Journal* 80(5), 1121–1134. https://doi.org/10.2136/sssaj2016.06.0179.

Trenkel, M. (1997). *Controlled-Release and Stabilized Fertilizers in Agriculture*. Paris, France.

Trenkel, M. (2010). *Slow-and Controlled-Release and Stabilized Fertilizers: An Option of Enhancing Nutrient Use Efficiency in Agriculture*. Paris, France: International Fertilizer Industry Association (IFA).

Venterea, R. T. (2007). Nitrite-driven nitrous oxide production under aerobic soil conditions: Kinetics and biochemical controls. *Global Change Biology* 13(8), 1798–1809. https://doi.org/10.1111/j.1365-2486.2007.01389.x.

Verburg, K., Bristow, K. L., Mclachlan, G. D., Muster, T. H., Poole, J. M., Mardel, J. I., Zhao, Z., Wittwer-Schmid, K. and Wong, M. T. F. (2021). Understanding soil water effects on nitrogen release from controlled-release fertilizers. *Soil Science Society of America Journal* 85(1), 59–72. https://doi-org.ezproxy.library.uq.edu.au/10.1002/saj2.20141.

Wang, W., Park, G., Reeves, S., Zahmel, M., Heenan, M. and Salter, B. (2016a). Nitrous oxide emission and fertiliser nitrogen efficiency in a tropical sugarcane cropping system applied with different formulations of urea. *Soil Research* 54(5), 572–584. https://doi.org/10.1071/SR15314.

Wang, W. J., Reeves, S. H., Salter, B., Moody, P. W. and Dalal, R. C. (2016b). Effects of urea formulations, application rates and crop residue retention on N_2O emissions from sugarcane fields in Australia. *Agriculture, Ecosystems and Environment* 216, 137–146. https://doi.org/10.1016/j.agee.2015.09.035.

Wu, P., Chen, G., Liu, F., Cai, T., Zhang, P. and Jia, Z. (2021). How does deep-band fertilizer placement reduce N_2O emissions and increase maize yields? *Agriculture, Ecosystems and Environment* 322, 107672. https://doi.org/10.1016/j.agee.2021.107672.

Wu, S. F., Wu, L. H., Shi, Q. W., Wang, Z. Q., Chen, X. Y. and Li, Y. S. (2007). Effects of a new nitrification inhibitor 3,4-dimethylpyrazole phosphate (DMPP) on nitrate and potassium leaching in two soils. *Journal of Environmental Sciences* 19(7), 841–847.

Xie, Y., Tang, L., Yang, L., Zhang, Y., Song, H., Tian, C., Rong, X. and Han, Y. (2020). Polymer-coated urea effects on maize yield and nitrogen losses for hilly land of southern China. *Nutrient Cycling in Agroecosystems* 116(3), 299–312. https://doi.org/10.1007/s10705-019-10038-9.

Ying, J., Li, X., Wang, N., Lan, Z., He, J. and Bai, Y. (2017). Contrasting effects of nitrogen forms and soil pH on ammonia ozidizing microorganisms and their responses to long-term nitrogen fertilization in a typical steppe ecosystem. *Soil Biology and Biochemistry* 107, 10–18.

Zantua, M. I. and Bremner, J. M. (1977). Stability of urease in soils. *Soil Biology and Biochemistry* 9(2), 135–140. https://doi.org/10.1016/0038-0717(77)90050-5.

Chapter 9

Dynamic models for addressing complexities of nitrogen management in maize production

Harold M.van Es and Jeff Melkonian, Cornell University, USA; and Rebecca Marjerison, Yara North America, USA

1 Introduction

Improved N use efficiency in cropping systems has become a compelling issue with concerns about cost, energy use, and multiple environmental impacts. N is most challenging to manage among essential crop nutrients due to strong influences of climate, soil, seasonal weather, cropping system and management factors. Also, conventional soil testing approaches are of limited utility for N management because of dynamic transformations of N in cropping systems. Crop yield is related to crop N needs, and higher yields with recent advances in crop genetics and crop and soil management typically necessitate greater N inputs. But a large fraction of cropland areas have N rates applied that substantially exceed the N needed by the crop (McLellan et al., 2018). This generally results in high potential N losses to the environment as evidenced by persistent water quality concerns, including nitrate contamination of groundwater in agricultural regions and hypoxia in estuaries (van Grinsven et al., 2015). Also, gaseous forms of N are the largest contributor to agricultural greenhouse gas emissions (EPA, 2023) and contribute to fine particulate air pollution (Pinder et al., 2007).

Maize (*Zea mays* L.) is the most widely grown commodity crop worldwide and requires high N fertilization levels. A C_4 plant, it is physiologically more

http://dx.doi.org/10.19103/AS.2024.0135.14

efficient at utilizing N (more yield per unit N crop accumulation) than most other non-leguminous crops, which are generally C_3 plants. But paradoxically, maize production systems generally have low fertilizer N use recovery efficiencies (Cassman et al., 2002). Maize is also a major focal crop for concerns related to water and air quality and climate, because it is a dominant source of water-dissolved and aerial losses to the environment (McLellan et al., 2018). Studies have reported rapid increases in nitrate leaching and nitrous oxide gaseous losses when N rates for maize production exceed the optimum (Randall et al., 1997; Millar et al., 2010). This highlights the importance of minimizing excessive N applications which in turn requires correct estimation of maize fertilizer N needs. But the optimum N rate recommendation has remained elusive due to complex N dynamics that vary by location and time.

2 Nitrogen management complexity

2.1 4R decision support

The 4R concept is widely promoted as a way to increase nutrient use efficiency (Snyder et al., 2009). This approach promotes field application of nutrients from the right source, at the right place, right time, and right rate (TFI, 2018). The 4R approach is a straightforward concept and has the potential to steer growers toward managing crop N more efficiently. However, under the complexity of real production environments in farmer fields, effective implementation of the 4R concept is challenging for nitrogen (N). The 'right' answer to each R factor, especially for N *rate*, remains difficult to define as it varies spatially among or within fields, as well as temporally for the same field location. Moreover, the right rate depends in part on how the other 3Rs are implemented, as well as other aspects of a complex production environment (tillage system, rotation and cover crops, organic amendments, etc.), because they affect N transformations and losses (Sela and van Es, 2018). While most other crop nutrients can be managed based on soil testing, the dynamic nature of soil N requires a different approach to estimating the correct N rate. Multiple N decision tools have been developed to aid growers, many of them static and highly generalized in space and time (Morris et al., 2018). In this chapter, we focus on dynamic simulation models as a more integrative approach to N management.

2.2 Nitrogen rate and precision

The ability to estimate the economically optimal N rate (EONR, providing maximum economic return to N input) has remained challenging, because crops like maize show high variability in N response (Scharf et al., 2005; Sawyer et al., 2006). Many factors impact N dynamics in fields and should be considered in the process of determining an EONR. These vary from location to location and season to season, depending on soil texture, weather, N mineralization rates, prior and future N losses, organic N inputs, and yield potential, among others. In fact, the optimum recommendation may vary as the growing season progresses with in-season weather events (van Es et al., 2007).

The concept of 'N balance' (OECD, 2013) can serve as a simple field- and farm-level indicator of N excess and the potential for loss. N balance is defined as the difference between N inputs to, and N removed in products from, an agricultural system. It can provide insights on the impact of farm management on environmental outcomes and can be calculated from records of inorganic and organic nutrient applications and crop yields. Although the N balance concept has limitations, studies have shown that high N rates and N balances are associated with increased N losses (McLellan et al., 2018; Sela et al., 2019; Hoben et al., 2011). The focus therefore should be on reducing N inputs while not jeopardizing yield or causing soil depletion, and in this context Zhang et al. (2015) suggested a critical N balance value of 78 kg ha^{-1} as the limit for sustainable agricultural production. The ability to achieve this target depends on regional factors, like climate, soils, management, and the correct estimation of the optimum N rate (Sela et al., 2019).

2.3 Computing optimum rates

Most current crop N rate recommendations are based on a simple mass-balance equation involving crop N demand (from yield goals) and N credits from soil-specific criteria and crop management practices such as rotation and manure application (Stanford, 1973):

$$N_{Fert} = (N_{Crop} - N_{Soil})/N_{eff}$$

where N_{Fert} is the recommended N rate, N_{Crop} is the N rate required to reach optimum yield, N_{Soil} is the N rate supplied by the soil (including contributions from manure and crop rotations), and N_{eff} is an efficiency factor that accounts for N losses during the production period. These factors are typically based on either user input (expected yield for N_{Crop}) or assumed values (N_{Soil} and N_{eff}).

As an alternative, a generalized empirical approach was developed for maize by averaging the results of numerous N response field trials on a broad regional basis (Sawyer et al., 2006). This maximum return to N (MRTN) approach not only incorporates fertilizer and grain prices into EONR calculations but also inherently assumes that (1) the EONR is not related to yield, and (2) biophysical factors (soil, weather, etc.) and management factors are minimally relevant. Both the generalized empirical approach and the mass-balance approach may account for N credits from rotations and amendments in generalized ways.

The main concern with these static and generalized corn N recommendation methodologies is their limited ability to manage variability in N response associated with different production environments, especially related to weather, and their insufficient consideration of crop management factors (Ransom et al., 2023).

2.4 Factors affecting optimum nitrogen rate

In order to achieve more accurate N rate recommendations for a wide range of production environments, N decision tools need to incorporate the most relevant processes that affect N dynamics and, therefore, crop available N based on a field's biophysical environment and management practices.

2.4.1 Soil and climate factors

Mineralization of soil organic matter (SOM) can potentially supply a significant proportion of the required crop N, with typical values ranging from 60 kg N ha^{-1} to 130 kg N ha^{-1} (Ketterings et al., 2003; Cassman et al., 2002), primarily depending on climate, soil type, and especially organic matter content. The difference between the crop requirement and the soil N supply is ideally provided by organic or synthetic N fertilizer applications (Stanford, 1973; van Es et al., 2007). In general, the soil N supply is a function of readily mineralizable SOM which in turn can be estimated using information on the organic matter content and texture (Melkonian et al., 2008). Soils with higher clay content generally provide less mineralizable N for the same amount of SOM. The labile SOM pool can be enhanced through cropping and agronomic management practices that also improve soil health (Magdoff and van Es, 2021). Soil N supply is furthermore a function of both soil physics and chemistry where temperature, hydrology, and pH among others significantly impact N transformation processes including SOM mineralization (Kyveryga et al., 2004).

Climate (temperature and precipitation) can impact general N availability from soils. Soil biological activity is temporarily reduced in regions that experience either extended periods of cold temperatures or drought. This mostly preserves the labile SOM pool from mineralization until the soil warms in spring or wets at the onset of a rainy season, after which rapid SOM mineralization occurs. When this is in synchrony with the high N uptake growth phase of a new crop, it may provide a significant portion of the crop N required to achieve yield. Conversely, warm and humid regions experience more continuous mineralization of organic N, which implies that the gains and losses of reactive N are more even throughout the year.

2.4.2 Agronomic management factors

Cropping system: The soil N supply is enhanced when the crop is preceded by one that contributes residual biomass (Correndo et al., 2021a). This N credit is especially significant when it involves a perennial legume (e.g. alfalfa [*Medicago sativa*]) or a cropping system that involves common application of significant amounts of manure or compost. Benefits may extend over multiple years with decreasing N contribution (Klausner et al., 1994). N availability is also strongly impacted by management practices like timing of application and soil incorporation, especially involving manure with a significant ammonium fraction (affecting volatile losses; Jokela and Meisinger, 2008). Similarly, cover crops may contribute significant amounts of N to the following maize crop depending on type, biomass amount, and C:N ratio, as well as timing and method of termination (Melkonian et al., 2017).

Synthetic or organic N amendments: Application of N like fertilizer, manure, and compost directly impacts soil N availability. Synthetic N sources like urea, anhydrous ammonia, and ammonium and nitrate salts are generally available for crop uptake within a few days. Organic sources like manure also have rapidly available N as ammonium and nitrate (urea-N component), but the organic N pool releases at a slow rate through mineralization (Klausner et al., 1994). This in turn is impacted by weather conditions and fertilizer placement. Urea-based fertilizers, similar to manure, experience high ammonia losses when applied on the surface under certain soil (high pH) and weather (no subsequent rain) conditions. Enhanced efficiency compounds (EECs), such as urease and nitrification inhibitors, can delay N transformations and result in reduced loss potential of soil N (Abalos et al., 2014; Halvorson et al., 2014) and also lower the EONR (Sela et al., 2019). The benefits of using EECs are expected to be higher in production environments that are more prone to N losses, such as very coarse or fine-textured soils in climates that experience high rainfall events during the early growing season.

Nitrogen placement: The location of fertilizer N application can affect N losses and crop availability. Incorporated applications, for example, are expected to increase crop N availability compared to surface broadcasting (see Nkebiwe et al. (2016) for a recent review), especially when the formulation is susceptible to losses through volatilization such as with urea or high ammonium organic sources (Rochette et al., 2009). Banding of N application, e.g. through injectors or dribblers can also enhance N availability by applying the product closer to the crop roots. However, there is uncertainty regarding the quantitative gains of such practices in different production environments.

Timing of nitrogen application: The scheduling of N application affects the ability to synchronize crop N availability with crop N needs, which is known to minimize environmental losses. In the US, only about a third of US maize growers apply fertilizer N within the active growing season, with the remainder applying either in the fall, in the spring, or both (USDA ERS, 2010). Livestock-based farm operations often apply manure to fields well before crop uptake, typically for logistical reasons. These early fertilizer or manure applications also increase the risk of N losses, especially under humid climate conditions. In a multi-year simulation study including 19 midwestern locations, split fertilizer applications where the majority of N is applied within the growing season required on average 50% and 40% less N than fall and spring preplant N applications, respectively (McLellan et al., 2018). Conversely, other studies found mixed or no agronomic or environmental benefits in shifting N application from spring to in-season (Kovács et al., 2015; Pittelkow et al., 2017). In general, it is challenging for growers to understand the field and climate conditions that benefit from modifying the timing of N application in the context of logistical and economic considerations.

Other 'R' practices: The 4R concept could be extended to include more factors. These include (1) proper irrigation and drainage management, which impact yield as well as the potential for N losses and associated crop N availability; (2) soil health management, affecting among others soil structure and organic matter content, and thereby N availability; (3) appropriate crop management through precision seeding and location-adapted crop genetics; and (4) optimized use of other agronomic inputs that affect crop growth, including management of pH, other nutrients, pest control, etc. These factors all affect the ability to optimize the management of N, especially to determine the right N rate. Together, the broader use of best N management practices may be referred to as 4R+ (Sela and van Es, 2018).

2.4.3 Weather factors

In addition to climate factors, shorter-term weather events are significant sources of variability for the EONR (Sogbedji et al., 2001; Ransom et al., 2021). Fields that received high precipitation in the time period around sidedressing were found to have greater N response (higher EONR) than those receiving low precipitation (Tremblay et al., 2012). From an extensive analysis of 481 site-years of maize N response curves in North America, Correndo et al. (2021b) determined that weather parameters were by far the most relevant (72% to 81% of variability) to predicting N response. But a unique challenge with weather factors is that they are not predictable nor manageable, and they interact strongly with the inherent (soil) and management (crop rotation, tillage, nutrient applications, etc.) factors (Kay et al., 2006). For most crops, including maize, *early*-season events are the strongest determinant for optimum N rates (Sogbedji et al., 2000; Xie et al., 2013). This is related to the positive soil water balance generally experienced in early parts of the growing season in rainfed systems, i.e. precipitation rates higher than low evapotranspiration (ET) rates driving greater N leaching and denitrification losses. These N loss risks generally decline when the crop reaches higher ET rates as the leaf area increases and soils dry out. The strong early-season signal of precipitation provides opportunities for in-season supplemental N applications without yield loss (Nafziger et al., 2021) through adaptive tools (van Es et al., 2007). Moreover, the yield potential of maize is in part impacted by early-season conditions (Abendroth et al., 2011), which allows for adjustment of yield estimates and N rates.

2.4.4 Risk affecting optimum nitrogen rate

Correndo et al. (2021b) concluded that improved N rate recommendations should be dynamic using weather, soil, and agronomic information but should also consider the uncertainty of yield response to N as a risk component (risk = probability × cost of losses). In fact, the uncertainty with the static recommendation systems affects an underappreciated aspect of N management, *asymmetric risk* (Fig. 1). From a farmer perspective, the uncertainty in optimum N rate poses risk for profit losses. It is related to either unnecessary fertilizer expense when the EONR is overestimated, or yield loss when the EONR is underestimated (Fig. 1a). But fertilizer N cost, although often a major crop input expense, typically is still low compared to the costs from yield losses. When the EONR is uncertain, producers tend to manage this asymmetric risk by hedging against insufficient N. For a typical N response function based in US Midwest maize N rate experiments, the integration of risk with N response implies that the EONR is approximately 34 kg ha^{-1} higher (Fig. 2) than when it is solely based on N response and fertilizer and grain prices (like with the MRTN approach; Sawyer et al., 2006). Although farmers are often criticized for overfertilizing, the use of excessive (insurance) fertilizer is a rational economic response to uncertainty

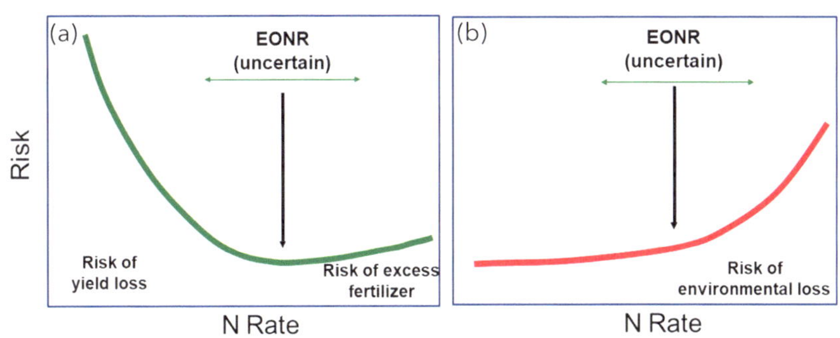

Figure 1 Asymmetric risk with uncertain optimum N rates encourages excessive N rates with producers (a), while social costs associated with environmental losses favor reduced N rates (b). Source: Adapted from: van Es et al. (2002); Hoben et al. (2010); Sela et al. (2019).

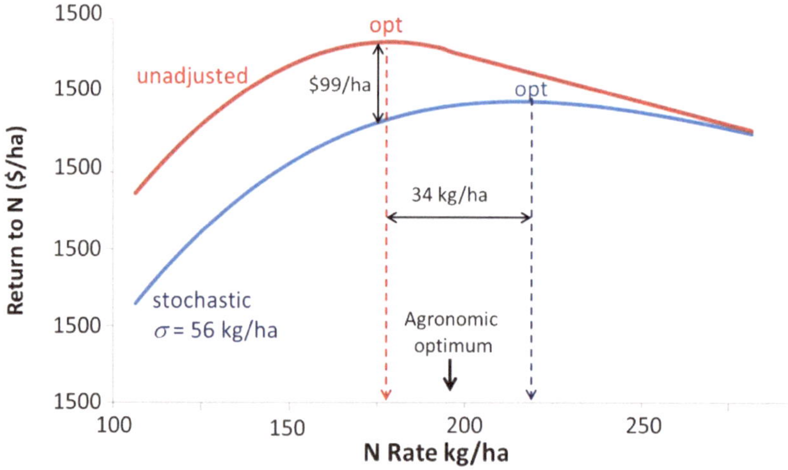

Figure 2 Optimum N rates are increased by 34 kg ha^{-1} and economic returns decrease by \$99 ha^{-1} when asymmetric risk is incorporated into calculation of optimum N rates based on N response and fertilizer/grain price ratios. (Based on N response curve from typical Midwest US maize field following soybean and EONR uncertainty of σ = 56 kg ha^{-1}

around the actual EONR for a specific field and growing season. Conversely, concerns about environmental impacts of excess N are pressuring farmers to reduce their N balance and limit those impacts, i.e. social (environmental impacts) risk is asymmetric toward higher N rates while farmer risk is toward lower N rates (Fig. 1a and b). These opposing interests are best addressed by providing more accurate location- and time-specific recommendations that increase precision and explicitly address the risk to farmers.

3 Improved nitrogen rate prediction technologies

3.1 Adaptive approaches

Due to the complex interconnections between soil-climate and management factors that in turn interact with unpredictable weather factors, an adaptive and dynamic N management approach is needed that comprehensively considers the soil-crop N system (Jin et al., 2017). This is best done in mid–late vegetative stages of the maize crop when potential N losses from early spring rain events have passed. In such cases, opportunities to apply supplemental N that optimizes yield outcomes still exist with high-clearance field equipment. In-season soil testing (e.g. pre-sidedress N test; Magdoff et al., 1984) provides insights into the combined effects of management practices and weather to refine N fertilizer applications. Despite some adoption, there are challenges in using these tests as they are labor intensive and somewhat imprecise (Klausner et al., 1993), especially when there have been banded/injected nutrient applications.

Plant N status can be measured from leaf samples but is generally more efficiently assessed through remote or proximal sensing of crop canopies from different platforms (satellite, aircraft, or field equipment mounted; Colaço and Bramley, 2018). Current challenges with the sensing approaches to N management include required reference strips with high N application rates, detection problems at adequate plant N levels due to biomass or water stress, and the unknown risk and economic factors (Hatfield, 2013). Satellite-based sensing is often limited by relatively long passover intervals (3–5 days) combined with common cloud cover so that actionable sensing information during critical N management periods is not available. Aircraft-based imaging is generally costly, and unoccupied aerial vehicles (drones) still have limited coverage.

3.2 Dynamic simulation models incorporating complexity of production environments

Location- and time-specific N recommendations can be made through the application of dynamic simulation models of the soil-crop system (Kersebaum, 1995; Tremblay et al., 2018). Recent advances in data-driven computational and information technologies stimulated a new generation of dynamic-adaptive N recommendation tools based on the application of mechanistic or stochastic models (Melkonian et al., 2008). These tools aim for real-time monitoring of soil and crop N status and, in most cases, recommend N application rates. As a management approach, simulation models are generally well suited to integrating 4R+ factors and account for different soil, weather, and management elements to derive optimum N rates (Sela and van Es, 2018). There have been several model-based tools that have been commercialized for use by farmers, including the tool discussed in the next section (Adapt-N). However, with most models there is little known about the mechanics or performance validation for different production environments.

Use of simulation models involves data inputs and includes associated uncertainty with inputs and outputs. However, the use cost can be low due to high scalability of the technology. Once the model and data framework have been built, the marginal cost of additional users or land areas tends to decline as upscaling mostly involves low-cost additional cloud computer resources. Model-driven recommendation systems also allow for continuous (daily) monitoring of field N conditions throughout the growing season. They can provide alerts when threshold conditions are reached, while sensing and sampling approaches are generally based on one-time or very infrequent assessments.

In order to adequately simulate N dynamics in a soil-crop system, Sela and van Es (2018) suggested data inputs from five different factor groupings: (1) inherent resources (soil type and health, climate, etc.), (2) agronomic management (rotations, cultivar, cover crops, tillage, irrigation/drainage, etc.), (3) the 3Rs other than rate (source, placement, and timing of previous N applications), (4) weather (precipitation, temperature, solar radiation at daily interval or less), and (5) prices and risk information. The modeling system needs to be able to simulate all interacting factors to derive estimates of the N status of the soil and crop system and ultimately develop an N fertilizer recommendation. Model simulations could be augmented with external data inputs from either sampled or sensed field observations (Sela and van Es, 2018).

Data-driven tools that integrate these management concepts and dynamically adapt N management to the localized conditions of the production environment can play a central role in unifying the fragmented 4R+ components into a coherent N management approach. It also allows for N decision tools to adapt to evolving multi-objective (yield, water quality, GHG emissions, energy) production environments, i.e. where the factors affecting the EONR change over time as a result of environmental, technological, and policy changes. This includes changing weather patterns due to climate change, enhanced soil health, adoption of innovative agronomic practices like no-tillage and cover cropping, greater cycling of bionutrients from organic waste materials, new crop genetics, enhanced efficiency products, and new types of application equipment. Without dynamic models, the impacts of numerous factorial combinations of practices and the changes in technology cannot be identified (Stanford, 1973; Sawyer et al., 2006).

4 Case study: Adapt-N

4.1 Framework

The Adapt-N tool (adapt-N.com) is a cloud-based N decision technology that includes virtual monitoring of soil N, crop N, and other agronomic and environmental elements using daily model simulations. It incorporates the main factors affecting EONR discussed earlier in this chapter (soil climate, agronomic management, weather, and economic risk). Adapt-N provides real-time N recommendations for maize-based systems using daily high-resolution climate data and regionally adapted soil parameters under a diverse range of management scenarios (Melkonian et al., 2008). It accounts for different maize systems (grain, silage, sweet corn); cultivars; fertilizer formulations; tillage systems and residue levels; crop rotations; cover crops; organic material

Table 1 Adapt-N inputs

Weather	High-resolution daily precipitation, temperature, and solar radiation data
Soil	Texture/soil type, slope, rooting depth, % organic matter, pH, from which hydrologic and N transformation parameters are derived
N fertilizer applications	Rate, type, timing, placement, stabilizers
Maize cultivar	Silage, grain, or sweet corn; planting date; maturity class; population; expected yield
Tillage system	Tillage system and residue management; timing
Organic inputs (manure, etc.)	Date, rate, N analysis, incorporation date and type
Rotations	Multiple preceding annual or perennial crops - last 3 years, % legume, surface killed or incorporated
Cover crops	Types, mixes, termination date, stage, and method, stand quality
Irrigation	Amounts and dates
Economic and risk	Fertilizer and grain prices and EONR uncertainty
Field observations	PSNT result; date of emergence

applications; and enhanced efficiency products (stabilizers, etc.; Table 1). It also accommodates field observations to enhance prediction accuracy (date of crop emergence, in-season soil nitrate test results, irrigation, and soil pH). In addition to daily tracking of soil and crop N status, the model provides estimates of crop stage and soil water status (facilitating irrigation planning). Adapt-N's primary client group is fertilizer retailers and crop consultants. Users are generally accessing it with either the Adapt-N interface or through application programming interface (API) technology with third-party software.

4.2 Adapt-N rate calculator

Adapt-N incorporates both deterministic and stochastic model components that are used to derive optimum N recommendations at either field, zone, or grid (20 × 20 m) scale. N rate recommendations calculated by Adapt-N are based on a mass-balance equation (Melkonian et al., 2008):

$$N_{Fert} = Exp(NCrop_{Harvest}) - NCrop_t - NSoil_t - Exp(NSoil_t)$$
$$Exp(NLeg) + Exp(NLoss_t) - Nprofit$$

Equation components that include Exp involve expected future values based on either user input or a stochastic assessment. Components that include the subscript $_t$ refer to a real-time estimate for the date of simulation. $Exp(NCrop_{Harvest})$ is the expected total uptake of maize N at physiological maturity and is based on expected yield and an adjustable N uptake multiplier. $NCrop_t$ is the total crop uptake of N up to the stage of development at time *t*, the date the simulation. $NSoil_t$ is the amount of plant-available N in the soil, also at time *t*. Exp($NSoil_t$) refers to the expected additional plant-available N in the soil from mineralization after time *t* through the end of the season. This is a stochastic value that is estimated from simulations based on 30-year climate data but is adjusted for the date of simulation (more post-simulation mineralized N is expected earlier in the season). The Exp(NLeg) term represents a credit of plant-available N from a previous legume crop (soybean). The Exp($NLoss_t$) term is an estimate of the applied N that will be lost from the soil–plant system after a fertilizer application is made, also from simulations based on 30-year climate data. The final term, Nprofit, adjusts the recommended N rate based on the prices of fertilizer and grain, into which an asymmetric economic risk factor is integrated based on residual uncertainties of the EONR (Fig. 2).

4.3 Model simulations

The Precision Nitrogen Management (PNM) model (Melkonian et al., 2007) is the core of the Adapt-N technology and is used to estimate components of the mass-balance equation. It is a set of process-based computational routines that simulate soil hydrology, chemical transformations and movement of N in the soil, as well as crop growth and yield. The PNM model was developed from two earlier models, the LEACHN soil model (Hutson, 2010) and a maize growth model by Sinclair and Muchow (1995). Transformations of C and N are primarily modeled using functions modulated by soil temperature and soil water content as predicted from high-resolution climate data (Melkonian et al., 2008). The model also reports environmental N losses in the forms of gases or leachate.

The historical evolution of the Adapt-N technology (discussed in detail by van Es et al., 2020) followed a series of development steps: (1) model development, (2) model calibration and validation efforts based on detailed multi-year field trials, (3) development of an integrated decision tool and user interface, and (4) independent validation on commercial farms in broad geographical regions in the US.

4.4 Model inputs

Table 1 summarizes data inputs for the Adapt-N tool, which are generally available from either field management software, public databases, or Adapt-N's internal data dictionaries. Some routines incorporate adaptations to regional differences in production environments based on climate, soil, and agronomic management practices. For North America, Adapt-N includes eighteen soil-climate regions. As an example, more detailed cover crop input requirements are provided in Table 2. It represents the most important information that is

Table 2 Adapt-N cover crop user input options

Cover crop type	*Grass*	*Legume*	*Mix: mostly grass*	*Mix: 50-50*	*Mix: Mostly legume*
Stage at termination	*Tillering/ early vegetative*	*Stem elongation/ mid vegetative*	*Boot-head/ bud-flower*	*Anthesis*	
Quality of stand	*Good*	*Medium*		*Poor*	
Termination date	*Calendar date*				
Incorporation	*None*	*Light*	*Full*		

required to achieve good estimates of net N mineralization from a previous cover crop (Melkonian et al., 2017).

An important feature of Adapt-N is its dynamic access to gridded high-resolution (4 × 4 km) weather data (maximum and minimum temperature, precipitation, solar radiation; Table 1), which allows for field-specific and timely adjustments. The high resolution weather database are derived from public or commercial temperature models and operational Doppler radars (DeGaetano and Belcher, 2007; DeGaetano and Wilks, 2009). Soils data used in Adapt-N are adapted from NRCS SSURGO datasets (http://soildatamart.nrcs.usda.gov/).

4.5 User experience

Adapt-N provides N recommendations at the field, zone, and grid (20 × 20 m) scale (Fig. 3). Users receive tabular and graphical simulation results to gain insights and understanding of the N recommendation. This includes crop stage and N uptake, N in the soil (root zone and top 30 cm), N needed at crop harvest, N mineralized from organic sources (e.g. SOM, previous crop biomass, and organic amendments), N that was recently lost to the environment, predicted future N losses, rainfall received, and soil water status and irrigation needs. In addition, more detailed graphical outputs (by date) are provided for N leaching, N gaseous losses, N mineralization, inorganic N in the soil, precipitation (cumulative, daily), and crop stage development (Fig. 3). The interface has review and editing features for input data, map displays, aggregated field information (e.g. N rates and total field N needs), and export utilities tailored to products (including prescription shapefiles). Users can further adjust the tool by specifying fertilizer and grain prices, crop N content at crop maturity, and threshold levels for daily alerts by text or email.

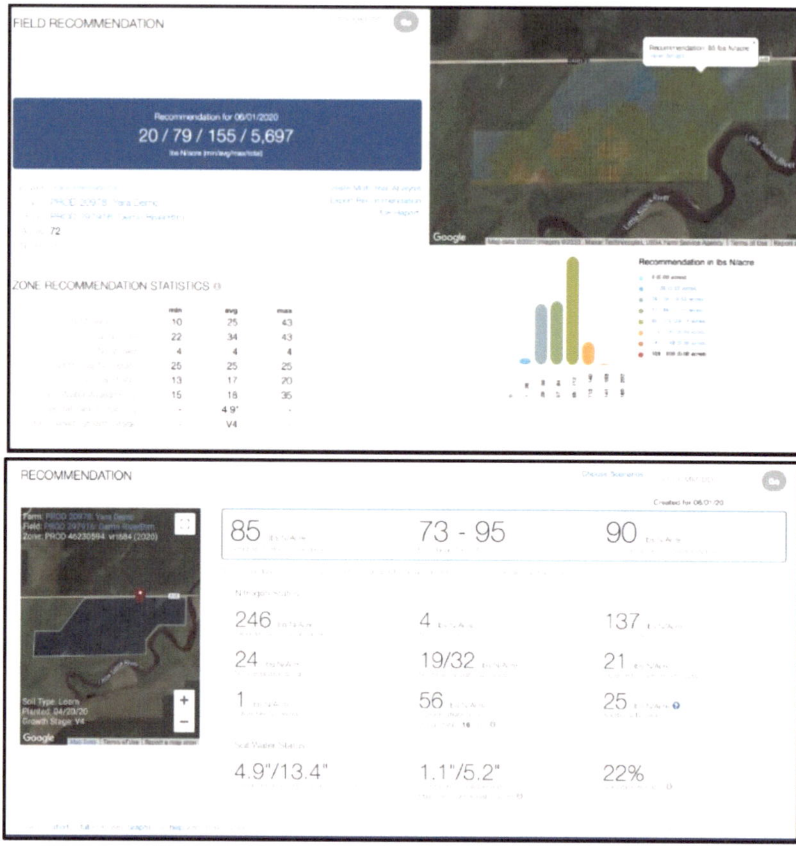

Figure 3 User interface for a field recommendation and detailed output of a single management unit.

4.6 Technology performance

Agronomic performance: Adapt-N recommendations were compared with grower practice under a variety of management scenarios through on-farm strip trials in New York and Iowa (n = 113; Sela et al., 2016). N inputs were on average 45 kg ha^{-1} lower and marginal profits \$64 ha^{-1} higher with Adapt-N rates. The monetary benefits were associated with fertilizer savings when Adapt-N rates were lower (83% of cases) and averted yield reductions from N deficits when the tool recommended higher rates (17% of cases; mostly in wet years). Another 16 on-farm multi-N rate trials in New York and 23 trials in the Midwest (Indiana, Ohio and Wisconsin; Sela et al., 2017) compared Adapt-N recommendations with estimated *ex-post facto* EONR values. In the New York trials Adapt-N achieved RMSE values of 33 kg ha^{-1} and an average bias of −7 kg ha^{-1}. For the Midwest trials Adapt-N also had an RMSE of 33 kg ha^{-1} and bias of −10 kg ha^{-1}. Profits were \$64 ha^{-1} higher for Adapt-N for the New York trials and \$25 ha^{-1} higher for the Midwest trials compared to official static N rate calculators based on a Stanford-type equation and MRTN, respectively. Recent refinements of model parameters based on trial results (e.g. N uptake efficiency) have improved the precision and reduced bias in N recommendations.

In on-farm strip trials in North Carolina, Adapt-N on average recommended 13 kg N ha^{-1} less than experienced crop consultants in 38 site-years with identical profits (Osmond et al., 2018). Recent multi-year independent evaluations of the Adapt-N technology in Nebraska, USA, demonstrated that the technology achieves higher N use efficiencies with the same yields and partial profits as conventional grower practice or canopy sensor-based recommendations (University of Nebraska, 2022, 2023).

Environmental performance: The environmental impacts were assessed from some of the above trials by simulating leaching (beyond bottom of root zone) and gaseous losses (from the soil surface). For the Adapt-N vs. Grower trials, estimated post-sidedress N losses averaged 36% lower leaching and 39% lower gaseous losses for Adapt-N (Sela et al., 2016). In the multi-N rate studies, Adapt-N reduced simulated N leaching losses by 53% and gaseous losses by 54% compared to the conventional N calculator in New York trials and by 3% and 22%, respectively, in Midwest trials (Sela et al., 2017). A separate field lysimeter study corroborated these results (van Es et al., 2020) with mean leached NO_3 concentrations reduced by 37% and 41% for a clay loam and loamy sand soil, respectively, while yields were not negatively impacted.

4.7 4R+ analysis features

The abovementioned studies only addressed N rate differences. Larger gains may be achieved when multiple 4R+ practices are optimized (Sela and van Es, 2018). The ability to gain insights into the effects of management practices and seasonal weather conditions is an additional feature of model-based frameworks, offering learning opportunities for farmers and consultants on how 4R+ factors impact N gains and losses. Features explorable in Adapt-N that have been shown to contribute to enhanced cropping system sustainability include

- estimates of reduced N inputs and environmental losses associated with improved N application timing;
- estimates of the benefits of improved fertilizer technology and enhanced efficiency products; and
- estimates of the benefits of organic inputs, alternative rotations, and cover crops, as well as increased soil health.

Table 3 Adapt-N sidedress recommendations, total N requirements, and estimated total N losses (since October 1 of previous year) for June 30 for a maize field in Missouri, USA, that received either no previous N applications or 112 kg ha^{-1} anhydrous N in the previous fall (November 25)

Growing season	30 June Adapt-N rate rec (change from 2019); total N (kg ha^{-1})		Total N loss since October 1 (kg ha^{-1})		Crop availability of fall fertilizer
	No previous N	Fall fertilizer N	No previous N	Fall fertilizer N	
2022	146 (−4%); 146	108 (−4%); 220	66	151	34%
2021	140 (−8%); 140	96 (-12%); 208	62	140	40%
2020	146 (−4%); 146	108 (−4%); 220	66	152	33%
2019	151 (0%); 151	113 (0%); 225	65	151	33%
2018	123 (−19%); 123	57 (−40%); 169	50	110	57%
2017	140 (−8%); 140	102 (−8%); 214	71	156	34%
2016	129 (−16%); 129	57 (−40%); 169	57	115	58%
2015	140 (−8%); 140	91 (−16%); 203	59	132	45%
2014	140 (−8%); 140	96 (−20%); 208	56	123	50%
2013	146 (−4%); 146	108 (−4%); 220	63	145	36%
2012	123 (−19%); 123	46 (−48%); 158	55	105	65%

Planting date: May 13; previous crop: soybean; SOM: 2.4%; yield target: 15.8 Mg ha^{-1}. Change from 2019 is calculated as it required the highest N rate.

Also, simulations for up to 10 preceding years can be performed to understand seasonal variability on the estimated optimum N rate. Table 3 shows the results of a year-by-year comparison of the estimated optimum N rate for the years 2022 back to 2012 for a maize field in Missouri, USA. N recommendations for a scenario involving mainly in-season N applications (i.e. the N recommendation provides the total amount of needed N) was compared with one where a previous fall N fertilizer application of 112 kg ha^{-1} occurred (the Adapt-N recommendation is supplemental). The next season's crop availability of fall applied N fertilizer was estimated to range from as low as 33% for the wet year 2019 to 65% in the dry year 2012. This is related to higher N losses when most of N fertilizer is applied in the previous fall, resulting in considerably higher average total N requirements (63 kg ha^{-1}) with the fall applied N scenario compared to the scenario with mainly in-season applications. Even in the dry years 2012 and 2016, the total N requirements were 35 kg ha^{-1} and 40 kg ha^{-1} higher, respectively, when N fertilizer was applied in the fall. In addition, annual variability in EONR increases when more N is applied outside the growing season. For example, in the dry seasons of 2018 and 2012 the estimated EONR was only 19% lower compared to the highest N recommendation in the wet year 2019 when all N is applied in-season. However, when there is a fall N application, EONR for 2018 and 2012 was 40% and 48%, respectively, lower than for 2019. Therefore, when N fertilizer applications are not timed close to crop uptake, both N use efficiency and uncertainty are increased.

5 Conclusion

Correct estimation of crop N requirements is challenging due to the dynamic processes that impact N availability and optimum N rates under field conditions. This is especially the case with increasing complexity of crop production environments as farmers adopt sustainable management practices and use more organic inputs. Dynamic simulation models offer opportunities for data-driven N management through simulations of diverse sets of crop production systems, but other investments are required to turn a modeling framework into an effective and commercially viable technology for use by farmers and advisors. These include (1) well-conceived, parameterized, and validated simulation routines and data inputs, (2) consistent dynamic access to real-time climate data, (3) regional adaptation to production environments (soil, climate, crops, management practices), (4) adaptive testing and refinement of model features and parameters through on-farm testing and user feedback, and (5) options for integration into third-party field management software products through API links. In all, model-based N decision technologies can provide significant producer and societal benefits and are inherently well suited to allow farmers to adapt to technological advances and new production practices. These technologies, when transparent about mechanics and performance validation for different production environments, need greater consideration as part of climate, water, and energy initiatives.

6 Conflicts of interest

The authors confirm that there was no conflict of interest writing this chapter.

7 References

Abalos, D., Jeffery, S., Sanz-Cobena, A., Guardia, G. and Vallejo, A. 2014. Meta-analysis of the effect of urease and nitrification inhibitors on crop productivity and nitrogen use efficiency. *Agriculture, Ecosystems and Environment* 189, 136–144.

Abendroth, L. J., Elmore, R. W., Boyer, M. J. and Marlay, S. K. 2011. *Corn Growth and Development*. . Iowa State University Extension PMR 1009.

Cassman, K. G., Dobermann, A. and Walters, D. T. 2002. Agroecosys- tems, nitrogen-use efficiency, and nitrogen management. *Ambio* 31(2), 132–140. https://doi.org/10.1579/0044-7447-31.2.132.

Published by Burleigh Dodds Science Publishing Limited, 2024.

Colaço, A. F. and Bramley, R. G. V. 2018. Do crop sensors promote improved nitrogen management in grain crops? *Field Crops Research* 218, 126–140.

Correndo, A. A., Rotundo, J. L., Tremblay, N., Archontoulis, S., Coulter, J. A., Ruiz-Diaz, D., Franzen, D., Franzluebbers, A. J., Nafziger, E., Schwalbert, R., Steinke, K., Williams, J., Messina, C. D. and Ciampitti, I. A. 2021a. Assessing the uncertainty of maize yield without nitrogen fertilization. *Field Crops Research* 260, 107985.

Correndo, A. A., Tremblay, N., Coulter, J. A., Ruiz-Diaz, D., Franzen, D., Nafziger, E., Prasad, V., Rosso, L. H. M., Steinke, K., Du, J., Messina, C. D. and Ciampitti, I. A. 2021b. Unraveling uncertainty drivers of the maize yield response to nitrogen: a Bayesian and machine learning approach. *Agricultural and Forest Meteorology* 311, 108668.

DeGaetano, A. T. and Belcher, B. N. 2007. Spatial interpolation of daily maximum and minimum air temperature based on meteorological model analyses and independent observations. *Journal of Applied Meteorology and Climatology* 46(11), 1981–1992. https://doi.org/10.ll75/2007JAMC1536.1.

DeGaetano, A. T. and Wilks, D. S. 2009. Radar-guided interpolation of climatological precipitation data. *International Journal of Climatology* 29(2), 185–196. https://doi.org/10.1002/joc.1714.

EPA. 2023. *Inventory of U. S. Greenhouse Gas Emissions and Sinks: 1990–2021*. U.S. Environmental Protection Agency, EPA 430-R-23-002. Available at: https://www.epa.gov/ghgemissions/inventory-us-greenhouse-gas-emissions-and-sinks-1990-2021.

Halvorson, A. D., Snyder, C. S., Blaylock, A. D. and Del Grosso, S. J. 2014. Enhanced-efficiency nitrogen fertilizers: potential role in nitrous oxide emission mitigation. *Agronomy Journal* 106(2), 715–722.

Hatfield, J. L. 2013. Precision nutrient management and crop sensing. In: Kumar, J., et al. (Eds). *Phenomics in Crop Plants: Trends, Options and Limitations*. India: Springer pp. 207-222.

Hoben, J. P., Gehl, R. J., Millar, N., Grace, P. R. and Robertson, G. P. 2011. Nonlinear nitrous oxide (N2O) response to nitrogen fertilizer in on-farm corn crops of the US Midwest. *Global Change Biology* 17(2), 1140–1152.

Hutson, J. L. 2010. *LEACHM: Leaching Estimation and Chemistry Model: A Process-Based Model of Water and Solute Movement, Trans- Formations, Plant Uptake and Chemical Reactions in the Unsaturated Zone Version 4*. Research Series R03-l. Ithaca, NY: Cornell University.

Jin, Z. N., Prasad, R., Shriver, J. and Zhuang, Q. L. 2017. Crop model- and satellite imagery-based recommendation tool for variable rate N fer¬tilizer application for the US Corn system. *Precision Agriculture* 18(5), 779–800.

Jokela, W. E. and Meisinger, J. J. 2008. Ammonia emissions from land-applied manure: environmental and economic benefits. Proceedings of the Wisconsin Fertilizer, Aglime and Pest Management Conference, pp. 199–208.

Kay, B. D., Mahboubi, A. A., Beauchamp, E. G. and Dharmakeerthi, R. S. 2006. Integrating soil and weather data to describe variability in plant available nitrogen. *Soil Science Society of America Journal* 70(4), 1210-1221. https://doi.org/10.2 l 36/sssaj2005.0039.

Kersebaum, K. C. 1995. Application of a simple management model to simulate water and nitrogen dynamics. *Ecological Modelling* 81(1–3), 145–156.

Ketterings, Q. M., Klausner, S. and Czymmek, K. J. 2003. *Nitrogen Guidelines for Field Crops in New York. Second Rel*. Ithaca, NY: Cornell University.

Klausner, S. D., Kanneganti, V. R. and Bouldin, D. R. 1994. An approach for estimating a decay series for organic nitrogen in animal manure. *Agronomy Journal* 86(5), 897-903.

Klausner, S. D., Reid, W. S. and Bouldin, D. R. 1993. Relationship between late spring soil nitrate concentrations and corn yields in New York. *Journal of Production Agriculture* 6(3), 350-354.

Kovács, P., Van Scoyoc, G. E., Doerge, T. A., Camberato, J. J. and Vyn, T. J. 2015. Anhydrous ammonia timing and rate effects on maize nitrogen use efficiencies. *Agronomy Journal* 107(4), 1205-1214.

Kyveryga, P. M., Blackmer, A. M., Ellsworth, J. W. and Isla, R. 2004. Soil pH effects on nitrification of fall-applied anhydrous ammonia. *Soil Science Society of America Journal* 68(2), 545-551.

Magdoff, F. R. and van Es, H. M. 2021. *Building Soils for Better Crops: Managing for Soil Health*. Handbook Series Book 10. College Park, MD: Sustainable Agriculture Research and Education, 374 pp.

Magdoff, F., Ross, D. and Amadon, J. 1984. A soil test for nitrogen availability to corn. *Soil Science Society of America Journal* 48, 1301-1304.

McLellan, E. L., Cassman, K. G., Woodbury, P. B., Sela, S., Eagle, A. J., Tonitto, C., Marjerison, R. D. and van Es, H. M. 2018. The nitrogen balancing act: tracking the environmental performance of food supply chains. *BioScience* 68(3), 194-203.

Melkonian, J. J., van Es, H. M., DeGaetano, A. T. and Joseph, L. 2008. ADAPT-N: adaptive nitrogen management for maize using high-resolution climate data and model simulations. In: Kosla, R. (Ed.). *Proceedings of the 9th International Conference on Precision Agriculture*. Denver, CO: (CD-ROM), July 20-23.

Melkonian, J., van Es, H. M., DeGaetano, A. T., Sogbedji, J. M. and Joseph, L. 2007. Application of dynamic simulation modeling for nitrogen management in maize. In: Bruulsema, T. (Ed.). *Managing Crop Nutrition for Weather* (pp. 14-22). Peachtree Corners, GA: International Plant Nutrition Institute Publishing.

Melkonian, J., Poffenbarger, H. J., Mirsky, S. B., Ryan, M. R. and Moebius-Clune, B. N. 2017. Estimating nitrogen mineralization from cover crop mixtures using the precision nitrogen management model. *Agronomy Journal* 109, 1-13.

Millar, N., Robertson, G. P., Grace, P. R., Gehl, R. J. and Hoben, J. P. 2010. Nitrogen fertilizer management for nitrous oxide (N2O) mitigation in intensive corn (Maize) production: an emis-sions reduction protocol for US Midwest agriculture. *Mitigation and Adaptation Strategies for Global Change* 15(2), 185-204.

Morris, T. F., Murrell, T. S., Beegle, D. B., Camberato, J. J.,Ferguson, R. B., Grove, J., Ketterings, Q., Kyveryga, P. M., Laboski, C. A. M., McGrath, J. M., Meisinger, J. J., Melkonian, J., Moebius-Clune, B. N., Nafziger, E. D., Osmond, D., Sawyer, J. E., Scharf, P. C., Smith, W., Spargo, J. T., van Es, H. M. and Yang, H. 2018. Strengths and limitations of nitrogen rate recommendations for corn and opportunities for improvement. *Agronomy Journal* 110(1), 1-37.

Nafziger, E. D. B., Yoder, B., Mathesius, J. and Carter, P. 2021 Nitrogen deficiency and corn yield with delayed N application. *Agronomy Journal* 113(4), 3665-3674.

Nkebiwe, P. M., Weinmann, M., Bar-Tal, A. and Muller, T. 2016. Fertilizer placement to improve crop nutrient acquisition and yield: a review and meta-analysis. *Field Crops Research* 196, 389-401.

OECD (Organisation for Economic Co-operation and Development). 2013. *Compendium of Agri-environmental Indicators*. Paris: OECDPublishing. Available at: https://

oecd-ilibrary.org/agriculture-and-food/oecdcompendium-of-agri-environmental-indicators_9789264186217-en (Accessed: 23 May 2023).

Osmond, D. E., Shelton, S., Autin, R., van Es, H. and Sela, S. 2018. Evaluation of Adapt-N and realistic yield expectation approaches for maize management in North Carolina. *Soil Science Society of America Journal* 82, 1449–1458.

Pinder, R. W., Adams, P. J. and Pandis, S. N. 2007. Ammonia emission controls as a cost-effective strategy for reducing atmospheric particulate matter in the eastern United States. *Environmental Science and Technology* 41(2), 380–386.

Pittelkow, C. M., Clover, M. W., Hoeft, R. G., Nafziger, E. D., Warren, J. J., Gonzini, L. C. and Greer, K. D. 2017. Tile drainage nitrate losses and corn yield response to fall and spring nitrogen management. *Journal of Environmental Quality* 46(5), 1057–1064.

Randall, G. W., Huggins, D. R., Russelle, M. P., Fuchs, D. J., Nelson, W. W. and Anderson, J. L. 1997 Nitrate losses through subsurface tile drainage in conservation reserve program, alfalfa, and row crop systems. *Journal of Environmental Quality* 26(5), 1240–1247.

Ransom, C. J., Kitchen, N. R., Camberato, J. J., Carter, P. R., Ferguson, R. B., Fernández, F. G., Franzen, D. W., Laboski, C. A. M., Myers, D. B., Nafziger, E. D., Sawyer, J. E. and Shanahan, J. F. (2023). Combining corn N recommendation tools for an improved economical optimal nitrogen rate estimation. *Soil Science Society of America Journal* 87, 902–917.

Ransom, C. J., Kitchen, N. R., Sawyer, J. E., Camberato, J. J., Carter, P. R., Ferguson, R. B., Fernandez, F. G., Franzen, D. W., Laboski, C. A. M., Meyers, D. B., Nafziger, E. D. and Shanahan, J. F. 2021. Improving publicly available corn nitrogen rate recommendation tools with soil and weather measurements. *Agronomy Journal* 113(2), 2068–2090.

Rochette, P., Angers, D. A., Chantigny, M. H., MacDonald, J. D., Bissonnette, N. and Bertrand, N. 2009. Ammonia volatilization following surface application of urea to tilled and no-till soils: a laboratory comparison. *Soil and Tillage Research* 103(2), 310–315.

Sawyer, J., Nafziger, E., Randall, G., Bundy, L., Rehm, G. and Joern, B. 2006. *Concepts and Rationale for Regional Nitrogen Rate Guide- Lines for Corn (PM-2015).* Ames: Iowa State University Extension. Available at: https://store.extension.iastate.edu/product/Conce pts-and-Rationale-for-Regional- Nitrogen-Rate-Guide!ines-for-Corn (Accessed: 23 May 2023).

Scharf, P. C., Kitchen, N. R., Sudduth, K. A., Davis, J. G., Hubbard, V. C. and Lory, J. A. 2005. Field-scale variability in optimal nitrogen fertilizer rate for corn. *Agronomy Journal* 97(2), 452–461.

Sela, S. and van Es, H. M. 2018. Dynamic tools unify fragmented 4Rs into an integrative nitrogen management approach. *Journal of Soil and Water Conservation* 73(4), 107A–112A.

Sela, S., Woodbury, P. B., Marjerison, R. and van Es, H. M. 2019. Towards applying N balance as a sustainability indicator for the US cornbelt: realistic achievable targets, spatio-temporal variability and policy implications. *Environmental Research Letters* 14(6), 064015.

Sela, S., van Es, H. M., Moebius-Clune, B. N., Marjerison, R., Melkonian, J. J., Moebius-Clune, D., Schindelbeck, R. and Gomes, S. 2016. Adapt- N outperforms grower-selected nitrogen rates in Northeast and Midwest USA strip trials. *Agronomy Journal* 108, 1726–1734.

Sela, S., van Es, H. M., Moebius-Clune, B. N., Marjerison, R., Moebius- Clune, D., Schindelbeck, R., Severson, K. and Young, E. 2017. Dynamic model improves agronomic and environmental outcomes for maize N management over static approach. *Journal of Environmental Quality* 46(2), 311–319.

Sinclair, T. R. and Muchow, R. C. 1995. Effect of nitrogen sup- ply on maize yield. I. Modeling physiological responses. *Agronomy Journal* 87(4), 632-641.

Snyder, C. S., Bruulsema, T. W., Jensen, T. L. and Fixen, P. E. 2009. Review of greenhouse gas emissions from crop production systems and fertilizer management effects. *Agriculture, Ecosystems and Environment* 133(3–4), 247–266.

Sogbedji, J. M., van Es, H. M., Klausner, S. D., Bouldin, D. R. and Cox, W. J. 2001. Spatial and temporal processes affecting nitrogen availability at the landscape scale. *Soil and Tillage Research* 58(3–4), 233–244.

Sogbedji, J. M., van Es, H. M., Yang, C. L., Geohring, L. D. and Magdoff, F. R. 2000. Nitrate leaching and N budget as affected by maize N fertilizer rate and soil type. *Journal of Environmental Quality* 29(6), 1813–1820.

Stanford, G. 1973. Rationale for optimum nitrogen fertilization in corn production. *Journal of Environmental Quality* 2(2), 159–166.

TFI (The Fertilizer Institute). 2018. The 4R of nutrient stewardship. Available at: https://nutrientstewardship.com/4rs/ (Accessed: 23 May 2023).

Tremblay, N., Bouroubi, Y. M., B´elec, C., Mullen, R. W., Kitchen, N. R., Thomason, W. E., Ebelhar, S., Mengel, D. B., Raun, W. R., Francis, D. D., Vories, E. D. and Ortiz-Monasterio, I. 2012. Corn response to nitrogen is influenced by soil texture and weather. *Agronomy Journal* 104(6), 1658–1671.

University of Nebraska. 2022. 2021 On-farm research results. Available at: https://on-farm-research.unl.edu/pdfs/research/result-publications/2021research-results.pdf.

University of Nebraska. 2023. 2022 On-farm research results. Available at: https://on-farm-research.unl.edu/pdfs/research/result-publications/2022research-results.pdf.

USDA ERS (United States Department of Agriculture Economic Research Service). 2010. *Tailored Reports: Crop Production Practices*. Washington, DC: USDA Economic Research Service. Available at: https://data.ers.usda.gov/reports. aspx?ID=17883.

van Es, H. M., Czymmek, K. J. and Ketterings, Q. M. 2002. Management effects on N leaching and guidelines for an N leaching index in New York. *Journal of Soil and Water Conservation* 57, 499–504.

van Es, H. M., Kay, B. D., Melkonian, J. J. and Sogbedji, J. M. 2007. Nitrogen management under maize in humid regions: case for a dynamic approach. In: Bruulsema, T. (Ed.). *Managing Crop Nutrition for Weather* (pp. 6–13). Peachtree Corners, GA: International Plant Nutrition Institute Publishing.

van Grinsven, H. J. M., Bouwman, L., Cassman, K. G., van Es, H. M., McCrackin, M. L. and Beusen, A. H. W. 2015. Losses of ammonia and nitrate from agriculture and their effect on nitrogen recovery in the European Union and the United States between 1900 and 2050. *Journal of Environmental Quality* 44(2), 356–367.

Xie, M., Tremblay, N., Tremblay, G., Bourgeois, G., Bouroubi, M. Y. and Wei, Z. 2013. Weather effects on corn response to in-season nitrogen rates. *Canadian Journal of Plant Science* 93(3), 407–417.

Zhang, X., Davidson, E. A., Mauzerall, D. L., Searchinger, T. D., Dumas, P. and Shen, Y. 2015. Managing nitrogen for sustainable development. *Nature* 528(7580), 51–59.

Chapter 10

The economics of nitrogen in farming systems and beyond[1]

David J. Pannell and Asjad Tariq Sheikh, University of Western Australia, Australia

1 Introduction

The use of nitrogen fertiliser contributes to global food security, but it causes pollution of rivers, lakes, and coastal waters around the world, and it contributes to emissions of greenhouse gases. It increases the profitability of individual farmers, but it is over-applied in many cases, wasting money and needlessly worsening environmental problems.

There is an extensive and diverse literature analysing these issues using the tools and frameworks of economics. In this chapter, we provide a review and synthesis of this literature. The economic insights that will be discussed are important for making sound decisions about farm-level management of nitrogen, for understanding farmers' behaviour in relation to nitrogen, for designing regional or national policies for nitrogen-related pollution, and for prioritising nitrogen-related research and extension.

[1] This chapter is an update and expansion of Pannell, D.J. (2017). Economic perspectives on nitrogen in farming systems: managing trade-offs between production, risk and the environment, *Soil Research* 55, 473-478.

http://dx.doi.org/10.19103/AS.2024.0135.20

Our objective is to identify and describe important economic principles, insights, and empirical results and to illustrate how technical aspects of nitrogen fit into the economic, social, and policy world. In this world, the technical issues are important, but they underpin a broader set of perspectives, encompassing profitability, risk, learning, human behaviour, human attitudes, human values, fairness, rights, markets, policy mechanisms, the role of government, and politics.

2 Economics of nitrogen as an input to production

The optimal level of nitrogen fertiliser for a farmer to apply to a crop or pasture is case-specific, depending on technical issues (e.g. the type of crop, soil type, and rainfall) and socio-economic issues (e.g. the sale price of grain, purchase price of fertiliser, and the objective of the farmer). The simplest economic model addressing this question (Fig. 1) is based on an assumption that the farmer's objective is to maximise the expected value of on-farm profit, ignoring factors such as production risk and off-farm impacts. The 'production function' in Fig. 1 shows the expected value of yield as a function of fertiliser rate. This function encompasses all of the relevant technical issues that determine the response of yield to fertiliser. Expected profit is maximised where the slope of the production function equals the fertiliser price divided by the grain sale price (Dillon and Anderson, 2012).

The production function for nitrogen fertiliser exhibits diminishing marginal returns - it flattens out at higher fertiliser rates. In certain situations (e.g. a drought late in the growing season), yield may decline at higher nitrogen rates, in which case the expected value of yield (based on the probability distribution of yields) will probably also decline at higher nitrogen rates. It is obvious that if

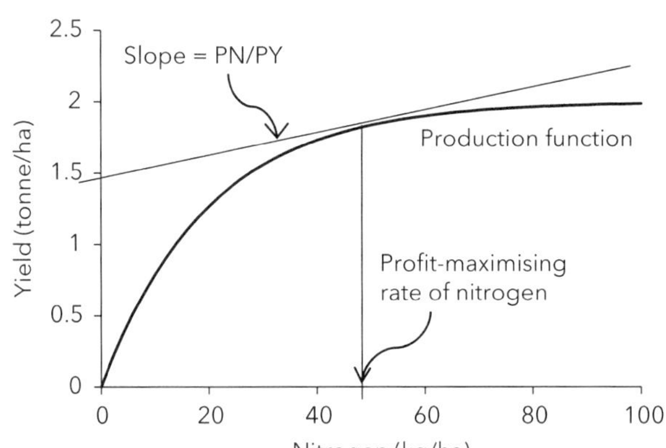

Figure 1 Simple economic model to optimise profit from application of nitrogen fertiliser. PN, price per unit of nitrogen ($/kg), PY, sale price of grain ($/tonne).

the fertiliser rate is so high that expected yield is decreasing, then the fertiliser rate is too high.

Perhaps slightly less obvious is the conclusion that the rate of nitrogen fertiliser that maximises expected profit is less than the rate that maximises expected yield. The gap between those two rates depends on the shape of the production function and the ratio of fertiliser price to grain price. In cases where the production function is relatively flat, and/or the ratio of nitrogen price to output price is relatively large, the gap can be large. Clearly, the optimal fertiliser rate depends not just on technical factors but also on prices. If the price of grain rises or the price of fertiliser falls, the optimal fertiliser rate increases (and vice versa), even in the absence of any change in the technical relationship between fertiliser and yield (Hendricks et al., 2014). Therefore, improving nitrogen use efficiency, if it is pursued by adjusting fertiliser rates, is not necessarily in the best interests of farmers. If it involves reducing the fertiliser rate below the profit-maximising rate, the farmer incurs an opportunity cost – he or she makes a lower expected profit than would have been possible at a higher nitrogen rate.

Aggregate fertiliser usage also depends on land use, which is also affected by output price (and, to a lesser extent, fertiliser price). For example, an increase in the price of a particular grain crop has dual effects: it leads to higher fertiliser rates per hectare of that crop and a larger area of production of that crop (assuming the prices of other agricultural outputs are unchanged). Whether such a price increase leads to increased nitrogen usage overall depends on which land use is displaced by the expanding area of this crop, and the usage of nitrogen on the displaced land use.

3 Nitrogen and economic risk

Maximisation of expected profit is a reasonable approximation of the management objective of some farmers (Abadi Ghadim and Pannell, 2003), but many are prepared to trade off some expected profit to reduce risk (Binswanger, 1980; Antle, 1987). If the riskiness of cropping is different at different rates of nitrogen fertiliser, then this would influence the rate preferred by these 'risk-averse' farmers. Risk aversion can vary from mild to extreme, depending on the psychological make-up and the circumstances of different farmers. Economists generally measure risk as the variance or standard deviation of the probability distribution of some measure of profit. The greater the variance of profit for a practice (e.g. a fertiliser rate), the less attractive that practice is to risk-averse farmers. The use of variance as the measure of riskiness reflects that there are both downside and upside aspects to risk. Risk-averse farmers give greater weight to the downside, but the upside is not irrelevant to them.

There are various sources of risk that can affect farmers' decisions about nitrogen fertiliser. One is unpredictability of the price at which the output will be sold. If this price is subject to risk, then increasing crop yield by applying nitrogen fertiliser will increase the farmer's overall risk, by increasing the variance of income from the crop. This means that, for risk-averse farmers, greater output-price risk results in a lower optimal fertiliser rate. Conversely, policies or contracts that reduce output-price risk may increase fertiliser use, depending on what those policies or contracts do to the expected value of output price.

The other main source of risk relevant to nitrogen fertiliser decisions is the riskiness of crop production. The consequence of yield risk for optimal fertiliser rate is less clear-cut than for price risk. It has sometimes been suggested that the application of nitrogen fertiliser reduces farmers' risks, either by reducing the variance of income or by acting as a kind of insurance policy that reduces the probability of bad outcomes (e.g. Sheriff, 2005). Pearce and Koundouri (2003, p.144) stated that 'Fertilisers can be expected to be overused due to risk aversion among farmers'. This statement implies that farmers see nitrogen fertiliser as being a risk-reducing input. SriRamaratnam et al. (1987) surveyed grain sorghum producers in Texas and found that they perceived the variance of profit to decrease at higher nitrogen fertiliser rates. Houser (2022) found that some farmers in the Midwest of the United States over-applied nitrogen in the belief that it is a risk-mitigation strategy. Asci et al. (2015) revealed that risk-averse potato growers in northern Florida prefer higher nitrogen fertiliser application rates compared with risk-neutral growers, again implying that they believe nitrogen fertiliser to be a risk-reducing input.

However, the weight of empirical evidence about the effect of nitrogen fertiliser on risk contradicts this view, both in developing countries (e.g. Mazid and Bailey, 1992) and developed countries (Roosen and Hennessy, 2003; Rajsic et al., 2009; Paulson and Babcock, 2010; Monjardino et al., 2015). Studies that have demonstrated an increase in the variance of profit at higher nitrogen rates include Just and Pope (1979) (using experimental data for corn and oats in Mississippi), Nelson and Preckel (1989) (using farm data for corn in Iowa), Love and Buccola (1991) (using farm data for corn in Iowa), Roosen and Hennessy (2003) (using experimental data for corn in Iowa), Rajsic et al. (2009) (using experimental data for corn in Ontario, Canada), and Meyer-Aurich and Karatay (2019) (using experimental data for wheat in Germany).

It appears that misperceptions amongst farmers about the effects of nitrogen fertiliser on risk may be contributing to over-application, leading to environmental problems. There may be opportunities for a behavioural economics approach (e.g. Byerly et al., 2018) to contribute to reducing the apparent bias in farmers' perceptions.

Another nitrogen management practice that relates to risk is applying nitrogen in two or more applications, rather than applying all of the fertiliser at the start of the production season. The decision about the second application can be adjusted depending on the predicted crop yield at that point in the season – a better-informed prediction than would be possible at seeding time. However, if the later application is delivered by machinery driven in the field, there is a trade-off inherent in using split applications because high rainfall may mean that the field is not accessible due to the risk of bogging. Gramig et al. (2017) presented a decision support tool for evaluating this trade-off, showing that the strategy of splitting nitrogen applications may or may not be worthwhile in different circumstances.

4 Economics of nitrogen fixation from legumes

The rotation of cereal crops with legumes is an ancient practice that continues to be widely practiced. Legume rotations (or legumes grown in an intercropping system) provide various benefits to cereal producers, potentially including risk reduction through diversification of income sources, reduced cereal crop disease, diversification of weed management strategies, provision of high-quality livestock feed and fixation of atmospheric nitrogen (Pannell, 1995).

There is much less research on the economics of nitrogen fixation from legumes than there is on the economics of nitrogen fertiliser. There are various studies that have examined the economics of including legumes in farming systems, including nitrogen-related benefits (e.g. Pannell et al., 2014; Preissel et al., 2015) but these usually do not separate out nitrogen from other factors.

An exception is the work of Khataza et al. (2017), who estimated the 'shadow price' of nitrogen fixed by legumes in a maize/legume rotation in Malawi. The shadow price indicates the change in farm profit for a one unit change in fixed nitrogen. They found that these shadow prices were, on average, higher than the subsidised price of commercial nitrogen fertiliser, suggesting that the existing subsidies are reducing the incentive for farmers to include legumes in their rotations.

Pannell and Falconer (1988) identified two components of the benefits of nitrogen fixation: (1) reduction in the economically optimal rate of nitrogen fertiliser and (2) contributing to an increase in yield due to biologically fixed nitrogen that does not act as a direct substitute for nitrogen fertiliser, perhaps because it becomes available to crops more slowly as biological material breaks down. The latter is difficult to distinguish from yield boosts due to other factors, such as reductions in crop disease. The existence of these two components means that the value of fixed nitrogen is not equal to the cost of a similar amount of nitrogen fertiliser, because some fixed nitrogen does not actually substitute for nitrogen fertiliser.

For the broadacre farming system of south-western Australia, Pannell and Falconer (1988) found that the value of the yield boost due to legumes (including the yield boost due to nitrogen fixation) was far greater than the value of a reduced need for nitrogen fertiliser. They also found that the contribution to profits from legumes in this farming system at that time exceeded the contribution from nitrogen fertiliser.

The economics of nitrogen fixation and legumes more broadly are highly case-specific (Schilizzi and Pannell, 2001). Their contributions to profitability can vary greatly between regions and farming systems and even between different farms in the same region (e.g. Pannell et al., 2014). In some cases, nitrogen fixation is a relatively minor factor in a decision to grow legumes (e.g. where they are grown primarily for livestock feed), whereas in other cases it is amongst the main reasons.

Huang and Uri (1993) compared continuous corn with a corn-soybean rotation. For a particular modelled scenario, they found that the legume rotation delivered both higher profits and a lower level of excess nitrogen. Similarly, Cox et al. (2010) simulated a range of rotations in Queensland, Australia, and found that cereal/legume rotations were substantially more profitable than cereal-only rotations and could profitably reduce nitrogen fertiliser use. MacWilliam et al. (2014) explored the economic and environmental consequences of including legume crops (dry pea or lentils) in an oilseed-cereal rotation in Western Canada. As well as finding that either legume was a profitable addition to the rotation, they also identified benefits in terms of climate change mitigation, ecosystem quality, and human health. Although the relative profit results for legumes depend on a variety of factors (Pannell, 1995), the widespread use of legumes in rotations (e.g. corn-soybean rotations in the USA and parts of South America, chick pea-cereal rotations in India, lupin-wheat rotations in Australia, and various legume-cereal rotations in Africa) indicate that the strategy is profitable in many situations.

5 Flat pay-off functions

An under-recognised but important aspect of the economics of nitrogen is the frequent occurrence of flat pay-off functions for nitrogen fertiliser (Pannell, 2006). Near to the fertiliser rate that maximises profit (or any other measure of the pay-off), there are always other fertiliser rates that are only slightly less profitable (i.e. there is a range of rates where the pay-off function is relatively flat) and that flat range can be wide. This is also true if the farmer's objective includes an allowance for risk aversion; the pay-off function (including a negative risk premium) is flat within the vicinity of its highest point.

For example, Fig. 2 shows profit as a function of nitrogen application rate for several soil types in the central wheatbelt of Western Australia, as represented in the whole-farm bioeconomic model, MIDAS (Morrison et al., 1986; Kingwell and Pannell, 1987). For sandy loam over clay, fertiliser rates

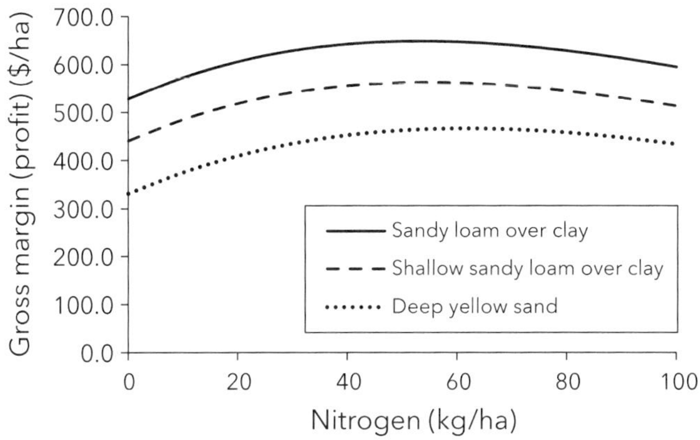

Figure 2 Gross margin (revenue minus nitrogen fertiliser cost) as a function of nitrogen application rate in the central wheatbelt of Western Australia. Production functions, prices, and costs from MIDAS model 2015. Source: Adapted from: Morrison et al. (1986).

within +77% to -51% of the optimal fertiliser rate provide revenue net of nitrogen costs within 5% of the optimum (i.e. any rate between 24 kg/ha and 88 kg/ha of N gives almost the same profit). The profit functions for other soils are similarly flat. Equivalent ranges for the other soils are +75% to -46% for shallow sandy loam over clay and +55% to -42% for deep yellow sand. Results broadly similar to this are typical almost everywhere that nitrogen fertiliser is applied. Jardine (1975) noted that on presenting evidence to agronomists about flat profit curves for fertilisers, he 'observed such reactions as complete disbelief, blank incomprehension, incipient terror, and others less readily categorized'. It remains an under-appreciated issue that deserves a higher profile.

The existence of flat pay-off functions has a number of management implications. They mean that the farmer has some flexibility in choosing the fertiliser rate. If a lower rate would better satisfy another objective (e.g. risk reduction), the farmer can choose that rate with little sacrifice of profit. If a farmer wants to adopt a simple strategy of applying the same nitrogen rate each year, saving effort but foregoing potentially beneficial adjustments in response to variations in grain price or yield potential, this can be done with relatively little economic sacrifice. If regulators require a moderate reduction in fertiliser rate below the farmer's economic optimum, the cost to the farmer will be small.

A second implication of flat pay-off functions is that the benefits of precision agriculture technologies that spatially adjust fertiliser rates within a field will usually be small. Well before these technologies had been developed, Anderson (1975) argued that 'In pursuing ... optimal levels of decision variables, precision is pretence and great accuracy is absurdity'. Unless the required rate adjustments within a field are very large, a standard rate is likely to fall within the flat range of the pay-off function in most parts of the field, meaning that a failure to adjust rates spatially does not result in a large loss of profits. That is why economists evaluating this type of precision technology have typically found that their benefits are not large. Paz et al. (1999), Babcock and Pautsch (1998), Thrikawala et al. (1999), and Walters et al. (2017) all estimated small benefits from variable-fertiliser-rate technology - too small to cover realistic costs of the technology in most cases. Variable-rate technology can result in reduced environmental impacts due to reduced losses of nitrogen from the field (e.g. Walters et al., 2017), but private financial benefits from the technology are unlikely to be a powerful motivation for its adoption by farmers. This raises interesting questions about what has motivated the slowly growing adoption of variable-fertiliser-rate technology in various locations.

A third implication of flat pay-off functions is for agronomists developing recommendations for fertiliser rates: they should not recommend a single fertiliser rate for any particular farming context. Much more relevant and helpful to farmers would be to identify the range of rates that will deliver profits within a certain percentage of the maximum. Farmers can then bring in their own preferences when selecting which rate to apply.

6 Over-application of nitrogen fertilisers

It is sometimes observed that a group of farmers is applying nitrogen fertiliser at rates that exceeds local recommended rates. For example, Yadav et al. (1997) (for corn in Minnesota), Villacis et al. (2020) (for Corn in Colorado), and Biggs et al. (2021) (for sugarcane in Queensland, Australia) found that farmers' average nitrogen rate exceeded the profit-maximising levels.

In some cases, the issue is that the rates recommended by agronomists are less than the profit-maximising rates (e.g. Rajsic and Weersink, 2008), so farmers are actually in line with profit maximisation. It may be that government programmes increase the profit-maximising rate by increasing the expected price of grain (Preckel et al., 2000), and that this is not factored into agronomists' recommendations. Or it may be that farmers are accounting for uncertainty about the availability of nitrogen in the soil, which has been shown to increase the optimal fertiliser application rate (Paulson and Babcock, 2010). Using data from the USDA's Agricultural Resource Management Survey, Williamson (2011) found that farmers who tested their soil for nitrogen (thereby reducing their uncertainty about nitrogen availability) applied 33.6 lbs/ha less nitrogen on average relative to non-testers.

It may also be that farmers are applying nitrogen at rates above their own economic optimum. One likely contributor is the flat-pay-off-functions phenomenon, discussed above. Flat pay-off functions mean that there is little economic incentive pushing farmers toward the economic optimum. If they are using excessive rates for whatever reason (e.g. historical precedent or misinformation), they can continue to do so without sacrificing much profit.

Stuart et al. (2014) found that limited access to information was contributing to fertiliser overuse in the USA. Huang et al. (2015) found that in-field training could prompt farmers to reduce their nitrogen rates, suggesting that misinformation or lack of information was causing over-application. This misinformation could include the mistaken view that fertilisers reduce the riskiness of crop production (Sheriff, 2005; Houser, 2022). Or it may be that farmers are using a heuristic to manage risk that differs from the logical approach used by economists. For example, Berbel and Martínez-Dalmau (2021) suggest that it may be due to farmers observing that the cost of over-application is lower than the cost of under-application.

In contrast to the concerns about farmers' over-application of nitrogen fertiliser in developed countries, a concern in some developing countries has been under-application, detracting from food security. However, this concern may be diminishing. For example, Sheahan et al. (2013) found that maize farmers in Kenya had been increasing their usage of nitrogen fertiliser over the previous 13 years, and that their application rates sometimes exceeded profit-maximising rates.

7 Nitrogen pollution

Economists recognise that nitrogen fertiliser, as well as increasing agricultural production, is a cause of 'external costs' such as water pollution (e.g. Dabrowski et al., 2009), greenhouse gas emissions (e.g. Ribaudo et al., 2011), and health impacts (e.g. Wang and Lu, 2020). Farmers who generate these costs do not have to bear the costs themselves, so they lack the appropriate incentives to prevent or repair the costs. As a result, an external cost can result in 'market failure', where the decisions made by unregulated individuals in a free market lead to outcomes that are not in the overall best interests of society. This potentially justifies government intervention of some kind.

7.1 Farm-level economic implications

External costs due to nutrients or greenhouse gases from agriculture arise as a result of activities that are undertaken to generate economic benefits for the farmers. This means that reducing emissions of these pollutants usually comes at a cost. That cost may be borne by farmers (e.g. if they are required by regulation to reduce emissions), by consumers of farm products (to the extent that the additional costs are passed on), or by the broader community (e.g. if farmers are paid to reduce emissions as part of an agri-environmental programme). Economists have studied consequences of bringing these external costs into farmers' production decisions. Studies have identified various potential responses by farmers to the resulting economic incentives, including reductions in fertiliser rates (Nkonya and Featherstone, 2000), timing nitrogen applications to better match the nitrogen needs of crops (Huang et al., 1998; Shankar et al., 2000), switching to organic fertiliser (Vermersch et al., 1993), changing crops (Weersink et al., 1998), switching to break crops requiring little or no nitrogen fertiliser (England, 1986), intentional use of wetlands to retain nitrogen (Byström, 1998), reduced application of irrigation water (Knapp and Schwabe, 2008), and improved management of manure (Andréasson, 1990). There have been numerous estimates made of the cost in terms of lost agricultural output as a result of adopting such practices. Typically, the marginal cost of nitrogen emissions abatement is low for low levels of abatement but increases at an increasing rate as the required level of abatement increases (e.g. Helin, 2014; Johnson et al., 1991; Doole and Romero, 2015; Mandrini et al., 2022). As a result, modest targets for abatement can often be achieved at low cost, but ambitious targets can be extremely costly.

Another strand of work examines the impacts on emissions from changes in the farm system or the farming context. The prices of farming outputs have been one focus of study, given that higher output prices lead to higher application of nitrogen fertiliser (Section 2). For example, Hendricks et al. (2014) examined the impacts of US mandates for use of corn ethanol in fuel. They found that increases in use of nitrogen fertiliser as a result of this policy would expand the area of the hypoxic zone in the Gulf of Mexico by roughly 30 square miles on average. Another example of a policy which had an unintended consequence for nitrogen pollution was the Gross Revenue Insurance Program in Canada. This increased the relative support for corn, which has higher leachate potential than other crops (Weersink et al., 1998).

7.2 Policy-level economics

The largest body of nitrogen-related economic literature is concerned with government policies for reducing nitrogen pollution. A number of studies have quantified in monetary terms the benefits of policies to reduce nitrogen pollution (e.g. Morgan and Owens, 2001; Poor et al., 2007; Mandrini et al., 2022) or undertaken Benefit: Cost Analysis of pollution-reduction programmes (e.g. Hyytiäinen et al., 2015).

There have been many studies evaluating the economic efficiency of particular policy mechanisms to reduce nitrogen pollution, including taxes or levies on fertiliser (Lansink and Peerlings, 1997; Brady, 2003; Finger, 2012; Bostian et al., 2015), easements (Wu and Tanaka, 2005), a deposit-refund system (Hansen, 1999), a market to trade permits to pollute (Hanson and McConnell,

2008), incentive payments to encourage pollution reductions (Cooper and Keim, 1996), insurance-style schemes to underwrite losses from adoption of low-nitrogen practices (Metcalfe et al., 2007), quotas on the use of nitrogen fertiliser (Moxey and White, 1994), land-use regulations (Brady, 2003), and provision of information to farmers (Fleming et al., 1998).

These studies often provide insights into the design of a particular policy mechanism to achieve a particular policy target. For example, Whittaker et al. (2003) compared the use of an input tax for nitrogen fertiliser with a mandated reduction in fertiliser use in the Columbia River Basin (in the northwest of the United States). They found that a 300% input tax would be needed to provide the same overall result (in terms of fertiliser usage) as a 25% reduction applied uniformly to all farmers.

Economists have undertaken studies to select the most effective policy mechanism for reducing nitrogen pollution (Drucker and Latacz-Lohmann, 2003) and developed general frameworks for doing that (Pannell, 2008). The selection of the most appropriate policy mechanism can make a large difference to the cost-effectiveness of a policy. For example, Semaan et al. (2007) compared several policy approaches for reducing nitrate leaching from irrigated agriculture in the Apulia Region of Southern Italy. For a target of reducing nitrate leaching by 40%, they found that the net social cost would be 95 €/ha for a policy based on subsidies to encourage adoption of a set of efficient practices (including land levelling, water metres, and low fertiliser rates), 183 €/ha for a tax on fertiliser and 269 €/ha for a policy of imposing a price on irrigation water.

Mandrini et al. (2022) compared four policy scenarios to reduce nitrate leaching for the US Midwest: higher nitrogen prices, a tax on nitrogen leaching, a tax on the nitrogen 'balance' (nitrogen applied minus nitrogen removed), and voluntary reduction of nitrogen use by land managers. Interestingly, they found that the costs of reducing nitrate leaching were broadly similar for the four policy approaches.

There have been economic studies of the interactions between nitrogen pollution and other seemingly unrelated policies. Policies intended to increase farmers' incomes can have the unintended consequence of increasing nitrogen pollution by increasing the incentive to apply fertiliser (Weersink et al., 1998). Conversely, the need to maintain or improve water quality can have implications for the operation of other policies, such as a water market intended to allocate water efficiently to different irrigators (Dabrowski et al., 2009).

There have been many studies to identify efficient strategies for reducing external costs from nitrogen application (e.g. Vatn et al., 1997; Ribaudo et al., 2001; Fröschl et al., 2008; Ribaudo et al., 2011), including studies of the optimal spatial allocation of abatement measures within a region or catchment

(Doole and Romero, 2014; Konrad et al., 2014). On a larger spatial scale, there has been work on the design of systems for internationally coordinated pollution abatement for countries that share an affected water body, such as the countries around the Baltic Sea (Gren, 2001; Hasler et al., 2014). Spatial targeting of abatement effort (both at the regional and international scales) can generate much larger benefits than untargeted policies, although these additional benefits are likely to be offset to some degree by increased costs required to run a targeted programme (costs of information and administration).

8 Nitrogen research, development and extension

Economics provides information that can contribute to improved evaluation and prioritisation of agricultural research, development, and extension. Alston et al. (1995) provided a framework for integrating information about the benefits and costs of agricultural research to inform decisions about research priorities. There are numerous examples of the framework being applied to actual or potential agricultural research projects, although there are relatively few for research on nitrogen. Related work has investigated the role of new agricultural technologies within the farming system using bio-economic modelling (e.g. Pannell, 1995). This provides scientists with insights into the ways in which the technology is likely to be integrated into the farming system, including the scale of adoption within the farm, interactions with other parts of the farming system, and the contribution to farm profits.

Economists have also contributed to a large literature on the adoption of new practices by farmers (e.g. Abadi Ghadim and Pannell, 1999; Chavas and Nauges, 2020; Weersink and Fulton, 2020; Pannell and Claassen, 2020). Much of this work aims to quantify adoption and explain past patterns of adoption as influenced by a wide range of economic and non-economic factors (e.g. Feder and Umali, 1993; Pannell et al., 2006), but in recent work, Kuehne et al. (2017) have developed a tool to assist scientists and extension agents make predictions about future adoption of new practices. Complementing this tool, Llewellyn et al. (2005) presented a systematic and evidence-based approach to the planning of specific extension messages and the targeting of extension effort. A recent development is the growth of Behavioral Economics, a partnership between economics and psychology that is providing new understandings of human behaviour and developing tools to influence it (e.g. Byerly et al., 2018; Streletskaya et al., 2020; Balmford et al., 2021; Palm-Forster and Messer, 2022). There may be untapped opportunities for these insights and approaches to be utilised to influence farmers' management of nitrogen in future.

9 The market for nitrogen fertiliser

There is a global market for nitrogen fertiliser, with the largest producers being (largest first) China, the United States, India, Russia, and Indonesia,[2] and the largest consumers being China, India, the United States, Brazil, and Indonesia.[3] Given the similarities in those lists, much of the nitrogen fertiliser produced is consumed domestically. On the other hand, some countries are major importers to overcome their lack of domestic consumption, including Brazil, France, and Thailand, and despite their large domestic production, India and the United States are major importers as well.[4]

At any point in time, the world prices of nitrogen fertilisers depend on a variety of supply factors and demand factors. In the short term, demand is particularly influenced by the prices of agricultural outputs, since higher crop output prices increase the area of crop production and make higher nitrogen fertiliser rates more profitable. Typically, the demand curve for nitrogen fertiliser has been found to be 'inelastic' (Whittaker et al., 2003; Hansen, 2004), meaning that a given percentage change in fertiliser price causes a smaller percentage change in the demand for fertiliser. As a result, nitrogen fertiliser prices can rise sharply during times of shortage. This was observed during 2021 and 2022, due in part to restrictions on exports imposed by the Chinese government.

On the supply side of the market, the key factor in the short term is the price of energy, since manufacture of nitrogen fertiliser is an energy-intensive process. High energy prices in 2022 imposed downward pressures on supply of nitrogen fertilisers, contributing to high fertiliser prices.

Given the high energy intensity of nitrogen fertiliser production (Lim et al., 2021), if the energy source is fossil fuels, the process results in high emissions of greenhouse gasses. As a result, there has been growing interest in the feasibility of a variety of technologies that can produce nitrogen fertiliser with fewer emissions (e.g. Puértolas et al., 2022; Ravi and Makepeace, 2022).

10 Conclusion

The insights, evidence and principles outlined here are of central relevance to management, policy, and research related to nitrogen. However, they are relatively underutilised in practice. Many recommendations for nitrogen application rates are made without adequate consideration of highly relevant economic principles and evidence. The influence of nitrogen application on farm business risk is often perceived inaccurately. The very common occurrence of flat pay-off functions is often not recognised. The available economic tools

[2] https://www.statista.com/statistics/1252656/nitrogen-fertilizer-production-by-country/ accessed 23 December 2022.
[3] https://www.statista.com/statistics/1252671/nitrogen-fertilizer-consumption-by-country/ accessed 23 December 2022.
[4] https://www.nationmaster.com/nmx/ranking/nitrogen-fertilizer-imports accessed 23 December 2022.

for spatially targeting reductions in nitrogen-related externalities are often unutilised in situations where they could contribute greatly. Greater utilisation of these economic insights and tools has the potential to substantially increase the benefits to farmers and to society as a whole. Economists face the challenge of ensuring that their research on nitrogen is widely known and well understood.

11 References

Abadi Ghadim, A. K. and Pannell, D. J. (1999). A conceptual framework of adoption of an agricultural innovation. *Agricultural Economics* 21, 145-154.

Abadi Ghadim, A. K. and Pannell, D. J. (2003). Risk attitudes and risk perceptions of crop producers in Western Australia. In Babcock, B. A., Fraser, R. W. and Lekakis, J. N. (Eds), *Risk Management and the Environment: Agriculture in Perspective*, Kluwer Publishers, Dordrecht, pp. 113-133.

Alston, J. M., Norton, G. W. and Pardey, P. G. (1995). *Science Under Scarcity: Principles and Practice for Agricultural Research Evaluation and Priority Setting*, Cornell University Press, Ithaca.

Anderson, J. R. (1975). One more or less cheer for optimality. *Journal of the Australian Institute of Agricultural Science* 41, 195-197.

Andréasson, I. M. (1990). Costs for reducing farmers' use of nitrogen in Gotland, Sweden. *Ecological Economics* 2(4), 287-299.

Antle, J. M. (1987). Econometric estimation of producers' risk attitudes. *American Journal of Agricultural Economics* 69(3), 509-522.

Asci, S., Borisova, T. and VanSickle, J. J. (2015). Role of economics in developing fertilizer best management practices. *Agricultural Water Management* 152, 251-261.

Babcock, B. A. and Pautsch, G. R. (1998). Moving from uniform to variable fertilizer rates on Iowa corn: effects on rates and returns. *Journal of Agricultural and Resource Economics* 23(2), 385-400.

Balmford, A., Bradbury, R. B., Bauer, J. M., Broad, S., Burgess, G., Burgman, M., Byerly, H., Clayton, S., Espelosin, D., Ferraro, P. J., Fisher, B., Garnett, E. E., Jones, J. P. G., Marteau, T. M., Otieno, M., Polasky, S., Ricketts, T. H., Sandbrook, C., Sullivan-Wiley, K., Trevelyan, R., van der Linden, S., Veríssimo, D. and Nielsen, K. S. (2021). Making more effective use of human behavioural science in conservation interventions. *Biological Conservation* 261, 109256.

Berbel, J. and Martínez-Dalmau, J. (2021). A simple agro-economic model for optimal farm nitrogen application under yield uncertainty. *Agronomy* 11(6), 1107.

Biggs, J. S., Everingham, Y., Skocaj, D. M., Schroeder, B. L., Sexton, J. and Thorburn, P. J. (2021). The potential for refining nitrogen fertiliser management through accounting for climate impacts: an exploratory study for the Tully region. *Marine Pollution Bulletin* 170, 112664.

Binswanger, H. P. (1980). Attitudes toward risk: experimental measurement in rural India. *American Journal of Agricultural Economics* 62(1), 395-407.

Bostian, M., Whittaker, G., Barnhart, B., Färe, R. and Grosskopf, S. (2015). Valuing water quality tradeoffs at different spatial scales: an integrated approach using bilevel optimization. *Water Resources and Economics* 11, 1-12.

Brady, M. (2003). The relative cost-efficiency of arable nitrogen management in Sweden. *Ecological Economics* 47(1), 53-70.

Byerly, H., Balmford, A., Ferraro, P. J., Wagner, C. H., Palchak, E., Polasky, S., Ricketts, T. H., Schwartz, A. J. and Fisher, B. (2018). Nudging pro-environmental behavior: evidence and opportunities. *Frontiers in Ecology and the Environment* 16(3), 159–168.

Byström, O. (1998). The nitrogen abatement cost in wetlands. *Ecological Economics* 26(3), 321-331.

Chavas, J.-P. and Nauges, C. (2020). Uncertainty, learning and technology adoption in agriculture. *Applied Economic Perspectives and Policy* 42(1), 42–53.

Cooper, J. C. and Keim, R. W. (1996). Incentive payments to encourage farmer adoption of water quality protection practices. *American Journal of Agricultural Economics* 78(1), 54–64.

Cox, H. W., Kelly, R. M. and Strong, W. M. (2010). Pulse crops in rotation with cereals can be a profitable alternative to nitrogen fertiliser in central Queensland. *Crop and Pasture Science* 61(9), 752–762.

Dabrowski, J. M., Murray, K., Ashton, P. J. and Leaner, J. J. (2009). Agricultural impacts on water quality and implications for virtual water trading decisions. *Ecological Economics* 68(4), 1074–1082.

Dillon, J. L. and Anderson, J. R. (2012). *The Analysis of Response in Crop and Livestock Production* (3rd ed.), Pergamon Press, Oxford.

Doole, G. J. and Romero, A. J. (2014). Cost-effective regulation of nonpoint emissions from pastoral agriculture: a stochastic analysis. *Australian Journal of Agricultural and Resource Economics* 58(3), 471–494.

Doole, G. J. and Romero, A. J. (2015). Trade-offs between profit, production, and environmental footprint on pasture-based dairy farms in the Waikato region of New Zealand. *Agricultural Systems* 141, 14–23.

Drucker, A. G. and Latacz-Lohmann, U. (2003). Getting incentives right? A comparative analysis of policy instruments for livestock waste pollution abatement in Yucatán, Mexico. *Environment and Development Economics* 8(2), 261–284.

England, R. A. (1986). Reducing the nitrogen input on arable farms. *Journal of Agricultural Economics* 37(1), 13–24.

Feder, G. and Umali, D. L. (1993). The adoption of agricultural innovations: a review. *Technological Forecasting and Social Change* 43(3–4), 215–239.

Fleming, R. A., Adams, R. M. and Ervin, D. E. (1998). The role of soil test information in reducing groundwater pollution. *Journal of Agricultural and Resource Economics* 23(1), 20–38.

Fröschl, L., Pierrard, R. and Schönbäck, W. (2008). Cost-efficient choice of measures in agriculture to reduce the nitrogen load flowing from the Danube River into the Black Sea. An analysis for Austria, Bulgaria, Hungary and Romania. *Ecological Economics* 68(1–2), 96–105.

Finger, R. (2012). Nitrogen use and the effects of nitrogen taxation under consideration of production and price risks. *Agricultural Systems* 107, 13–20.

Gramig, B. M., Massey, R. and Do-Yun, S. D. (2017). Nitrogen application decision-making under climate risk in the US Corn Belt. *Climate Risk Management* 15, 82–89.

Gren, I. M. (2001). International versus national actions against nitrogen pollution of the Baltic Sea. *Environmental and Resource Economics* 20(1), 41–59.

Hansen, L. G. (1999). A deposit-refund system applied to non-point nitrogen emissions from agriculture. *Environmental Economics and Policy Studies* 2(4), 231–247.

Hansen, L. G. (2004). Nitrogen fertilizer demand from Danish crop farms: regulatory implications of heterogeneity. *Canadian Journal of Agricultural Economics* 52(3), 313-331.

Hanson, J. C. and McConnell, K. E. (2008). Simulated trading for Maryland's nitrogen loadings in the Chesapeake Bay. *Agricultural and Resource Economics Review* 37(2), 211-226.

Hasler, B., Smart, J. C. R., Fonnesbech-Wulff, A., Andersen, H. E., Thodsen, H., Blicher-Mathiesen, G., Smedberg, E., Göke, C., Czajkowski, M., Was, A., Elofsson, K., Humborg, C., Wolfsberg, A. and Wulff, F. (2014). Hydro-economic modelling of cost-effective transboundary water quality management in the Baltic Sea. *Water Resources and Economics* 5, 1-23.

Helin, J. A. (2014). Reducing nutrient loads from dairy farms: a bioeconomic model with endogenous feeding and land use. *Agricultural Economics* 45(2), 167-184.

Hendricks, N. P., Sinnathamby, S., Douglas-Mankin, K., Smith, A., Sumner, D. A. and Earnhart, D. H. (2014). The environmental effects of crop price increases: nitrogen losses in the U. S. Corn Belt. *Journal of Environmental Economics and Management* 68(3), 507-526.

Houser, M. (2022). Farmer motivations for excess nitrogen use in the U. S. Corn Belt. *Case Studies in the Environment* 6(1), 1688823.

Huang, J., Huang, Z., Jia, X., Hu, R. and Xiang, C. (2015). Long-term reduction of nitrogen fertilizer use through knowledge training in rice production in China. *Agricultural Systems* 135, 105-111.

Huang, W. Y., Hewitt, T. I. and Shank, D. (1998). An analysis of on-farm costs of timing N applications to reduce N losses. *Journal of Agricultural and Resource Economics* 23(2), 445-467.

Huang, W. Y. and Uri, N. D. (1993). The effect of crop rotation on reducing excess nitrogen fertilizer use. *Applied Mathematical Modelling* 17(3), 141-148.

Hyytiäinen, K., Ahlvik, L., Ahtiainen, H., Artell, J., Huhtala, A. and Dahlbo, K. (2015). Policy goals for improved water quality in the Baltic Sea: when do the benefits outweigh the costs? *Environmental and Resource Economics* 61(2), 217-241.

Jardine, R. (1975). Two cheers for optimality. *Journal of the Australian Institute of Agricultural Science* 41, 30-34.

Johnson, S. L., Adams, R. M. and Perry, G. M. (1991). The on-farm costs of reducing groundwater pollution. *American Journal of Agricultural Economics* 73(4), 1063-1073.

Just, R. E. and Pope, R. D. (1979). Production function estimation and related risk considerations. *American Journal of Agricultural Economics* 61(2), 276-284.

Khataza, R. R. B., Hailu, A., Kragt, M. E. and Doole, G. J. (2017). Estimating shadow price for symbiotic nitrogen and technical efficiency for legume-based conservation agriculture in Malawi. *Australian Journal of Agricultural and Resource Economics* 61(3), 462-480.

Kingwell, R. S. and Pannell, D. J. (Eds). (1987). *MIDAS, A Bioeconomic Model of a Dryland Farm System*, Pudoc, Wageningen, 207pp.

Knapp, K. C. and Schwabe, K. A. (2008). Spatial dynamics of water and nitrogen management in irrigated agriculture. *American Journal of Agricultural Economics* 90(2), 524-539.

Konrad, M. T., Andersen, H. E., Thodsen, H., Termansen, M. and Hasler, B. (2014). Cost-efficient reductions in nutrient loads; identifying optimal spatially specific policy measures. *Water Resources and Economics* 7, 39-54.

Kuehne, G., Llewellyn, R., Pannell, D. J., Wilkinson, R., Dolling, P., Ouzman, J. and Ewing, M. (2017). Predicting farmer uptake of new agricultural practices: a tool for research, extension and policy. *Agricultural Systems* 156, 115–125.

Lansink, A. O. and Peerlings, J. (1997). Effects of N-surplus taxes: combining technical and historical information. *European Review of Agricultural Economics* 24(2), 231–247.

Lim, J., Fernández, C. A., Lee, S. W. and Hatzell, M. C. (2021). Ammonia and nitric acid demands for fertilizer use in 2050. *ACS Energy Letters* 6(10), 3676–3685.

Llewellyn, R. S., Lindner, R. K., Pannell, D. J. and Powles, S. B. (2005). Targeting key perceptions when planning and evaluating extension. *Australian Journal of Experimental Agriculture* 45(12), 1627–1633.

Love, H. A. and Buccola, S. T. (1991). Joint risk preference-technology estimation with a primal system. *American Journal of Agricultural Economics* 73(3), 765–774.

MacWilliam, S., Wismer, M. and Kulshreshtha, S. (2014). Life cycle and economic assessment of Western Canadian pulse systems: the inclusion of pulses in crop rotations. *Agricultural Systems* 123, 43–53.

Mandrini, G., Pittelkow, C. M., Archontoulis, S., Kanter, D. and Martin, N. F. (2022). Exploring trade-offs between profit, yield, and the environmental footprint of potential nitrogen fertilizer regulations in the US Midwest. *Frontiers in Plant Science* 13, 852116.

Mazid, A. and Bailey, E. (1992). Incorporating risk in the economic analysis of agronomic trials: fertilizer use on barley in Syria. *Agricultural Economics* 7(2), 167–184.

Metcalfe, T., Bosch, D. J., Pease, J. W., Alley, M. M. and Phillips, S. B. (2007). Yield reserve program costs in the Virginia Coastal plain. *Agricultural and Resource Economics Review* 36(2), 197–212.

Meyer-Aurich, A. and Karatay, Y. N. (2019). Effects of uncertainty and farmers' risk aversion on optimal N fertilizer supply in wheat production in Germany. *Agricultural Systems* 173, 130–139.

Monjardino, M., McBeath, T., Ouzman, J., Llewellyn, R. and Jones, B. (2015). Farmer risk-aversion limits closure of yield and profit gaps: a study of nitrogen management in the southern Australian wheat belt. *Agricultural Systems* 137, 108–118.

Morgan, C. and Owens, N. (2001). Benefits of water quality policies: the Chesapeake Bay. *Ecological Economics* 39(2), 271–284.

Morrison, D. A., Kingwell, R. S., Pannell, D. J. and Ewing, M. A. (1986). A mathematical programming model of a croplivestock farm system. *Agricultural Systems* 20(4), 243–268.

Moxey, A. and White, B. (1994). Efficient compliance with agricultural nitrate pollution standards. *Journal of Agricultural Economics* 45(1), 27–37.

Nelson, C. H. and Preckel, P. V. (1989). The conditional beta distribution as a stochastic production function. *American Journal of Agricultural Economics* 71(2), 370–378.

Nkonya, E. M. and Featherstone, A. M. (2000). Determining socially optimal nitrogen application rates using a delayed response model: the case of irrigated corn in Western Kansas. *Journal of Agricultural and Resource Economics* 25(2), 453–467.

Palm-Forster, L. and Messer, K. (2022). Behavioral and experimental economics to inform agri-environmental programs and policies. Chapter 6. In Barrett, C. B. and Just, D. R. (Eds), *Handbook of Agricultural Economics* (vol. 5), Elsevier, Amsterdam. 4331–4406.

Pannell, D. J. (1995). Economic aspects of legume management and legume research in dryland farming systems of southern Australia. *Agricultural Systems* 49(3), 217–236.

Pannell, D. J. (2006). Flat-earth economics: the far-reaching consequences of flat payoff functions in economic decision making. *Review of Agricultural Economics* 28(4), 553–566.

Pannell, D. J. (2008). Public benefits, private benefits, and policy intervention for land-use change for environmental benefits. *Land Economics* 84(2), 225–240.

Pannell, D. J. and Claassen, R. (2020). The roles of adoption and behavior change in agricultural policy. *Applied Economic Perspectives and Policy* 42(1), 31–41.

Pannell, D. J. and Falconer, D. A. (1988). The relative contributions to profit of fixed and applied nitrogen in a croplivestock farm systems. *Agricultural Systems* 26(1), 1–17.

Pannell, D. J., Llewellyn, R. S. and Corbeels, M. (2014). The farm-level economics of conservation agriculture for resource-poor farmers. *Agriculture, Ecosystems and Environment* 187(1), 52–64.

Pannell, D. J., Marshall, G. R., Barr, N., Curtis, A., Vanclay, F. and Wilkinson, R. (2006). Understanding and promoting adoption of conservation practices by rural landholders. *Australian Journal of Experimental Agriculture* 46(11), 1407–1424.

Paulson, N. D. and Babcock, B. A. (2010). Readdressing the fertilizer problem. *Journal of Agricultural and Resource Economics* 35(3), 368–384.

Paz, J. O., Batchelor, W. D., Babcock, B. A., Colvin, T. S., Logsdon, S. D., Kaspar, T. C. and Karlen, D. L. (1999). Model-based technique to determine variable rate nitrogen for corn. *Agricultural Systems* 61(1), 69–75.

Pearce, D. and Koundouri, P. (2003). Diffusion pollution and the role of agriculture. In Helm, D. (Ed.), *Water Sustainability and Regulation: The Next Periodic Review and Beyond*, OXERA Press, Oxford. 125–153.

Poor, P. J., Pessagno, K. L. and Paul, R. W. (2007). Exploring the hedonic value of ambient water quality: a local watershed-based study. *Ecological Economics* 60(4), 797–806.

Preckel, P. V., Shively, G. E., Baker, T. G., Chu, M. C. and Burrell, J. E. (2000). Contract incentives and excessive nitrogen use in agriculture. *Journal of Agricultural and Resource Economics* 25(2), 468–484.

Preissel, S., Reckling, M., Schläfke, N. and Zander, P. (2015). Magnitude and farm-economic value of grain legume pre-crop benefits in Europe: a review. *Field Crops Research* 175, 64–79.

Puértolas, B., Comesaña-Hermo, M., Besteiro, L. V., Vázquez-González, M. and Correa-Duarte, M. A. (2022). Challenges and opportunities for renewable ammonia production via plasmon-assisted photocatalysis. *Advanced Energy Materials* 12(18), 2103909.

Rajsic, P. and Weersink, A. (2008). Do farmers waste fertilizer?: A comparison of ex post optimal nitrogen rates and ex ante recommended rates by model, site and year. *Agricultural Systems* 97(1-2), 56–67.

Rajsic, P., Weersink, A. and Gandorfer, M. (2009). Risk and nitrogen application levels. *Canadian Journal of Agricultural Economics* 57(2), 223–239.

Ravi, M. and Makepeace, J. W. (2022). Lithium-nitrogen-hydrogen systems for ammonia synthesis: exploring a more efficient pathway using lithium nitride-hydride. *Chemical Communications* 58(41), 6076–6079.

Ribaudo, M. O., Delgado, J. and Livingston, M. (2011). Preliminary assessment of nitrous oxide offsets in a cap and trade program. *Agricultural and Resource Economics Review* 40(2), 266–281.

Ribaudo, M. O., Heimlich, R., Claassen, R. and Peters, M. (2001). Least-cost management of nonpoint source pollution: source reduction versus interception strategies for controlling nitrogen loss in the Mississippi Basin. *Ecological Economics* 37(2), 183-197.

Roosen, J. and Hennessy, D. A. (2003). Tests for the role of risk aversion on input use. *American Journal of Agricultural Economics* 85(1), 30-43.

Schilizzi, S. and Pannell, D. J. (2001). The economics of nitrogen fixation. *Agronomie* 21(6-7), 527-537.

Semaan, J., Flichman, G., Scardigno, A. and Steduto, P. (2007). Analysis of nitrate pollution control policies in the irrigated agriculture of Apulia Region (Southern Italy): A bio-economic modelling approach. *Agricultural Systems* 94(2), 357-367.

Shankar, B., DeVuyst, E. A., White, D. C., Braden, J. B. and Hornbaker, R. H. (2000). Nitrate abatement practices, farm profits, and lake water quality: a central Illinois case study. *Journal of Soil and Water Conservation* 55(3), 296-303.

Sheahan, M., Black, R. and Jayne, T. S. (2013). Are Kenyan Farmers under-utilizing fertilizer? Implications for input intensification strategies and research. *Food Policy* 41, 39-52.

Sheriff, G. (2005). Efficient waste? Why farmers over-apply nutrients and the implications for policy design. *Review of Agricultural Economics* 27(4), 542-557.

SriRamaratnam, S., Bessler, D. A., Rister, M. E., Matocha, J. E. and Novak, J. (1987). Fertilization under uncertainty: an analysis based on producer yield expectations. *American Journal of Agricultural Economics* 69(2), 349-357.

Streletskaya, N. A., Bell, S. D., Kecinski, M., Li, T., Banerjee, S., Palm-Forster, L. H. and Pannell, D. J. (2020). Agriculture adoption and behavioral economics: bridging the gap. *Applied Economic Perspectives and Policy* 42(1), 54-66.

Stuart, D., Schewe, R. L. and McDermott, M. (2014). Reducing nitrogen fertilizer application as a climate change mitigation strategy: understanding farmer decision-making and potential barriers to change in the US. *Land Use Policy* 36, 210-218.

Thrikawala, S., Weersink, A., Kachanoski, G. and Fox, G. (1999). Economic feasibility of variable-rate technology for nitrogen on corn. *American Journal of Agricultural Economics* 81(4), 914-927.

Vatn, A., Bakken, L. R., Lundeby, H., Romstad, E., Rørstad, P. Kr., Vold, A. and Botterweg, P. (1997). Regulating nonpoint-source pollution from agriculture: an integrated modelling analysis. *European Review of Agricultural Economics* 24(2), 207-229.

Vermersch, D., Bonnieux, F. and Rainelli, P. (1993). Abatement of agricultural pollution and economic incentives: the case of intensive livestock farming in France. *Environmental and Resource Economics* 3(3), 285-296.

Villacis, A. H., Ramsey, A. F., Delgado, J. A. and Alwang, J. R. (2020). Estimating economically optimal levels of nitrogen fertilizer in no-tillage continuous corn. *Journal of Agricultural and Applied Economics* 52(4), 613-623.

Walters, C. G., Shumway, C. R. and Huggins, D. R. (2017). Impacts of terrain attributes on economics and the environment: costs of reducing potential nitrogen pollution in wheat production. *Agricultural Economics* 48(2), 143-152.

Wang, Y. and Lu, Y. (2020). Evaluating the potential health and economic effects of nitrogen fertilizer application in grain production systems of China. *Journal of Cleaner Production* 264, 121635.

Weersink, A., Dutka, C. and Goss, M. (1998). Crop price and risk effects on farm abatement costs. *Canadian Journal of Agricultural Economics* 46(2), 171-190.

Weersink, A. and Fulton, M. (2020). Limits to profit maximization as a guide to behavior change. *Applied Economic Perspectives and Policy* 42(1), 67-79.

Whittaker, G., Färe, R., Srinivasan, R. and Scott, D. W. (2003). Spatial evaluation of alternative nonpoint nutrient regulatory instruments. *Water Resources Research* 39(4).

Williamson, J. M. (2011). The role of information and prices in the nitrogen fertilizer management decision: new evidence from the agricultural resource management survey. *Journal of Agricultural and Resource Economics*, 36(3), 552-572.

Wu, J. and Tanaka, K. (2005). Reducing nitrogen runoff from the upper Mississippi River basin to control hypoxia in the Gulf of Mexico: easements or Taxes? *Marine Resource Economics* 20(2), 121-144.

Yadav, S. N., Peterson, W. and Easter, K. W. (1997). Do farmers overuse nitrogen fertilizer to the detriment of the environment? *Environmental and Resource Economics* 9(3), 323-340.

Part 3

Organic sources of nitrogen

Chapter 11

Optimizing livestock manure as a source of nitrogen and other nutrients

Samantha Glaze-Corcoran and Masoud Hashemi, University of Massachusetts Amherst, USA

1 Introduction

Manure traditionally refers to both liquid and solid excreta of farm animals and has been used as a source of fertility to support crop growth. Additionally, manure provides organic matter, and both stimulates and supplements soil biology. For these reasons, manure is quite valuable. However, efficient use of manure and its nutrients requires careful management. This chapter will discuss the effective management of nutrients applied in a manure matrix collected in a barn or feedlot in temperate climates.

It is critical to understand that manure nutrients are not equivalent to fertilizers like mined rock phosphate or manufactured nitrogen (N) fertilizer using the Haber–Bosch process; fertilizers obtained from these processes have clear origins, well-established quality control, and can be viewed explicitly as additive fertility. Conversely, manure nutrients should be understood as nutrients in one stage of an active, dynamic cycle observable in a human timescale, not the geological timescale applicable to mined nutrients or atmospheric N. Moreover, manure is not a primary source of revenue for farmers. Therefore, it is often viewed as a chore to be dealt with rather than a lucrative fertility source to be carefully managed.

http://dx.doi.org/10.19103/AS.2024.0135.15

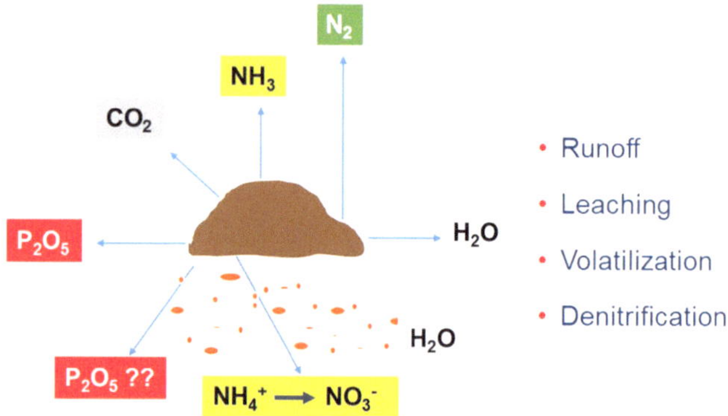

Figure 1 Nitrogen (N) and phosphorus (P) in manure will be lost in different ways, if not managed properly. While N is lost through volatilization, denitrification, and leaching, the P in manure is often lost by runoff and sometimes by leaching, the latter occurring when the soil P level is very high.

Manure can be thought of as a train station terminal for nutrients in transition as they arrive from many places and leave in many directions. Manure-N is a particularly mobile traveler as it is readily converted from one form to another, volatilizes into the atmosphere, leaches into the soil, and leaves with runoff (Fig. 1).

The nutrients in manure come from the plants consumed by the animal; the nutrients are transferred when the undigested part of the feed is excreted as manure. The source of nutrients in those plants comes from the soil that the plants were grown in. Therefore, the nutrients in manure, such as N, phosphorus (P), and potassium, originate from the soil where the feed or forage was grown.

The sources of field fertility for the said feed and forage include elements from both atmospheric N fixation by legume crops, including alfalfa and clovers, and soil parent materials that are liberated into plant-available forms through physical, chemical, and biological processes. Nutrients in the soil also come from the addition of inorganic chemical fertilizers and manure or other organic amendments. Those organic fertility sources again hark back to a field origin and fertility source of their own. Carbon in plants, which will pass into manure, comes from the atmosphere and from the process of photosynthesis. Thus, the nutrients and carbon found in manure are an amalgamation of nutrients from many origins, and manure is a significant component of nutrient and carbon cycling.

A portion of the nutrients and carbon consumed in feed and forage will be assimilated into animals' bodies or transferred to milk and eggs. The remainder of nutrients and carbon that are indigestible or unassimilated will pass through the animal into the excreta (with the exception of methane lost through enteric emissions, A.K.A., burps). The digestive process of animals expedites the liberation of elements from organic molecules back into plant-available inorganic forms - including the N fixed by legumes - and can accelerate the conversion process of plant-assimilated carbon into more stable forms of soil organic carbon.

To optimize manure as a source of nutrients and organic matter, we must approach the concept with two minds. First, the multiple steps of manure generation, collection, storage, and field application must be considered for efficiency at each stage. Second, the practical matters of manure management must be acknowledged, including storage capacity, equipment access, and farm nutrient management plans.

2 Manure as a source of nutrients

2.1 Nitrogen in manure

The amount of manure and nutrient profiles of manure, based on type of animal, are presented in Table 1.

The manure-N is in the inorganic form of ammonium, NH_4^+, and organic-N. However, we frequently discuss ammonia (NH_3) volatilization, N gas emissions, and nitrate (NO_3^-) leaching from manure without context. These leaps in discussion are natural for some but leave gaps for others.

When plant-organic-N is consumed, animal bodies convert it to forms of animal-organic-N. However, not all ingested feed and forage is digested. The organic-N in the undigested portion passes into the feces without further transformation.

Manure-organic-N is decomposed into proteins, then amino acids, and then NH_3 in the process of ammonification. This NH_3 then reacts with water to

Table 1 Average manure production and nutrient contents of various animal types

Animal type	Manure lbs/AU	N P_2O_5 K_2O (lbs/ton)		
Dairy - solid	82	10	4	8
Dairy (liq.) (lb/K gal)	-	28	13	25
Veal	63	8	2	11
Beef cattle	60	11	7	10
Pigs	65	14	11	11
Sheep	40	23	8	20
Horse	45	12	5	9
Poultry	61	30	20	10

Source: Adapted from The Penn State Agronomy Guide (2021-2022). Presented values exclude bedding.
AU = animal unit = 1000 lb.

form NH_4^+. This conversion of organic-N to NH_4^+-N is mineralization. Having a positive charge, the ammonium ion is stable in soils and is plant available. However, NH_4^+ does not accumulate in the soil because it is converted into NO_3^- in the nitrification process.

Urea, CH_4N_2O, is the primary source of N in urine. The liver metabolizes proteins into amino acids, which subsequently form NH_3. Ammonia is toxic in the body, and enzymes in the liver convert NH_3 into urea that is later excreted in urine. Once excreted, the urease enzyme found in feces and soil converts urea to NH_3, which then reacts with water to form NH_4^+.

If manure is not incorporated in soil, significant amounts of NH_3 are likely to volatilize before conversion to NH_4^+. The volatilization of NH_3 increases with temperature and pH.

Nitrate is present in small amounts in raw manure and is thus rarely measured in commercial manure analyses, despite the interest that the scientific community may have in NO_3^-. For context, forage plants rarely contain significant amounts of NO_3^-, with the exception of nitrate accumulation in grasses under certain conditions. Even when NO_3^- is present, it is found in the bottom of the plant in the stems that are not harvested. Alas, N in the form of NO_3^- is minimally ingested in the first place with proper forage management. If ingested, nitrate is converted to nitrite in the animal body by bacteria, which results in what is known as nitrate toxicity. Evolution has selected against NO_3^- and NO_2^- accumulation in the body, and therefore it is not passed into the liquid or solid fraction of manure. The eventual presence of NO_3^- from manure sources is due to mineralization and nitrification achieved in-field or during composting.

Denitrification results in the conversion of NO_3^- to N gas, including nitrous oxide, N_2O, and dinitrogen gas, N_2. The process is mediated by denitrifying bacteria. This N loss from manure becomes more significant with time as the N cycle progresses.

2.2 Sources of variation in manure composition

2.2.1 Initial composition

The initial composition of manure, both within and among species, is affected by four primary factors:

1. Type of animal. The ruminant biology of cattle, sheep, and goats versus the nonruminant biology of swine, equines, and poultry impacts the digestion process and the digestion products. For example, phytate is a form of stored P found in seeds, including feed grains, that can comprise 75% of the total P in the seeds (Raboy, 1997). Ruminants can digest and assimilate P in the form of phytate (Nelson et al., 1976) but non-ruminants cannot

(Cowieson et al., 2011). While phytate will be excreted by non-ruminants, less than 5% of ingested phytate is found in ruminant excretions (Morse et al., 1992). This is one reason for the lower P concentrations in ruminant manure compared to non-ruminants. Even animal breed has been shown to impact manure composition within a species (Aguerre et al., 2011).

2. Animal diet and dietary supplements. The animal type will inherently impact the animal diet. Ruminants and equines thrive on diets high in plant fiber and cannot tolerate excessive levels of protein (Constable et al., 2016), while by contrast poultry requires high-protein diets. This is in part responsible for the higher N and P concentration in poultry manure compared to ruminant manure. Dietary differences among a single species - such as a diet high in grain versus a diet high in forage - may impact manure-N and carbon content (Powell et al., 2006), but this outcome is not consistent (Aguerre et al., 2011; Knowlton et al., 2010). Feed supplements can reliably increase or reduce the nutrients in manure. For example, phytase can be added to the feed of non-ruminants to increase phytate absorption and decrease the P content of manure (Biehl et al., 1995). Zinc has been shown to increase the N content of poultry manure (Kim and Patterson, 2005).
3. Life stage. Manure composition has been shown to vary relative to animal age, milk production, and egg production. Young animals assimilate more nutrients into growing bodies, and lactating and egg-laying animals release nutrients in their eggs and milk thereby diverting some nutrients from excreta (Nennich et al., 2005; Ashworth et al., 2020; Collins et al., 1999).
4. Bedding material. Bedding material will add substantial amounts of carbon to a manure product and alter the carbon-to-N (C:N) ratio. The impact will be relative to the bedding material itself, such as woodchips versus straw. Bedding material can also absorb and reintroduce urine nutrients to the final composite product that could otherwise be lost (Woodward et al., 2019; Misselbrook and Powell, 2005).

2.2.2 Collection and storage

Livestock manure can be solid, semisolid, or liquid depending on animal housing and the collection, storage, and transportation strategy of the operation. An exception is an equine manure, which is found in a solid form due to the high carbon content from eating higher fiber forage, a comparatively low stocking rate of animals in barns, large amounts of bedding used in stalls, and the potential use of woodchips in sacrifice areas. Once collected, manure can be applied directly to fields as a fresh product or stored for later use. The latter is more common. How often manure is collected, how manure is collected, and

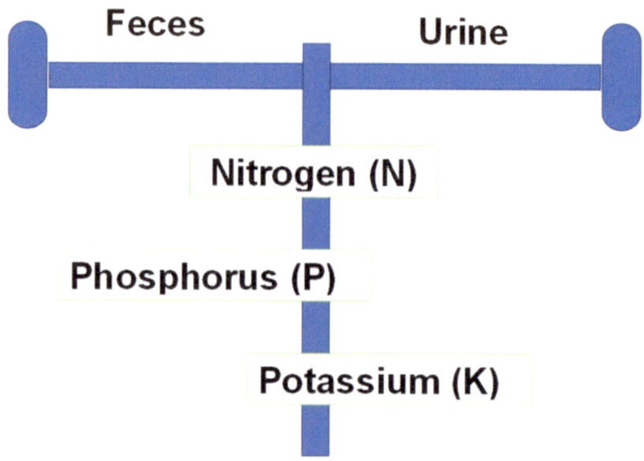

Figure 2 Distribution of major nutrients in feces and urine of manure.

the storage method will impact the nutrient composition. This is of particular importance for N.

Manure accumulated in open feedlots and in barns can be scraped and handled as a solid. Nutrients are excreted in different amounts in urine and feces. If an operation utilizes bedding, significant amounts of urine nutrients can be captured and contained within the bedding that will become a part of the final manure product (Aguerre et al., 2011; Mroz et al., 2000). However, in open feedlots, nutrients in both urine and feces can be lost.

Nutrient loss from open feedlots can be mitigated by the use of a runoff containment system. Once in the containment system, solids can be separated from liquids in a settling basin before the liquid fraction is released. Once solids are separated and liquids are released, the solids will dry prior to collection and handling. If solids are collected, most P, which is mainly in the feces, is retained but K, which is mainly in the urine (Fig. 2), and any remaining N are lost in the liquid. However, if solids are not separated from the liquid fraction, P will be lost as well.

After manure solids are collected – whether from a feedlot, barn, or basin – they will be transported to storage. Commonly, manure solids are stacked on grassed areas or concrete pads (Fig. 3). The presence of the pad limits some nutrient loss via leaching. Solid manures can be stockpiled or composted. Stockpiling is a storage method. By contrast, composting is a management practice that encourages microbial activity and the breakdown of organic materials. Composting results in a more stable product and less volume to transport compared to stockpiled manure but not without tradeoffs.

N and C are lost in significant amounts during both the composting and stockpiling processes. In fact, 30–40% of manure-N can be lost (Bai et al., 2020; Shah et al., 2012; Onema et al., 2007; Eghball et al., 1997), but N losses are greater with composting. About 90% of the N loss during composting is due

Figure 3 Manure stockpiled on a grassed area (left) and on a concrete pad (right).

to ammonia (NH_3) volatilization (Bai et al., 2020; Eghball et al., 1997). The loss of N results in a lower ratio of N to P in the final product which will impact the ability of the product to meet crop needs as desired. Stockpiling, compared to composting, can reduce NH_3 volatilization by nearly 80% (Bai et al., 2020) and help maintain a higher ratio of N to P. The simple addition of plastic sheeting to a compacted stockpile (Fig. 3) can also reduce N loss by 75–90% compared to compost and over 30% compared to uncovered stockpiles (Shah et al., 2012; Chadwick, 2005).

For C, fresh manure has the highest content, but results are mixed on the effects of composting versus stockpiling. In either case, 50% or more C can be lost during the composting or stockpiling process. However, composting increases the C to N ratio due to higher N loss compared to C (Bai et al., 2020; Shah et al., 2012; Cooperband et al., 2002; Eghball et al., 1997). The loss of C is due to microbial decomposition.

Environmental factors, the initial C to N ratio, and how often manure is turned (aerated), if composted, will impact the final C and N content (Ogunwande et al., 2008). The C to N ratio of either stockpiled or composted manure will affect when manure-N becomes available for crop uptake (Lazicki et al., 2020).

In some livestock operations, including dairy production and swine finishing, animals are housed under a roof. These operations may choose not to use bedding so that both liquid and solid excrement can be easily collected in a slurry form (Fig. 4). Minimal water is added in this process but is introduced when animals, equipment, and facilities are cleaned. Many operations have slatted floors with collection pits below the barn. The below barn pit may be the long-term storage location, or slurry may be pumped from this pit into an in-ground pit or above ground tank exterior to the barn. Exterior storage is preferred to increase in-barn air quality for both animals and humans.

In concentrated animal feeding operations, flushing systems are used for the rapid and effective cleaning of facilities. This process introduces substantial

Figure 4 Slurry manure storage; earthen lagoon (left) and slurry tank (right).

amounts of water, resulting in a liquid manure product rather than a slurry. Lagoons are favored for the storage of this liquid manure.

Exterior tanks or in-ground pits can be covered or uncovered, and lagoons are typically uncovered. Uncovered collection areas - including lagoons and slurry pits or tanks - are subject to precipitation which will influence the final liquid composition, nutrient concentration, and the pH of the manure product. Slurry storage results in 5–40% N loss, which is comparable to but can be less than N loss from solid manure. By contrast, lagoons typically result in 70–85% N loss (Sutton et al., 1994).

P loss from manure is typically 5-15%, depending on collection and storage; P loss from lagoons can reach 85%. While this P loss is certainly an environmental concern, the ratio of N to P in manure is often much lower than desired to meet crop needs. Therefore, while P losses are not without serious consequence, N receives the most attention when fertilizer efficiency is the primary focus. The loss of K is intermediate to N and P (Oenema et al., 2007; Sutton et al., 1994). Although K is valuable, the lack of environmental concern regarding K and the persistence of K in the soil results in far less attention to K preservation in manure compared to N or P.

2.2.3 Manure analysis

Due to the many factors influencing manure nutrient content, the true composition of manure may well deviate from the book values found in Table 1. For example, in an analysis of 16 poultry manures collected from different manure stockpiles and poultry houses, the average total-N was 53.3 g kg^{-1}. Although the standard error of the mean was only 4.4, the sample standard deviation was 17.5 (Bitzer and Sims, 1988).

Even when errors or deviations seem small, this variation can quickly result in nutrient excesses or deficits when magnified on the field scale especially when applications span multiple years. When possible, manure should be sampled and analyzed by a laboratory. However, well-mixed samples can be challenging to obtain. As with any sampling protocol, the data generated is only as good as the sample collected. In the absence of a properly collected manure sample, book values are adequate but should be paired with routine soil testing to monitor macro- and micronutrient accumulations as well as crop monitoring for signs of N imbalances.

3 Key considerations of manure fertility

3.1 Nutrient availability

It is necessary to know when nutrients will be released from manure in order to meet the crop needs. Ammonium, nearly all of the K, and most of the P is available during the growing season. More than half of the calcium and magnesium are available, while less than half of the micronutrients are immediately available (Eghball et al., 2002). Although NH_4^+ represents an average of 70% of total N in liquid manure from lagoons, NH_4^+ represents just 30–35% of total N in solids and slurries at the time of application. Additionally, lagoon liquid manure total-N is only 0.619 kg 1000 L^{-1}, while slurry total-N is 4.3 kg 1000 L^{-1} and solid total-N is 10 kg $megagram^{-1}$ (Sutton et al., 1994). Therefore, there is a substantial quantity of N that can be released from the organic fraction of slurry and solid manure via mineralization.

Mineralization refers to the conversion of organically bound nutrients into inorganic forms. The former is not plant-available while the latter may be, depending on the form. The mineralization of organic-N into inorganic-N is mediated by soil microorganisms. Some converted N is released during manure decomposition and is available to plants, while some is temporarily immobilized in microbial bodies. When the microorganisms die, the N is again released and can continue the mineralization process.

Decomposition and mineralization are influenced by temperature, moisture, soil type, soil macrofauna, microorganisms, and the manure C:N ratio (Bhogal et al., 2016; Sistani et al., 2008; Aira et al., 2007; Peacock et al., 2001; Sims, 1986). Of these, only the C:N and soil type presently serve as reliable parameters for predicting mineralization rates in-field, as the other parameters are variable within seasons and across years.

Manures with low C:N ratios, 20:1 or less, are associated with the greatest amount of N-mineralization in the short term (within 10 weeks of application). Manures with C:N ratios above 20:1 demonstrate reduced short-term mineralization (Qian and Schoenau, 2002; Probert et al., 2005). While manures with high C:N still provide valuable inorganic-N at the time of application, the organic-N may not be mineralized in time to be of value during critical crop growth periods. High C:N of manure may also induce significant

N immobilization and may necessitate supplemental N-fertility to meet crop needs.

3.2 Nitrogen- or phosphorus-based application decisions

The N:P ratio (total N to P_2O_5) in manure generally ranges from 1:1 to 4:1 (Sileshi et al., 2017; Cooperband et al., 2002; Eghball 2002; Sutton et al., 1994), depending on the animal, collection method, and storage type. A ratio of 2:1 is often considered to be the 'rule of thumb'. The ratio of N:P uptake by agronomic crops typically ranges from 4:1 to 9:1 (Seidel et al., 2019; Xie et al., 2017; Cooperband et al., 2002; Hons et al., 1986). If manure is applied based on N needs, then it is likely P will be overapplied and accumulated in the soil (Karimi et al., 2018; Sadeghpour et al., 2016; Eghball and Power, 1999). While P is not particularly mobile in soil compared to N, increased soil test P levels are associated with an increased risk of surface runoff (Daverede et al., 2004; Kleinman et al., 2002; Gilley and Eghball, 1998). As a response, P-index informed applications are widely used (Nelson and Shober, 2012).

3.3 Manure form and incorporation methods

For this discussion, the slurry will be addressed as a liquid.

All forms of manure are subject to N-loss through both leaching and volatilization as well as N, P, and K loss through surface runoff and soil erosion. Both liquid and solid manures are incorporated into the soil to limit ammonia volatilization and surface loss of nutrients. Incorporating manure can reduce ammonia loss by as much as 75% (Sadeghpour et al., 2015) compared with surface manure application. However, the efficacy of mechanical incorporation can be greatly improved if one operator spreads manure while another operator follows immediately behind to incorporate manure. Both solid and liquid manures can be incorporated with tillage or light disking, depending on the overall goals and management practices of the operation.

In a 3-year study, Sadeghpour et al. (2015) concluded that regardless of the tillage system, the majority of N in the form of ammonia will be lost in the first 24 h after land application. However, while immediately disking manure into the soil on average retained 70% of N, the use of an aerator after manure application or before manure application retained ammonia N by 41% and 22%, respectively. Manure incorporation with tillage also reduces total P loss due to runoff by 55–93% (Kaiser et al., 2009; Allen and Mallarino, 2008; Daverede et al., 2004).

Liquid manure can be injected into the soil, which is compatible with no-till systems. However, the injector is costly for small operations, and the presence of rocks in the soil may limit the use of manure injectors.

3.4 Timing of application

The timing of the application is influenced by three factors.

1. Crop needs: manure spread prior to crop growth is used as a source of fertility for cash crops. This typically occurs in the spring prior to the primary growing season.
2. Storage capacity: farms are often limited by their ability to store manure. This is particularly true for farms that rely on slurry and liquid storage facilities. When they are full, the contents simply must be spread. Ideally, this occurs in the spring prior to cash crop growth and in the fall prior to cover crop or winter grain planting. In both events, there is a crop to uptake manure nutrients. Sometimes, manure will be spread in the winter due to storage issues. This can be problematic. Even if the ground is not frozen and manure can infiltrate the soil, crop growth and thus manure uptake are reduced. Nutrients are more likely to be wasted via environmental loss and are at greater risk for runoff, with P 2.5–3.6 times more likely to run off compared to non-winter applications (Liu et al., 2018; Vadas et al., 2017; Owens et al., 2011).
3. Time and crop rotation: in New England, manure application and cover planting should occur in September for maximum manure-nutrient capture, especially N (Hashemi et al., 2013). However, farmers may still be harvesting silage corn at this time. Even though some silage fields are harvested early, the continual harvest of remaining silage fields takes priority over timely manure spreading and cover crop planting. The temperature and soil moisture at the time of manure application will significantly increase the amount of N-mineralization. At cooler temperatures, mineralization is reduced. As soil moisture increases, mineralization increases (Agehara and Warncke, 2005; Eghball, 2000). Generally, spring and fall conditions are cool but wet, resulting in mixed effects on mineralization.

3.5 Safety of manure

It is well known that manure can contain pathogens that are harmful to human health, including *E. coli*, *Salmonella,* and *Listeria* (You et al., 2006; Kudva et al., 1998; Renterghem et al., 1991). Improper use of manure in fresh produce production systems can lead to foodborne illness. The United States Food and Drug Administration (FDA) and the United States Department of Agriculture (USDA) regulate manure application to vegetable production. Heat, adjustment to alkaline pH, and composting can be used to treat manure for pathogens; in the absence of these methods, the manure is considered untreated.

At this time, the FDA requires that untreated manure is applied 'in a manner that does not contact covered produce during application and minimizes the potential for contact with covered produce after application'. Covered produce refers to a list of raw produce covered under the Food Safety and Modernization Act. The FDA does not currently impose an application interval from when untreated or treated manure is applied until a covered crop can be harvested, but the specification is forthcoming. In the interim, the FDA 'does not object to' farmers following the organic management standards set by the USDA (FDA, N.D.; Biological Soil Amendments of Animal Origin and Human Waste, 2023).

The guidelines for manure use in organic vegetable production, set by the USDA, require that animal manure must be composted before it is applied to cropland. The exceptions are crops that are not for human consumption, manure incorporated into the soil at least 120 days prior to harvesting a product with edible portions that are in contact with the soil surface or soil particles, or manure incorporated into the soil at least 90 days prior to harvesting a product with edible portions that are not in contact with the soil surface or soil particles. Composted manure can be applied if it has met certain N-to-C ratio requirements and temperature requirements during composting, and the producer has appropriate records documenting these two parameters (Soil fertility and crop nutrient management practice standard, 2023).

4 Challenges and responses in manure management

4.1 Minimizing nitrogen loss

4.1.1 Acidification

Ammonium is the conjugate acid of NH_3. Below a pH of 7, NH_3 is protonated to form NH_4^+. This results in significantly more NH_4 than NH_3 at low pH values, thus reducing NH_3 volatilization. This chemistry has been capitalized on via the acidification of manure to minimize NH_3 loss. The addition of sulfuric acid (H_2SO_4) significantly reduces NH_3 loss from raw manure in the barn, during storage, and after field application (Kai et al., 2008) as well as from composted manure (Cao et al., 2020). Recently, acetic acid (CH_3COOH) was shown to be slightly more effective than sulfuric acid in preventing NH_3 loss and was even more effective in preventing methane (CH_3) emission (Fuchs et al., 2021). This method has the added benefits of eliminating odorous hydrogen sulfide (H_2S) production associated with sulfuric acid and being less corrosive to equipment.

Lactic acid ($C_3H_6O_3$) has the capacity to acidify manure as well and equally conveys the benefit of reduced corrosion and hydrogen sulfide production. However, this comparatively weak acid is required in large, cost-prohibitive amounts to achieve the desired pH. A recent study demonstrated the potential to manipulate the microbial community to favor a substantial population of lactic acid-producing bacteria that successfully lowered the pH to favor NH_4^+ formation. Additionally, urease-producing bacteria were inhibited. This limited the amount of urea conversion to NH_3, further reducing the potential for NH_3 volatilization. In total, NH_3 loss was reduced by nearly 96% (Liu et al., 2023). This experiment demonstrated proof of concept and offers an interesting opportunity to use microbiology to drive the desired chemical reactions.

4.1.2 Alkalization and ammonia stripping

Treating manure to increase the pH and thus the ammonia content can kill microbes (Park and Diez-Gonzalez, 2003; Vincini et al., 1994) and increase the safety of manure (Islam et al., 2004). This is why alkalinization is a recommendation to treat manure to reduce pathogens, including *E. coli*, *Salmonella*, and *Listeria*. Acidification for ammonium retention is obviously in conflict with alkaline treatments. However, alkalinization can be an element of N preservation.

Increasing the pH of a manure solution will facilitate the conversion of NH_4^+ to NH_3 and concurrently increasing temperature encourages NH_3 volatilization from the solution. The NH_3 can be captured, thermally destroyed, released into the atmosphere, or converted into a solid. The process is widely cited in wastewater treatment systems to remove N from the water before it is released (Guštin and Marinšek-Logar, 2011). This same logic can be applied to manure management.

Anaerobic manure digesters capture methane (CH_4) and use it for energy production. The addition of an ammonia stripping system enables digesters to also convert NH_3 gas into a solid N form, such as ammonium sulfate ($(NH_4)_2SO_4$) (Bi et al., 2020; Mohammed-Nour et al., 2019; Jiang et al., 2014). Ammonia stripping can improve the anaerobic digestion process as well. The rate of digestion increases at higher temperatures but so does the accumulation of NH_3 gas that can kill methanogen microbes. When the NH_3 is stripped from the system, digestion can occur at a higher temperature (Yao et al., 2017). Likewise, manure high in inorganic-N – such as poultry manure – has been limited by the accumulation of excessive levels of NH_3 (Bi et al., 2020). Ammonia stripping can increase the value of these manures for anaerobic digestion and correspondingly improve their N management by converting NH_3 into a solid N fertilizer.

4.1.3 Adsorbents

The successful use of amendments to reduce NH_3 and N_2O has been demonstrated in many studies. Adsorbent materials that hold onto NH_4 and NH_3, including bentonite (Wang et al., 2016; Li et al., 2012), zeolite (Li et al., 2008; Lefcourt and Meisinger, 2001), biochar (Awasthi et al., 2020; Dougherty et al., 2017), and fly ash (Zhao et al., 2023; Usmani et al., 2019; Bhattacharya et al., 2004), have clear physical and chemical effects that result in reduced N loss. However, recent research has found downstream effects of these adsorbents on microbial community structure and function; these microbial changes are also responsible for improvements to N retention.

In one study, the addition of bentonite resulted in a 19% and 73% reduction of NH_3 and N_2O emissions from composted manure. After 43 days, the inorganic-N content (NH_4 and NO_3) was nearly seven times greater than composted manure that did not receive bentonite. The adsorbent properties of bentonite were partially credited with the retention of NH_3 and NH_4. In addition, the physical and chemical changes induced by bentonite altered the microbial community during the composting process. Nitrification gene abundance was significantly lower and nitrite reductase gene abundance was significantly higher, favoring the accumulation of NH_4 (Guo et al., 2020).

This study highlighted the importance of investigating not just physical and chemical changes with adsorption technology, but the subsequent effects on the microbiology involved in the N cycle. Studies have taken similar approaches to better understand how biochar can expedite the humification of composted manure (Zhao et al., 2022) or how diatomite can decrease N_2O and NH_3 losses and increase total N in the final product (Ren et al., 2021).

Improving our understanding of how manure amendments - including adsorbents as well as lime, bulking agents, and micronutrients - affect microbial diversity will support improvements to manure management strategies. The 'omics' research suite (genomics, transcriptomics, proteomics, metabolomics) continues to grow more affordable, accessible, and powerful. These tools stand to be significant resources for years to come.

4.1.4 Cover crops

Manure must often be spread at times in the year when there is no actively growing cash crop. This necessity is driven by limited storage capacity. In the absence of a crop to capture N and store it as organic plant-N, much of the N will be lost to leaching and denitrification. Cover crops can be planted to uptake manure-N that could otherwise be lost. The cover crops serve as a temporary savings bank of nutrients that will eventually be returned to the field as residues decompose and mineralize.

Cover crops can reduce soil-N (Everett et al., 2019) thus limiting the potential for N loss. Compared to fallow conditions, a rye cover crop has been shown to reduce nitrate leaching up to 95% in the laboratory (Parkin et al., 2006) and up to 60% in modeling (Malone et al., 2017), but in field and meta-analysis, the results range from as little as 13% to over 85% (Nouri et al., 2022; Sedghi and Weil, 2022; Adeli et al., 2021; Hanrahan et al., 2021; Strock et al.,

2004). In one widely cited study, a rye cover crop had no significant effect on preventing nitrate loss. The failure was attributed to soil type (Ritter et al., 1998), but the timing of manure application and cover cropping may have also played a role. The time between manure application and cover crop planting date can significantly affect nitrate leaching. In one study, an early planting reduced leaching by 86%, while a delayed planting reduced leaching by 45% (Sedghi and Weil, 2022).

Likewise, nitrous oxide (N_2O) emissions were reduced in the lab by an average of 52% (Parkin et al., 2006). However, a meta-analysis found that cover crops were associated with N_2O emissions only 40% of the time (Basche et al., 2014). The analysis included but did not specifically focus on manured systems. However, it again reveals the variation in the literature. In both the case of nitrate and N_2O loss prevention, further work is needed to optimize cover crops for in-field N retention.

Cover crops harvested as a supplemental source of forage may also contribute to P management. When harvested, organic-P in plants is removed from the field. It can re-enter the on-farm nutrient cycle if fed to livestock or exit the farm's nutrient system if the forage is sent off-farm. Such an approach can help limit P accumulation in soils (Glaze-Corcoran et al., 2022).

4.1.5 Injection

Subsurface injection of manure is an efficient method of eliminating ammonia volatilization. Manure injection can also enhance nutrient cycling and reduce the loss of nutrients (Battaglia et al., 2021). For example, the runoff of P in manure could be minimized by injection, more so after heavy precipitation (Jahanzad et al., 2019). One major benefit of manure injection is compatibility with no-tillage systems.

However, there are several challenges with manure injection into agricultural lands. The main challenges with the injection include the cost of the equipment, higher fuel consumption, increased time of application, and when the field contains a significant amount of rocks.

4.1.6 Urease and nitrification inhibitors

Urea is the primary source of ammonia volatilization from manure. Urease inhibitors limit the conversion of urea to ammonia and ammonium. Liu et al. (2023) demonstrated a microbiology management practice to limit populations of ureolytic bacteria in manure. By contrast, urease inhibitors function to inactivate the urease enzyme. *N*-(*n*-butyl)-thiophosphoric triamide (NBPT) has long been a popular urease inhibition molecule. NBPT can control urea loss from feedlots when applied directly to pens (Varel et al., 1999) and when applied to collected manure (Singh et al., 2009).

The effects of NBPT last only a few days without reapplication (Parker et al., 2005), and novel chemistries have been investigated to improve efficacy (Hagenkamp-Korth et al., 2015). *N*-(2-nitrophenyl) phosphoric triamide (2-NPT) has recently shown promise in extending urease inhibition effects longer than NBPT. Ammonia volatilization measured 29 days after urease application was significantly lower with 2-NPT compared to NBPT, and 2-NPT continued to reduce volatilization up to 59 days after application. However, the results were specific to soil type (Adhikari et al., 2019). Urease chemistries and the effect of soil type remain valuable avenues of investigation.

Nitrification inhibitors contain compounds that suppress *Nitrosomonas* and *Nitrobacter*; the former is responsible for the conversion of NH_4^+ to NO_2^- and the latter is responsible for the conversion of NO_2^- to NO_3^-. Nitrification inhibition arrests mineralization. Ammonium can accumulate at higher levels for a longer time as a result, allowing plant uptake while reducing NO_3^- leaching or denitrification.

N inhibitors are associated with the reduction of NO_3^- leaching by over 75% and N2O emissions by over 80%. However, successful nitrification inhibition is relative to soil type, temperature, and moisture. Sandy soils, warm temperatures, and increased soil moisture reduce the success of nitrification inhibition. While N_2O reductions are largely conserved, NO_3^- leaching reductions may be insignificant (Lin and Hernandez-Ramirez, 2020; Nair et al., 2020; Smith et al., 2002; Di and Cameron, 2002; Beckwith et al., 1998). Future research into the optimized use of nitrification inhibitors can improve plant N uptake and reduce N loss. N inhibitors used in combination with biochar, straw, or other organic amendments may provide increased success (Wang et al., 2019).

4.1.7 Modification of carbon-to-nitrogen ratio

Many studies have investigated co-composting manure with plant-based additives to manipulate the C:N ratio. Higher, initial C:N ratios are associated with reduced N loss and increased inorganic-N in the finished compost, in part due to temporary N immobilization that prevents N loss (Jiang et al., 2011; Gao et al., 2010; Zhu, 2007; Calderón et al., 2005). Recently, higher, initial C:N ratios have been shown to promote the relative abundance of non-symbiotic N-fixing bacteria and decrease bacteria associated with denitrification and ureolysis (Liu et al., 2022; Zhang et al., 2020), adding an explanation to the observed phenomenon induced by high C:N ratios. This once again demonstrates the prominent role that omics and microbiology can currently play in improving our understanding of manure nutrient dynamics, especially that of N.

4.2 Mitigating negative impacts to soil quality

4.2.1 Salt accumulation

Long-term application of nutrient-rich manure can result in soluble salt build up (Chang et al., 1991). Efforts have been made to identify safe salt concentrations of manure to avoid negative effects on crop production (Shapiro et al., 2005). However, salt build up from manure over many years is likely to occur when manure is applied at maximum rates (Hao and Chang, 2003). This is particularly true in areas with inadequate precipitation to facilitate salt leaching as well as in high tunnels. Despite risks, maximum application serves the purpose of meeting crop-N needs as well as clearing manure storage facilities.

Calcium amendments have been shown to increase crop performance in soils with elevated salts. Following this principle, calcium-fortified composted manure was investigated. This manure modification resulted in increased plant performance despite elevated salt conditions (Niamat et al., 2019). While leaching can be induced in high tunnels with drenching, this cannot reasonably be performed on a field scale. Although the modification did not alleviate salt accumulation, it demonstrated a potential management practice to minimize the effects associated with salt stress.

4.2.2 Non-uniform application

Lack of nutrient uniformity in agricultural soils is a common issue. The variability of nutrient concentrations in different parts of agricultural fields is partly due to the history of cropping since nutrient removal varies with crop species. Another reason for the lack of uniformity is related to different soil types in a single field. However, one major reason for various levels of nutrients in different parts of fields could be the non-uniform application of manure. Traditionally, the fields that are closer to the barn receive more manure than those in the distance. A three-year study conducted by the University of Connecticut, the University of Massachusetts, and the University of Maine revealed that there is considerable variability in P level in corn fields with a history of manure application (Fig. 5).

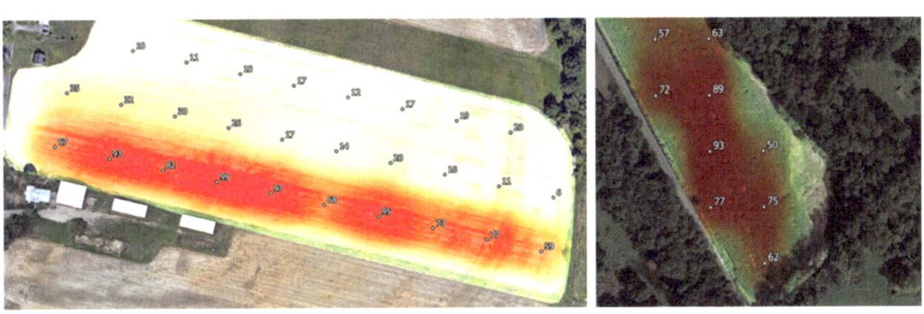

Figure 5 Ununiform P level in different parts of the manured fields. The darker red sections demonstrate a higher P level.

The non-uniform application of nutrient resources can result in decreased crop productivity. This is because while excess nutrients in one section do not necessarily produce a higher yield, the areas with less than optimum level of nutrients do not reach the potential yield. Areas with nutrient deficits may not reach maximum yields if manure is the only source of fertility. Moreover, the non-uniform application will result in a composite soil sample that may not be correct, exacerbating fertility shortcomings.

4.3 Integration with whole farm management

4.3.1 Synchrony of nitrogen mineralization and crop uptake

The portion of N immediately available to crops at the time of manure application can be predicted with a manure test and estimates of mineralization have been assessed. Laboratory experiments have offered valuable insight into mineralization under controlled conditions (Cassity-Duffey et al., 2020; Lazicki et al., 2020). Despite this knowledge, the only condition that influences mineralization that can presently be controlled in the field is the C:N ratio of the selected manure. The ratio can be selected based on manure product, or it may be altered during co-composting or co-digesting to reach more ideal conditions. Even so, it can be difficult to achieve optimum mineralization timing that aligns with patterns of crop N uptake in-season.

Both first-year and second-year N availability are greatest with properly handled fresh manure, followed by dried manure, with composted manure having the least availability. Fresh manure with low C:N ratios does provide yield increases compared to crops grown with no supplemental fertility. However, composted manures with a high C:N ratio may not provide yield increases compared to unfertilized plots in the first 2 years (Muñoz et al., 2008). These composts may be limited by mineralization rates as well as N immobilization. Significant contributions of these manures may not be seen until 3 years after application (Pang and Letey, 2000).

Despite the potential for delayed mineralization in the early years of application, a field with a long history of manure application can result in twice as much in-season N mineralization compared to a field with no history of manure application. Yet, the application of supplemental N to a field with a history of manure application may reduce the total N availability compared to a field with a history of manure and no supplemental N application (Mallory and Griffin, 2007).

Supplemental N is recommended to meet crop needs when manure applications are based on P levels (Moshia et al., 2014). Supplemental N fertility can induce changes in microbial populations (Zeng et al., 2016; Ramirez et al., 2010). Inorganic-N applications can have a stronger effect on microbial populations than organic-N sources when considered in a short-term timescale

(Zhang et al., 2021). These changes may stimulate the growth of populations that can compete with crops for N and decrease bacteria that support mineralization. The extent of these effects is largely influenced by soil type and environmental conditions.

Research that improves our understanding of how, why, and when N-mediating microbial communities shift is of great importance. This is especially true when the confounding effect of cover crop residue is introduced. In the absence of our ability to control the weather, our ability to optimize mineralization and crop uptake may be limited or expanded relative to our ability to intentionally and consistently influence soil microbial populations. In the meantime, the pre-sidedress nitrate test remains the most effective method to optimize in-season supplement N fertility in manured systems and field crop production.

4.3.2 No-till and organic systems

Manure injectors are valuable in no-till systems. Injection prevents N loss as NH_3 and prevents P stratification associated with surface manure application, all without tillage-based incorporation. This provides required compliance with the use of manure in organic vegetable production. However, these systems are not suitable for most small farmers due to cost.

4.3.3 Storage and transportation limitations

Many farms, especially small-scale operations, do not have manure storage, therefore they have to apply manure daily. The main challenges with the daily application of manure include loss of ammonia to the atmosphere due to volatilization and non-uniform rate of application, thus creating variable nutrient levels in the fields, loss of nutrients by runoff, and in some areas prohibition of manure application to snow-covered lands.

When a farm has manure storage, often the capacity of storage is limited and can hold only 6 months of manure generated on the farm. Traditionally, on farms that keep animals and produce crops, farmers empty their storage once in the fall after crop harvest, and in spring prior to planting. Manure applied in fall must be protected by integrating cover crops into the system, otherwise, a significant amount of manure nutrients will be lost to the environment through leaching, volatilization, and runoff. In a study, by Hashemi et al. (2013) roughly 160 kg N ha^{-1} was captured by the winter rye cover crop planted in the first week of September otherwise, this amount of N would have been lost to the environment.

Farmers may also not prioritize the time and fuel required to apply manure to fields farther from the barn. Manure with low nutrient concentrations is also prohibitive to move long distances. The expense of fuel and labor to transport and apply manure can easily undermine the value of the nutrients in manure. This logistical challenge does not have a clear scientific or economic solution, which is partially responsible for the over application of manure in fields closer to manure storage facilities and nutrient deficiencies in far fields. Policy that prevents over application is a mechanism to overpower these practical challenges. While the policy may solve the problem of overapplication, it does not solve the economic or logistical challenges.

5 Case studies

5.1 Case study 1: managing high carbon horse manure

Management of manure on horse farms is a challenge for horse owners and equine facility managers. This is of particular concern at farms where horses are kept in stalls and land availability for manure spreading is limited. An average size horse generates about 45 lb. of manure per day, almost 10 tons per year plus a considerable amount of soiled bedding. On average horse waste that includes manure and woody bedding often possesses a C:N ratio of >75:1 depending on the amount and frequency of bedding replacement. This explains why a pile of stable waste cannot noticeably decompose over time. Other than the proper C:N ratio, successful composting requires adequate moisture and oxygen for optimum microbial activity.

In a sanctuary for draft horses with only 8 acres of land, 40 horses were kept mainly in the barn. The barn waste was not proper to be used as fertilizer due to a very high C:N ratio. As a result, there were several piles of waste accumulated on the farm causing serious threats to the nearby water bodies. A static aerated system was used to convert the waste into a useful material (Fig. 6). The system consists of 1-4 perforated PVC pipes, 1-HP air pump, and a geotextile fabric (Hashemi et al., 2017).

Figure 6 One of the waste piles at the facility (left) and a single-line aerated composting pile (right).

The perforated PVC pipe was placed in a groove, and the composting materials were placed on top and then covered with a geotextile fabric that is water resistant but air can freely move in and out. The PVC pipe was connected to the pump, using a PVC elbow. The pump was connected to a timer and was set to work for 2 min per hour. The pile was 5 feet wide and 5 feet tall. The temperature of the composting material was monitored using a long-stem thermometer. The compost was finished in 3 months with a C:N ratio of 26:1. Composting could have accelerated by adding about an inch of well-finished compost or addition of N-rich materials including grass clippings or kitchen scraps. More PVC pipes could have been connected to the pump, when a larger amount of materials are available or for harvesting the finished compost at different stages.

5.2 Case study 2: reducing ammonia volatilization

Reducing ammonia volatility is a major step in manure-N management for retention and sustainable N cycling in a farming system. Traditionally, planting winter grain cover crops, more specifically winter annual rye, has been considered an efficient strategy for capturing N released from manure mineralization. However, in a no-till planting system or when planting cover crops is delayed, surface application of slurry manure results in significant N loss through volatilization. In this case, two strategies can be considered for mitigating ammonia volatility. The first strategy is the use of an aerator to create holes, thus capturing a major portion of the applied manure N before it is lost to the atmosphere as ammonia. Sadeghpour et al. (2015) demonstrated that the use of an aerator before manure application saved as much as 41% of ammonia compared to the conventional surface application. The second strategy is delaying manure application until late fall when the temperature is low (Akbari et al., 2020). This strategy could be efficient if planting cover crops also is delayed, otherwise, high cover crop biomass will prevent contact between manure and soil.

6 Future research directions for manure management

An excellent body of research has investigated manure dynamics for decades and has well identified the challenges in manure management for optimum fertility and environmental health. From the chemistry of leaching and volatilization to the practical necessities to store and spread, both the scientific community and farmers are aware of the limitations. To advance the field, some further investigatory work is needed. However, an emphasis must be placed on identifying solutions rather than precisely cataloging challenges in all possible scenarios one study at a time.

6.1 Microbiology

The recent work investing microbial population dynamics in digesters, compost, and in the field provides valuable insight as well as potential solutions. This research has pushed past our physical and chemical understanding of manure management techniques. It has offered evidence of downstream effects on microbial communities that in turn influence the outcomes of N form and preservation. This research has also cast new light on why nitrification inhibitors and other manure amendments have variable field results.

There is tremendous value to the omics tools currently available, and cross-disciplinary research to advance this area can result in new or improved management techniques.

6.2 Agricultural engineering

Precision soil sampling and manure application technology have the potential to overcome the challenges of non-uniform application and the heterogeneity it creates. The concept of precision manure management is not a new one. However, technology is rapidly evolving to improve this practice and make it more affordable for smaller operations.

Improvements to manure injectors can increase the adoption of this method. This would overcome hurdles with no-till management, including nutrient stratification as well as volatilization. The current implements are slow and costly, limiting their adoption.

The use of injectors does not provide an avenue for solid manure application in no-till systems. Liquid injection should be prioritized in no-till systems to limit nutrient losses associated with surface-applied manure. Thus, solid manure is relegated to only tilled systems when optimum nutrient management is the goal.

Increasing ammonia scrubbers on-farm that convert ammonia to solid N-fertilizer offers multiple benefits. These systems reduce N loss and move N into an easily manageable form. As biogas production continues to expand in the energy sector, there is significant potential to increase the use of such systems.

6.3 Phosphorus solutions

Policy exists to mitigate the over-application of P. However, this does not solve the problem of the farmers. Programs that offset the cost of long-distance manure transport should be considered. Such programs are just as valuable as funding the planting of cover crops. Moreover, on-farm nutrient cycling can be zoomed out to a state or regional scale. If a larger area is viewed as a connected system of nutrient cycling, multiple farmers could benefit.

It is known that liquid and solid separators capture a significant amount of P in the solids, thus reducing P in the liquid fraction. Increased access to this method can help farmers apply more volume and more N to their crops. This also makes it possible to export P from livestock farms to operations deficient in P. However, this significantly changes to the C:N ratio of the end products and will alter soil organic matter accumulation. Improvements would be required for both economic and logistical practicality.

Management practices that remove P from the field are limited. The use of cover crops harvested for forage should be further investigated. Such an approach provides practical incentive to plant cover crops as well as environmental benefits.

6.4 Coordinating nitrogen mineralization with crop uptake

Many researchers are trying to untangle in-field mineralization dynamics from organic inputs in order to create reliable predictions. The impacts of environmental parameters have been characterized. The fact is temperature and soil moisture are uncontrollable in a field crop. Moisture may be influenced in an irrigated system to an extent. Annual adjustments could be made using long-range weather forecasts. Despite our scientific endeavors to influence soil microbes in the field, this is a substantial challenge that will take years to achieve.

Emphasis should be placed on developing reasonable predictions given soil type, manure form, C:N, long-range forecasts, and field history. Acceptance that some variation is simply unpredictable will help develop responses to this variation. In-field soil-N monitoring should be emphasized. The pre-sidedress nitrate test remains a reliable tool for field crops. The use of in-field N-monitoring sensors as well as remote crop monitoring can further improve the precision of supplemental-N application decisions.

6.5 Public perception

Agricultural production is too often perceived as occurring in remote, rural areas. While this is true in some parts of the United States, there is a significant amount of agricultural production occurring near cities and integrated into residential areas. The undeveloped spaces and idyllic farm scenes are desirable for many homeowners.

However, the odor of manure can easily result in complaints from neighbors. Perceptions that manure is not being applied safely or sustainably can also draw community attention. Concerns about complaints can significantly impact farmer decisions regarding manure application and may influence farmers to choose non-optimum management practices. For example, a farmer may choose to apply manure in late fall or early winter when ammonia volatilization will be slower and thus have less smell. However, applying in late summer or early fall is preferable so that a growing cover crop can capture manure nutrients.

Extension programs and agricultural service agencies can participate in public education. Local policymakers and regulatory agencies can support the installation of digesters, which have the added benefit of reduced ammonia in the applied product. These social benefits have a role to play in effective manure management.

7 Conclusion

Historically, manure has been valued as a vital component in farming systems. It contains numerous recyclable elements, including macro and micronutrients, organic matter, and fiber. However, its fertilizer value can be significantly diminished if not utilized efficiently. Nutrient loss from manure represents not only an economic setback for growers but also poses a potential environmental liability for animal farming operations, affecting air quality and surface as well as groundwater resources. Effective manure management is crucial in this context. This involves strategies such as appropriate application rates and timing, sufficient storage capacity, method of application, use of cover crops, and maintaining the proper carbon-to-nitrogen ratio. These practices play a crucial role in optimizing nutrient value while minimizing environmental pollution risks. The use of manure in no-till systems is an area that requires further research. Another important aspect of manure management is better synchronizing the nutrient release from manure with the nutrient uptake needs of crops.

8 Where to look for further information

- Esteves, E. M. M., Herrera, A. M. N., Esteves, V. P. P. and Morgado, C. D. R. V. (2019). Life cycle assessment of manure biogas production: A review. *Journal of Cleaner Production* 219, 411–423. doi.org/10.1016/j.jclepro.2019.02.091.
- Nugrahaeningtyas, E., Lee, D. J., Song, J. I., Kim, J. K. and Park, K. H. (2022). Potential application of urease and nitrification inhibitors to mitigate emissions from the livestock sector: a review. *Journal of Animal Science and Technology* 64(4), 603–620. doi.org/10.5187%2Fjast.2022.e5.
- The Penn State Agronomy Guide. https://extension.psu.edu/the-penn-state-agronomy-guide.
- Sommer, S. G., Zhang, G. Q., Bannink, A., Chadwick, D., Misselbrook, T., Harrison, R., Hutchings, N. J., Menzi, H., Monteny, G. J., Ni, J. Q., Oenema, O. and Webb, J. (2006). Algorithms determining ammonia emission from

buildings housing cattle and pigs and from manure stores. *Advances in Agronomy* 89, 261–335. doi.org/10.1016/S0065-2113(05)89006-6.

- University of Massachusetts Crop. Dairy, livestock, and equine extension program website. https://ag.umass.edu/crops-dairy-livestock-equine.
- Westendorf, M. L., Williams, C. A., Murphy, S., Kenny, L. and Hashemi, M. (2020). Generation and management of manure from horses and other equids. In Waldrip, H. M., Pagliari, P. H. and He, Z. (Eds), *Animal Manure: Production, Characteristics, Environmental Concerns, and Management*. ASA Special Publication, USA. Vol. 67.
- Zalewska, M., Błażejewska, A., Czapko, A. and Popowska, M. (2021). Antibiotics and antibiotic resistance genes in animal manure-consequences of its application in agriculture. *Frontiers in Microbiology* 12, 640. doi.org/10.3389/fmicb.2021.610656.

9 References

Adeli, A., Brooks, J. P., Read, J. J., Miles, D. M., Shankle, M. W. and Jenkins, J. N. (2021). Impact of cover crop on nutrient losses in upland soil. *Communications in Soil Science and Plant Analysis* 52(5), 536–550. doi.org/10.1080/00103624.2020.1862154.

Adhikari, K. P., Saggar, S., Hanly, J. A. and Guinto, D. F. (2019). Comparing the effectiveness and longevity of the urease inhibitor N-(2-nitrophenyl) phosphoric triamide (2-NPT) with N-(n-butyl) thiophosphoric triamide (nBTPT) in reducing ammonia emissions from cattle urine applied to dairy-grazed pasture soils. *Soil Research* 57(7), 719–728. doi.org/10.1071/SR18337.

Agehara, S. and Warncke, D. D. (2005). Soil moisture and temperature effects on nitrogen release from organic nitrogen sources. *Soil Science Society of America Journal* 69(6), 1844–1855.

Aguerre, M. J., Wattiaux, M. A., Powell, J. M., Broderick, G. A. and Arndt, C. (2011). Effect of forage-to-concentrate ratio in dairy cow diets on emission of methane, carbon dioxide, and ammonia, lactation performance, and manure excretion. *Journal of Dairy Science* 94(6), 3081–3093. doi.org/10.3168/jds.2010-4011.

Aira, M., Monroy, F. and Domínguez, J. (2007). Eisenia fetida (Oligochaeta: Lumbricidae) modifies the structure and physiological capabilities of microbial communities improving carbon mineralization during vermicomposting of pig manure. *Microbial Ecology* 54(4), 662–671. doi.org/10.1007/s00248-007-9223-4.

Akbari, P., Herbert, S., Hashemi, M., Barker, A., Zandvakili, O. R. and Bistgani, Z. E. (2020). Winter annual rye seeding date influence on nitrogen recovery and ammonia volatilization from late fall surface-applied manure. *Agronomy* 10(7), 931; doi.org/10.3390/agronomy10070931.

Allen, B. L. and Mallarino, A. P. (2008). Effect of liquid swine manure rate, incorporation, and timing of rainfall on phosphorus loss with surface runoff. *Journal of Environmental Quality* 37(1), 125–137. doi.org/10.2134/jeq2007.0125.

Ashworth, A. J., Chastain, J. P. and Moore, Jr, P. A. (2020). Nutrient characteristics of poultry manure and litter. *Animal Manure: Production, Characteristics, Environmental Concerns, and Management* 67, 63–87. doi.org/10.2134/asaspecpub67.c5.

Awasthi, M. K., Duan, Y., Awasthi, S. K., Liu, T. and Zhang, Z. (2020). Influence of bamboo biochar on mitigating greenhouse gas emissions and nitrogen loss during poultry manure composting. *Bioresource Technology* 303, 122952. doi.org/10.1016/j.biortech.2020.122952.

Bai, M., Flesch, T., Trouvé, R., Coates, T., Butterly, C., Bhatta, B., Hill, J. and Chen, D. (2020). Gas emissions during cattle manure composting and stockpiling. *Journal of Environmental Quality* 49(1), 228–235. doi.org/10.1002/jeq2.20029.

Basche, A. D., Miguez, F. E., Kaspar, T. C. and Castellano, M. J. (2014). Do cover crops increase or decrease nitrous oxide emissions? A meta-analysis. *Journal of Soil and Water Conservation* 69(6), 471–482. Doi.org/10.2489/jswc.69.6.471.

Battaglia, M. L., Ketterings, Q. M., Godwin, G. and Czymmek, K. J. (2021). Conservation tillage is compatible with manure injection in corn silage systems. *Agronomy Journal* 113(3), 2900–2912. doi.org/10.1002/agj2.20604.

Beckwith, C. P., Cooper, J., Smith, K. A. and Shepherd, M. A. (1998). Nitrate leaching loss following application of organic manures to sandy soils in arable cropping. I. Effects of application time, manure type, overwinter crop cover, and nitrification inhibition. *Soil Use and Management* 14(3), 123–130. doi.org/10.1111/j.1475-2743.1998.tb00135.x.

Bhattacharya, S. S. and Chattopadhyay, G. N. (2004). Transformation of nitrogen during vermicomposting of fly ash. *Waste Management and Research* 22(6), 488–491.

Bhogal, A., Williams, J. R., Nicholson, F. A., Chadwick, D. R., Chambers, K. H. and Chambers, B. J. (2016). Mineralization of organic nitrogen from farm manure applications. *Soil Use and Management* 32(S1), 32–43. doi.org/10.1111/sum.12263.

Bi, S., Qiao, W., Xiong, L., Mahdy, A., Wandera, S. M., Yin, D. and Dong, R. (2020). Improved high solid anaerobic digestion of chicken manure by moderate in situ ammonia stripping and its relation to metabolic pathway. *Renewable Energy* 146, 2380–2389 doi.org/10.1016/j.renene.2019.08.093.

Biehl, R. R., Baker, D. H. and DeLuca, H. F. (1995). 1α-hydroxylated cholecalciferol compounds act additively with microbial phytase to improve phosphorus, zinc, and manganese utilization in chicks fed soy-based diets. *The Journal of Nutrition* 125(9), 2407–2416. doi.org/10.1093/jn/125.9.2407.

Biological Soil Amendments of Animal Origin and Human Waste 21 CFR § 112.51 - § 112.60 (2023).

Bitzer, C. C. and Sims, J. T. (1988). Estimating the availability of nitrogen in poultry manure through laboratory and field studies. *Journal of Environmental Quality* 17(1), 47–54. doi.org/10.2134/jeq1988.00472425001700010007x.

Calderón, F. J., McCarty, G. W. and Reeves, J. B. (2005). Analysis of manure and soil nitrogen mineralization during incubation. *Biology and Fertility of Soils* 41(5), 328–336. doi.org/10.1007/s00374-005-0843-x.

Cao, Y., Wang, X., Liu, L., Velthof, G. L., Misselbrook, T., Bai, Z. and Ma, L. (2020). Acidification of manure reduces gaseous emissions and nutrient losses from subsequent composting process. *Journal of Environmental Management* 264, 110454. doi.org/10.1016/j.jenvman.2020.110454.

Cassity-Duffey, K., Cabrera, M., Gaskin, J., Franklin, D., Kissel, D. and Saha, U. (2020). Nitrogen mineralization from organic materials and fertilizers: predicting N release. *Soil Science Society of America Journal* 84(2), 522–533. doi.org/10.1002/saj2.20037.
Chadwick, D. R. (2005). Emissions of ammonia, nitrous oxide and methane from cattle manure heaps: effect of compaction and covering. *Atmospheric Environment* 39(4), 787–799. doi.org/10.1016/j.atmosenv.2004.10.012.
Chang, C., Sommerfeldt, T. G. and Entz, T. (1991). Soil chemistry after eleven annual applications of cattle feedlot manure. *Journal of Environmental Quality* 20(2), 475-480. doi.org/10.2134/jeq1991.00472425002000020022x.
Collins, J. C., Jr., Barker, L. E., Carr, L. E., Brodie, H. L. and Martin, J. H., Jr. (1999). *Poultry Waste Management Handbook (NRAES-132)*. Natural Resource, Agriculture, and Engineering Service, Cooperative Extension, Ithaca, NY.
Constable, P. D., Gruenberg, W., Hinchcliff, K. W., Done, S. H. and Gruenberg, W. (2016). *Veterinary Medicine: A Textbook of the Diseases of Cattle, Horses, Sheep, Pigs and Goats*, Elsevier Health Sciences, USA.
Cooperband, L., Bollero, G. and Coale, F. (2002). Effect of poultry litter and composts on soil nitrogen and phosphorus availability and corn production. *Nutrient Cycling in Agroecosystems* 62(2), 185–194. doi.org/10.1023/A:1015538823174.
Cowieson, A. J., Wilcock, P. and Bedford, M. R. (2011). Super-dosing effects of phytase in poultry and other monogastrics. *World's Poultry Science Journal* 67(2), 225–236. Doi .org/10.1017/S0043933911000250.
Daverede, I. C., Kravchenko, A. N., Hoeft, R. G., Nafziger, E. D., Bullock, D. G., Warren, J. J. and Gonzini, L. C. (2004). Phosphorus runoff from incorporated and surface-applied liquid swine manure and phosphorus fertilizer. *Journal of Environmental Quality* 33(4), 1535–1544. doi.org/10.2134/jeq2004.1535.
Di, H. J. and Cameron, K. C. (2002). The use of a nitrification inhibitor, dicyandiamide (DCD), to decrease nitrate leaching and nitrous oxide emissions in a simulated grazed and irrigated grassland. *Soil Use and Management* 18(4), 395–403. doi.org /10.1111/j.1475-2743.2002.tb00258.x.
Dougherty, B., Gray, M., Johnson, M. G. and Kleber, M. (2017). Can biochar covers reduce emissions from manure lagoons while capturing nutrients? *Journal of Environmental Quality* 46(3), 659–666. doi.org/10.2134/jeq2016.12.0478.
Eghball, B. (2000). Nitrogen mineralization from field-applied beef cattle feedlot manure or compost. *Soil Science Society of America Journal* 64(6), 2024–2030. doi.org/10 .2136/sssaj2000.6462024x.
Eghball, B. (2002). Soil properties as influenced by phosphorus-and nitrogen-based manure and compost applications. *Agronomy Journal* 94(1), 128–135. doi.org/10 .2134/agronj2002.1280.
Eghball, B. and Power, J. F. (1999). Phosphorus-and nitrogen-based manure and compost applications corn production and soil phosphorus. *Soil Science Society of America Journal* 63(4), 895-901. doi.org/10.2136/sssaj1999.634895x.
Eghball, B., Power, J. F., Gilley, J. E. and Doran, J. W. (1997). Nutrient, carbon, and mass loss during composting of beef cattle feedlot manure. *Journal of Environmental Quality* 26(1), 189–193. doi.org/10.2134/jeq1997.00472425002600010027x.
Eghball, B., Wienhold, B. J., Gilley, J. E. and Eigenberg, R. A. (2002). Mineralization of manure nutrients. *Journal of Soil and Water Conservation* 57(6), 470–473.

Everett, L. A., Wilson, M. L., Pepin, R. J. and Coulter, J. A. (2019). Winter rye cover crop with liquid manure injection reduces spring soil nitrate but not maize yield. *Agronomy* 9(12), 852. doi.org/10.3390/agronomy9120852.

Federal Drug Administration (n.d.). Produce safety rule (21 CFR 112). *Biological Soil Amendments of Animal Origin. Fact Sheet*. Retrieved 4/24/2023. www.fda.gov/media/114775/download.

Fuchs, A., Dalby, F. R., Liu, D., Kai, P. and Feilberg, A. (2021). Improved effect of manure acidification technology for gas emission mitigation by substituting sulfuric acid with acetic acid. *Cleaner Engineering and Technology* 4, 100263. doi.org/10.1016/j.clet.2021.100263.

Gao, M., Liang, F., Yu, A., Li, B. and Yang, L. (2010). Evaluation of stability and maturity during forced-aeration composting of chicken manure and sawdust at different C/N ratios. *Chemosphere* 78(5), 614–619. doi.org/10.1016/j.chemosphere.2009.10.056.

Gilley, J. E. and Eghball, B. (1998). Runoff and erosion following field application of beef cattle manure and compost. *Transactions of the ASAE* 41(5), 1289–1294. doi.org/10.13031/2013.17301.

Glaze-Corcoran, S., Smychkovich, A. and Hashemi, M. (2022). Use of dual-purpose winter-grain cover crops as emergency forage and for management of high soil phosphorous in manured fields. *Agronomy* 12(10), 2334. doi.org/10.3390/agronomy12102334.

Guo, H., Gu, J., Wang, X., Nasir, M., Yu, J., Lei, L., Wang, J., Zhao, W. and Dai, X. (2020). Beneficial effects of bacterial agent/bentonite on nitrogen transformation and microbial community dynamics during aerobic composting of pig manure. *Bioresource Technologyy* 298, 122384. doi.org/10.1016/j.biortech.2019.122384.

Guštin, S. and Marinšek-Logar, R. (2011). Effect of pH, temperature, and air flow rate on the continuous ammonia stripping of the anaerobic digestion effluent. *Process Safety and Environmental Protection* 89(1), 61–66. doi.org/10.1016/j.psep.2010.11.001.

Hagenkamp-Korth, F., Haeussermann, A., Hartung, E. and Reinhardt-Hanisch, A. (2015). Reduction of ammonia emissions from dairy manure using novel urease inhibitor formulations under laboratory conditions. *Biosystems Engineering* 130, 43–51. doi.org/10.1016/j.biosystemseng.2014.12.002.

Hanrahan, B. R., King, K. W., Duncan, E. W. and Shedekar, V. S. (2021). Cover crops differentially influenced nitrogen and phosphorus loss in tile drainage and surface runoff from agricultural fields in Ohio, USA. *Journal of Environmental Management* 293, 112910. doi.org/10.1016/j.jenvman.2021.112910.

Hao, X. and Chang, C. (2003). Does long-term heavy cattle manure application increase salinity of clay loam soil in semi-arid southern Alberta? *Agriculture, Ecosystems and Environment* 94(1), 89–103. doi.org/10.1016/S0167-8809(02)00008-7.

Hashemi, M., Farsad, A., Sadeghpour, A., Weis, S. A. and Herbert, S. J. (2013). Cover-crop seeding-date influence on fall nitrogen recovery. *Journal of Plant Nutrition and Soil Science* 176(1), 69–75. doi.org/10.1002/jpln.201200062.

Hashemi, M., Kadi, A. and Kraemer, K. (2017). *Manure Composting Bin for Small Livestock Operation*. [Fact Sheet] UMass Extension Crops, Dairy, Livestock, Equine, 17–23. https://ag.umass.edu/sites/ag.umass.edu/files/fact-sheets/pdf/manure_composting_for_small_livestock_operation_0.pdf.

Hons, F. M., Moresco, R. F., Wiedenfeld, R. P. and Cothren, J. T. (1986). Applied nitrogen and phosphorus effects on yield and nutrient uptake by high-energy sorghum produced for grain and biomass 1. *Agronomy Journal* 78(6), 1069–1078. doi.org/10.2134/agronj1986.00021962007800060026x.

Islam, M., Morgan, J., Doyle, M. P., Phatak, S. C., Millner, P. and Jiang, X. (2004). Persistence of Salmonella enterica serovar Typhimurium on lettuce and parsley and in soils on which they were grown in fields treated with contaminated manure composts or irrigation water. *Foodborne Pathogens and Disease* 1(1), 27–35. doi.org/10.1089/153531404772914437.

Jahanzad, E., Saporito, L. S., Karsten, H. D. and Kleinman, P. J. A. (2019). Varying influence of dairy manure injection on phosphorus loss in runoff over four years. *Journal of Environmental Quality* 48(2), 450–458 doi.org/10.2134/jeq2018.05.0206.

Jiang, A., Zhang, T., Zhao, Q. B., Li, X., Chen, S. and Frear, C. S. (2014). Evaluation of an integrated ammonia stripping, recovery, and biogas scrubbing system for use with anaerobically digested dairy manure. *Biosystems Engineering* 119, 117–126. doi.org/10.1016/j.biosystemseng.2013.10.008.

Jiang, T., Schuchardt, F., Li, G., Guo, R. and Zhao, Y. (2011). Effect of C/N ratio, aeration rate, and moisture content on ammonia and greenhouse gas emission during the composting. *Journal of Environmental Sciences* 23(10), 1754–1760. doi.org/10.1016/S1001-0742(10)60591-8.

Kai, P., Pedersen, P., Jensen, J. E., Hansen, M. N. and Sommer, S. G. (2008). A whole-farm assessment of the efficacy of slurry acidification in reducing ammonia emissions. *European Journal of Agronomy* 28(2), 148–154. doi.org/10.1016/j.eja.2007.06.004.

Kaiser, D. E., Mallarino, A. P., Haq, M. U. and Allen, B. L. (2009). Runoff phosphorus loss immediately after poultry manure application is influenced by the application rate and tillage. *Journal of Environmental Quality* 38(1), 299–308. doi.org/10.2134/jeq2007.0628.

Karimi, R., Akinremi, W. and Flaten, D. (2018). Nitrogen-or phosphorus-based pig manure application rates affect soil test phosphorus and phosphorus loss risk. *Nutrient Cycling in Agroecosystems* 111(2–3), 217–230. doi.org/10.1007/s10705-018-9924-8.

Kim, W. K. and Patterson, P. H. (2005). Effects of dietary zinc supplementation on hen performance, ammonia volatilization, and nitrogen retention in manure. *Journal of Environmental Science and Health. Part. B, Pesticides, Food Contaminants, and Agricultural Wastes* 40(4), 675–686. doi.org/10.1081/PFC-200061598.

Kleinman, P. J., Sharpley, A. N., Moyer, B. G. and Elwinger, G. F. (2002). Effect of mineral and manure phosphorus sources on runoff phosphorus. *Journal of Environmental Quality* 31(6), 2026–2033. doi.org/10.2134/jeq2002.2026.

Knowlton, K. F., Wilkerson, V. A., Casper, D. P. and Mertens, D. R. (2010). Manure nutrient excretion by Jersey and Holstein cows. *Journal of Dairy Science* 93(1), 407–412. doi.org/10.3168/jds.2009-2617.

Kudva, I. T., Blanch, K. and Hovde, C. J. (1998). Analysis of Escherichia coli O157: H7 survival in ovine or bovine manure and manure slurry. *Applied and Environmental Microbiology* 64(9), 3166–3174. doi.org/10.1128/AEM.64.9.3166-3174.1998.

Lazicki, P., Geisseler, D. and Lloyd, M. (2020). Nitrogen mineralization from organic amendments is variable but predictable. *Journal of Environmental Quality* 49(2), 483–495. doi.org/10.1002/jeq2.20030.

Lefcourt, A. M. and Meisinger, J. J. (2001). Effect of adding alum or zeolite to dairy slurry on ammonia volatilization and chemical composition. *Journal of Dairy Science* 84(8), 1814–1821. doi.org/10.3168/jds.S0022-0302(01)74620-6.

Li, H., Xin, H., Liang, Y. and Burns, R. T. (2008). Reduction of ammonia emissions from stored laying hen manure through topical application of zeolite, Al+ Clear, Ferix-3, or poultry litter treatment. *Journal of Applied Poultry Research* 17(4), 421–431. doi.org/10.1016/j.wasman.2018.01.015.

Li, R., Wang, J. J., Zhang, Z., Shen, F., Zhang, G., Qin, R., Li, X. and Xiao, R. (2012). Nutrient transformations during composting of pig manure with bentonite. *Bioresource Technology* 121, 362–368. doi.org/10.1016/j.biortech.2012.06.065.

Lin, S. and Hernandez-Ramirez, G. (2020). Nitrous oxide emissions from manured soils as a function of various nitrification inhibitor rates and soil moisture contents. *Science of the Total Environment* 738, 139669. doi.org/10.1016/j.scitotenv.2020.139669.

Liu, C., Yan, J., Huang, Q., Liu, H., Qiao, C., Li, R., Shen, B. and Shen, Q. (2022). The addition of sawdust reduced the emission of nitrous oxide in pig manure composting by altering the bacterial community structure and functions. *Environmental Science and Pollution Research International* 29(3), 3733–3742. doi.org/10.1007/s11356-021-15786-2.

Liu, J., Kleinman, P. J. A., Aronsson, H., Flaten, D., McDowell, R. W., Bechmann, M., Beegle, D. B., Robinson, T. P., Bryant, R. B., Liu, H., Sharpley, A. N. and Veith, T. L. (2018). A review of regulations and guidelines related to winter manure application. *Ambio* 47(6), 657–670. doi.org/10.1007/s13280-018-1012-4.

Liu, J., Li, X., Xu, Y., Wu, Y., Wang, R., Zhang, X., Hou, Y., Qu, H., Wang, L., He, M., Kupczok, A. and He, J. (2023). Highly efficient reduction of ammonia emissions from livestock waste by the synergy of novel manure acidification and inhibition of ureolytic bacteria. *Environment International* 172, 107768. doi.org/10.1016/j.envint.2023.107768.

Mallory, E. B. and Griffin, T. S. (2007). Impacts of soil amendment history on nitrogen availability from manure and fertilizer. *Soil Science Society of America Journal* 71(3), 964–973. doi.org/10.2136/sssaj2006.0244.

Malone, R. W., Kersebaum, K. C., Kaspar, T. C., Ma, L., Jaynes, D. B. and Gillette, K. (2017). Winter rye as a cover crop reduces nitrate loss to subsurface drainage as simulated by HERMES. *Agricultural Water Management* 184, 156–169. doi.org/10.1016/j.agwat.2017.01.016.

Misselbrook, T. H. and Powell, J. M. (2005). Influence of bedding material on ammonia emissions from cattle excreta. *Journal of Dairy Science* 88(12), 4304–4312. doi.org/10.3168/jds.S0022-0302(05)73116-7.

Mohammed-Nour, A., Al-Sewailem, M. and El-Naggar, A. H. (2019). The influence of alkalization and temperature on ammonia recovery from cow manure and the chemical properties of the effluents. *Sustainability* 11(8), 2441. doi.org/10.3390/su11082441.

Morse, D., Head, H. H. and Wilcox, C. J. (1992). Disappearance of phosphorus in phytate from concentrates in vitro and from rations fed to lactating dairy cows. *Journal of Dairy Science* 75(7), 1979–1986. doi.org/10.3168/jds.S0022-0302(92)77957-0.

Moshia, M. E., Khosla, R., Longchamps, L., Reich, R., Davis, J. G. and Westfall, D. G. (2014). Precision manure management across site-specific management zones: grain yield and economic analysis. *Agronomy Journal* 106(6), 2146–2156. doi.org/10.2134/agronj13.0400.

Mroz, Z., Moeser, A. J., Vreman, K., van Diepen, J. T. M., Van Kempen, T., Canh, T. T. and Jongbloed, A. W. (2000). Effects of dietary carbohydrates and buffering capacity on nutrient digestibility and manure characteristics in finishing pigs. *Journal of Animal Science* 78(12), 3096–3106. doi.org/10.2527/2000.78123096x.

Muñoz, G. R., Kelling, K. A., Rylant, K. E. and Zhu, J. (2008). Field evaluation of nitrogen availability from fresh and composted manure. *Journal of Environmental Quality* 37(3), 944–955. doi.org/10.2134/jeq2007.0219.

Nair, D., Baral, K. R., Abalos, D., Strobel, B. W. and Petersen, S. O. (2020). Nitrate leaching and nitrous oxide emissions from maize after grass-clover on a coarse sandy soil: mitigation potentials of 3, 4-dimethylpyrazole phosphate (DMPP). *Journal of Environmental Management* 260, 110165.

Nelson, N. O. and Shober, A. L. (2012). Evaluation of phosphorus indices after twenty years of science and development. *Journal of Environmental Quality* 41(6), 1703–1710. doi.org/10.2134/jeq2012.0342.

Nelson, T. S., Daniels, L. B., Hall, J. R. and Shields, L. G. (1976). Hydrolysis of natural phytate phosphorus in the digestive tract of calves. *Journal of Animal Science* 42(6), 1509–1512. doi.org/10.2527/jas1976.4261509x.

Nennich, T. D., Harrison, J. H., Vanwieringen, L. M., Meyer, D., Heinrichs, A. J., Weiss, W. P., St-Pierre, N. R., Kincaid, R. L., Davidson, D. L. and Block, E. (2005). Prediction of manure and nutrient excretion from dairy cattle. *Journal of Dairy Science* 88(10), 3721–3733. doi.org/10.3168/jds.S0022-0302(05)73058-7.

Niamat, B., Naveed, M., Ahmad, Z., Yaseen, M., Ditta, A., Mustafa, A., Rafique, M., Bibi, R., Sun, N. and Xu, M. (2019). Calcium-enriched animal manure alleviates the adverse effects of salt stress on growth, physiology, and nutrient homeostasis of *Zea mays* L. *Plants* 8(11), 480. doi.org/10.3390/plants8110480.

Nouri, A., Lukas, S., Singh, S., Singh, S. and Machado, S. (2022). When do cover crops reduce nitrate leaching? A global meta-analysis. *Global Change Biology* 28(15), 4736–4749. doi.org/10.1111/gcb.16269.

Oenema, O., Oudendag, D. and Velthof, G. L. (2007). Nutrient losses from manure management in the European Union. *Livestock Science* 112(3), 261–272. doi.org/10.1016/j.livsci.2007.09.007.

Ogunwande, G. A., Osunade, J. A., Adekalu, K. O. and Ogunjimi, L. A. O. (2008). Nitrogen loss in chicken litter compost as affected by carbon to nitrogen ratio and turning frequency. *Bioresource Technology* 99(16), 7495–7503. doi.org/10.1016/j.biortech.2008.02.020.

Owens, L. B., Bonta, J. V., Shipitalo, M. J. and Rogers, S. (2011). Effects of winter manure application in Ohio on the quality of surface runoff. *Journal of Environmental Quality* 40(1), 153–165. doi.org/10.2134/jeq2010.0216.

Pang, X. P. and Letey, J. (2000). Organic farming challenge of timing nitrogen availability to crop nitrogen requirements. *Soil Science Society of America Journal* 64(1), 247–253. doi.org/10.2136/sssaj2000.641247x.

Park, G. W. and Diez-Gonzalez, F. (2003). Utilization of carbonate and ammonia-based treatments to eliminate Escherichia coli O157: H7 and Salmonella Typhimurium DT104 from cattle manure. *Journal of Applied Microbiology* 94(4), 675–685. doi.org/10.1046/j.1365-2672.2003.01899.x.

Parker, D. B., Pandrangi, S., Greene, L. W., Almas, L. K., Cole, N. A., Rhoades, M. B. and Koziel, J. A. (2005). Rate and frequency of urease inhibitor application for minimizing ammonia emissions from beef cattle feed yards. *Transactions of the ASAE* 48(2), 787–793. doi.org/10.13031/2013.18321.

Parkin, T. B., Kaspar, T. C. and Singer, J. W. (2006). Cover crop effects on the fate of N following soil application of swine manure. *Plant and Soil* 289(1–2), 141–152. doi.org/10.1007/s11104-006-9114-3.

Peacock, A. D., Mullen, M. D., Ringelberg, D. B., Tyler, D. D., Hedrick, D. B., Gale, P. M. and White, D. C. (2001). Soil microbial community responses to dairy manure or ammonium nitrate applications. *Soil Biology and Biochemistry* 33(7–8), 1011–1019. doi.org/10.1016/S0038-0717(01)00004-9.

Powell, J. M., Wattiaux, M. A., Broderick, G. A., Moreira, V. R. and Casler, M. D. (2006). Dairy diet impacts on fecal chemical properties and nitrogen cycling in soils. *Soil Science Society of America Journal* 70(3), 786–794. doi.org/10.2136/sssaj2005.0286.

Probert, M. E., Delve, R. J., Kimani, S. K. and Dimes, J. P. (2005). Modeling nitrogen mineralization from manures: representing quality aspects by varying C: N ratio of sub-pools. *Soil Biology and Biochemistry* 37(2), 279–287. doi.org/10.1016/j.soilbio.2004.07.040.

Qian, P. and Schoenau, J. J. (2002). Availability of nitrogen in solid manure amendments with different C:N ratios. *Canadian Journal of Soil Science* 82(2), 219–225. doi.org/10.4141/S01-018.

Raboy, V. (1997). Accumulation and storage of phosphate and minerals. In *Cellular and Molecular Biology of Plant Seed Development*, 441–477. Kluwar Academic Publishers, The Netherlands.

Ramirez, K. S., Lauber, C. L., Knight, R., Bradford, M. A. and Fierer, N. (2010). Consistent effects of nitrogen fertilization on soil bacterial communities in contrasting systems. *Ecology* 91(12), 3463–3470. doi.org/10.1890/10-0426.1.

Ren, X., Wang, Q., Chen, X., He, Y., Li, R., Li, J. and Zhang, Z. (2021). Pathways and mechanisms of nitrogen transformation during co-composting of pig manure and diatomite. *Bioresource Technology* 329, 124914. doi.org/10.1016/j.biortech.2021.124914.

Renterghem, B. V., Huysman, F., Rygole, R. and Verstraete, W. (1991). Detection and prevalence of Listeria monocytogenes in the agricultural ecosystem. *Journal of Applied Bacteriology* 71(3), 211–217.

Ritter, W. F., Scarborough, R. W. and Chirnside, A. E. M. (1998). Winter cover crops as a best management practice for reducing nitrogen leaching. *Journal of Contaminant Hydrology* 34(1–2), 1–15. doi.org/10.1016/S0169-7722(98)00087-4.

Sadeghpour, A., Hashemi, M., Weis, S. A., Spargo, J. T., Mehrvarz, S. and Herbert, S. J. (2015). Assessing tillage systems for reducing ammonia volatilization from spring-applied slurry manure. *Communications in Soil Science and Plant Analysis* 46(6), 724–735. doi.org/10.1080/00103624.2015.1005223.

Sadeghpour, A., Ketterings, Q. M., Vermeylen, F., Godwin, G. S. and Czymmek, K. J. (2016). Soil properties under nitrogen-vs. phosphorus-based manure and compost management of corn. *Soil Science Society of America Journal* 80(5), 1272–1282. doi.org/10.2136/sssaj2016.03.0086.

Sedghi, N. and Weil, R. (2022). Fall cover crop nitrogen uptake drives reductions in winter-spring leaching. *Journal of Environmental Quality* 51(3), 337–351. doi.org/10.1002/jeq2.20342.

Seidel, S. J., Gaiser, T., Kautz, T., Bauke, S. L., Amelung, W., Barfus, K., Ewert, F. and Athmann, M. (2019). Estimation of the impact of precrops and climate variability on soil depth-differentiated spring wheat growth and water, nitrogen and phosphorus uptake. *Soil and Tillage Research* 195, 104427. doi.org/10.1016/j.still.2019.104427.

Shah, G. M., Groot, J. C. J., Oenema, O. and Lantinga, E. A. (2012). Covered storage reduces losses and improves crop utilisation of nitrogen from solid cattle manure. *Nutrient Cycling in Agroecosystems* 94(2–3), 299–312. doi.org/10.1007/s10705-012-9543-8.

Shapiro, C. A., Kranz, W. L. and Wortmann, C. S. (2005). Salt thresholds for liquid manure applied to corn and soybean. *Transactions of the ASAE* 48(3), 1005–1013. doi.org/10.13031/2013.18512.

Sileshi, G. W., Nhamo, N., Mafongoya, P. L. and Tanimu, J. (2017). Stoichiometry of animal manure and implications for nutrient cycling and agriculture in sub-Saharan Africa. *Nutrient Cycling in Agroecosystems* 107(1), 91–105. doi.org/10.1007/s10705-016-9817-7.

Sims, J. T. (1986). Nitrogen transformations in a poultry manure amended soil: temperature and moisture effects. *Journal of Environmental Quality* 15(1), 59–63. doi.org/10.2134/jeq1986.00472425001500010014x.

Singh, A., Casey, K. D., King, W. D., Pescatore, A. J., Gates, R. S. and Ford, M. J. (2009). Efficacy of urease inhibitor to reduce ammonia emission from poultry houses. *Journal of Applied Poultry Research* 18(1), 34–42. doi.org/10.3382/japr.2008-00046.

Sistani, K. R., Adeli, A., McGowen, S. L., Tewolde, H. and Brink, G. E. (2008). Laboratory and field evaluation of broiler litter nitrogen mineralization. *Bioresource Technology* 99(7), 2603–2611. doi.org/10.1016/j.biortech.2007.04.069.

Smith, K. A., Beckwith, C. P., Chalmers, A. G. and Jackson, D. R. (2002). Nitrate leaching following autumn and winter application of animal manures to grassland. *Soil Use and Management* 18(4), 428–434.

Soil fertility and crop nutrient management practice standard 7 CFR §205.203 (2023).

Strock, J. S., Porter, P. M. and Russelle, M. P. (2004). Cover cropping to reduce nitrate loss through subsurface drainage in the northern US Corn Belt. *Journal of Environmental Quality* 33(3), 1010–1016. doi.org/10.2134/jeq2004.1010.

Sutton, A. L., Jones, D. D., Joern, B. C. and Huber, D. M. (1994). *Animal Manure as a Plant Nutrient Resource*. Purdue University Press and Cooperative Extension Service, USA.

Usmani, Z., Kumar, V., Gupta, P., Gupta, G., Rani, R. and Chandra, A. (2019). Enhanced soil fertility, plant growth promotion and microbial enzymatic activities of vermicomposted fly ash. *Scientific Reports* 9(1), 10455. doi.org/10.1038/s41598-019-46821-5.

Vadas, P. A., Good, L. W., Jokela, W. E., Karthikeyan, K. G., Arriaga, F. J. and Stock, M. (2017). Quantifying the impact of seasonal and short-term manure application decisions on phosphorus loss in surface runoff. *Journal of Environmental Quality* 46(6), 1395–1402. doi.org/10.2134/jeq2016.06.0220.

Varel, V. H., Nienaber, J. A. and Freetly, H. C. (1999). Conservation of nitrogen in cattle feedlot waste with urease inhibitors. *Journal of Animal Science* 77(5), 1162–1168. doi.org/10.2527/1999.7751162x.

Vincini, M., Carini, F. and Silva, S. (1994). Use of alkaline fly ash as an amendment for swine manure. *Bioresource Technology* 49(3), 213–222. doi.org/10.1016/0960-8524(94)90042-6.

Wang, D., Guo, L., Zheng, L., Zhang, Y., Yang, R., Li, M., Ma, F., Zhang, X. and Li, Y. (2019). Effects of nitrogen fertilizer and water management practices on nitrogen leaching from a typical open field used for vegetable planting in northern China. *Agricultural Water Management* 213, 913–921. doi.org/10.1016/j.agwat.2018.12.015.

Wang, Q., Li, R., Cai, H., Awasthi, M. K., Zhang, Z., Wang, J. J., Ali, A. and Amanullah, M. (2016). Improving pig manure composting efficiency employing Ca-bentonite. *Ecological Engineering* 87, 157–161. doi.org/10.1016/j.ecoleng.2015.11.032.

Xie, J., Schoenau, J. and Warkentin, T. D. (2017). Yield and uptake of nitrogen and phosphorus in soybean, pea, and lentil and effects on soil nutrient supply and crop yield in the succeeding year in Saskatchewan, Canada. *Canadian Journal of Plant Science* 98(1), 5–16. doi.org/10.1139/cjps-2016-0342.

Yao, Y., Yu, L., Ghogare, R., Dunsmoor, A., Davaritouchaee, M. and Chen, S. (2017). Simultaneous ammonia stripping and anaerobic digestion for efficient thermophilic conversion of dairy manure at high solids concentration. *Energy* 141, 179–188. doi.org/10.1016/j.energy.2017.09.086.

You, Y., Rankin, S. C., Aceto, H. W., Benson, C. E., Toth, J. D. and Dou, Z. (2006). Survival of Salmonella enterica serovar Newport in manure and manure-amended soils. *Applied and Environmental Microbiology* 72(9), 5777–5783. doi.org/10.1128/AEM.00791-06.

Zeng, J., Liu, X., Song, L., Lin, X., Zhang, H., Shen, C. and Chu, H. (2016). Nitrogen fertilization directly affects soil bacterial diversity and indirectly affects bacterial community composition. *Soil Biology and Biochemistry* 92, 41–49. doi.org/10.1016/j.soilbio.2015.09.018.

Zhang, M., Zhang, X., Zhang, L., Zeng, L., Liu, Y., Wang, X., He, P., Li, S., Liang, G., Zhou, W. and Ai, C. (2021). The stronger impact of inorganic nitrogen fertilization on soil bacterial community than organic fertilization in short-term conditions. *Geoderma* 382, 114752. doi.org/10.1016/j.geoderma.2020.114752.

Zhang, W., Yu, C., Wang, X. and Hai, L. (2020). Increased abundance of nitrogen transforming bacteria by higher C/N ratio reduces the total losses of N and C in chicken manure and corn stover mix composting. *Bioresource Technology* 297, 122410. doi.org/10.1016/j.biortech.2019.122410.

Zhao, X., Xu, K., Wang, J., Wang, Z., Pan, R., Wang, Q., Li, S., Kumar, S., Zhang, Z. and Li, R. (2022). The potential of biochar integrated manganese sulfate for promoting pig manure compost humification and its biological mechanism. *Bioresource Technology* 357, 127350. doi.org/10.1016/j.biortech.2022.127350.

Zhao, Y., Luan, H., Yang, B., Li, Z., Song, M., Li, B. and Tang, X. (2023). Adsorption of low-concentration ammonia nitrogen from water on alkali-modified coal fly ash: characterization and mechanism. *Water* 15(5), 956. doi.org/10.3390/w15050956.

Zhu, N. (2007). Effect of low initial C/N ratio on aerobic composting of swine manure with rice straw. *Bioresource Technology* 98(1), 9–13. doi.org/10.1016/j.biortech.2005.12.003.

Chapter 12

Characterizing soil nitrogen availability to improve nitrogen fertilizer recommendations

Alan J. Franzluebbers, USDA – Agricultural Research Service, USA

1 Introduction

Soil management for agricultural production has enormous implications for the water quality of streams, lakes, groundwater and estuaries. Water bodies in agricultural areas are susceptible to nutrient enrichment, primarily from nitrogen (N) leaching and run-off, which is not necessarily due to a lack of desire for environmental stewardship by farmers. Rather, conditions outside of a farmer's control may be highly influential in these nutrient loss processes. Such conditions include extreme weather events (drought, flooding etc.), shifting economic conditions that promote excess fertilizer input in order to strive for high production targets, lack of scientific understanding of specific nutrient dynamics under a diversity of environmental and management conditions and social behaviour that influences management decisions (e.g. tradition, peer pressure, agribusiness marketing and news media).

Although N is one of the most studied elements of modern agricultural production systems in the USA, knowledge of its behaviour in agricultural systems is limited. For example, in the *Agronomy Journal* (https://acsess.onlinelibrary.wiley.com/journal/14350645), the number of publications focused on N as a topic has continued to increase, perhaps due to greater emphasis on environmental considerations rather than production aspects (Fig. 1). Sources of available N to growing crops need to be quantified to assure full accountability, as knowing the quantity and flux rates from these sources enables the more accurate calculation of external N fertilizer input. A key source

http://dx.doi.org/10.19103/AS.2024.0135.16

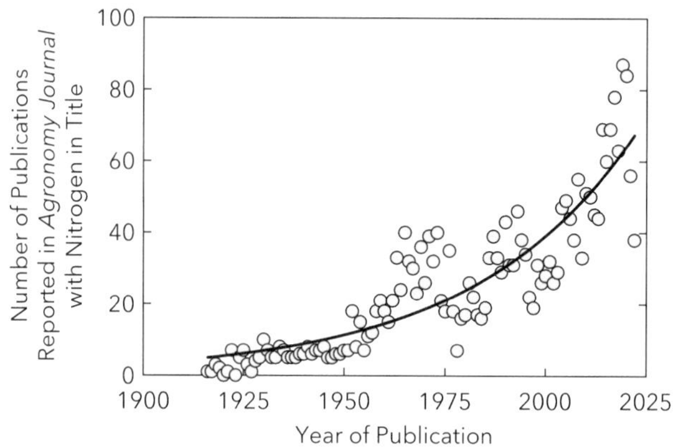

Figure 1 Number of articles appearing in the *Agronomy Journal* during the past century with 'nitrogen' in the title.

of plant-available N that has not been adequately characterized across diverse environmental and management conditions in the USA is mineralizable soil N, also known as biologically active N. Nitrogen mineralization from soil organic matter is essentially a biologically driven process requiring soil microorganisms to consume organic matter and release N as inorganic N. This biological component is essential to estimate the balance of external N required to optimize crop production. Research to improve N fertilizer recommendations to achieve high productivity and preserve environmental quality is vital for agriculture to be competitive. Importantly, contemporary farmers must efficiently manage resources to be effective production specialists, smart business operators and sensitive land stewards of our abundant, yet vulnerable, natural resources.

1.1 Research need

The Chesapeake Bay Watershed is located in the Mid-Atlantic region of the eastern USA and occupies 16.5 Mha of land, has 150 major rivers and streams, and supports 17 million human inhabitants in New York, Pennsylvania, Maryland and Virginia. Contaminant loading into the Chesapeake Bay has been, and continues to be, a major concern (EPA, 2010). Contaminants include nutrients from agricultural run-off and urban settings, chemicals of emerging concern from industrial operations, and suspended sediment from agriculture and development. Other Eastern Seaboard estuaries have similar nutrient loading issues, and therefore, reducing N loading from agricultural run-off to these estuaries is of high priority to stakeholders throughout the eastern USA. The Albermarle-Pamlico Sound Watershed, which is directly south of the Chesapeake Bay Watershed in North Carolina, occupies 8.0 Mha of land area, has 10 major rivers, supports ~3 million people and is also highly impacted by nutrient loading (http://www.fws.gov/nc-es/edout/albewhatwrong.html). Nitrogen and phosphorus loadings are not only through agricultural run-off, as atmospheric deposition accounts for 5–44% of the total N loading to bays and estuaries along the Atlantic Coast (Valigura et al., 1996).

Agriculture in the eastern USA is diverse not only in types of crops and animals but also in how they are managed. Farmland occupies 27–33% of total land area in Pennsylvania, Maryland, Virginia and North Carolina. Corn is one of the most demanding crops for N and therefore often requires a high rate of N fertilizer to achieve high productivity. Corn is grown on 1.3 Mha of land in these four states (USDA Census of Agriculture, 2017). In 2010, N fertilizer was applied at an average of 157 kg N ha^{-1} to corn throughout the USA, with corn in North Carolina receiving an average of 143 kg N ha^{-1}, while corn in Pennsylvania received an average of 97 kg N ha^{-1} (USDA-NASS, 2013). A significant portion of farmland in these states receives animal manure as a nutrient supplement (e.g. 9% of all farmland in these states compared with only 3% of all farmland in the USA as a whole (USDA Census of Agriculture, 2017)) and this complicates fertilizer recommendations due to manure composition, timing and method of application, as well as frequency and rate of application that can control nutrient release and availability to crops (Griffin and Honeycutt, 2000).

The cost of N fertilizer has risen dramatically over the past few years primarily due to the spike in natural gas price, which is necessary for fertilizer manufacturing (Halecki and Bedla, 2022). Further pressures were induced by limitations on the manufacture and distribution of ammonia caused by the Russian invasion of Ukraine, both global leaders in N fertilizer manufacturing. Grain prices have also risen, so the margin of return has allowed expansion of land planted to corn. Keys to success in the future decline of grain prices will involve producers acquiring increased knowledge of (1) how much N fertilizer is needed, (2) when and where to apply it to be most effective for plant uptake and minimize loss to the environment, and (3) how to choose the right source of N to optimize economic and environmental considerations. It would not be safe to assume that the cost of N fertilizer will precipitously decline in the future, given the volatile and likely rising energy costs that control its manufactured value.

1.2 Overall objectives

Research over the past decade has focused on the following goals (NRC, 2010):

- Satisfying human food, feed and fibre needs and contributing to biofuel needs;
- Enhancing environmental quality and the natural resource base;
- Sustaining the economic viability of agriculture; and
- Enhancing the quality of life for farmers, farm workers and society as a whole.

Hypotheses included:

- Supplying just enough N to cereal crops would yield abundant grain (addressing goal 1), would not pollute the environment and would contribute to improved soil quality (addressing goal 2);
- Optimizing N fertilizer recommendations on a site-specific basis would improve economic investment and return for producers (addressing goal 3); and
- Reducing N loading to rivers and estuaries along the Atlantic Coast would lead to improved quality of life for all stakeholders as (1) the economic viability of agricultural operators and consultants would increase, (2) water quality would improve, (3) biodiversity would rebound and (4) the friction between conservative farm organizations and environmental groups would disappear after realizing their goals are compatible and necessary to achieve sustainability (addressing goal 4).

Specifically, the research team at the Soil Ecology and Management Lab in Raleigh, North Carolina, wanted to develop a biologically based tool to improve N management in current agricultural systems of the region. One of the recommendations of the National Research Council Committee on Twenty-First Century Systems Agriculture was to measure progress toward sustainability (NRC, 2010). Reducing N losses to the environment, maintaining high yield and increasing economic return to the producer were targets to achieve via the development of a biologically based soil test to predict N mineralization during the growing season using pre-season soil samples. The research team envisioned this was possible not in isolation but in combination with a multitude of other well-described analytical tools already being deployed by agronomic consultants. A simple, rapid, robust biologically based soil test was considered a missing tool in the path to achieve sustainability of N use, utilization and fate in agricultural systems.

The hypotheses detailed that a biologically based soil N test could:

- Increase N fertilizer recommendations in soils with low N-supplying capacity to improve productivity; and
- Decrease N fertilizer recommendations in soils with high N-supplying capacity to alleviate economic loss and environmental threats.

These ideas have been previously presented by Franzluebbers (2016). The low N-supplying capacity was considered to be due to:

- Excessive tillage that historically depleted soil organic matter;
- Soil physical conditions that could be overcome with nutrient amelioration; and
- Long-term history of exploitive cropping.

The high N-supplying capacity was considered to be due to:

- Long-term deployment of conservation management (e.g. no-till and cover cropping);
- Recent manure N additions;
- N-rich cover crops; and
- Crop rotations that enhanced the biologically active component of soil organic matter.

The high N-supplying capacity of the soil appears to have been the 'norm' in a survey of corn fields analysed using the basal stalk nitrate test (BSNT) in Virginia. Data were collected in 2012 (a very good growing season) from a cooperative project with the Virginia Department of Conservation and Recreation (Fig. 2). Data suggested that 57% of corn fields should have received a lower N fertilizer application rate and 20.5% should have received a higher N fertilizer application rate than the current N recommendation.

On-farm research was undertaken to develop a solid foundational knowledge base in fundamental and applied agricultural science that could help solve the persistent water quality issues experienced throughout the estuary regions of the Eastern Seaboard. The fundamental issue was to understand and describe the process of soil N mineralization and the applied issue was to develop an improved N fertilizer recommendation system that

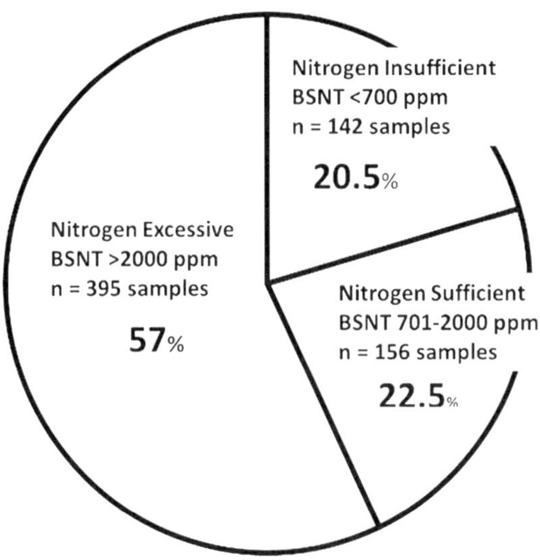

Figure 2 Distribution of corn fields that were below sufficiency, at sufficiency and above sufficiency based on the end-of-season basal stalk nitrate test (BSNT) in Virginia in 2012. Source: data courtesy of Alec Lipscomb at Virginia Department of Conservation and Recreation.

accounted for soil biological activity. Although intended specifically for the regional stakeholders in the Chesapeake Bay and Albermarle-Pamlico Sound Watersheds, the research could have practicality and utilization in other regions of the USA, as well as other regions of the world.

2 Understanding nitrogen availability in soil

Soil organic matter is a large source and sink of nutrients that can significantly influence plant nutrition (Doran and Smith, 1987). Nutrients are made available to plants from soil organic matter at a relatively slow but continual rate primarily through the activity of soil microorganisms. How soil microbial activity affects soil fertility has been investigated for more than a century (Russell, 1905). Organically bound forms of nutrients within soil organic matter generally occur in a ratio of 140 C/10 N/1.3 P/1.3 S (Stevenson, 1986). Marked geographical variation in the quantity of soil organic matter across the USA is largely due to temperature and moisture gradients, which influence plant production and microbial activity (Jenny, 1930). Locally within a particular landscape position, biological activity exerts considerable influence on soil organic matter accumulation. Biological activity is a measure of both plant production of fixed C and uptake of nutrients and subsequent decomposition of these organic materials by soil fauna and microorganisms. Increasing plant production (e.g. increasing plant growth via fertilizer application or adding winter cover cropping) increases the amount of fixed C and the subsequent availability of organic C for decomposition by microorganisms. Relative rates of organic C additions via plant residues and disappearance through decomposition control the steady-state level of soil organic matter.

Several pools of soil organic matter have been proposed, e.g. active, slow and passive (Parton et al., 1987). Paul and Juma (1981) divided soil organic matter into four pools:

- The living microbial biomass;
- The non-biomass active N;
- A residual fraction with a half-life of ~30 years; and
- A very old pool.

The fast turnover rate of the microbial biomass gives importance to this pool in controlling nutrient dynamics. Soil microbial biomass methodology blossomed with the work of Jenkinson and Powlson (1976) following years of investigation of soil microbial activity as an index of soil fertility (Waksman and Starkey, 1924).

Nitrogen transformations in soil are closely coupled with C dynamics (McGill et al., 1975). The active fraction of soil organic N that participates in short-term biological cycling consists of the soil microbial biomass, microbial products and labile soil organic N (Duxbury et al., 1991). The portion of active organic N as microbial biomass was estimated to be 23–77%, depending on the calculation method of microbial biomass N. The portion of total soil N as active organic N in the upper 30 cm of soil was 10–12% (Duxbury et al., 1991) and in the upper 10 cm of soil was 5–6% (Follett and Schimel, 1989).

Stanford and Smith (1972) developed a method to determine potentially mineralizable N through successive leaching of inorganic N over a 30-week period under aerobic conditions. They found that the portion of total soil N as potentially mineralizable N was 5–41%, depending on soil type and management - soils cropped to pasture or with animal manure applied tended to have greater potentially mineralizable N. Bonde et al. (1988) observed that microbial biomass N was 55–89% of the potentially mineralizable N fraction released during a 40-week incubation. During the first 6 weeks of incubation, rapid N mineralization occurred, but a lower and nearly constant rate of mineralization occurred during the final 34 weeks. Microbial biomass C and N declined proportionally to the rate of N mineralization during incubation - being reduced to 7–14% of original levels after 47 weeks of successive leaching and incubation.

Close coupling of C and N occurs in the turnover of soil organic matter (McGill et al., 1975). The possibility of measuring soil CO_2 evolution to estimate potentially mineralizable N was effectively demonstrated using a C:N ratio of 11 (Robertson et al., 1988). In the diverse cropping systems of Texas, strong correlations were observed between mineralizable N and total organic C and N, microbial biomass C and N, and mineralizable C (Franzluebbers et al., 1994a). More detailed seasonal analyses of cropping systems also suggested a strong correlation between mineralizable C and N but only in surface samples (0–5 cm) and when wheat was fertilized but not when unfertilized (Franzluebbers et al., 1994b). Strong, positive relationships were also observed between C and N mineralization rates following the harvesting of sorghum, wheat and soybean crops, reflecting the decomposition of crop residues and the release of nutrients (Franzluebbers et al., 1995).

2.1 Flush of CO_2 following rewetting of dried soil

The flush of CO_2 following rewetting of dried soil reflects the initial non-linear rate of C mineralization during an aerobic incubation. Soils are dried and rewetted in nature, sometimes to a greater extreme in semi-arid climates but also in humid climates between extended periods without rainfall in summer. As early as Birch (1959), associations between flushes of CO_2 following rewetting of dried soil and soil fertility have been investigated. During the growth of wheat, the flush of CO_2 exhibited seasonal variations representative of active organic matter via rhizodeposition (Franzluebbers et al., 1995b). In more intensive cropping systems involving soybean double cropped with wheat, a similar seasonal pulse in active organic matter was detected (Franzluebbers et al., 1996b). This study revealed that the flush of CO_2 was a good predictor of the greater yearly mean N mineralization potential under long-term no-tillage compared with conventional-tillage cropping, but that coupling of C and N was not always in synchrony during monthly evaluations of these processes. During periods of very high C mineralization following crop residue incorporation, N mineralization declined temporarily - followed by increasing N mineralization a few months later. This result suggested that very high C mineralization can cause the temporary immobilization of N.

During experimentation to test if microbial biomass C methodology could be developed for dried rather than field-moist soil, it was noted that the flush of CO_2 following rewetting of dried soil was indeed highly related to soil microbial biomass C and net N mineralization among a diversity of soils (Fig. 3). Drying soil kills part of the soil microbial biomass (Jenkinson, 1966); therefore, the flush of CO_2 during the first day reflects the contribution of the soil microbial biomass and active soil organic matter fractions that are readily mineralizable. The prediction of net N mineralization via CO_2 evolution was suggested for plant residues added to soil (Gilmour et al., 1985), but relationships were

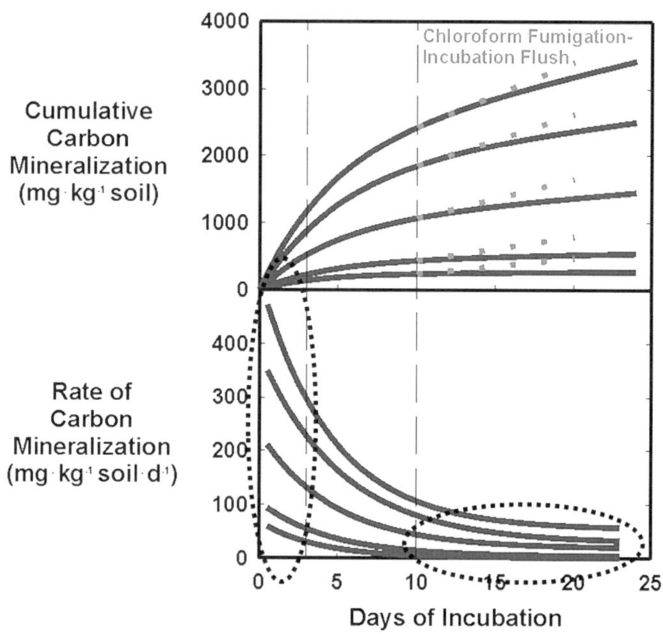

Figure 3 Cumulative (top panel) and daily rate (bottom panel) of C mineralization over time in five soils with different inherent biological capacity. Source: Adapted from: Franzluebbers et al. (1996a).

dependent on the initial C:N ratio of plant residues as well as the C:N ratio of residues remaining after decomposition of the rapidly mineralizable fraction. In soil with a relatively stable C:N ratio of 8–12 (Alexander, 1991), variations in relationships between C and N mineralization among soils would be more of a function of the active fraction of soil organic matter, including the soil microbial biomass, its metabolic by-products and recently introduced organic residues.

By manipulating the microbial biomass of a single soil with a gradient of amino acid additions during a 20-day incubation, Haney and Franzluebbers (2009) demonstrated that the flush of CO_2 following rewetting of dried soil clearly reflected differences in soil microbial biomass C. Since soil microbial biomass is highly related to N mineralization among soils managed in the long term with widely differing soil organic matter contents (Alef et al., 1988; Franzluebbers et al., 1999a), it was reasonable to connect the flush of CO_2 directly with potential N mineralization. In fact, 44 samples from 10 different soil series throughout the southern USA were evaluated for N mineralization indices, and the flush of CO_2 during 3 days of incubation was the best predictor of long-term net N mineralization potential (Schomberg et al., 2009).

Excess C mineralization (i.e. levels approaching 1000 mg kg^{-1} 3 d^{-1}) can cause significant N immobilization and a temporary breakdown of the relationship between the flush of CO_2 and N mineralization (Franzluebbers et al., 1996b; Franzluebbers, 1999b; Franzluebbers and Brock, 2007). Such high levels of mineralizable C can occur at the surface of long-term no-till cropland soils, with high application rates of readily digestible organic matter such as immature compost or during vigorous rooting of grasses – all with high soil moisture to activate soil microbial activity. However, under normal field conditions such high levels of mineralizable C are not typical. The possibility of fine-tuning the flush of CO_2 to predict N mineralization was at the core of research conducted in the early 2010s, with a particular focus on soluble organic C:N (Haney et al., 2012). This led to the development of the Haney Soil Health Tool to assess soil biological activity and potentially predict soil N availability (Haney et al., 2018).

The relationship between the flush of CO_2 and net N mineralized during a growing season relies on a stable relationship between C mineralized in the short term and N mineralized in the long term. Since C is mineralized at ~10 times the rate of N, sensitivity of C to subtle soil changes caused by management is greater than for N. With peak demand for N by corn at 60–90 days from planting, the amount of N mineralized during 24 days of aerobic incubation under ideal temperature and moisture conditions in the laboratory approximates the amount of N mineralized during 60–90 days in the field. For example, the combined temperature × precipitation coefficient scaled to optimum was ~0.3–0.7 during the summer growing season in southcentral Texas, suggesting that N mineralized during 24 days under optimum conditions would require 34–80 days under these field conditions (Franzluebbers et al., 1998).

2.2 Are chemical extractions suitable?

For many years, the sequential leaching and incubation method developed by Stanford and Smith (1972) was the standard of determining potentially mineralizable N (N_0). However, various issues make the method unsuitable for routine soil testing, including the several months necessary to obtain results, depletion of C and N substrates during weekly leaching that does not reflect natural soil systems and analytical complexity in rate calculations. Various soil microbial biomass methods were developed and evaluated for the relationship with potential N mineralization, including chloroform fumigation-incubation (Jenkinson and Powlson, 1976; Voroney and Paul, 1984), chloroform fumigation-extraction (Brookes et al., 1985; Vance et al., 1987), substrate-induced respiration (Anderson and Domsch, 1978; Smith et al., 1985), adenosine triphosphate (Webster et al., 1984), arginine ammonification (Alef et al., 1988) and phospholipid fatty acids (Zelles et al., 1997). Although many of these techniques could be considered useful in varying conditions, none have been robust enough to predict net N mineralization from a variety of soils due to complicating soil physical and chemical conditions. In addition, all require complex laboratory techniques and considerable time for soil preparation and analysis.

Short-term aerobic N incubations have been used to measure the net N mineralization potential, with success in relating microbial activity and microbial biomass across diverse conditions (Franzluebbers, 1999a,b). Shorter term aerobic N incubations correlate well to long-term incubations (Schomberg et al., 2009) but still require one to several weeks of incubation in addition to analysis time for the extraction of inorganic N from samples.

Various chemical extraction techniques have been proposed as rapid, simple approaches to estimate net N mineralization from a diversity of soils (Schomberg et al., 2009), including hot-KCl extraction of inorganic N (Gianello and Bremner, 1986), NaOH distilled N (Stanford, 1978), phosphate-borate distilled N (Gianello and Bremner, 1988), 7-day anaerobic N mineralization (Keeney and Bremner, 1966), calcium hypochlorite oxidation (Picone et al., 2002) and permanganate oxidizable C (Weil et al., 2003). These procedures can generally be accomplished within a day, depending on equipment availability and number of samples to be analysed. All chemical extraction methods (except permanganate oxidizable C, which was not evaluated) were reasonably correlated with potentially mineralizable N (the statistical plateau obtained from a 41-week incubation) (Schomberg et al., 2009). However, the best predictors of potentially mineralizable N were obtained with total organic

C, total soil N, 7-day anaerobic N mineralization and net N mineralized during 24 days of aerobic incubation. In an evaluation of 44 samples from 10 different soil series in southern USA, Schomberg et al. (2009) concluded that combining total soil N and the flush of CO_2 to predict potentially mineralizable N would rely on relatively simple methods, which were a logical combination of indices defining available substrate and microbial biomass. Most evaluations of chemical extractants remain limited to validation against long-term incubations to determine N mineralization potential. Field studies with actively growing plants and changing environmental conditions were generally lacking to verify the robustness of these methods.

Permanganate oxidizable C has been evaluated across a wide diversity of soils under different management conditions in the USA (Culman et al., 2012). Strong correlations were obtained with soil microbial biomass C (r = 0.7), particulate organic C (r = 0.5) and total organic C (r = 0.8). This method is simple and rapid and, therefore, is considered a potential soil testing tool. The relationship with density and size fractions of particulate organic matter suggested that it was related to a greater degree to heavier and smaller particulate organic matter - an indication that it reflects a more processed, stabilized pool of labile soil C (Hurisso et al., 2016).

Culman et al. (2013) evaluated permanganate oxidizable C, 7-day anaerobic N mineralization and the flush of CO_2 following rewetting of dried soil several times during the growing season of corn under three management systems (conventional, integrated and compost) and two rotation sequences (continuous corn and corn-soybean-wheat). All three of these soil tests exhibited significant sensitivity to management system, rotation sequence and time of sampling, although permanganate oxidizable C was more sensitive than 7-day anaerobic N mineralization. Compared with corn grain yield, the flush of CO_2 was the only test with a significant relationship at pre-plant sampling. At the fifth-leaf growth stage of corn, the flush of CO_2 and soil nitrate concentration were significantly related to grain yield. At the tenth-leaf growth stage, grain yield was significantly related to the flush of CO_2, soil nitrate concentration and 7-day anaerobic N mineralization. Grain yield was not significantly related to permanganate oxidizable C or leaf chlorophyll content. Total aboveground corn biomass and total aboveground N uptake in corn, however, were significantly predicted by all measures of analysis. A field study in Texas used the flush of CO_2 as a predictor of N uptake by forage (Haney et al., 2001). Forage N uptake was strongly related (r^2 = 0.93 in 1995 and r^2 = 0.91 in 1996) to the flush of CO_2, averaging 7.1 kg N uptake ha^{-1} per unit of CO_2 flush (mg kg^{-1} day^{-1}).

Although chemical extractants can be rapid (important for soil testing) and simple to use and interpret, the reliability of the tests to make robust N fertilizer recommendations has not been established. At the beginning of a focused research effort in 2013, available evidence suggested that the flush of

CO_2 would be even more robust than proposed chemical extractants. The flush of CO_2 is rapid (1–3 days, with sensitivity increasing over a longer incubation time), simple (container, soil, water, alkali trap and acid dispensed in a burette, and even simpler with a colorimetric approach offered by the Solvita CO_2 Burst Test; https://solvita.com/co2-burst/) and potentially robust (evidence correlated against net N mineralization under controlled laboratory conditions was strong); however, sufficient evidence to predict N availability in the field was lacking.

3 Correlation and calibration of soil-test biological activity with soil nitrogen availability

3.1 Laboratory incubations

A strong association between the flush of CO_2 following rewetting of dried soil and net N mineralization during 24 days of aerobic incubation had been previously observed from individual studies of a single soil or a few soils with a wide diversity of soil organic C and N content induced by management and soil depth differences (Franzluebbers et al., 2000, 2007; Franzluebbers and Brock, 2007; Franzluebbers and Stuedemann, 2008). This focused research effort, initiated in 2013, attempted to clarify some of the discrepancies and discussions that had begun emerging in the wider scientific community about the short-term C mineralization approach. Early research indicated that net N mineralization over 21 days was associated with the flush of CO_2 during the first day following the rewetting of dried soil (Franzluebbers et al., 1996). There was a close correlation between CO_2 released during 1 day and 3 days (Franzluebbers et al., 2000), but the lead investigator chose to focus subsequent studies on the response during 3 days as part of incubations that would extend to 24 days to determine net N mineralization. A co-investigator, Richard Haney, focused his subsequent research programme on what he called '1-day CO_2', which eventually led to collaborative development with Will Brinton of the Solvita CO_2 Burst Test (Haney et al., 2008) and with other colleagues for the Haney Soil Health Tool to assess general indications of life in soil (Haney et al., 2018).

Testing of the Haney Soil Health Tool by other investigators led to questions as to the applicability of this technique beyond the original testing domain. In the western US region, Sullivan and Granatstein (2015) questioned the usefulness of the Haney Soil Health Tool in modifying standard N fertilizer recommendations due to the lack of field validation studies. In Minnesota, 1-day CO_2 was not strongly associated with an economically optimum N rate (Tu, 2016). In Tennessee, 1-day CO_2 and 7-day anaerobic N mineralization were weakly associated (Chu et al., 2019). These authors suggested 1-day CO_2 was not applicable in Tennessee without further field calibration. In North Carolina, the Haney Soil Health Tool was not strongly associated with crop yield, and authors suggested soil health testing does not differentiate among long-term management systems in North Carolina (Roper et al., 2017). However, at 17 sites in 8 midwestern US states, 1-day CO_2 as a specific component of the Haney Soil Health Tool explained 55% of the variation in the economically optimum N rate in corn (Yost et al., 2018).

Soils from 47 corn production fields in North Carolina and Virginia were collected for testing in the Soil Ecology and Management Lab in Raleigh, North Carolina (Franzluebbers et al., 2018b). This study illustrated that soil N mineralization during 24 days of aerobic incubation could be predicted with the flush of CO_2 during 3 days following the rewetting of dried soil (and hence termed 'soil-test biological activity'). Soil texture did not alter the association between soil-test biological activity and aerobic N mineralization. Soils were clay loam (32% of sites), loam (21% of sites), silty clay loam (13% of sites), sandy loam (13% of sites) and a smaller number of sites were comprised of sandy clay loam, clay, silt loam and silty clay. Soil depth had a considerable effect on net N mineralization, with an interquartile range of 54 to 114 mg N kg^{-1} 24 day^{-1} at 0–10 cm depth, 22 to 41 mg N kg^{-1} 24 day^{-1} at 10–20 cm depth and 12 to 22 mg N kg^{-1} 24 day^{-1} at 20–30 cm depth. This study concluded that soil-test biological activity was a rapid and reliable indicator of soil N availability. Across the 541 samples evaluated, there was a strong coefficient of determination (r^2 = 0.77, $P < 0.001$).

In an evaluation of additional medium-term (3–10 years old) and long-term (>10 years old) research sites, as well as various on-farm and research station management systems under cropping and pasture management, climate zone but not land use (i.e. cropland versus pastureland) was found to modify the association between soil-test biological activity and net N mineralization over 24 d (Franzluebbers and Pershing, 2020). More variable associations occurred in western soils (few observations) than in eastern soils (multiple observations). This study concluded that holistic soil fertility analyses should be determined from a combination of chemical (i.e. routine soil chemical analyses) and biological components, as reflected by the strong association between soil-test biological activity and other soil organic C and N fractions.

Finally, a comprehensive evaluation of how soil-test biological activity associated with net N mineralization, basal soil respiration, cumulative C mineralization during 24 days, soil microbial biomass C and total and particulate organic C and N across a diversity of studies illustrated the robust nature of this relatively simple soil assay (Franzluebbers, 2018a). The review demonstrated the strong association of soil-test biological activity with soil microbial biomass C and net N mineralization, but weaker associations with total and particulate organic C and N. The association between soil-test biological activity and basal soil respiration was consistent across selected soils from diverse environmental conditions. Consistency across these recent evaluations was likely due to the use of a standard protocol for both C and N analyses. Other investigators have chosen a variety of different protocols, and this may cause discrepancies, such as using 7-day or 14-day anaerobic incubation to determine soil N availability rather than 24-day aerobic incubation or using 1-day, 2-day or 4-day incubation with variable soil mass, water addition and temperature rather than a standard scoop of soil (59 mL), water to 50% water-filled pore space and 25 $\pm$ 0.5°C incubation for 3 days, as comprehensively described in the methods chapter (Franzluebbers, 2021).

Variations in protocol lead to differences in obtaining short-term C mineralization values and how they might be interpreted. Incubation temperature and method of water delivery were noted as key variables leading to discrepancies between soil-test biological activity and a 4-day incubation following the Cornell Assessment of Soil Health (Franzluebbers and Veum, 2020). Close association between 1-day and 3-day estimates of CO_2 was found along a soil textural gradient, but values at 1 day were more variable and sandy loam soil CO_2 emissions were suppressed when water was supplied to near saturation via capillary action (Franzluebbers and Haney, 2018). Wetting soil with excess water via capillary action has also led to CO_2 suppression in other studies (Wade et al., 2016; Castro Bustamante and Hartz, 2016). Additionally, the extent of sieving or pulverizing soil prior to incubation can lead to suppression or enhancement of C mineralization (Franzluebbers, 1999b; Wade et al., 2018). Variable soil volume or large differences in soil mass may cause bias in quantitative estimates, and using a small mass could lead to larger analytical variations when coarsely sieving soil (Franzluebbers, 2020a).

3.2 Greenhouse growth bioassays

A second level in correlation analysis and a preliminary step in calibration of soil-test biological activity with soil N availability was undertaken in standardized greenhouse growth bioassays. A series of studies grew sorghum-sudangrass for 6 weeks as an unfertilized test crop in greenhouse growth trials to relate plant N uptake and dry matter accumulation to a suite of inherent soil properties. Greenhouse growth conditions were semi-controlled for temperature (24 ± 4°C) and moisture (water supplied via capillary action every few days to keep within a range of 40% to 100% water-holding capacity). Soils from 47 corn production fields in North Carolina and Virginia at three sampling depths (0–10, 10–20 and 20–30 cm) and four field replicates were evaluated (Franzluebbers and Pershing, 2018). This study showed that soil-test biological activity was a key indicator of how much N was taken up by plants under these semi-controlled conditions in the greenhouse. Among all soil properties

Published by Burleigh Dodds Science Publishing Limited, 2024.

measured, soil-test biological activity, residual inorganic N and total soil N were the most important variables explaining plant dry matter production and N uptake. Soils with low biological activity were effectively limited in growth by the supply of mineralized N.

In soils from croplands and pasturelands in the eastern USA and a few samples from Oklahoma and Nebraska, plant N uptake in greenhouse growth trials was strongly associated with a suite of biologically active components of soil organic matter, such as soil-test biological activity, basal soil respiration, soil microbial biomass C, particulate organic C, particulate organic N and cumulative C and net N mineralization during 24 days of aerobic incubation (Franzluebbers and Pershing, 2020). Plant N uptake was explained in a multiple regression analysis by soil-test biological activity (62.0% of variation), extractable Ca (6.5% of variation), plant-available N (9.4% of variation) and soil microbial biomass C (2.4% of variation). Although significant variations in association caused by climatic zone and physiographic region were detected, the relative effects were small (5% of variation) compared with the overall strong relationship between plant N uptake and soil-test biological activity. Some of the variation may have been caused by the extremely shallow sampling depths that contained concentrated plant litter inputs in some of the samples, as shallow-depth sampling addressed previous research objectives and these samples were analysed from archived storage. Excess C availability can cause temporary N immobilization, depending on the total amount of N present in the soil. If the intent were to predict soil N availability from soil-test biological activity, then it would be advisable to avoid extremely shallow surface sampling depths (<5 cm). Difference in land use between annual cropland and perennial pasture was not a significant variable affecting the relationship between plant N uptake and soil-test biological activity. These results indicated that soil-test biological activity reasonably predicted plant N uptake in a greenhouse growth bioassay broadly across land uses and physiographic regions. However, some adjustments were suggested when the climatic conditions varied substantially, although this needs additional testing.

In a variety of soils (0–10 cm depth only) under 57 tall fescue pastures in North Carolina and surrounding states, plant dry matter production and N uptake in two greenhouse growth bioassays were highly associated with soil-test biological activity (72% and 77% of total variation) with minor additional association with total soil N, residual inorganic N, extractable Mn and particulate organic C (Pehim Limbu and Franzluebbers, 2022). Plant N uptake and soil-test biological activity were greater in soils managed under rotational stocking than under hayed management and under medium-textured than under coarse-textured or fine-textured soils.

3.3 Yield response trials in the field

Calibration of soil-test biological activity with the supply of mineralizable soil N during the growing season was investigated in corn and fall-stockpiled tall fescue. Both crops were expected to have large N fertilizer requirements, so a variable supply of mineralizable soil N among fields could be effectively assessed with multiple N fertilizer rates to satisfy plant demand. Current fertilizer N recommendations for corn are independent of management and only slightly adjusted for soil type (i.e. physiographic region) in North Carolina (16.1 to 17.9 kg N Mg^{-1} grain; Osmond et al., 2020) and Virginia (17.9 kg N Mg^{-1} grain; Virginia DCR, 2014). Recently conducted research was undertaken to challenge this standard approach and suggest that management (i.e. site-specific adjustment) could alter the extent of fertilizer N required to optimize production. A conceptual framework is illustrated in Fig. 4.

During 2014 to 2016, 47 corn N trials conducted in North Carolina and Virginia demonstrated that relative corn yield was positively associated with plant available N (i.e. residual inorganic + net N mineralization during 24 days), as well as with soil-test biological activity as a strong surrogate (Franzluebbers,

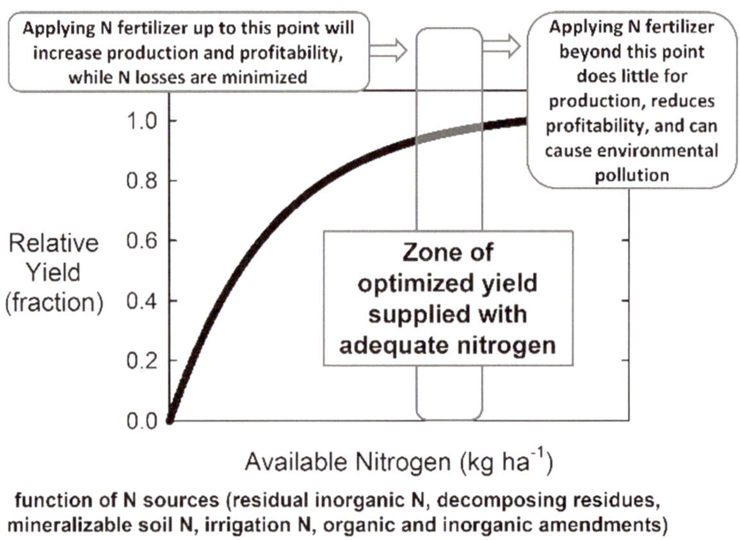

Figure 4 Idealized response of any non-leguminous crop in response to available N. The zone of optimized yield is less than maximum yield because the cost of N fertilizer can be substantial and environmental loss of N needed to achieve maximum yield can also be substantial. The zone of optimized yield may not be so narrow, as the cost-to-value threshold may fluctuate during the growing season, causing significant uncertainty in achieving absolute optimum. The goal of soil testing is to measure the residual inorganic N and estimate the supply of mineralizable soil N so that the rate of inorganic or organic amendments can be subsequently supplied to optimize yield. Source: Adapted from: Franzluebbers et al. (2022).

2018b). Economically optimum N fertilizer requirement (EONR) at sidedress declined with increasing levels of plant available N and soil-test biological activity. The agronomically optimum N rate to maximize yield would typically be achieved at higher N rates than EONR. The cost of N fertilizer is considered in EONR, and therefore, EONR can typically be 30–60 kg N ha^{-1} lower than the N rate to maximize yield, i.e. agronomically optimum N rate. The external N input needed to optimize production (i.e. to achieve EONR) was as high as 20 kg N Mg^{-1} of grain with very low soil biological activity and declined linearly to no external N required when soil-test biological activity was 600 mg CO_2-C kg^{-1} (depth of 0–10 cm). Plant available N determined at 0–20 and 0–30 cm depths did not improve associations with corn yield, suggesting that changes in the surface 10 cm were the most important. Soil-test biological activity determined in spring at, or prior to, planting corn was considered an ideal soil-test indicator due to its simple, rapid and reliable characteristics related to potential soil N mineralization and corn yield responses to applied N fertilizer. Soil-test biological activity can vary seasonally, particularly in response to active crop root growth, although the effect was measured as 53 $\pm$ 25% under tightly constrained greenhouse growth conditions (Franzluebbers and Assmann, 2020). The intense input of roots under these constrained conditions was thought to provide an upper limit of variation expected over a short period of time.

Extending field evaluations to an additional 111 corn N trials conducted throughout the Carolinas and Virginia in 2017 to 2018, plant available N (residual inorganic N + net N mineralization during 24 days) was inversely proportional to EONR scaled to grain production level (r^2 = 0.47, P < 0.001) (Franzluebbers, 2020b). Soil-test biological activity was also predictive of EONR (r^2 = 0.46, P < 0.001) and validated previous trials. Greater soil-test biological activity was obtained from private farms than from research stations (259 versus 172 mg CO_2-C kg^{-1} 3 $days^{-1}$, respectively; P < 0.001). This curious result was probably due to the aggressive soil-health building practices deployed by private farmers in the region, compared with more mild or reserved conservation approaches on research stations. This region has relatively warm (mean annual temperature of 14–18°C) and wet (mean annual precipitation of 1100–1400 mm) conditions, although periodic drought can occur in the summer with consistently high daytime temperatures (28–34°C). At a depth of 0–10 cm, soil managed with minimal compared with severe soil disturbance had greater total organic C (20.9 versus 15.8 g kg^{-1}, P = 0.04), greater net N mineralization (78 versus 41 mg N kg^{-1} soil 24 $days^{-1}$, P < 0.001) and greater soil-test biological activity (254 versus 142 mg CO_2-C kg^{-1} 3 $days^{-1}$, P < 0.001). Overall, these properties were not affected by crop rotation. Sorting trials by recent use of cover cropping led to differences in yield, net N mineralization and soil-test biological activity at 0–10 cm depth, and a change in economic

Sensitivity to cover cropping

Property	No Cover		Single-Species Cover Crop		Multi-Species Cover Crop
Number of observations	39		18		42
Soil-Test Biological Activity (mg CO_2-C kg^{-1} 3 d^{-1})$_{0\text{-}4'' \text{ depth}}$	189		209	<	325
Net N Mineralization (mg N kg^{-1} 24 d^{-1})$_{0\text{-}4'' \text{ depth}}$	62		67	<	85
Corn Grain Yield (bu/acre)	129	<	175		166
Return over STD ($/acre)	13.61*		5.62		27.57***

Franzluebbers (2020) Agronomy Journal, 112, 2188-2204; Franzluebbers & Shoemaker (2020) Agronomy Journal, 112, 2848-2865

Figure 5 Soil, plant and economic responses to on-farm management conditions as sorted by the use of cover cropping in 99 field trials conducted in the Carolinas and Virginia from 2017 to 2018. Soil and plant data are from Franzluebbers (2020b) and economic calculations are from Franzluebbers and Shoemaker (2020). Single-species cover crop was typically cereal rye or wheat, while multi-species cover crops were various combinations of annual legumes (e.g. crimson clover, hairy vetch), grasses (e.g. rye, triticale, ryegrass) and broadleaves (e.g. turnip, rape, radish). Economic return is the difference between a soil-based approach and a standard (STD) N fertilizer recommendation based on 21.4 kg N Mg^{-1} grain. *Indicates value with soil-based N recommendation was greater ($P < 0.05$) than with the standard approach. ***Indicates value with soil-based N recommendation was greater ($P < 0.001$) than that with the standard approach. < between columns indicates significance at $P < 0.05$.

return by adopting a novel adjustment of fertilizer N required based on soil testing (Fig. 5). Fields historically receiving animal manure had greater soil-test biological activity (317 versus 206 mg CO_2-C kg^{-1} 3 $days^{-1}$, $P < 0.001$) and net N mineralization (101 versus 61 mg N kg^{-1} soil 24 $days^{-1}$, $P < 0.001$) at a depth of 0–10 cm. The results imply that some farmers are making choices to improve soil health conditions, which can lead to a lower requirement for N fertilizer input. A shift towards site-specific N management should therefore focus on soil biological activity and its association with N mineralization as an indication of improved soil health to increase profit and reduce environmental impacts.

From 55 tall fescue field trials conducted in North Carolina and surrounding states from 2015 to 2016, forage dry matter production, forage C and N uptake and forage tissue N concentration and moisture were generally positively influenced by fertilizer N application at the end of summer prior to autumn stockpiling of tall fescue stands (Franzluebbers et al., 2018a). To achieve a low cost-to-value threshold of 5 kg dry matter kg^{-1} N, EONR was nil for 75% of the field trials, <45 kg N ha^{-1} for 16% of the field trials and >45 kg N ha^{-1} for only 9% of the field trials. However, standard N fertilizer recommendations for autumn-stockpiled pastures in the region suggest an application of 50–100 kg N ha^{-1}. Plant available N (residual inorganic + mineralizable N) and soil-test biological activity were key indicators that helped determine responsive and nonresponsive trials.

Additional autumn-stockpile field trials ($n = 37$) were conducted in the region in 2018 to target more low-fertility sites (Franzluebbers and Poore, 2020). Soil-test biological activity was validated in these trials as an excellent indicator of soil N availability, as significant quantities of N are recycled in well-managed pastures via the mineralization of organic matter. These additional trials also investigated forage mass response to P fertilizer, but none of the sites responded significantly to P fertilizer application, irrespective of the Mehlich-3 soil-test P level. Autumn-stockpile trials suggested: (1) fertilizer N is needed to optimize economic return only when soil N mineralization is low as indicated by soil-test biological activity, (2) EONR can be expressed as a non-linear function of soil-test biological activity and (3) the cost of fertilizer N and value of forage should be used to adjust N fertilizer recommendations in a logical manner so that livestock producers can balance economic returns on their farm and preserve water, air and soil resources for themselves and surrounding communities.

The preponderance of evidence over this past decade of research correlating and calibrating soil-test biological activity with soil N availability is robust and consistent, suggesting that fertilizer N recommendations should be adjusted on a site-specific basis. Soil health is crucial in the supply of mineralizable soil N. Conservation management approaches that foster surface residue cover, promote active rooting throughout the year and limit soil disturbance are essential to improve soil health conditions in many soils in the eastern USA, and most probably around the world. There may be some exceptions to this general description, but those should be identified and fully characterized with additional research. A recent summary of this approach illustrates that N fertilizer recommendations of crops tested so far should be adjusted based on (1) soil-test biological activity and (2) cost-to-value threshold (Fig. 6).

4 Conclusion

Soil health can be modified by conservation agricultural approaches, especially when implemented over more than a few years. Available evidence suggests that soil-test biological activity can reasonably predict site-specific mineralizable soil N, which has largely been ignored in standard N fertilizer recommendation systems. Improved economic and ecological outcomes could be achieved using optimized N fertilizer rates based on realistic yield goals and knowledge of soil-test biological activity as a proxy for mineralizable soil N. The cost of N fertilizer and agricultural product value (i.e. cost-to-value threshold) is another important attribute that has not received enough attention in modifying N fertilizer recommendations. Both fundamental and applied investigations

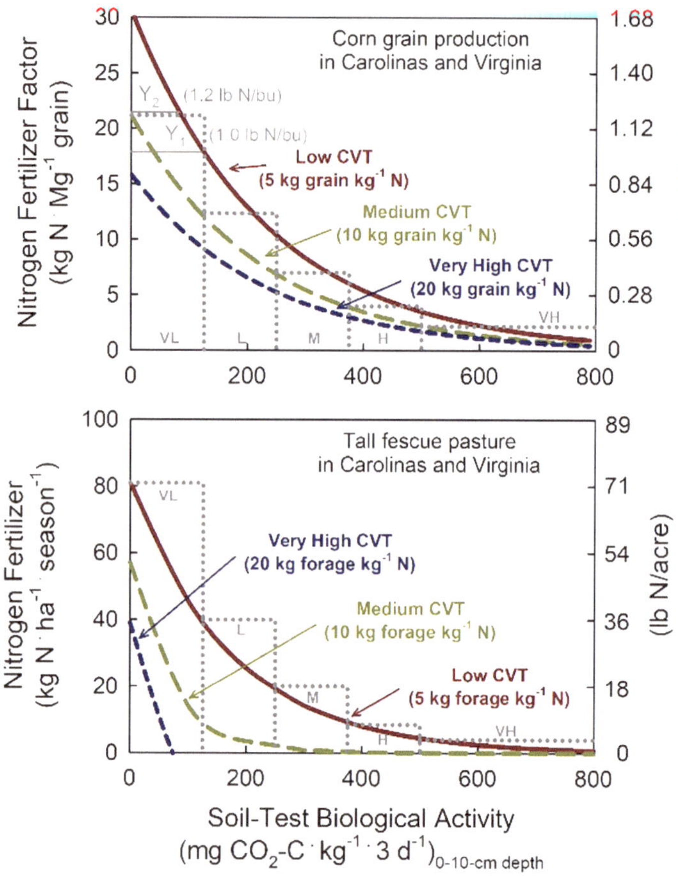

Figure 6 Recommended N fertilizer factor to optimize yield of corn grain (top panel) and tall fescue pasture (bottom panel) as a function of cost-to-value threshold (CVT) and soil-test biological activity as a proxy for mineralizable soil N. Lines associated with Y1 and Y2 indicate static N fertilizer factors of 1.0 and 1.2 lb N/bu, which are reasonable only for soils with very low soil-test biological activity and are insensitive to CVT and soil-test biological activity levels. The N fertilizer factor in the upper panel is the recommended N fertilizer input needed to optimize corn grain on a per unit basis based on regression analysis from 158 field trials reported in Franzluebbers (2018, 2020b). The recommended N fertilizer application rate to optimize the yield of tall fescue pastures was based on regression analysis of 92 field trials reported in Franzluebbers and Poore (2020). Boxes outlined in grey dashed lines indicate very low (VL), low (L), medium (M), high (H) and very high (VH) categories of soil-test biological activity. English units on the right side of the vertical axes are a direct translation from metric units on the left side of the upper and lower panels. Source: Adapted from: Franzluebbers et al. (2022).

conducted over numerous years and trials provide a sound basis for this shift in how N fertilizer rates should be recommended. A final effort on full-scale implementation and assessment of the impacts of such an implementation on farm economics, change in ecological condition of receiving water bodies and other environmental quality responses will be needed to complete this new approach to N fertilizer recommendations.

5 References

Alef, K., Beck, T., Zelles, L. and Kleiner, D. (1988). A comparison of methods to estimate microbial biomass and N-mineralization in agricultural and grassland soils. *Soil Biology and Biochemistry* 20(4), 561–565.

Alexander, M. (1991). *Introduction to Soil Microbiology* (2nd edn.), Krieger Publishing Co., Malabar, FL.

Anderson, J. P. E. and Domsch, K. H. (1978). A physiological method for the quantitative measurement of microbial biomass in soils. *Soil Biology and Biochemistry* 10(3), 215–221.

Birch, H. F. (1959). Further observations on humus decomposition and nitrification. *Plant and Soil* 11(3), 262–286.

Bonde, T. A., Schnuerer, J. and Rosswall, T. (1988). Microbial biomass as a fraction of potentially mineralizable nitrogen in soils from long-term field experiments. *Soil Biology and Biochemistry* 20(4), 447–452.

Brookes, P. C., Landman, A., Pruden, G. and Jenkinson, D. S. (1985). Chloroform fumigation and the release of soil nitrogen: A rapid direct extraction method to measure microbial biomass nitrogen in soil. *Soil Biology and Biochemistry* 17(6), 837–842.

Castro Bustamante, S. J. and Hartz, T. K. (2016). Carbon mineralization and water extractable organic carbon and nitrogen as predictors of soil health and nitrogen mineralization potential. *Communications in Soil Science and Plant Analysis* 47(sup1), 46–53.

Chu, M., Singh, S., Walker, F. R., Eash, N. S., Buschermohle, M. J., Duncan, L. A. and Jagadamma, S. (2019). Soil health and soil fertility assessment by the Haney soil health test in an agricultural soil in West Tennessee. *Communications in Soil Science and Plant Analysis* 50(9), 1123–1131.

Culman, S. W., Snapp, S. S., Freeman, M. A., Schipanski, M. E., Beniston, J., Lal, R., Drinkwater, L. E., Franzluebbers, A. J., Glover, J. D., Grandy, A. S., Lee, J., Six, J., Maul, J. E., Mirksy, S. B., Spargo, J. T. and Wander, M. M. (2012). Permanganate oxidizable carbon reflects a processed soil fraction that is sensitive to management. *Soil Science Society of America Journal* 76(2), 494–504.

Culman, S. W., Snapp, S. S., Green, J. M. and Gentry, L. E. (2013). Short- and long-term labile soil carbon and nitrogen dynamics reflect management and predict corn agronomic performance. *Agronomy Journal* 105(2), 493–502.

Doran, J. W. and Smith, M. S. (1987). Organic matter management and utilization of soil and fertilizer nutrients. In: Follett, R. F., Stewart, J. W. B. and Cole, C. V. (Eds), *Soil Fertility and Organic Matter as Critical Components of Production Systems*. Soil Science Society of America Spec. Publ. No. 19, Madison, WI, pp. 53–72.

Duxbury, J. M., Lauren, J. G. and Fruci, J. R. (1991). Measurement of the biologically active soil nitrogen fraction by a 15N technique. *Agriculture, Ecosystems and Environment* 34(1–4), 121–129.

EPA (Environmental Protection Agency). (2010). *Chesapeake Bay Total Maximum Daily Load for Nitrogen, Phosphorus and Sediment*. Washington, DC. Available at: http://

www.epa.gov/reg3wapd/pdf/pdf_chesbay/FinalBayTMDL/CBayFinalTMDLExecSumSection1through3_final.pdf.

Follett, R. F. and Schimel, D. S. (1989). Effect of tillage practices on microbial biomass dynamics. *Soil Science Society of America Journal* 53(4), 1091–1096.

Franzluebbers, A. J. (1999a). Microbial activity in response to water-filled pore space of variably eroded southern Piedmont soils. *Applied Soil Ecology* 11(1), 91–101.

Franzluebbers, A. J. (1999b). Potential C and N mineralization and microbial biomass from intact and increasingly disturbed soils of varying texture. *Soil Biology and Biochemistry* 31(8), 1083–1090.

Franzluebbers, A. J. (2016). Should soil testing services measure soil biological activity? *Agricultural and Environmental Letters* 1(1), 150009.

Franzluebbers, A. J. (2018a). Short-term C mineralization (aka the flush of CO_2) as an indicator of soil biological health. *CAB Reviews* 13, No. 017.

Franzluebbers, A. J. (2018b). Soil-test biological activity with the flush of CO_2: III. Corn yield responses to applied nitrogen. *Soil Science Society of America Journal* 82(3), 708–721.

Franzluebbers, A. J. (2020a). Soil mass and volume affect soil-test biological activity estimates. *Soil Science Society of America Journal* 84(2), 502–511.

Franzluebbers, A. J. (2020b). Soil-test biological activity with the flush of CO_2: V. Validation of predicted nitrogen for corn production. *Agronomy Journal* 112(3), 2188–2204.

Franzluebbers, A. J. (2021). Assessment and interpretation of soil-test biological activity. In: Karlen, D. R., Stott, D. E. and Mikha, M. M. (Eds), *Soil Health Series: Volume 2 Laboratory Methods for Soil Health Analysis*, Soil Science Society of America, John Wiley & Sons, New Jersey.

Franzluebbers, A. J. and Assmann, T. S. (2020). Soil-test biological activity with short- and long-term carbon contributions. *Agricultural and Environmental Letters* 5, e20035.

Franzluebbers, A. J. and Brock, B. G. (2007). Surface soil responses to silage cropping intensity on a Typic Kanhapludult in the piedmont of North Carolina. *Soil and Tillage Research* 93(1), 126–137.

Franzluebbers, A. J. and Haney, R. L. (2018). Evaluation of soil processing conditions on mineralizable C and N across a textural gradient. *Soil Science Society of America Journal* 82(2), 354–361.

Franzluebbers, A. J., Haney, R. L., Honeycutt, C. W., Schomberg, H. H. and Hons, F. M. (2000). Flush of carbon dioxide following rewetting of dried soil relates to active organic pools. *Soil Science Society of America Journal* 64(2), 613–623.

Franzluebbers, A. J., Haney, R. L., Hons, F. M. and Zuberer, D. A. (1996a). Determination of microbial biomass and nitrogen mineralization following rewetting of dried soil. *Soil Science Society of America Journal* 60(4), 1133–1139.

Franzluebbers, A. J., Hons, F. M. and Zuberer, D. A. (1994a). Long-term changes in soil carbon and nitrogen pools in wheat management systems. *Soil Science Society of America Journal* 58(6), 1639–1645.

Franzluebbers, A. J., Hons, F. M. and Zuberer, D. A. (1994b). Seasonal changes in soil microbial biomass and mineralizable C and N in wheat management systems. *Soil Biology and Biochemistry* 26(11), 1469–1475.

Franzluebbers, A. J., Hons, F. M. and Zuberer, D. A. (1995b). Soil organic carbon, microbial biomass, and mineralizable carbon and nitrogen in sorghum. *Soil Science Society of America Journal* 59(2), 460–466.

Franzluebbers, A. J., Hons, F. M. and Zuberer, D. A. (1996b). Seasonal dynamics of active soil carbon and nitrogen pools under intensive cropping in conventional and no tillage. *Zeitschrift fuer Pflanzernnaehrung und Bodenkunde* 159(4), 343–349.

Franzluebbers, A. J., Hons, F. M. and Zuberer, D. A. (1998). In situ and potential CO_2 evolution from a Fluventic Ustochrept in southcentral Texas as affected by tillage and cropping intensity. *Soil and Tillage Research* 47(3–4), 303–308.

Franzluebbers, A. J., Pehim-Limbu, S. and Poore, M. H. (2018a). Soil-test biological activity with the flush of CO_2: IV. Fall-stockpiled tall fescue yield response to applied nitrogen. *Agronomy Journal* 110(5), 2033–2049.

Franzluebbers, A. J. and Pershing, M. R. (2018). Soil-test biological activity with the flush of CO_2: II. Greenhouse growth bioassay from soils in corn production. *Soil Science Society of America Journal* 82(3), 696–707.

Franzluebbers, A. J. and Pershing, M. R. (2020). Soil-test biological activity with the flush of CO_2: VIII. Soil type and management diversity. *Soil Science Society of America Journal* 84(5), 1658–1674.

Franzluebbers, A. J., Pershing, M. R., Crozier, C., Osmond, D. and Schroeder-Moreno, M. (2018b). Soil-test biological activity with the flush of CO_2: I. C and N characteristics of soils in corn production. *Soil Science Society of America Journal* 82(3), 685–695.

Franzluebbers, A. J. and Poore, M. H. (2020). Soil-test biological activity with the flush of CO_2: VII. Validation of nitrogen prediction for fall-stockpiled tall fescue. *Agronomy Journal* 112(3), 2240–2255.

Franzluebbers, A. J., Schomberg, H. H. and Endale, D. M. (2007). Surface-Soil responses to paraplowing of long-term no-tillage cropland in the Southern Piedmont USA. *Soil and Tillage Research* 96(1–2), 303–315.

Franzluebbers, A. J. and Shoemaker, R. (2020). Soil-test biological activity with the flush of CO_2: VI. Economics of optimized N inputs for corn. *Agronomy Journal* 112(4), 2848–2865.

Franzluebbers, A. J., Shoemaker, R., Cline, J., Lipscomb, B., Stafford, C., Farmaha, B. S., Waring, R., Lowder, N., Thomason, W. E. and Poore, M. H. (2022). Adjusting the N fertilizer factor based on soil health as indicated by soil-test biological activity. *Agricultural and Environmental Letters* 7(2), e20091.

Franzluebbers, A. J. and Stuedemann, J. A. (2008). Early response of soil organic fractions to tillage and integrated crop-livestock production. *Soil Science Society of America Journal* 72(3), 613–625.

Franzluebbers, A. J. and Veum, K. S. (2020). Comparison of two alkali trap methods for measuring the flush of CO_2. *Agronomy Journal* 112(2), 1279–1286.

Franzluebbers, A. J., Zuberer, D. A. and Hons, F. M. (1995a). Comparison of microbiological methods for evaluating quality and fertility of soil. *Biology and Fertility of Soils* 19(2–3), 135–140.

Gianello, C. and Bremner, J. M. (1986). Comparison of chemical methods of assessing potentially available organic nitrogen in soil. *Communications in Soil Science and Plant Analysis* 17(2), 215–236.

Gianello, C. and Bremner, J. M. (1988). A rapid steam distillation method of assessing potentially available organic nitrogen in soil. *Communications in Soil Science and Plant Analysis* 19(14), 1551–1568.

Gilmour, J. T., Clark, M. D. and Sigua, G. C. (1985). Estimating net nitrogen mineralization from carbon dioxide evolution. *Soil Science Society of America Journal* 49(6), 1398–1402.

Griffin, T. S. and Honeycutt, C. W. (2000). Using growing degree days to predict nitrogen availability from livestock manures. *Soil Science Society of America Journal* 64(5), 1876–1882.

Halecki, W. and Bedla, D. (2022). Global wheat production and threats to supply chains in a volatile climate change and energy crisis. *Resources* 11(12), 118.

Haney, R. L. and Franzluebbers, A. J. (2009). Soil CO_2 evolution: response from arginine additions. *Applied Soil Ecology* 42(3), 324–327.

Haney, R. L., Franzluebbers, A. J., Jin, V. L., Johnson, M., Haney, E. B., White, M. J. and Harmel, R. D. (2012). Soil organic C:N vs. water-extractable organic C:N. *Open Journal of Soil Science* 2(3), 269–274.

Haney, R. L., Haney, E. B., Smith, D. R., Harmel, R. D. and White, M. J. (2018). The soil health tool: theory and initial broad-scale application. *Applied Soil Ecology* 125, 162–168.

Haney, R. L., Hons, F. M., Sanderson, M. A. and Franzluebbers, A. J. (2001). A rapid procedure for estimating nitrogen mineralization in manured soil. *Biology and Fertility of Soils* 33(2), 100–104.

Hurisso, T. T., Culman, S. W., Horwath, W. R., Wade, J., Cass, D., Beniston, J. W., Bowles, T. M., Grandy, A. S., Franzluebbers, A. J., Schipanski, M. E., Lucas, S. T. and Ugarte, C. M. (2016). Comparison of permanganate-oxidizable carbon and mineralizable carbon for assessment of organic matter stabilization and mineralization. *Soil Science Society of America Journal* 80(5), 1352–1364.

Jenkinson, D. S. (1966). Studies on the decomposition of plant material in soil. II. Partial sterilization of soil and the soil biomass. *Journal of Soil Science* 17(2), 280–302.

Jenkinson, D. S. and Powlson, D. S. (1976). The effects of biocidal treatments on metabolism in soil. V. A method for measuring soil biomass. *Soil Biology and Biochemistry* 8(3), 209–213.

Jenny, H. (1930). A study on the influence of climate upon the nitrogen and organic matter content of the soil. *Missouri Agricultural Experiment Station Bulletin* 152. 66 pp.

Keeney, D. R. and Bremner, J. M. (1966). Comparison and evaluation of laboratory methods of obtaining an index of soil nitrogen availability. *Agronomy Journal* 58(5), 498–503.

McGill, W. B., Shields, J. A. and Paul, E. A. (1975). Relation between carbon and nitrogen turnover in soil organic fractions of microbial origin. *Soil Biology and Biochemistry* 7(1), 57–63.

NRC (National Research Council). (2010). *Toward Sustainable Agricultural Systems in the 21st Century*. Committee on Twenty-First Century Systems Agriculture, National Academies Press, Washington, DC.

Osmond, D., Gatiboni, L., Kulesza, S., Austin, R. and Crouse, D. (2020). North Carolina realistic yield expectations and nitrogen fertilizer decision making. NC extension, AG-878. Available at: https://content.ces.ncsu.edu/north-carolina-realistic-yield-expectations-and-nitrogen-fertilizer-decision-making.

Parton, W. J., Schimel, D. S., Cole, C. V. and Ojima, D. S. (1987). Analysis of factors controlling soil organic matter levels in Great Plains grasslands. *Soil Science Society of America Journal* 51(5), 1173–1179.

Paul, E. A. and Juma, N. G. (1981). Mineralization and immobilization of nitrogen by microorganisms. In Clark, F. E. and Rosswall, T. (Eds), *Terrestrial Nitrogen Cycles. Ecological Bulletins (Stockholm)* 33, 179–195.

Pehim Limbu, S. and Franzluebbers, A. J. (2022). Greenhouse growth bioassay confirms soil nitrogen availability indicated by the flush of CO_2. *Biology and Fertility of Soils* 58(1), 91–103.

Picone, L. I., Cabrera, M. L. and Franzluebbers, A. J. (2002). A rapid method to estimate potentially mineralizable nitrogen in soil. *Soil Science Society of America Journal* 66(6), 1843–1847.

Robertson, K., Schnuerer, J., Clarholm, M., Bonde, T. A. and Rosswall, T. (1988). Microbial biomass in relation to C and N mineralization during laboratory incubations. *Soil Biology and Biochemistry* 20(3), 281–286.

Roper, W. R., Osmond, D. L., Heitman, J. L., Wagger, M. G. and Reberg-Horton, S. C. (2017). Soil health indicators do not differentiate among agronomic management systems in North Carolina soils. *Soil Science Society of America Journal* 81(4), 828–843.

Russell, E. J. (1905). Oxidation of soils and its connection with fertility. *Journal of Agricultural Science* 1(3), 261–279.

Schomberg, H. H., Wietholter, S., Griffin, T. S., Reeves, D. W., Cabrera, M. L., Fisher, D. S., Endale, D. M., Novak, J. M., Balkcom, K. S., Raper, R. L., Kitchen, N. R., Locke, M. A., Potter, K. N., Schwartz, R. C., Truman, C. C. and Tyler, D. D. (2009). Assessing indices for predicting potential nitrogen mineralization in soils under different management systems. *Soil Science Society of America Journal* 73(5), 1575–1586.

Smith, J. L., McNeal, B. L. and Cheng, H. H. (1985). Estimation of soil microbial biomass: an analysis of the respiratory response of soils. *Soil Biology and Biochemistry* 17(1), 11–16.

Stanford, G. (1978). Evaluation of ammonium release by alkaline permanganate extraction as an index of soil nitrogen availability. *Soil Science* 126(4), 244–253.

Stanford, G. and Smith, S. J. (1972). Nitrogen mineralization potentials of soils. *Soil Science Society of America Proceedings* 36(3), 465–472.

Stevenson, F. J. (1986). *Cycles of Soil: Carbon, Nitrogen, Phosphorus, Sulfur, Micronutrients*, John Wiley and Sons, Inc., New York.

Sullivan, D. M. and Granatstein, D. (2015, Summer). Are "Haney tests" meaningful indicators of soil health and estimators of nitrogen fertilizer credits? *Nutrient Digest* 7(2), 1-2.

Tu, X. (2016). Solvita and other measures as indicators of soil N availability in Minnesota soils. Master's thesis, University of Connecticut, 1175. Available at: http://opencommons.uconn.edu/gs_theses/1175.

USDA Census of Agriculture. (2007). *2007 Census*, Volume 1, Chapter 2: State Level Data. Available at: http://www.agcensus.usda.gov/Publications/2007/Full_Report/Volume_1,_Chapter_2_US_State_Level/.

USDA-NASS (National Agricultural Statistics Service). (2013). *Agricultural Chemical Use Program*. Available at: http://www.nass.usda.gov/Surveys/Guide_to_NASS_Surveys/Chemical_Use/index.asp.

Valigura, R., Luke, W., Artz, R. and Hicks, B. (1996). *Atmospheric Nutrient Inputs to Coastal Areas: Reducing the Uncertainties*. NOAA Coastal Ocean Program Decision Analysis Series No. 9, National Oceanic and Atmospheric Administration Coastal Ocean Office, Silver Spring, MD.

Vance, E. D., Brookes, P. C. and Jenkinson, D. S. (1987). An extraction method for measuring soil microbial biomass C. *Soil Biology and Biochemistry* 19(6), 703–707.

Virginia DCR (Department of Conservation and Recreation). (2014). Virginia nutrient management standards and criteria. Available at: https://www.dcr.virginia.gov/document/standardsandcriteria.pdf.

Voroney, R. P. and Paul, E. A. (1984). Determination of k_C and k_N in situ for calibration of the chloroform fumigation-incubation method. *Soil Biology and Biochemistry* 16(1), 9–14.

Wade, J., Culman, S. W., Hurisso, T. T., Miller, R. O., Baker, L. and Horwath, W. R. (2018). Sources of variability that compromise mineralizable carbon as a soil health indicator. *Soil Science Society of America Journal* 82(1), 243–252.

Wade, J., Horwath, W. R. and Burger, M. B. (2016). Integrating soil biological and chemical indices to predict net nitrogen mineralization across California agricultural systems. *Soil Science Society of America Journal* 80(6), 1675–1687.

Waksman, S. A. and Starkey, R. L. (1924). Microbiological analysis of soil as an index of soil fertility. VII: Carbon dioxide evolution. *Soil Science* 17(2), 141–162.

Webster, J. J., Hampton, G. J. and Leach, F. R. (1984). ATP in soil: A new extractant and extraction procedure. *Soil Biology and Biochemistry* 16(4), 335–342.

Weil, R. R., Islam, K. R., Stine, M. A., Gruver, J. B. and Samson-Liebig, S. E. (2003). Estimating active carbon for soil quality assessment: A simplified method for laboratory and field use. *American Journal of Alternative Agriculture* 18(1), 3–17.

Yost, M. A., Veum, K. S., Kitchen, N. R., Sawyer, J. E., Camberato, J. J., Carter, P. R., Ferguson, R. B., Fernández, F. G., Franzen, D. W., Laboski, C. A. and Nafziger, E. D. (2018). Evaluation of the Haney soil health tool for corn nitrogen recommendations across eight Midwest states. *Journal of Soil and Water Conservation* 73(5), 587–592.

Zelles, L., Palojaervi, A., Kandeler, E., von Luetzo, M., Winter, K. and Bai, Q. Y. (1997). Changes in soil microbial properties and phospholipid fatty acid fractions after chloroform fumigation. *Soil Biology and Biochemistry* 29(9–10), 1325–1336.

Chapter 13

Service crops as a source of nitrogen in temperate Europe

Iris Vogeler, Aarhus University, Denmark and Christian-Albrechts University, Germany; Peter Sørensen and Ingrid K. Thomsen, Aarhus University, Denmark; and Friedhelm Taube, Christian-Albrechts University, Germany

1 Introduction

The use of service crops (SCs), defined as crops not used for harvest of food, feed, or fibre but for protecting and improving the soil and impacts on the environment, is gaining attention worldwide. One approach is an increased use of cover crops (or catch crops) grown as SCs between times of regular annual main crop production. This has been promoted through various governmental legislations and programs that provide financial assistance for growing SCs (Wallander et al., 2021). Another approach is to replace inorganic nitrogen (N) fertiliser use by targeted cultivation of leguminous SCs in cereal cropping systems.

The use of SCs to increase soil fertility is an ancient practice and dating back in China as far as 3000 years, when grasses and weeds were used for green manuring (Pieters, 1927). The value of SCs for providing N has been documented as early as 1896, when a legume winter SC was used as an N source for cotton (Mitchell et al., 2008). Studies on the role of SCs for reducing N leaching started in the 1930s (Meisinger et al., 1991), and considerable work has been done to determine the role of SCs for improving ecosystem

http://dx.doi.org/10.19103/AS.2024.0135.17

services beyond crop production, such as the reduction of runoff, soil erosion and N leaching, soil improvement and soil carbon (C) sequestration, and weed suppression (Gardarin et al., 2022; Schipanski et al., 2014). Until the 1970s, SCs were grown in mixed farming systems in Europe both for delivering primary production services to meet the feed requirements of ruminants and for ecosystem services. With the shift to specialised arable systems within the last 40 years, SCs for green manuring are now mainly grown in less intensive cropping systems as in organic farming and for providing regulating services in general.

Legal obligations in Europe from the Common Agricultural Policy (CAP) and the Nitrates Directive, as well as the 'greening direct payment strategy', has incentivised the use of SCs in the form of cover crops (Fendrich et al., 2023). For instance, in Denmark and France, the use of cover crops in the autumn/winter period is mandatory for obtaining payment, and in many European countries, the use of cover crops is directly subsidised (e.g. Norway, Sweden, Finland, France, and the Netherlands). These obligations and incentives have resulted in an increase in the use of SCs from 6.5% to 8.9% of all arable lands from 2010 to 2016 (Borrelli and Panagos, 2020). Fendrich et al. (2023) estimated, based on satellite data and statistical surveys, that the extent of SCs in Europe is generally below 5% of the land under arable cultivation. Higher proportions, with up to 20–50% in some regions have been identified for Northern France, Germany, and Denmark.

Another option which has been promoted as an option towards 'sustainable intensification' is intercropping, or bi-cropping, where two or more crops are grown simultaneously over a certain period of time. Intercropping cereals with annual legumes such as pea (*Pisum sativum* L.) or common vetch (*Vicia sativa* L.) is one of the oldest agricultural practices for both forage and grain production (Mihailović et al., 2004). Traditionally intercropping has been practiced in low-input, labour-intensive, and low-yield systems, often on small parcels of land. While it is still very common in many parts of the world, including Africa, Latin America, and China, this practice has been lost in mainstream systems in Europe. This is partly due to practical management problems, including sowing, application of disease controls and fertilisers to physiological differently developed crops, and harvesting. Only recently has intercropping gained attention, not only for simultaneously growing two cash crops but also intercropping a cash crop with an SC. This has been driven by the increasing awareness of the negative environmental impacts of general agricultural farming practices, including land and water degradation and biodiversity loss.

In this review, we provide recent advances on SCs as a source for N in arable crop rotations based on data from the literature and unpublished studies from Northern Europe. SCs are defined here as crops grown in an agroecosystem for ecosystem services, including direct services via N transfer to the main cash crop, and a range of other multiple and diverse ecosystem services (Garcia et al., 2018). As such SCs include cover crops and catch crops which are specifically grown to catch post-harvest soil mineral N and to reduce N leaching losses and crops which are mainly grown as an additional source of N. They can be grown in the time between two main crops, undersown into a main crop, or intercropped with a main crop. Excluded are crops which are grown for forage production. Furthermore, our review focuses on SC studies and findings for arable cropping systems in regions with a temperate climate. Discussed are benefits and challenges regarding the use of SCs for reducing N leaching losses and for providing N to cash crops, and research gaps that require further investigations.

2 Service crops in rotation with cash crops

SCs are commonly sown after harvest of the main cash crop in autumn, mainly to prevent a period of bare soil between two successive cash crops and its associated benefits. This can be a challenge when harvest occurs late in autumn and when the growing season is too short in autumn for sufficient establishment, and thus SCs are often undersown into the main cash crop for better establishment. Ideally, to accomplish the various ecosystem services, SCs should have the ability to easily establish, develop quickly, develop a substantial biomass with high N uptake, service the soil, resist diseases, be easily terminated, be economically viable, and not act as a host for cash crop diseases (Reeves, 1994). The major commonly grown SCs in temperate regions are leguminous (i.e. clover and vetches) and non-leguminous brassicas (i.e. radishes and turnip), grasses, and cereals. Legumes contribute N through symbiotic N fixation, brassicas generally grow rapidly under warm conditions and potentially take up large quantities of N, and their residues mineralise N rapidly. In contrast, residues from grasses and cereals are relatively resistant to decomposition. SCs can either be sown as single species or mixtures or be annual, bi-annual, or perennial species. Especially in low-input systems, the use of non-legume and legume mixtures has been promoted as an additional source of N. Provided that these mixtures are in balance (with similar proportions), they can be effective in reducing N leaching (Möller and Reents, 2009; Thapa et al., 2018; Tribouillois et al., 2016; Vogeler et al., 2019b) as well as increase the residual or fertiliser replacement value compared to pure non-legume SCs (Abdalla et al., 2019; Wittwer et al., 2017). To maintain a balance in mixtures requires a similar competition of the species for resources (e.g. light, water, and nutrients) apart from that of N (Yu et al., 2016). Studies have, however, also indicated that mixtures should be dominated by the legumes species for ensuring an adequate fertiliser replacement value in systems without the application of N (Kramberger et al., 2014).

While SCs also effect greenhouse gas emissions (GHGs) (Abdalla et al., 2014), this aspect is not included in the review, as the amounts of N lost via nitrous oxide emissions are small compared to those lost via leaching. Thus these losses can be neglected when evaluating the contribution of SCs for supplying N to cash crops. We, however, acknowledge that understanding the impacts of SCs on GHG is important for mitigating climate change, especially considering the discrepancies that have been reported. For instance, based on a meta data analysis, Basche et al. (2014) found a decrease in emission by 40% and an increase by 60% of the fields with SCs compared with bare soil, with higher emissions from legume SCs and when SC residues were incorporated into the soil.

Apart from these required characteristics of SCs, frost tolerance or winter hardiness is also important for their performance, especially regarding reductions in N leaching (Böldt et al., 2021; Gollner et al., 2020). Grasses are generally the most cold-tolerant SCs, and winter rye, brassicas, hairy vetch, red clover, and oats are good cool-season options with frost tolerance, whereas vetch and turnip often show limited winter survival.

3 Service crops intercropped with cash crops

Intercropping cash crops with SCs has been promoted for providing cash crops with N and for enhancing retention of N, and thereby reducing N leaching losses. Many studies have shown the benefits of intercropping, including crop yield (Bedoussac et al., 2015), weed suppression (Arlauskienė et al., 2021), reductions in pest and diseases (Boudreau, 2013), enhanced C and N input to the soil from residual biomass (Wichmann et al., 2005), and yield stability (Raseduzzaman and Jensen, 2017). While intercropping systems include either the use of two cash crops or the use of an SC and a cash crop, we focus on intercropping of SCs and a cash crop.

In organic, low-input systems, the use of legume SCs within the crop rotation is of significant interest to improve N input and a potential way towards delivering 'sustainable intensification' (Brooker et al., 2015; Jones and Clements, 1993). However, keeping the proportion of non-marketable SCs in the crop rotation low is important for economic benefits. In organic wheat cropping systems, green manure crops often occupy 20–40% of the rotation (Cuttle et al., 2003), and grass clover as a pre-crop before winter cereals is common. However, the high soil mineral N after ploughing of the grass clover in autumn increases the risk of N leaching during the following winter period (Hoffmann et al., 2018; Vogeler et al., 2021b). Intercropping of cereals and legumes, where a cash crop and an SC are grown in the same year, may present an alternative to a grass/clover SC within the rotation. The cash crops and SCs can either be sown at the same time – synchronous intercropping (Fig. 1), or

Figure 1 Example of synchronous intercropping system of winter wheat and white clover in a wide row system. Source: Adapted from: H. Neumann CAU Kiel.

the cash crop can be sown into a living mulch of a previously established SC – asynchronous intercropping (Gardarin et al., 2022).

Intercropping of cereals and legumes has been shown to reduce the competition for soil N compared with solely non-leguminous intercropping, and furthermore legumes can increase the soil N. This can have a positive effect on the yield of the non-legume crop through N transfer via i.e. rhizodeposition and indirect N transfers between plants through the promotion of arbuscular mycorrhizal fungi (Gardarin et al., 2022; Leigh et al., 2009). Intercropping of red clover/cereals has been shown to perform well, as the red clover can easily be established in cereals due to the tolerance of low light. Undersowing a winter cereal with a non-legume SC can be used as a strategy to reduce the risk of N leaching after N-intensive legume pre-crops (Mauscherning, 2008).

Living mulches are SCs sown either before or simultaneously with a main crop and maintained as a living ground cover throughout the growing season (Hartwig and Ammon, 2002). In temperate climates, most studies on living mulches have been done with perennial legumes, particularly not only white clover but also red clover (Bergkvist, 2003a; Neumann, 2005; Thorsted et al., 2002a). Benefits of such living mulches compared with terminated SCs include increased weed suppression, reduced soil erosion and N leaching, and in the case of a legume-living mulch a potential source of N for the cereal (Bhaskar et al., 2021; Lorin et al., 2015).

One of the main drawbacks of synchronous intercropping systems is the competition for water, nutrients, and light. The latter depends on early growth characteristics of the intercropped plants, the leaf area development, the height, and vertical growth and the spatial position, due to shading of the cash crop by the SC especially in the early stages (Carof et al., 2007).

4 Effects of service crops on yields

4.1 Undersown service crops and effects on the main crop

Undersowing of SCs into a main crop is often done to ensure adequate biomass growth after harvest for reducing the risk of N leaching (Wachendorf et al., 2006) and, in the case of a legume SC, to increase the potential N fixation and provision of N to the following crops. This can, however, also reduce the growth and yield of the main crop due to competition for resources (e.g. soil moisture, light, and plant nutrients). The competition can be reduced via delaying the time of undersowing, by reducing the seeding rate (Känkänen and Eriksson, 2007; Ohlander et al., 1996), or by enhancing the spatial distance between seed rows of the main crop and the SC (Wachendorf et al., 2006). Reductions in the yield of the cash crop due to competition by the SC are dependent on environmental conditions, management, and choice of the SC species. In a crop rotation experiment in an organic system, involving cereal and legume grain crops as sole crops or intercropped, and with or without undersowing of an SC (ryegrass/white clover), Hauggaard-Nielsen et al. (2012) found no significant effects of the undersown catch crop on the yield of the cash crop. Similarly, Ohlander et al. (1996) found only small reductions in grain yield of spring barley (up to 6%) due to undersown catch crops. Känkänen and Eriksson (2007) investigated the effect of a range of undersown SCs and seeding rates on the growth of spring barley, and generally found little effect of red and white clover on the yield, whereas grasses strongly competed with the spring barley. Other studies, in contrast, have found no effect of an undersown ryegrass on the yield of spring barley (Thomsen, 2005; Thomsen and Christensen, 1999). In order to reduce nitrate leaching, Wachendorf et al. (2006) established a ryegrass understory under forage maize resulting in negligible effects on yield performance of maize and a reduction in N leaching of around 50%, as long as the grass was drilled and a distance of 15 cm to the maize row was ensured in order to reduce the competition for light and nutrients.

4.2 Pre-emptive competition and effects on yield in the following main crop

SCs grown in rotation with the main crop take up water and N over the period from sowing to termination. Depending on environmental conditions, some of this water and N would not be lost through drainage and leaching without SCs, but would have been available for the following cash crop if not taken up by the SC. This phenomenon has been referred to as pre-emptive competition (Thorup-Kristensen et al., 2003). While termination in spring has been suggested to be more effective in reducing N leaching (Miranda-Vélez and Vogeler, 2022), pre-emptive competition occurs mainly in spring. As such, the termination time is a key management factor to avoid pre-emptive competition and yield reductions of the subsequent cash crop (Krueger et al., 2011; Wallgren and Lindén, 1994). The optimal timing depends on soil characteristics and climatic conditions. The rooting depth of the SC is another factor which influences pre-emptive competition, with shallow roots taking up N from the surface layers and thereby depleting the available N pool for the cash crop. In contrast, deep-rooted SCs can take up N from deeper soil layers, which is likely prone for leaching (Sapkota et al., 2012; Thorup-Kristensen and Nielsen, 1998). For water availability, pre-emptive competition depends on the balance between water uptake by the SC and the evaporation prevented by the residue mulch.

N limitation and yield reductions of the cash crop can also occur through N immobilisation, when following SCs with high C:N. Delaying termination time increases the C:N and fibre content of the residues, which are then more recalcitrant to decomposition, resulting in a more slow release of N (Alonso-Ayuso et al., 2014). Legume SCs may compensate for this by fixing more N when killed late.

4.3 Living mulches

Living mulches are SCs sown before the main crop and maintained as a living ground cover throughout the growing season. If perennial SCs are used, the mulch can be maintained for several years without the need of resowing. A major drawback of intercropping with living mulches is the potential competition between these and the cash crops, especially for water and light, as the SC has already formed an established cover with potential effects on cash crop establishment. Various studies have shown reductions in yield compared with sole cropping of cereals (Germeier, 2000). Especially, white clover as a living mulch shows strong competition for spring cereals due to the vigorous growth in spring (Jones and Clements, 1993; Williams and Hayes, 1991). Competition from white clover has also been shown for winter wheat, and Bergkvist (2003a) suggested that the white clover should be defoliated prior to the sowing of the winter wheat, and the clover growth should also be suppressed at the tillering stage of the wheat. Suppression of the living mulch by mechanical means is however labour intensive. Similarly, in an intercropping system with winter wheat directly drilled into pea stubble, white clover, or red clover in Lithuania, Sarunaite et al. (2013) found significant reduction in yields of winter wheat of ≥50% when compared to conventional sowing as a sole crop, which was partly due to bad establishment of the winter wheat. To overcome competition with white clover, different approaches have been tested, including the application of a herbicide in spring, the use of a less competitive white clover variety or subterranean clover (*Trifolium subterraneum* L.), and the application of a mineral N fertiliser dressing in spring (Baresel et al., 2018; Bergkvist, 2003b;

Thorsted et al., 2002b). Another option is the use of a frost-sensitive legume, which after functioning as a living mulch is killed by winter temperatures, potentially releasing their fixed N for the cereal (Lorin et al., 2015).

In agreement with these results, Neumann (2005) found in a 2-year experiment in an organic system in Northern Germany lower grain yields of winter wheat sown into a stand of white clover compared with sole winter wheat. In this study, two different intensities of tillage were investigated to suppress the growth of the clover, namely row rotary tillage for extensive suppression, and laminar rotary tillage for intensive suppression. Depending on the tillage system, the grain yield was reduced by 34% with intensive suppression and by 60% with extensive suppression. Whole crop silage was also significantly lower under intercropping with extensive suppression, while the crude protein was higher in both intercropping systems than in the sole wheat crop.

5 Residual effects of service crops

5.1 Yields and nitrogen uptake of subsequent cash crops

Contradictory results on the effects of SCs are reported in the literature, and an overview of the SC N use by the succeeding crop is provided in a recent review by Sieling (2019). Such an overview is thus not included here, but instead examples from various studies are given, which provide further insights into the complex interactions of N cycling, especially in crop rotations involving catch crops or intercropping. Reported residual effects of SC N in Sieling (2019) for Northern Europe range from −130 kg N ha^{-1} with the use of a winter rye as an SC to +67 kg N ha^{-1} with the inclusion of legumes. Differences can be linked to environmental and management factors, including the type and properties of SC, seeding time, establishment, development, and N uptake of the SCs, incorporation time, soil type, and weather conditions (Böldt et al., 2021; Komainda et al., 2016; Schipanski et al., 2014; Vogeler et al., 2019a). The N uptake by SCs depends on many factors, including the length of the growing season for the SC, the competition between the SC and the cash crop in an intercropping system, and the preceding crop and how much residual N (both mineral and organic N) is left in the soil after harvest. The SC's growth stage and biochemical properties (C:N ratio, N and biogeochemical composition) at termination affects the fate of N, resulting in either immobilisation or N mineralisation (Adetunji et al., 2020; Nicolardot et al., 2001; Sullivan and Andrews, 2012; Thomsen et al., 2016). N mineralisation from SC residues within the short term (up to 1 year) after residue incorporation has been reported to range between 20% and 50% of the SC N, although varying results have been reported. If this N mineralisation is synchronised with the need for the subsequent cash crop N use efficiency is increased, but if the timing is not good, the risk of losses via leaching is increased. Some studies also indicate that effects of SC on the subsequent crop might only be positive after a long-term use of SC (Gabriel et al., 2016; Hansen and Djurhuus, 1997; Kuo and Jellum, 2000).

In the organic crop rotation study by Hauggaard-Nielsen et al. (2012) with cereal and legume grain crops with or without undersowing of an SC (ryegrass/white clover), the yield of a subsequent spring cereal was significantly reduced by the SC, while the yield of a subsequent winter cereal was significantly increased. These contrasting effects can be explained by the availability of soil mineral N, pre-emptive competition between the SC and the spring sown cash crop for N, and the provision of N to the winter cereal via mineralisation of the SC residue.

In an organic crop rotation, Böldt et al. (2021) found no effect of SCs on the grain N yield of peas following a cereal pre-crop and a cover crop, whereas the grain of oats following pea as a pre-crop and a cover crop had on average about 20 kg ha^{-1} additional N uptake compared with those grown with a bare soil over autumn/winter. The additional N uptake can be explained by a higher soil mineral N after harvest of the pea pre-crop, resulting in high N uptake of the SCs (ranging from about 90–180 kg N ha^{-1}). In the study by Vogeler et al. (2021a) involving increasing rates of N fertilisation to a spring barley with a previous SC of either ryegrass or fodder radish additional grain N was only about 4 kg N ha^{-1}, with no effect on the fertilisation rate. Based on these studies, supplemented with data from the literature, there appears to be only a weak relationship between the N measured in the aboveground biomass (AGB) of the SCs and the additional N in the grain (Fig. 2a). Including measurements of the N in the belowground biomass (BGB), and separation by legume and non-legume SC species does not improve the relationship (Fig. 2b). A long-term simulation using APSIM also indicated no relationship between the uptake of N by ryegrass as an SC and a subsequent spring barley crop, and only a weak relationship when fodder radish was used as an SC (Fig. 3) (Vogeler et al., 2023). Similarly, Lorin et al. (2016) and Cadoux et al. (2015) found no relationship between N uptake in autumn by various frost-sensitive legumes, and the additional N uptake of intercropped winter oilseed rape in experiments carried out in various parts of France under different climatic conditions.

In an organic intercropping experiment by Neumann (2005), the residual effect of winter wheat, directly drilled into a living mulch of white clover was investigated in a subsequent cash crop of either oats or winter rye. The residual dry matter (DM) and N amounts in the residues of the winter wheat and white clover at the beginning of October prior to sowing of the cash crop was substantially higher in the intercropping system compared to solely grown winter wheat systems, potentially providing more than 200 kg N ha^{-1} of residual N and with a C:N between 20 and 25; some of this can be mineralised during the growth period of the following cash crop. These residual amounts are similar to

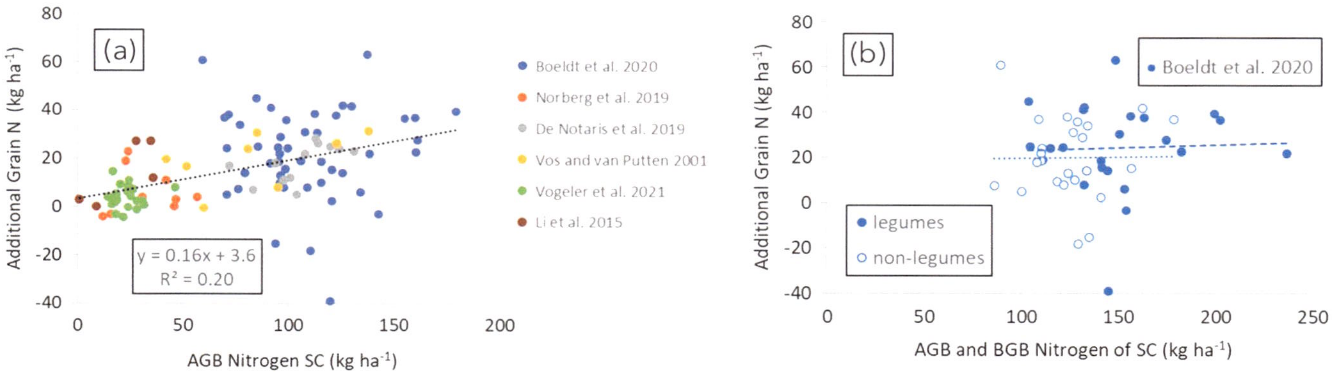

Figure 2 Additional grain N as a function of the (a) aboveground (ABG) N of SCs, and (b) for ABG and belowground (BGB) N separated for non-legume and legume SC.

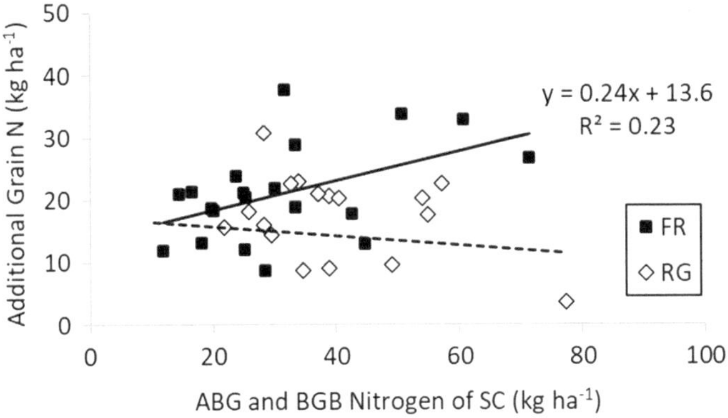

Figure 3 Additional grain N as a function of the (a) aboveground (ABG) and belowground (BGB) N in SC for spring barley based on long-term simulations by Vogeler et al. (2023) with the APSIM model and either fodder radish (FR) or ryegrass (RG) as a catch crop.

those that can be achieved from grass/clover crops, which are commonly used for green manure on organic farms (Mueller and Thorup-Kristensen, 2001). The intercropping of winter wheat and white clover resulted in higher total yields of both winter rye and spring oats compared to wheat grown as a sole crop following ploughing of white clover. The additional yield was about 2 tons ha^{-1} (+29.2%) for spring oats and 1.3 tons ha^{-1} (+30.7%) for winter rye. The study of Neumann (2005) further investigated how an intercropping system can be optimised within a crop rotation approach. For this, intercropping was carried out for a second year but restricted to the intercropping with intensive suppression of the white clover via laminar rotary tillage. To enable a comparison between the different systems with different crops, the measure of grain equivalent (GE) from the official tables of the German Federal Ministry of Food and Agriculture (BMELV, 2012) was used. Highest total GE yields over the 2 years were found when oats were grown as sole crops in year 2, with slightly higher yields following intercropping in year 1 (Table 1). Similar results, despite lower GE, were found for winter rye in year 2. Intercropping a second year reduced GE for both rye and oats. Results from this study suggest that the most successful biennial system consisted of wheat/clover intercropping in the first year, followed by a sole crop of oats after spring-ploughing in the second year.

5.2 Fertiliser replacement value of service crops

The residual N effect can be used to estimate a N fertiliser replacement value (NFRV) of SCs, based on the efficiency of fertiliser recovery of the crop/system. The NFRV indicates the equivalent amount of mineral N (kg N ha^{-1}) which the SC can replace and thus by how much the amount of N fertiliser applied to the succeeding cash crop can be reduced. As discussed earlier, residual

Table 1 Grain equivalent of different biennial cropping systems, either grown as a sole crop or intercropped with winter clover (IC)

System	Year 1		Year 2		Total GE
	Crop	Treatment	Crop	Treatment	
1	Wheat	Sole	Oats	Sole	83.5[a]
2	Wheat/clover	IC	Oats	Sole	88.5[a]
3	Wheat/clover	IC	Oats	IC	66.7[b]
4	Wheat	Sole	Rye	Sole	71.2[b]
5	Wheat/clover	IC	Rye	Sole	72.1[b]
6	Wheat/clover	IC	Rye	IC	58.7[c]

Values are means of two biennial cropping periods and four replicates. Different letters mark significant differences between factor levels.
GE, grain equivalent; IC, intercropping.
Source: Adapted from Neumann (2005).

effects of SCs are very variable, and so are fertiliser recovery efficiencies. N recovery efficiencies can be based on the N in the total AGB taken up within the first growing season, which reflects the capacity of the crop to acquire N. Alternatively recovery efficiencies can be based on the grain N, which reflects the physiological efficiency of the crop to use the acquired N to produce grain (Ladha et al., 2005). Based on studies from major cereal crops (e.g. maize, rice, and wheat) grown across large agroclimatic regions, Ladha et al. (2005) reported N fertiliser recovery efficiencies (based on the difference between fertilised and non-fertilised treatments) ranging from 20% to 90% for N in the total AGB and from 10% to 70% for grain N. Larger efficiencies were generally associated with low fertiliser N inputs. Average N fertiliser recoveries for Europe were 68% for the N in the total AGB and 41% for the grain N.

Based on a long-term modelling study with APSIM and a rotation of SC and spring barley (with the SC grown every year) in Denmark, Vogeler et al. (2023) calculated an average NFRV of 42 kg N ha^{-1} for fodder radish and of 29 kg N ha^{-1} for ryegrass, assuming an N fertiliser recovery of 50%. Year-to-year variability was large, with coefficients of variation of 40% for fodder radish and 70% for ryegrass (Fig. 4). The high variability can be explained by environmental conditions, with an average annual rainfall over the 20-year simulated period of 695 mm (ranging from 518 mm to 1083 mm) and an average annual temperature of 8.4°C (ranging from 6.2°C to 9.5°C). These simulated values are in line with measurements from various experiments conducted in Denmark, which also show large differences. Fontaine et al. (2020) reported NFRVs ranging from 61 kg N ha^{-1} to 85 kg N ha^{-1} for an SC mixture dominated by red clover. De Notaris et al. (2019) found NFRV ranging from 13 kg N ha^{-1} to 50 kg N ha^{-1} with early undersowing of an SC mixture consisting of legumes and non-legumes, with higher residual effects in the following crop when no fertiliser was applied to the main crop. Askegaard and Eriksen (2007) reported higher NFRV values for

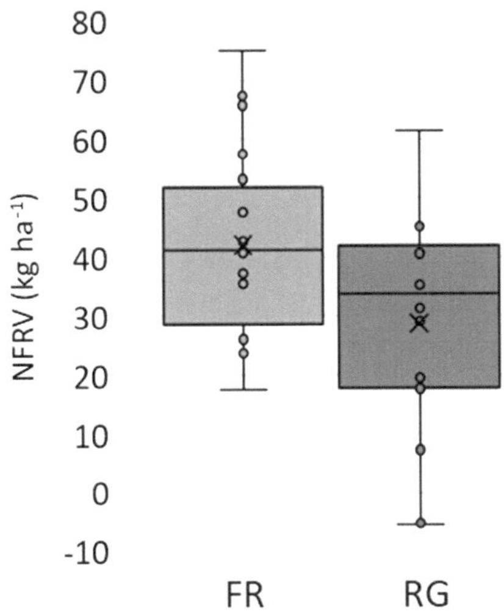

Figure 4 Nitrogen fertiliser replacement value of catch crops of fodder radish (FR) and ryegrass (RG) simulated by APSIM for a period of 20 years with SCs grown every year and calculated as additional grain N uptake above the treatment with spring barley (SB) as a sole crop and an annual N fertilisation of 100 kg N ha^{-1} (Vogeler et al., 2023).

SCs with legumes (ranging from 73 kg N ha^{-1} to 120 kg N ha^{-1}), compared to non-legume SCs (23 kg N ha^{-1}). For chicory as an SC, they even found a negative effect, which they explained by pre-emptive competition of N and N immobilisation.

6 Nitrate leaching

Nitrate leaching from intensive cropping systems is of concern and affected by the management, including tillage operations and application of N via inorganic and organic fertilisers. Even in low N input systems, N leaching can be elaborated due to high soil mineral N after grain legumes or grass clover leys (Biernat et al., 2020; Hauggaard-Nielsen et al., 2003), which are included in the rotation to improve fertility and N supply. The benefits of non-leguminous SCs in pure stand to reduce nitrate leaching over the autumn/winter period by taking up soil N has been shown in many studies (Justes et al., 1999; Thomsen, 2005; Thorup-Kristensen et al., 2001). N leaching can potentially be reduced by both the reduction of soil mineral N, and the uptake of water, which delays the percolation, although this latter has been found to be negligible under humid conditions (Vogeler et al., 2023). The use of mixtures of non-legume and legume SCs has recently also been investigated, as a way for reducing external N inputs via mineral fertilisers. However, as legume SCs are generally less effective in reducing N leaching (Gselman and Kramberger, 2008; Tonitto et al., 2006), studies indicate that SC mixtures with a critical biomass and balanced proportion of legumes and non-legumes need to be maintained throughout the leaching period for effective reduction in N leaching losses (De Notaris et al., 2021; Vogeler et al., 2019b).

Several studies have also shown that intercropping reduces N leaching compared with sole cropping (Hauggaard-Nielsen et al., 2003). This might be due to improved use of soil N and increases in C:N of crop residues compared with solely cropped grain legumes, which are less prone to mineralisation after harvest in autumn and thus losses via N leaching.

Literature data and some unpublished data indicate that the reduction in N leaching is associated with the amount of N accumulated in the SCs (Fig. 5). These data comprise both SCs grown between two cash crops (Böldt et al., 2021; Constantin et al., 2010; Munkholm et al., 2017; Norberg and Aronsson, 2020; Vogeler et al., 2021a; Vos and Van Der Putten, 2004) and SCs intercropped with a cash crop (Mauscherning, 2008; Neumann, 2005). However, the reduction in N leaching during the first winter following an SC is often larger than the aboveground N uptake by the SC. This difference can be due to that N uptake is generally only determined for the AGB determined in late autumn. The uptake by the roots and additional N uptake can occur during mild winters and in early spring, and this increases the total uptake of N by the SC. Data from Böldt et al. (2021) indicate that additional N uptake after late autumn is negligible, and instead standing DM (data not shown) and N in the AGB declines over the winter/ and spring period. However, N in the roots can contain 14–49% of the SCs' N (Fig. 6). Another reason for the high reduction in N leaching could be immobilisation of soil N by plant residues (Glendining et al., 2001; Zhao et al.,

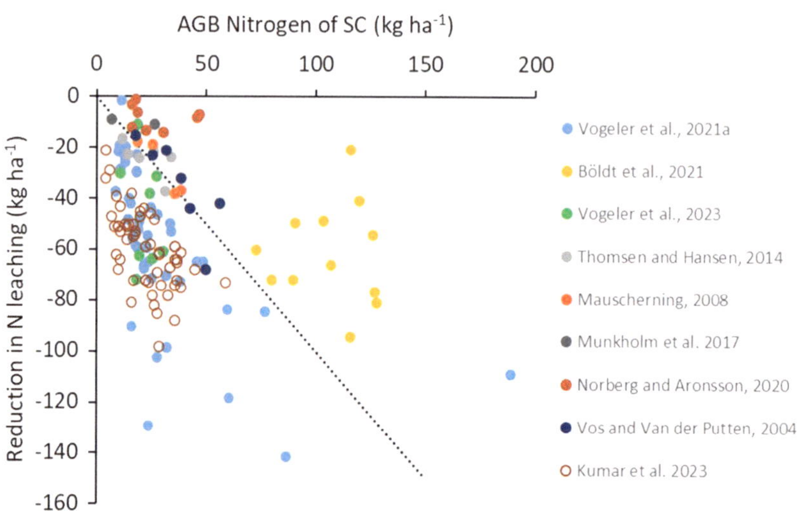

Figure 5 N leaching reduction as a function of the aboveground biomass (ABG) nitrogen of the service crop, either grown between two main cash crops or intercropped with the cash crop.

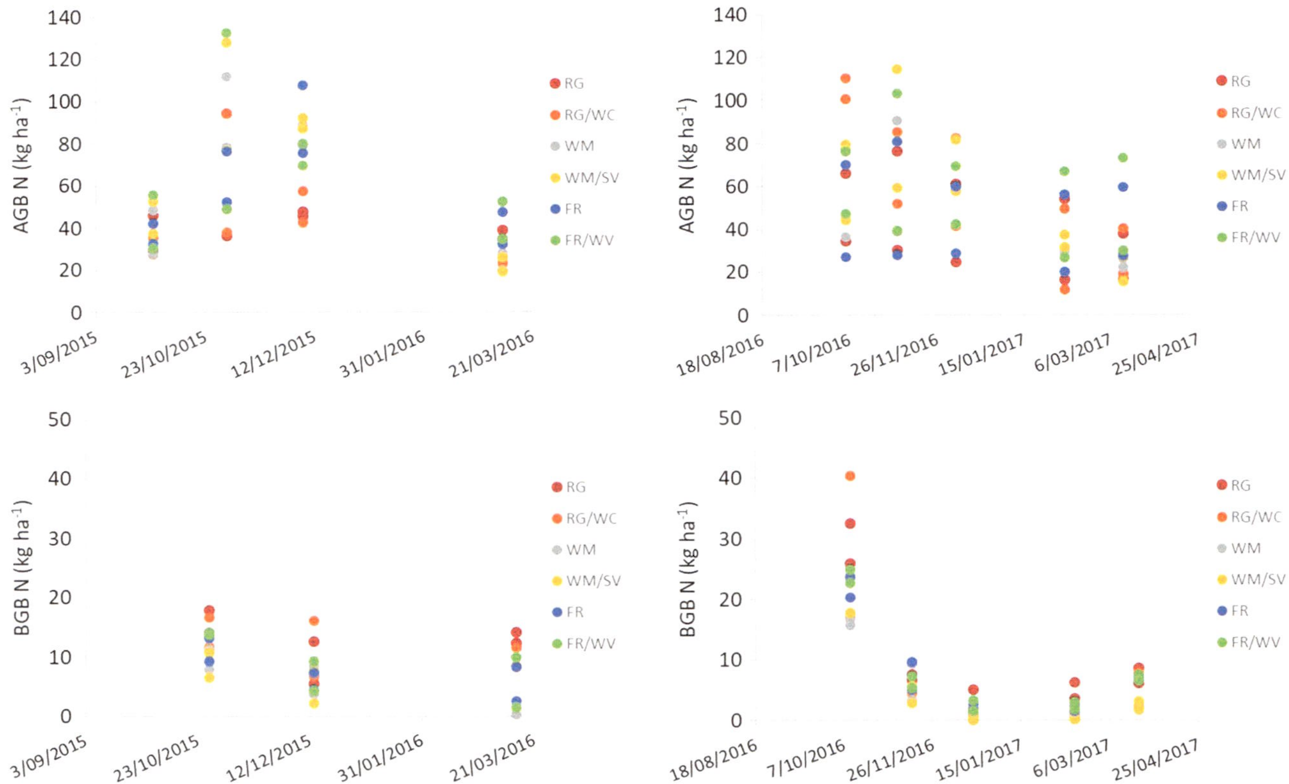

Figure 6 Nitrogen in the aboveground biomass (AGB) and belowground biomass (BGB) determined for different service crops at different times in 2 years Source: Adapted from the study by Böldt et al. (2021).

2022), and retention of the SC N in the soil organic N fraction (Harris et al., 1994). Also, some N may have been taken up that is not lost by leaching due to dry conditions.

In some studies, the reduction in N leaching is also lower than the aboveground N uptake of the SC, which can be due to N mobilisation from the residue over the leaching period, due to either tillage in autumn, and freeze-kill effects on the SC (Böldt et al., 2021; Gollner et al., 2020). Long-term effects of crop management and the use of SCs also influences N leaching, as shown in studies by Thomsen and Christensen (1999) and Hansen et al. (2000), where leaching was higher in treatments with previous ryegrass SCs compared to those without SCs. From this, they concluded that when assessing the effects of SCs for reducing N leaching in the long-term, legacy effects of N mineralisation from residues of SCs needs to be considered. Thus, mineral N fertilisation in a cropping system should be reduced by those amounts of additional mineralisation.

7 Conclusion

Introducing SCs into intensive, often simple cropping systems offers many benefits, including diversification, and as highlighted in this review potential reductions in external inputs of N fertilisers and N losses to the environment due to transfer of N to the subsequent cash crop.

Especially in intensive conventional systems with high use of N fertilisers, SCs should be used to reduce the risk of N leaching. However, the use of SCs is not easy to realise in pure winter cropping systems, as the time for a good biomass accumulation is generally not sufficient. Direct drilling into living mulches is also problematic, and intercropping of winter cereals with SCs can lead to high-yield suppressions of the main crop due to competition for resources from the SC. In rotations with spring cereals, SCs should be sown as early as possible, and the use of undersowing is often a good option, which generally has only small effects on the yield of the main cash crop. In environments, where SCs grow vigorously and provide too much competition with the cash crop, the time of undersowing should be delayed, or their seeding rate reduced.

In organic, low N input systems, the use of a legume SCs within the crop rotation is important for providing N into the systems, and the high risk of N leaching after a grain legume can be reduced with the use of an SC after the harvest. When sowing after harvest, it is important that the SCs are sown as early as possible to ensure a high biomass accumulation and N uptake, as the reduction in N leaching in general is positively related to the N uptake. SCs can also be beneficial for reducing N leaching in organic systems with often high organic matter contents and frequent grass clover leys within the rotation. In regions with cold temperatures and high precipitation over the winter, frost-killed SCs should be avoided to prevent N losses over winter. In areas with low and moderate precipitation, early termination can be beneficial to avoid pre-emptive competition.

Both the use of non-legume SCs and mixtures of legumes and non-legumes can be used. In the case of mixtures, it is important that a balance is kept between the legumes and non-legumes. This is important for both the reduction in N leaching from the residual N left after the harvest of the main crop, and of the SC residues, when tillage occurs in autumn. Low C:N ratios in legume species can result in fast mineralisation of N with subsequent leaching over the winter period. This reduces the potential N transfer to the subsequent crop.

Many studies have shown beneficial effects of SCs on spring cereals, due to N transfer from the SCs' residues, with increases in grain N. N fertiliser replacement values ranging from 23 kg N ha^{-1} to 85 kg N ha^{-1} have been observed, and with high year-to-year variability. This high variability highlights that further studies are warranted to understand the complex relationships between the characteristics of the residues and their mineralisation based on environmental conditions to better synchronise N release with the requirements of the subsequent crop. Management practices, especially tillage, can also influence the rate of mineralisation, by changing the soil structure and pore system characteristics, and thereby soil moisture and temperature.

Common arable rotations with SCs in temperate Europe include winter/spring cereal – SCs of either brassicas (oil radish, fodder radish) or ryegrass – spring cereal, with the SC either undersown, sown after harvest or spread into the main crop before harvest. In maize systems, undersowing of ryegrass or rye as an SC is common. Mandatory use of SCs is common for different purposes. For example in the Netherlands, farmers have to grow an SC after silage maize on sandy or loose soils to limit N leaching according to The Nitrates Directive. In various regions of Germany, France, and Belgium, SCs are also used to suppress harmful nematodes in cereal-sugar beet rotations.

Living mulches are also effective in reducing N leaching from winter cereals; however, they often show too strong competition with the cereal. Limiting the growth of these in springtime, either by labour-intensive mechanical means or by the use of herbicides is an unfavourable option. While living mulches have many benefits, including an increased biodiversity, prevention of erosion and suppression of weeds, further research should focus on how to limit the competition between the living mulch and the main cash crop to reduce yield losses.

Most studies on SCs have been conducted over only very short time periods, often 2 years. The few studies that have been carried out over long-term periods indicate legacy effects from SCs and fertiliser values to later crops in the crop rotation. The consequences of SCs on soil fertility, N transfer to subsequent crops and the N fertiliser replacement value, and N losses should be evaluated over a long time, at least a decadal time. The importance of long-term effects was shown in an organic cropping systems study in Denmark by Pullens et al. (2021), who found that the legacy effect of SCs on yield and grain N of spring barley was traceable for 5–8 years, as was also the legacy effect of crop residues and manure.

The use of process-based modelling is a cost-effective way to complement experimental studies to fully understand the implication of various management options, such as the use of SCs and intercropping under varying environmental conditions and for optimising cropping systems. Many studies have shown that most models have considerable difficulties in simulating crop growth and N dynamics in complex crop rotations. As such they need to be tested and refined based on long-term experiments, in order to achieve robust model outputs, which account for legacy effects from crop residues on crop yield and quality, as well as overall cropping system profitability.

8 Where to look for further information

Further information on service crops and their role in ecosystem services can be found in the recently published book by Blanco (2023). Information ranges from a detailed discussion of cover crop biomass production, soil erosion, greenhouse gas fluxes, nitrate leaching, soil C sequestration to effects of climate change and economics of cover crop farming. The role of policy for driving the adaptation of cover crops in Europe is discussed in Kathage et al. (2022). Based on farm surveys across Europe, they found that policy is the key factor for driving adoption rates and adoption intensities. Spatial coverage of cover crops for Europe and the United Kingdom, England, is provided in the high-resolution map, which is derived from satellite data combined with statistical surveys. The map can be accessed via https://esdac.jrc.ec.europa.eu/content/cover-crops-accross-europe, and a description is provided in Fendrich et al. (2023).

9 References

Abdalla, M., Hastings, A., Cheng, K., Yue, Q., Chadwick, D., Espenberg, M., Truu, J., Rees, R. M. and Smith, P. (2019). A critical review of the impacts of cover crops on nitrogen leaching, net greenhouse gas balance and crop productivity. *Global Change Biology* 25(8), 2530–2543.

Abdalla, M., Hastings, A., Helmy, M., Prescher, A., Osborne, B., Lanigan, G., Forristal, D., Killi, D., Maratha, P., Williams, M., Rueangritsarakul, K., Smith, P., Nolan, P. and Jones, M. B. (2014). Assessing the combined use of reduced tillage and cover crops for mitigating greenhouse gas emissions from arable ecosystem. *Geoderma* 223-225, 9-20.

Adetunji, A. T., Ncube, B., Mulidzi, R. and Lewu, F. B. (2020). Management impact and benefit of cover crops on soil quality: A review. *Soil and Tillage Research* 204, 104717.

Alonso-Ayuso, M., Gabriel, J. L. and Quemada, M. (2014). The kill date as a management tool for cover cropping success. *PLoS ONE* 9(10), e109587.

Arlauskienė, A., Jablonskytė-Raščė, D., Šarūnaitė, L., Toleikienė, M., Masilionytė, L., Gecaitė, V. and Kadžiulienė, Ž. (2021). Perennial forage legume cultivation and their above-ground mass management methods for weed suppression in arable organic cropping systems. *Chemical and Biological Technologies in Agriculture* 8.

Askegaard, M. and Eriksen, J. (2007). Growth of legume and nonlegume catch crops and residual-N effects in spring barley on coarse sand. *Journal of Plant Nutrition and Soil Science* 170(6), 773-780.

Baresel, J. P., Nichols, P., Charrois, A. and Schmidhalter, U. (2018). Adaptation of ecotypes and cultivars of subterranean clover (Trifolium subterraneum L.) to German environmental conditions and its suitability as living mulch. *Genetic Resources and Crop Evolution* 65(8), 2057-2068.

Basche, A. D., Miguez, F. E., Kaspar, T. C. and Castellano, M. J. (2014). Do cover crops increase or decrease nitrous oxide emissions? A meta-analysis. *Journal of Soil and Water Conservation* 69(6), 471-482.

Bedoussac, L., Journet, E.-P., Hauggaard-Nielsen, H., Naudin, C., Corre-Hellou, G., Jensen, E. S., Prieur, L. and Justes, E. (2015). Ecological principles underlying the increase of productivity achieved by cereal-grain legume intercrops in organic farming. A review. *Agronomy for Sustainable Development* 35(3), 911-935.

Bergkvist, G. (2003a). Effect of white clover and nitrogen availability on the grain yield of winter wheat in a three-season intercropping system. *Acta Agriculturae Scandinavica, Section B - Soil and Plant Science* 53(3), 97-109.

Bergkvist, G. (2003b). Effect of white clover and nitrogen availability on the grain yield of winter wheat in a three-season intercropping system. *Acta Agriculturae Scandinavica, Section B: Soil and Plant Science* 53(3), 97-109.

Bhaskar, V., Westbrook, A. S., Bellinder, R. R. and DiTommaso, A. (2021). Integrated management of living mulches for weed control: A review. *Weed Technology*, 35(5), 856-868.

Biernat, L., Taube, F., Vogeler, I., Reinsch, T., Kluß, C. and Loges, R. (2020). Is organic agriculture in line with the EU-nitrate directive? On-farm nitrate leaching from organic and conventional arable crop rotations. *Agriculture, Ecosystems and Environment* 298, 106964.

Blanco, H. (2023). *Cover Crops and Soil Ecosystem Services*. John Wiley & Sons.

BMELV (2012). *Getreideeinheitenschlüssel* (vol. 172). http://berichte.bmelv-statistik.de/SJT3120100-2012.pdf.

Böldt, M., Taube, F., Vogeler, I., Reinsch, T., Kluß, C. and Loges, R. (2021). Evaluating different catch crop strategies for closing the nitrogen cycle in cropping systems–field experiments and modelling. *Sustainability (Switzerland)* 13(1), 1-22.

Borrelli, P. and Panagos, P. (2020). An indicator to reflect the mitigating effect of Common Agricultural Policy on soil erosion. *Land Use Policy* 92, 104467.

Boudreau, M. A. (2013). Diseases in intercropping systems. *Annual Review of Phytopathology* 51, 499–519.

Brooker, R. W., Bennett, A. E., Cong, W. F., Daniell, T. J., George, T. S., Hallett, P. D., Hawes, C., Iannetta, P. P., Jones, H. G., Karley, A. J., Li, L., McKenzie, B. M., Pakeman, R. J., Paterson, E., Schöb, C., Shen, J., Squire, G., Watson, C. A., Zhang, C., Zhang, F., Zhang, J. and White, P. J. (2015). Improving intercropping: A synthesis of research in agronomy, plant physiology and ecology. *New Phytologist* 206(1), 107–117.

Cadoux, S., Sauzet, G., Morison, M. V., Pontet, C., Champolivier, L., Robert, C., Lieven, J., Flénet, F., Mangenot, O., Fauvin, P. and Landé, N. (2015). Intercropping frost-sensitive legume crops with winter oilseed rape reduces weed competition, insect damage, and improves nitrogen use efficiency. *OCL Oilseeds and Fats Crops and Lipids* 22(3).

Carof, M., De Tourdonnet, S., Coquet, Y., Hallaire, V. and Roger-Estrade, J. (2007). Hydraulic conductivity and porosity under conventional and no-tillage and the effect of three species of cover crop in northern France. *Soil Use and Management* 23(3), 230–237.

Constantin, J., Mary, B., Laurent, F., Aubrion, G., Fontaine, A., Kerveillant, P. and Beaudoin, N. (2010). Effects of catch crops, no till and reduced nitrogen fertilization on nitrogen leaching and balance in three long-term experiments. *Agriculture, Ecosystems & Environment* 135(4), 268–278.

De Notaris, C., Mortensen, E. Ø., Sørensen, P., Olesen, J. E. and Rasmussen, J. (2021). Cover crop mixtures including legumes can self-regulate to optimize N2 fixation while reducing nitrate leaching. *Agriculture, Ecosystems and Environment* 309, 107287.

De Notaris, C., Rasmussen, J., Sørensen, P., Melander, B. and Olesen, J. E. (2019). Manipulating cover crop growth by adjusting sowing time and cereal inter-row spacing to enhance residual nitrogen effects. *Field Crops Research* 234, 15–25.

Fendrich, A. N., Matthews, F., Van Eynde, E., Carozzi, M., Li, Z., d'Andrimont, R., Lugato, E., Martin, P., Ciais, P. and Panagos, P. (2023). From regional to parcel scale: A high-resolution map of cover crops across Europe combining satellite data with statistical surveys. *Science of the Total Environment* 873, 162300.

Fontaine, D., Eriksen, J. and Sørensen, P. (2020). Cover crop and cereal straw management influence the residual nitrogen effect. *European Journal of Agronomy* 118, 126100.

Gabriel, J. L., Alonso-Ayuso, M., García-González, I., Hontoria, C. and Quemada, M. (2016). Nitrogen use efficiency and fertiliser fate in a long-term experiment with winter cover crops. *European Journal of Agronomy* 79, 14–22.

Garcia, L., Celette, F., Gary, C., Ripoche, A., Valdés-Gómez, H. and Metay, A. (2018). Management of service crops for the provision of ecosystem services in vineyards: A review. *Agriculture, Ecosystems & Environment* 251, 158–170.

Gardarin, A., Celette, F., Naudin, C., Piva, G., Valantin-Morison, M., Vrignon-Brenas, S., Verret, V. and Médiène, S. (2022). Intercropping with service crops provides multiple services in temperate arable systems: A review. *Agronomy for Sustainable Development* 42(3), 39.

Germeier, C. U. (2000). Wide row spacing and living mulch: New strategies for producing high protein grains in organic cereal production. *Biological Agriculture & Horticulture* 18(2), 127–139.

Glendining, M. J., Poulton, P. R., Powlson, D. S., Macdonald, A. J. and Jenkinson, D. S. (2001). Availability of the residual nitrogen from a single application of 15N-labelled fertilizer to subsequent crops in a long-term continuous barley experiment. *Plant and Soil* 233(2), 231–239.

Gollner, G., Fohrafellner, J. and Friedel, J. K. (2020). Winter-hardy vs. freeze-killed cover crop mixtures before maize in an organic farming system with reduced soil cultivation. *Organic Agriculture* 10(S1), 5–11.

Gselman, A. and Kramberger, B. (2008). Benefits of winter legume cover crops require early sowing. *Australian Journal of Agricultural Research* 59(12), 1156–1163.

Hansen, E. M. and Djurhuus, J. (1997). Yield and N uptake as affected by soil tillage and catch crop. *Soil and Tillage Research* 42(4), 241–252.

Hansen, E. M., Djurhuus, J. and Kristensen, K. (2000). Nitrate leaching as affected by introduction or discontinuation of cover crop use. *Journal of Environmental Quality* 29(4), 1110–1116.

Harris, G. H., Hesterman, O. B., Paul, E. A., Peters, S. E. and Janke, R. R. (1994). Fate of legume and fertilizer nitrogen-15 in a long-term cropping systems experiment. *Agronomy Journal* 86(5), 910–915.

Hartwig, N. L. and Ammon, H. U. (2002). Cover crops and living mulches. *Weed Science* 50(6), 688–699.

Hauggaard-Nielsen, H., Ambus, P. and Jensen, E. S. (2003). The comparison of nitrogen use and leaching in sole cropped versus intercropped pea and barley. *Nutrient Cycling in Agroecosystems* 65(3), 289–300.

Hauggaard-Nielsen, H., Mundus, S. and Jensen, E. S. (2012). Grass-clover undersowing affects nitrogen dynamics in a grain legume-cereal arable cropping system. *Field Crops Research* 136, 23–31.

Hoffmann, M. P., Isselstein, J., Rötter, R. P. and Kayser, M. (2018). Nitrogen management in crop rotations after the break-up of grassland: Insights from modelling. *Agriculture, Ecosystems and Environment* 259, 28–44.

Jones, L. and Clements, R. O. (1993). Development of a low input system for growing wheat (Triticum vulgare) in a permanent understorey of white clover (Trifolium repens). *Annals of Applied Biology* 123(1), 109–119.

Justes, E., Mary, B. and Nicolardot, B. (1999). Comparing the effectiveness of radish cover crop, oilseed rape volunteers and oilseed rape residues incorporation for reducing nitrate leaching. *Nutrient Cycling in Agroecosystems* 55(3), 207–220.

Känkänen, H. and Eriksson, C. (2007). Effects of undersown crops on soil mineral N and grain yield of spring barley. *European Journal of Agronomy* 27(1), 25–34.

Kathage, J., Smit, B., Janssens, B., Haagsma, W. and Adrados, J. L. (2022). How much is policy driving the adoption of cover crops? Evidence from four EU regions. *Land Use Policy* 116, 106016.

Komainda, M., Taube, F., Kluß, C. and Herrmann, A. (2016). Above- and belowground nitrogen uptake of winter catch crops sown after silage maize as affected by sowing date. *European Journal of Agronomy* 79, 31–42.

Kramberger, B., Gselman, A., Kristl, J., Lešnik, M., Šuštar, V., Muršec, M. and Podvršnik, M. (2014). Winter cover crop: The effects of grass-clover mixture proportion and biomass management on maize and the apparent residual N in the soil. *European Journal of Agronomy* 55, 63–71.

Krueger, E. S., Ochsner, T. E., Porter, P. M. and Baker, J. M. (2011). Winter rye cover crop management influences on soil water, soil nitrate, and corn development. *Agronomy Journal* 103(2), 316–323.

Kuo, S. and Jellum, E. J. (2000). Long-term winter cover cropping effects on corn (Zea mays L.) production and soil nitrogen availability. *Biology and Fertility of Soils* 31(6), 470–477.

Ladha, J. K., Pathak, H., Krupnik, T. J., Six, J. and van Kessel, C. (2005). Efficiency of fertilizer nitrogen in cereal production: Retrospects and prospects. *Advances in Agronomy* 87, 85–156.

Leigh, J., Hodge, A. and Fitter, A. H. (2009). Arbuscular mycorrhizal fungi can transfer substantial amounts of nitrogen to their host plant from organic material. *New Phytologist* 181(1), 199–207.

Lorin, M., Jeuffroy, M.-H., Butier, A. and Valantin-Morison, M. (2015). Undersowing winter oilseed rape with frost-sensitive legume living mulches to improve weed control. *European Journal of Agronomy* 71, 96–105.

Lorin, M., Jeuffroy, M.-H., Butier, A. and Valantin-Morison, M. (2016). Undersowing winter oilseed rape with frost-sensitive legume living mulch: consequences for cash crop nitrogen nutrition. *Field Crops Research* 193, 24–33.

Mauscherning, I. (2008). *Interseeding Catch Crops in Organic Wheat and Rapeseed Production Systems Institut für Pflanzenbau u. Pflanzenzüchtung-Grünland und Futterbau*.

Meisinger, J., Hargrove, W., Mikkelsen, R., Williams, J. and Benson, V. (1991). Effects of cover crops on groundwater quality. *Cover Crops for Clean Water*, 46(3), 57–68.

Miranda-Vélez, J. F. and Vogeler, I. (2022). Autumn tillage reduces the effect of plant cover on topsoil nitrogen leaching. *Nitrogen* 3(2), 186–196.

Mitchell, C. C., Delaney, D. P. and Balkcom, K. S. (2008). A historical summary of Alabama's old rotation (circa 1896): The world's oldest, continuous cotton experiment. *Agronomy Journal* 100(5), 1493–1498.

Möller, K. and Reents, H. J. (2009). Effects of various cover crops after peas on nitrate leaching and nitrogen supply to succeeding winter wheat or potato crops. *Journal of Plant Nutrition and Soil Science* 172(2), 277–287.

Mueller, T. and Thorup-Kristensen, K. (2001). N-fixation of selected green manure plants in an organic crop rotation. *Biological Agriculture & Horticulture* 18(4), 345–363.

Munkholm, L. J., Hansen, E. M., Thomsen, I. K., Wahlström, E. M. and Østergaard, H. S. (2017). Nitrogen uptake, nitrate leaching and root development in winter-grown wheat and fodder radish. *Soil Use and Management* 33(2), 233–242.

Neumann, H. (2005). *Optimierungsstrategien für den Getreideanbau im ökologischen Landbau: System*. Weite Reiheund Direktsaat in ausdauernden Weißklee („Bi-cropping "). Schriftenreihe des Instituts für Pflanzenbau und Pflanzenzüchtung der Christian-Albrechts Universität zu Kiel, Heft 43.

Nicolardot, B., Recous, S. and Mary, B. (2001). Simulation of C and N mineralisation during crop residue decomposition: A simple dynamic model based on the C:N ratio of the residues. *Plant and Soil* 228(1), 83–103.

Norberg, L. and Aronsson, H. (2020). Effects of cover crops sown in autumn on N and P leaching. *Soil Use and Management* 36(2), 200–211.

Ohlander, L., Bergkvist, G., Stendahl, F. and Kvist, M. (1996). Yield of catch crops and spring barley as affected by time of undersowing. *Acta Agriculturae Scandinavica B-Plant Soil Sciences* 46(3), 161–168.

Pieters, A. J. (1927). *Green Manuring*. Willey: New York.

Pullens, J. W. M., Sørensen, P., Melander, B. and Olesen, J. E. (2021). Legacy effects of soil fertility management on cereal dry matter and nitrogen grain yield of organic arable cropping systems. *European Journal of Agronomy* 122, 129169.

Raseduzzaman, M. and Jensen, E. S. (2017). Does intercropping enhance yield stability in arable crop production? A meta-analysis. *European Journal of Agronomy* 91, 25–33.

Sapkota, T. B., Askegaard, M., Lægdsmand, M. and Olesen, J. E. (2012). Effects of catch crop type and root depth on nitrogen leaching and yield of spring barley. *Field Crops Research* 125, 129–138.

Sarunaite, L., Kadziuliene, Z., Deveikyte, I. and Kadziulis, L. (2013). Effect of legume biological nitrogen on cereals grain yield and soil nitrogen budget in double-cropping system. *Journal of Food, Agriculture and Environment* 11, 528–533.

Schipanski, M. E., Barbercheck, M., Douglas, M. R., Finney, D. M., Haider, K., Kaye, J. P., Kemanian, A. R., Mortensen, D. A., Ryan, M. R., Tooker, J. and White, C. (2014). A framework for evaluating ecosystem services provided by cover crops in agroecosystems. *Agricultural Systems* 125, 12–22.

Sieling, K. (2019). Improved N transfer by growing catch crops - A challenge. *Journal fur Kulturpflanzen* 71, 145–160.

Sullivan, D. M. and Andrews, N. (2012). *Estimating Plant-Available Nitrogen Release from Cover Crops.* A Pacific Northwest Extension Publication Oregon State University – Washington State University – University of Idaho.

Thapa, R., Mirsky, S. B. and Tully, K. L. (2018). Cover crops reduce nitrate leaching in agroecosystems: A global meta-analysis. *Journal of Environmental Quality* 47(6), 1400–1411.

Thomsen, I. K. (2005). Nitrate leaching under spring barley is influenced by the presence of a ryegrass catch crop: Results from a lysimeter experiment. *Agriculture, Ecosystems and Environment* 111(1-4), 21–29.

Thomsen, I. K. and Christensen, B. T. (1999). Nitrogen conserving potential of successive ryegrass catch crops in continuous spring barley. *Soil Use and Management* 15(3), 195–200.

Thomsen, I. K., Elsgaard, L., Olesen, J. E. and Christensen, B. T. (2016). Nitrogen release from differently aged Raphanus sativus L. nitrate catch crops during mineralization at autumn temperatures. *Soil Use and Management* 32(2), 183–191.

Thorsted, M. D., Olesen, J. E. and Koefoed, N. (2002). Effects of white clover cultivars on biomass and yield in oat/clover intercrops. *Journal of Agricultural Science* 138(3), 261–267.

Thorup-Kristensen, K., Magid, J. and Jensen, L. S. (2001). Catch crops and green manures as biological tools in nitrogen management in temperate zones. In *Advances in Agronomy*, Vol. 79, pp. 227–302.

Thorup-Kristensen, K., Magid, J. and Jensen, L. S. (2003). Catch Cropsand green manures as biological tools in nitrogen management in temperate zones. *Advances in Agronomy* 79, 227–302. Academic Press: UK .

Thorup-Kristensen, K. and Nielsen, N. E. (1998). Modelling and measuring the effect of nitrogen catch crops on the nitrogen supply for succeeding crops. *Plant and Soil* 203(1), 79–89.

Tonitto, C., David, M. B. and Drinkwater, L. E. (2006). Replacing bare fallows with cover crops in fertilizer-intensive cropping systems: A meta-analysis of crop yield and N dynamics. *Agriculture, Ecosystems and Environment* 112(1), 58–72.

Tribouillois, H., Cohan, J. P. and Justes, E. (2016). Cover crop mixtures including legume produce ecosystem services of nitrate capture and green manuring: Assessment combining experimentation and modelling. *Plant and Soil* 401(1-2), 347–364.

Vogeler, I., Cichota, R., Thomsen, I. K., Bruun, S., Jensen, L. S. and Pullens, J. W. M. (2019a). Estimating nitrogen release from Brassicacatch crop residues—comparison of different approaches within the APSIM model. *Soil and Tillage Research* 195, 104358.

Vogeler, I., Hansen, E. M., Thomsen, I. K. and Østergaard, H. S. (2019b). Legumes in catch crop mixtures: Effects on nitrogen retention and availability, and leaching losses. *Journal of Environmental Management* 239, 324–332.

Vogeler, I., Hansen, E. M. and Thomsen, I. K. (2023). The effect of catch crops in spring barley on nitrate leaching and their fertilizer replacement value. *Agriculture, Ecosystems & Environment* 343, 108282.

Vogeler, I., Jensen, J. L., Thomsen, I. K., Labouriau, R. and Hansen, E. M. (2021a). Fertiliser N rates interact with sowing time and catch crops in cereals and affect yield and nitrate leaching. *European Journal of Agronomy* 124, 126244.

Vogeler, I., Thomsen, I. K., Taube, F., Poulsen, H. V., Loges, R. and Hansen, E. M. (2021b). Effect of winter cereal sowing time on yield and nitrogen leaching based on experiments and modelling. *Soil Use and Management* 38(1), 663–675.

Vos, J. and Van Der Putten, P. E. L. (2004). Nutrient cycling in a cropping system with potato, spring wheat, sugar beet, oats and nitrogen catch crops. II. Effect of catch crops on nitrate leaching in autumn and winter. *Nutrient Cycling in Agroecosystems* 70(1), 23–31.

Wachendorf, M., Büchter, M., Volkers, K. C., Bobe, J., Rave, G., Loges, R. and Taube, F. (2006). Performance and environmental effects of forage production on sandy soils. V. Impact of grass understorey, slurry application and mineral N fertilizer on nitrate leaching under maize for silage. *Grass and Forage Science* 61(3), 243-252.

Wallander, S., Smith, D., Bowman, M. and Claassen, R. (2021). *Cover Crop Trends, Programs, and Practices in the United States*. EIB 222, U.S. Department of Agriculture, Economic Research Service, February 2021.

Wallgren, B. and Lindén, B. (1994). Effects of catch crops and ploughing times on soil mineral nitrogen. *Swedish Journal of Agricultural Research (Sweden)*. 24(2), 67-75

Williams, E. D. and Hayes, M. J. (1991). Growing spring cereals in a white clover (Trifolium repens) crop. *The Journal of Agricultural Science* 117(1), 23–37.

Wittwer, R. A., Dorn, B., Jossi, W. and van der Heijden, M. G. (2017). Cover crops support ecological intensification of arable cropping systems. *Scientific Reports* 7, 1–12.

Yu, Y., Stomph, T. J., Makowski, D., Zhang, L. and van der Werf, W. (2016). A meta-analysis of relative crop yields in cereal/legume mixtures suggests options for management. *Field Crops Research* 198, 269–279.

Zhao, J., Pullens, J. W. M., Sørensen, P., Blicher-Mathiesen, G., Olesen, J. E. and Børgesen, C. D. (2022). Agronomic and environmental factors influencing the marginal increase in nitrate leaching by adding extra mineral nitrogen fertilizer. *Agriculture, Ecosystems & Environment* 327, 107808.

Chapter 14

The role of crop rotations in optimizing nitrogen use efficiency in organic farming

Lucie Chmelíková[1] and Kurt-Jürgen Hülsbergen[1], Technical University of Munich, Germany; and Sebastian Wolfrum[1,2], Technical University of Munich, Germany, and Bavarian State Research Center for Agriculture, Germany

1 Introduction

The input of nitrogen (N) into the soil is a key factor for reliable yields and high productivity in modern agricultural systems. No synthetic N fertilizer is used in organic farming, making biological N_2 fixation, organic fertilization and nitrogen transfer within the crop rotation crucial for yield performance. High yield represents high N output from the system, as the N taken up from the soil by the plants stays only partially in the field in the form of straw, roots or other crop residues and is mostly taken away from the system in the form of harvested products. Balancing N inputs and outputs is a key challenge for farmers. This can reduce yield stability and affect the long-term sustainability of food production

[1] Chair of Organic Agriculture and Agronomy, TUM School of Life Sciences Weihenstephan,Technical University of Munich, Liesel-Beckmann-Str. 2, 85354 Freising, Germany.
[2] Bavarian State Research Center for Agriculture (LfL), Lange Point 6, 85354 Freising, Germany.

http://dx.doi.org/10.19103/AS.2024.0135.18

(Macholdt et al., 2019; Powlson et al., 2011; Tan et al., 2005). If the balance is too high, this can cause severe environmental issues due to loss of reactive N compounds, either by leaching into the groundwater or gaseous emissions into the air (Galloway et al., 2008; van Beek et al., 2003; van Groenigen et al., 2010). The N balance, and especially nitrogen use efficiency (NUE), which is the ratio of N output (crop yield) and N input (e.g. fertilizer N, symbiotically fixed N), are crucial indicators for optimizing the agronomic, economic and environmental performance of cropping systems (Fageria, 2014; Ladha et al., 2020).

Crop rotations are an important factor influencing NUE in plant production. They influence the N cycle in many ways and can control negative environmental effects and the contribution of agricultural systems to climate change (Barbieri et al., 2017). The composition of crop rotations influences N losses in the form of ammonia, nitrous oxide emissions and nitrate losses to groundwater (Behnke et al., 2018; Benoit et al., 2014; Holka et al., 2022; Jensen et al., 2012; Morari et al., 2012). Diversified crop rotations therefore have positive effects on soil organic matter and on farm nutrient cycles (Drinkwater et al., 1998), water balances, weed, pest and disease infestations (Kocira et al., 2020; Ratnadass et al., 2012) and erosion risks (Auerswald et al., 2003), thus improving the resilience of the overall system and soil fertility (Liu et al., 2022; Watson et al., 2002). Crop rotation can also increase NUE through the optimal integration of grain and forage legumes, the reduction of N losses through undersowing and catching crops, and the use of previous crop effects etc. (Ladha et al., 2022).

Agricultural intensification has both simplified the composition and reduced the number of crops in the rotation. The decrease in crop numbers and the shortening of crop rotations is associated with negative influences on the soil, such as a reduction in soil organic matter, erosion, decrease in soil fertility, loss of biodiversity and increased plant diseases. The proportion of different crop types has changed since 1950. Meyer et al. (2013) reported a decrease in spring cereals from 26% to 17% and an increase in winter cereals from 41% to 61% in ten study regions in Germany. In terms of the proportion of crop species, winter wheat (from 14% to 31%), winter rape (from 0% to 17%) and maize (from 1% to 9%) increased. Species such as potatoes, oat and beets decreased in their proportions (from 15% to 1%, 8% to 2% and 5% to 3%, respectively). Furthermore, the proportion of fodder legume crops decreased. Even in organic arable farming systems, a simplification of crop rotations and a decrease in legume proportions can be seen. On the other hand, there exist wide-ranging and versatile organized farming systems with low N input (Chmelíková et al., 2021).

Batool et al. (2022) modeled the annual long-term N balance for agricultural and non-agricultural soils across Europe for the period from 1850 to 2019. The highest values were calculated for Western and Central Europe. N surplus showed large temporal changes from 20 kg ha^{-1} $year^{-1}$ in 1930 to

Published by Burleigh Dodds Science Publishing Limited, 2024.

values of 60 kg ha^{-1} $year^{-1}$ in 1990. The four-fold improvement in global cereal production since 1961 was accompanied by a nine-fold increase in the use of fertilizer N (Ladha et al., 2022). While in some regions of the world, such as sub-Saharan Africa, inputs of N are still considerably less than required to achieve food security, countries such as China and India have large surpluses of N fertilizer (Ladha et al., 2020). Less than 50% of the applied N fertilizer is generally utilized by crops (Fageria and Baligar, 2005), increasing the risk of environmental pollution due to suboptimal application of fertilizer (Ladha et al., 2020). Discussions on yield and crop production, synthetic fertilizers and their long-term impact on soil organic matter (N and C) tend to be controversial (David et al., 2010; Mulvaney et al., 2009, 2010a,b; Powlson et al., 2010a,b). Nevertheless, most authors agree that long-term sustainability may require agricultural diversification, involving a gradual transition from intensive synthetic N inputs to more legume-based crop rotations to support food production, improve environmental outcomes and maintain ecosystem services (Jensen et al., 2012; Ladha et al., 2022).

2 Crop rotations and nitrogen flows in organic farming

Organic and conventional farming systems use N in different ways and from different sources. Organic farming is highly dependent on closed nutrient cycles, biological N_2 fixation and adapted crop rotations. This results in diverse crop rotations, intercropping schemes and the use of legumes as green manure (Bedoussac et al., 2015; Ladha et al., 2022). Conventional farming depends widely on mineral N fertilizers. The N fluxes of these systems are therefore different, with a noticeably increased level of complexity in organic farming with its large variety of fertilization options and their interactions in both time and space (Valenzuela, 2023). Furthermore, N flows differ not only between organic and conventional systems but also within organic farm types, e.g. between arable systems and dairy systems (Chmelíková et al., 2021).

Organic agriculture is known for promoting crop diversification and utilizing longer and more diversified crop rotations. At the global scale, organic rotations on average last for 4.5 ± 1.7 years (0.7 years longer than in conventional farming) and include 48% more crop types, indicating higher crop diversity over space and time (Barbieri et al., 2017). In their worldwide study, Barbieri et al. (2017) found that in organic crop rotations, the frequency of cereal intercropping with legume crops, fodder crops and catch crops as well as undersown cover crops is higher than in conventional systems (4.3 times, 2.8 times, 3.2 times and 12.1 times, respectively). Furthermore, the authors suggest that in organic rotations, leguminous grain pulses, fodder crops, catch and undersown cover crops supply 2.6 times more N to soils than in conventional rotations. Legumes play a pivotal role for the N input of crop rotations. N fixing crops can be introduced in crop rotation as cash crops, cover and catch crops, green manures and in association with other crops (Bedoussac et al., 2015; Liebman et al., 2012; Posner et al., 2008; Thorup-Kristensen et al., 2003). However, the net contributions of fixed N_2 by grain legumes to the overall N balance may be low due to the removal of large amounts of N harvested in the high-protein seed (Ladha et al., 2022), emphasizing the importance of non-food legumes in organic rotations to increase soil N supply (Barbieri et al., 2023).

An important indicator for evaluating the sustainability of the N cycle (an example of the N cycle of an organic dairy farm is given in Fig. 1) is the NUE. In addition to the analysis of the mean N fluxes of the farm presented here, crop-specific N fluxes must be taken into account. Different crops require different fertilizer amounts and have different nutrient uptakes.

Crops differ in terms of harvested biomass and leave different amounts of N in the soil after harvest. The utilization of organic sources of N such as senesced plant or green manure by the following crops can also differ greatly (Ladha et al., 2022). Combining crops spatially and temporally can promote efficient nutrient use (Wilkins, 2008). For this reason, it is important to take into account crop rotation, not just individual crops.

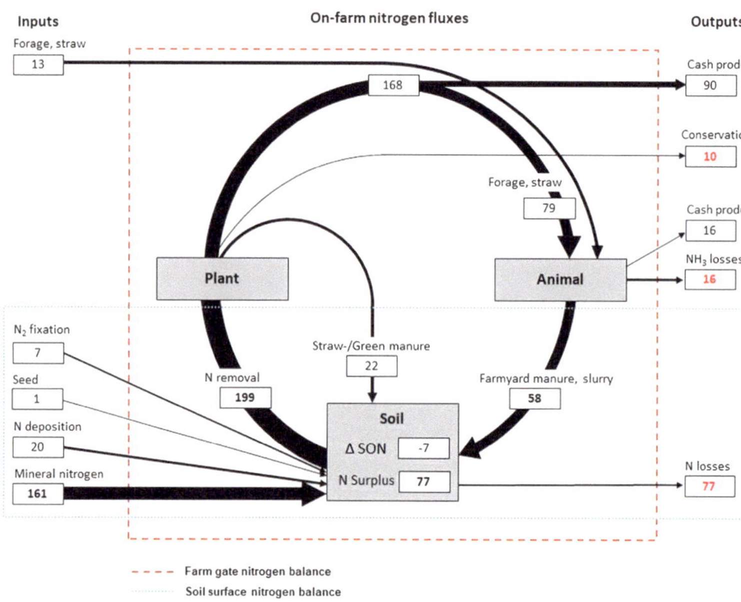

Figure 1 Example of a N cycle of an organic dairy farm (Chmelíková et al., 2021). The blue dotted box represents the N balance of the crop production system. N losses from soil are due to ammonia volatilization, denitrification and leaching (not specified). Changes in the stocks of feed and organic fertilizers (liquid manure, manure) are not shown. All values are in kg N ha^{-1} a^{-1}.

Published by Burleigh Dodds Science Publishing Limited, 2024.

Often, the crop rotation of different farm types and farm systems are compared for their NUE (Chmelíková et al., 2021; Kubota et al., 2018; Torstensson et al., 2006). Earlier farm investigations have shown that organic farms are structured very differently depending on site conditions and production direction. There are also large differences and a high individual variability of important crop rotation parameters within organic farming, such as crop species diversity, proportion of legumes etc. In organic farming, there is an increasing tendency toward specialization and simplified crop rotation, which raises the question of whether these systems are supplied with enough N to allow high yields, or whether N becomes a yield-limiting factor.

The influence of crop rotation on essential parameters of the N balance is examined in the following section. Based on the results, conclusions are drawn for the optimal design of crop rotations with regard to N-efficiency. A meta-analysis approach was used to analyze the effect of different crop rotation parameters on NUE. First, a comprehensive dataset of crop parameters from studies on NUE was created. The influences of the studied parameter on N output and N surplus, and afterward on NUE, were evaluated.

3 Assessing nitrogen flows and nitrogen use efficiency in organic farming: case study

Between 2017 and 2019, a comprehensive literature analysis comparing organic and conventional farming was conducted in the 'Impact of organic farming on environment and society' project (FKZ 2815OE009 / 2815OE034). The evaluation of scientific studies on public goods provided by organic farming was carried out on water protection, soil fertility, biodiversity, climate mitigation, climate adaptation, resource efficiency and animal welfare.

From 2020 to 2022, the joint project 'Remuneration for the environmental benefits of organic farming' (FKZ: 2819OE123) aimed to summarize and analyze the research and knowledge on the benefits of organic agriculture in areas such as water protection, soil fertility, climate protection and adaption, biodiversity and resource use efficiency. As part of a wider literature review, NUE in crop production under organic farming regulations was studied, taking into account N input and output as well as the N balance of the whole crop rotation and for wheat only.

3.1 Nitrogen cycle parameters

N fluxes were evaluated for the crop subsystem (soil surface balance). For more details on the soil surface balance, see Fig. 1. The soil surface balance was chosen to ensure comparability of the N cycle parameters (N inputs, N outputs, NUE and N balance). Equivalent input and output parameters enable the comparable analysis of N balance and NUE. In dairy farming, N fluxes of the livestock holdings were investigated to calculate the amount and N content of organic fertilizers (liquid manure, manure) which represent an input in crop production. Other N inputs and N outputs of animal husbandry (purchase of feed, animal products sold) are not necessary in the N balance of crop production. The N input included N fluxes from symbiotic N_2 fixation and N organic fertilizers and N atmospheric deposition. N output considered only the N contained in the harvested plant products. The N output corresponded to the N removal per hectare (dry matter yield × N content of the harvested products). N output is thus strongly related to the harvested product and a good indicator of yield. NUE was defined as N output in relation to N input (Halberg et al., 1995). Generally, efficiency is a relative measure and can be increased either by producing more output with the same input or by maintaining constant output with less input. In both ways, the farmer's profit will likely increase. NUE is thus a good indicator of profit. The difference between N inputs and N outputs is defined as the N balance (Halberg et al., 2005). Very low or negative balances indicate problems with long-term soil fertility whilst too high balances indicate an increased risk of negative environmental impacts like N loss to air or groundwater. N balance is therefore a good indicator of the sustainability of cropping systems. The three parameters were correlated to a certain degree but demonstrated distinct aspects relevant for targeted development of cropping systems and the potential adoption of these techniques. We therefore report results for all three parameters.

3.2 Literature review

A literature review utilizing the electronic database Web of Science was conducted with the following search string: TS= (organic* NEAR/1 (farm* OR agriculture* OR manag* OR practice* OR hortic* OR crop*)) OR (biologic* NEAR/1 (farm* OR agriculture* OR manag* OR practice* OR hortic* OR crop*)) OR (ecologic* NEAR/1 (farm* OR agriculture* OR manag* OR practice* OR hortic* OR crop*)) (nitrogen OR N OR energy) AND (efficienc* OR use-efficienc* OR balance* OR input* OR output* OR surplus* OR NUE OR EUE). This search was supplemented with a similar search in the organic eprints database (https://orgprints.org), which also stores project reports, conference papers and other unpublished documents, to reduce the impact of potential publication bias.

This combination of key terms resulted in 3107 papers. To reduce the number of papers included in the meta-analysis, the following criteria were applied: (1) publication period: 1990 to 2020; (2) region: temperate climates; (3) study design: at least one organic system; and (4) language: German or English. After reviewing all abstracts, 175 studies were selected and after additionally reviewing the data, a total of 91 crop rotation studies covering N parameters were entered into a database. Furthermore, a total of 47 studies considering wheat only were included in the database. Data from both conventional and organic crop production systems were recorded.

The selection of papers for this study focused on organic systems in the temperate climate zone because excessive nitrogen balances cause environmental problems and the need to improve NUE is a major concern, especially in temperate cropping systems. The initial goal of the project that this study was derived from was to develop a system of incentivizing organic farmers in Germany to foster ecosystem services and was targeted at an area in the temperate region.

The relevant variables suitable for analysis were selected from the database. All variables that were incomplete or for which no differentiation in the data could be identified were excluded. Furthermore, the exponential Shannon diversity index and the inverse Simpson diversity index (Jost, 2006) were calculated from the proportions of the crop types. Likewise, an aggregated variable for the soil texture was generated, distinguishing only sand and other soil types (silt and clay). In the next step, the data set was filtered so that only entries for organic farming remained in the table. In the last step, all incomplete data records (variables with NAs) were removed.

3.3 Preselection of explanatory variables

All data manipulations were conducted in R version 4.0.1 (R Core Team, 2020). In our basic dataset, we had three target variables and 34 explaining variables, of which some were highly correlated. In order to create stable and meaningful statistical models, it was necessary to avoid strongly correlated variables or those with a high variance inflation factor (VIF) and thus multicollinearity in the data set. To check the correlation of individual variables, all factor variables were first converted to numerical values, after which the Spearman rank correlations were calculated. Some of the variables were highly correlated and therefore removed from the data table. Furthermore, variables that were not relevant for further analysis or were already covered by other variables were removed. To check the correlation of the remaining variables, a correlation matrix was calculated using the Spearman rank correlations. Checking the VIF showed that none of the variables had a VIF above ten and the maximum value of 5.61 was found for the exponential Shannon Index of the crop types. However, N input and N output still showed a correlation of $r \geq 0.7$. Nevertheless, problems due to high correlations and thus multicollinearity for the final variable set of the models were not expected.

A new table was created from the raw data collected by the previous steps in which factors were consolidated, and the individual factor levels were renamed or adjusted. A far more complete dataset was therefore obtained for the analysis. Finally, an outlier check was carried out using a random forest

Table 1 Overview of source data (country, farming/cropping systems and their number)

Country	Number of studies	Farm type	Occurrence in studies
Denmark	7	Arable farm	4
		Mixed/dairy farm	5
Estonia	1	Arable farm	1
France	3	Arable farm	2
		Mixed/dairy farm	1
Germany	5	Arable farm	2
		Mixed/dairy farm	4
Italy	4	Arable farm	4
		Mixed/dairy farm	1
New Zealand	1	Mixed/dairy farm	1
Norway	2	Arable farm	2
		Mixed/dairy farm	2
Spain	1	Arable farm	1
		Mixed/dairy farm	1
Sweden	4	Arable farm	4
		Mixed/dairy farm	1
Switzerland	1	Mixed/dairy farm	1
USA	3	Arable farm	1
		Mixed/dairy farm	3
United Kingdom	2	Arable farm	2
Total	34		43

algorithm (Mayer et al., 2022) to identify and exclude remaining critical data points.

Following data preparation and cleaning, a data table with 115 data sets was available for the models and further analysis. The dataset contained three target values (N output, N balance and NUE) and 14 independent variables (Study ID, N input, farm type, soil texture category, tillage intensity, fertilizer category, duration of rotation, crop diversity (exp. Shannon Index), forage legumes (yes/no), cover/catch crops (yes/no), proportion of cereals in the crop rotation, proportion of root crops in the crop rotation, proportion of maize in the crop rotation and proportion of legumes in the crop rotation). The final data consisted of the results from 32 studies across 12 countries. Details of the data sources are shown in Table 1.

3.4 Statistical analysis

All statistical analysis was conducted in R version 4.0.1 (R Core Team, 2020). Log-linear models were used to model the N output and NUE. Due to potential negative values, only a linear model could be used for the N balance. First, variables were selected and then the results for the final models were analyzed. A stepwise approach based on the AIC value was used for variable selection. For this, first the minimal model, which contains only the ID of the studies, and then the full model with all variables was defined. The best model based on the AIC values was then determined using the base R step function with both directions. Model quality and validity was checked graphically. For the selected models, the significance of the individual variables was determined using an ANOVA. The value $P \leq 0.05$ (5%) was chosen as the significance level. A pairwise Tukey post hoc test was carried out for each of the factors from the ANOVA, correcting for multiple comparisons. A correction based on the Sidak method was used to calculate confidence intervals. For all effects, marginal means were calculated and visualized.

4 Assessing nitrogen flows and nitrogen use efficiency in organic farming: results

4.1 Nitrogen output

According to the final model, 85% of the variability of the N output can be explained by seven influencing variables (R2 = 0.85; adj. R2 = 0.77; F (38; 76) = 11.3; $P < 0.001$). Based on the results, the seven most important factors influencing N output were included in the following order - N input, fertilizer type, crop diversity, proportion of cereals in crop rotation, catch crops and tillage (Table 2). Minimum, maximum, mean values, median and standard deviation of all parameters are summarized in Table 3.

Table 2 ANOVA results for N output

Model: lm (log (N output)) ~ ID + N input + cover/catch crops + fertilizer category + crop diversity + proportion of cereals in the rotation + tillage intensity

	Df	Sum Sq	Mean Sq	*F*-value	*P*	ges
ID	31	10.90	0.35	4.34	< 0.001	0.639
N input	1	3.20	3.30	39.48	< 0.001	0.342
Cover/catch crops	1	0.37	0.37	4.50	0.037	0.056
Fertilizer category	2	0.95	0.47	5.84	0.004	0.133
Crop diversity	1	0.53	0.53	6.52	0.013	0.079
Proportion of cereals in the rotation	1	0.48	0.48	5.91	0.017	0.072
Tillage intensity	1	0.13	0.13	4.34	0.216	0.020
Residuals	76	6.16	0.08			

ID = study ID, df = degrees of freedom, sum sq = sum of squares, mean sq = mean sum of squares, *F*-value = *F* statistic, *P* = *P*-value Pr (>F), ges = effect size.

Table 3 Summary of statistics parameters of N input, N output, N balance, NUE, duration of rotation and proportion of crop groups in the crop rotation

	Minimum	Median	Mean value	Maximum	Standard deviation
N input	23.50	124.00	135.20	356.80	65.00
N output	24.00	80.25	98.17	262.31	57.29
N balance	−108.00	33.00	34.05	136.75	44.48
NUE	16.81	72.57	75.95	170.21	30.69
Duration of rotation	2.00	4.00	4.94	9.00	1.53
Exp. Shannon crop diversity	1.76	2.83	3.04	5.00	0.77
Proportion of cereals	0.00	0.50	0.45	0.75	0.13
Proportion of root crops	0.00	0.13	0.11	0,.25	0,.11
Proportion of maize	0.00	0.00	0.03	0.33	0.07
Proportion of legumes	0.00	0.25	0.26	0.64	0.13

N input, crop diversity and proportion of cereals in the crop rotation were continuous variables. Based on the results, a change in N input of 1 kg ha^{-1} caused a 0.5% increase in N output (Fig. 2a). An increase in crop species diversity by one effective species in the exponential Shannon Index resulted in a 20.1% reduction in N output. A 1% increase in the proportion of cereals in the rotation also resulted in a 1% decrease in N output.

Cover crops and fertilizer type were categorical variables with significant influence on N output. Significantly, higher N output was found in crop rotations without cover crops (92 kg ha^{-1}) compared to crop rotations with cover crops (76 kg ha^{-1}). Significant differences were also found between fertilizer type categories (organo-mineral fertilization - 109 kg ha^{-1}, organic fertilization - 84 kg ha^{-1} and green manure - 66 kg ha^{-1}).

N output correlated positively with NUE ($r(113) = 0.45$, $P < 0.001$) but not with N surplus ($r(113) = -0.12$, $P = 0.206$).

4.2 Nitrogen balance

When modeling N balance, 83% of the variability could be explained by eight explanatory variables (R2 = 0.83; adj. R2 = 0.75; F (39; 75) = 9.7; $P < 0.001$). Table 4 shows the parameters that had the greatest influence on the N balance - N input, duration of rotation, N output, cover crops, tillage, share of legumes in the crop rotation and type of fertilizer.

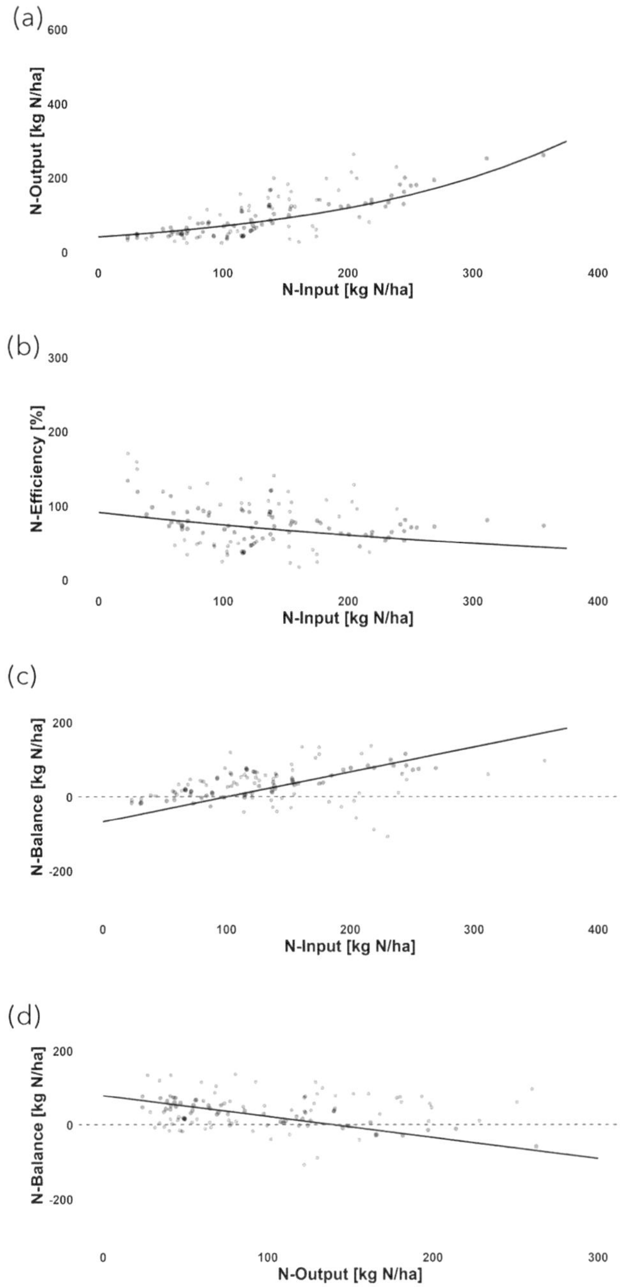

Figure 2 Overall relationship between (a) N input and N output, (b) N input and N efficiency, (c) N input and N balance, (d) N output and N balance.

Table 4 ANOVA results for N balance

Model: lm (N balance) ~ ID + N input + N output + duration of rotation + tillage intensity + cover/catch crops + proportion of legumes in the rotation + fertilizer category

	Df	Sum Sq	Mean Sq	*F*-value	*P*	ges
ID	31	45,684.20	1,473.68	2.95	<0.001	0.550
N input	1	27,296.93	27,296.93	54.69	<0.001	0.422
N output	1	15,827.33	15,827.33	31.71	<0.001	0.297
Duration of rotation	1	17,625.86	17,625.86	35.31	<0.001	0.320
Tillage Intensity	1	3,708.30	3,708.30	7.43	0.008	0.090
Cover/catch crops	1	4,150.43	4,150.43	8.32	0.005	0.100
Proportion of legumes in the rotation	1	3,044.08	3,044.08	6.10	0.016	0.075
Fertilizer category	2	3,025.09	1,512.55	3.03	0.054	0.075
Residuals	75	37,437.50	499.17			

ID = study ID, df = degrees of freedom, sum sq = sum of squares, mean sq = mean sum of squares, *F*-value = *F* statistic, *P* = *P*-value Pr (>*F*), ges = effect size.

N input, duration of rotation, N output and proportion of legumes in the crop rotation were evaluated as significant continuous influencing variables. A change in the N input of 1 kg ha^{-1} caused a 0.7% increase in the N balance (Fig. 2b) and a corresponding change in the N output caused a reduction in the N balance of 0.6% (Fig. 2c). Increasing the duration of the rotation by one year caused a 19% reduction in the N balance (Fig. 3a). An increase in the share of legumes by 1% also led to an increase in the N balance by 0.6%.

For N balance tillage, cover crops and fertilizer type were included in the model as categorical influencing variables. A significantly higher N balance was found in crop rotations with conventional tillage (45 kg ha^{-1}). In crop rotations with reduced tillage, the N balance amounted to 3 kg ha^{-1}. Cover crops had a significant influence on the N balance as well. In crop rotations with cover crops, the N balance was higher (34 kg ha^{-1}) than in crop rotations without cover crops (14 kg ha^{-1}). Green manure caused a significantly lower N balance (12 kg ha^{-1}) compared to other fertilization (27–32 kg ha^{-1}). N surplus correlated negatively with NUE ($r(113) = -0.78$, $P < 0.001$).

4.3 Nitrogen use efficiency

Finally, the model for NUE showed that 68% of the variability can be explained by ten explanatory variables (R2 = 0.68; adj. R2 = 0.49; F (41; 73) = 3 .72; $P <$ 0.001). The type of fertilizer had the greatest influence on NUE, followed by

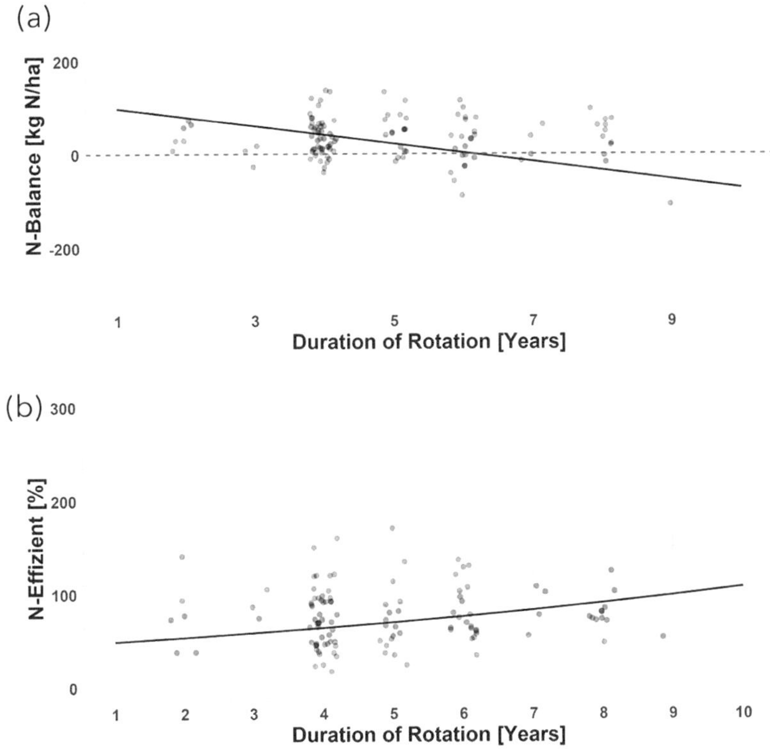

Figure 3 Overall relationship between (a) duration of rotation and N balance, (b) duration of rotation and N efficiency.

crop diversity, N input and the proportion of cereals in the crop rotation as well as the duration of the crop rotation (Table 5).

NUE was significantly influenced by the continuous variables of N input, proportion of row crops in the crop rotation and crop diversity. A 1 kg ha^{-1} change in N input resulted in a 0.2% reduction in N efficiency (Fig. 2d). A 1% increase in row crop percentage caused a 2.1% increase in NUE. An increase in crop diversity by one effective species in the exponential Shannon Index resulted in a 25.5% reduction in NUE. A 1% increase in the proportion of cereals in crop rotation also resulted in a 1% decrease in NUE. Increasing the duration of the crop rotation by one year increased NUE by 9.1% (Fig. 3b). For the categorical influencing variables, a significant effect on NUE was found for the fertilizer type. NUE was about 54% when only green manure was used for fertilization. For all other fertilizer types, NUE was in the range of 65–95%.

Table 5 ANOVA results for nitrogen use efficiency (NUE)

Model: lm (log (NUE)) ~ ID + fertilizer category + N input + cover/catch crops + proportion of row crops in the rotation + crop diversity + proportion of cereals in the rotation + duration of rotation + tillage intensity + soil texture dategory

	Df	Sum Sq	Mean Sq	*F*-value	*P*	ges
ID	31	11.88	0.38	3.98	<0.001	0.628
Fertilizer Category	2	0.85	0.42	4.40	0.016	0.108
N input	1	0.42	0.42	4.35	0.040	0.056
Cover/catch crops	1	0.19	0.19	2.01	0.161	0.027
Proportion of row crops in the rotation	1	0.38	0.38	3.97	0.050	0.052
Crop diversity	1	0.71	0.71	7.36	0.008	0.092
Proportion of cereals in the rotation	1	0.37	0.37	3.84	0.054	0.050
Duration of rotation	1	0.28	0.28	2.89	0.094	0.038
Tillage intensity	1	0.14	0.14	1.43	0.236	0.019
Soil texture category	1	0.13	0.13	1.32	0.255	0.018
Residuals	73	7.02	0.10			

ID = study ID, Df = degrees of freedom, sum sq = sum of squares, mean sq = mean sum of squares, *F*-value = *F* statistic, *P* = *P*-value Pr (>F), ges = effect size.

5 Improving the use of rotations to optimize nitrogen flows and nitrogen use efficiency

Summarizing and analyzing diverse studies in a meta-analysis is a great opportunity to study and evaluate the effect of different aspects of crop rotations on NUE. Most studies on N efficiency focus on the complete crop rotation when comparing organic and conventional or arable and dairy systems (Alaru et al., 2014; Aronsson et al., 2007; Benincasa et al., 2016; Torstensson et al., 2006). The results of these studies suggest that the effects are mainly due to differences in fertilizer management.

Furthermore, many studies focus on one aspect of crop rotation, such as the presence of legumes (Anglade et al., 2015; Stenberg et al., 2012; Swain et al., 2014) or cover crops (Abdalla et al., 2019; Basche et al., 2014; Sun and Shahrajabian, 2022) only. However, the design of field experiments or the type of management does not always accurately represent real farming system conditions (Mäder et al., 2002; Mayer et al., 2015; Stark et al., 2006). Therefore, in addition to conducting field experiments, studies of real farms, taking into account all production practices, is required (Dal Ferro et al., 2017). Even then, data from real farms is often highly variable and difficult to evaluate. A large number of studies describing aspects of whole rotations were used for this analysis. In the context of NUE, the relevant parameters of crop rotations identified were the duration of crop rotation, crop diversity (exp. Shannon Index), fodder legumes (yes/no), cover crops, proportion of cereals in rotation, proportion of row crops in rotation, proportion of cereals and maize and proportion of legumes in rotation. However, even with an extensive database comprising several studies, certain methodological problems complicate the analysis, including incomplete and inconsistent data on fertilization systems and crop rotations, which may not accurately reflect how farmers manage their farms. For example, our focus on organic farming led us to summarize fertilization with mineral fertilizers only and mixed fertilization as very few studies exist using mineral fertilizers. It also proved difficult to define the exact fertilization type and source as often only total amounts of N were reported. The results of such a meta-analysis approach should therefore be interpreted carefully.

A longer crop rotation duration can decrease the N balance. This is in agreement with the findings of Gaudino et al. (2014) who evaluated 22 indicators of inputs and environmental pressures in agricultural cropping systems. A greater number of different crop species showed a higher degree of sustainability. Furthermore, extended crop rotations were more agronomically sustainable than short-term rotations. The positive effect of a five-year rotation in comparison with a two-year rotation on maize yield (N output) was recorded by Stanger and Lauer (2008). The extended rotations involving forage crops (e.g. alfalfa) reduced N inputs and increased maize grain yields. Generally, longer crop rotations comprised of perennial forages and annual crops are used to reduce weed density in organic systems and increase crop yield and N supply (Anderson, 2010). Furthermore, diversified crop rotations improved yield stability (Sindelar et al., 2016).

Legumes are an important crop group in organic farming systems, whose cropping sequences are generally dominated by cereals. The N balance was significantly affected by the proportion of legumes in the crop rotation but not the NUE. An increase in the proportion of legumes caused an increase in the N balance. According to Anglade et al. (2015), forage legumes accounted for around 70% of total N inputs in organic farming systems. Nevertheless, in our analysis, legumes as fodder crops had no effect on N balance and NUE. In our study, forage legumes were not further differentiated (annual, perennial intercropped with grasses, etc.). However, this could influence the N input in the crop rotation (Carlsson and Huss-Danell, 2003). According to Swain et al. (2014), long-term nutrient balances are needed for the assessment of NUE of a rotation with legumes and organic amendments. Nutrients applied in one year may be stored in soil organic matter and released in subsequent years due to mineralization (Hartl and Erhart, 2005). Unfortunately, there was no detailed information regarding the different organic fertilizers used, making it impossible to take a closer look into these aspects in this study. Taking into consideration the accumulation or depletion of soil organic N strongly influences the results of NUE calculations (Chmelíková et al., 2021).

Attention should be paid to N losses in rotations with grass-clover and arable crops (Berntsen et al., 2006; Haynes, 1997). Furthermore, grain legumes (e.g. faba beans) and leguminous leys caused low uptake of N provided by biological N_2 fixation in subsequent crops, which caused higher N leaching than in the rotation without these crops in certain environments (Stenberg et al., 2012). Therefore, after faba beans, a crop with a long growth period is recommended as the ideal succeeding crop in order to make the most efficient use of N mineralized late in the growing season (Jensen et al., 2010).

Generally, forage and grain legume cover crops provide N and thus influence nitrate leaching. Grass species as non-legume cover crop supply only very limited amounts of N and reduce the risk of nitrate leaching (Benincasa et al., 2010). Our results suggested a significant influence of cover crops on N output and N balance. Many authors (Abdalla et al., 2019; Sun and Shahrajabian, 2022) have proposed cover crops to protect against the loss of nutrients, for improved nutrient recycling and increased soil organic carbon (SOC), thus contributing to the mitigation of climate change. Conversely, legume cover crops are commonly studied and discussed due to nitrous oxide emissions (Abdalla et al., 2019; Basche et al., 2014; Fiorini et al., 2020).

For further research, it is important to differentiate the cover crops with regard to crop group and species. According to Snapp et al. (2005), legume cover crops are slow growers, which are expensive to establish but enhance cash crop yield. According to Tonitto et al. (2006), legume-fertilized crops had 10% lower yields in comparison to conventional N-fertilization. The issue in conventional systems is that the inclusion of legumes in a rotation allows the rate of N fertilizer required to achieve the desired grain yields to be reduced and hence the environmental risks and implications of cereal production are also reduced (Jensen et al., 2012; Ladha et al., 2022). The nitrate released from legume crop residues and short duration green manures is currently only capable of satisfying part of the N demand of high-yielding cereal crops (Peoples et al., 1995).

It is worth noting that cereal cover crops increase soil organic matter, and Brassica cover crops can control pest infestations (Snapp et al., 2005). Wittwer et al. (2017) highlighted the importance of cover crops for maintaining a certain yield level when soil tillage intensity is reduced or when production is converted to organic agriculture. There is significant potential to improve the effects of cover crops depending on their management, selection of species, soil and proportion of legumes, timing of field operations and the application of N from manure (Doltra and Olesen, 2013).

Numerous authors have stressed the significant impact of the preceding crop for yield, particularly wheat and winter oilseed rape (Sieling and Christen, 2015). The combination of crops, and a better understanding of their root systems, can improve the NUE (Garnett et al., 2009).

Published by Burleigh Dodds Science Publishing Limited, 2024.

The importance of a suitable preceding crop cannot be stressed enough, e.g. yields of cereals (wheat) were higher following legumes (pea) (Babulicová, 2014). Results showed that the NUE of wheat was significantly affected by the preceding crop. The lowest NUEs were recorded for a wheat–wheat rotation and the highest in a potato–wheat rotation (Rahimizadeh et al., 2010). The high proportion of root crops with high yield can avoid economic or quality losses, which may be expected for these crops if too much N is applied.

In conclusion, the meta-analysis approach allowed us to summarize a larger body of evidence from different studies, resulting in interesting new insights into the influence of different factors on N output, N balances and NUE. Although this study did not address yield and profit explicitly, N output and NUE are good indicators for these parameters and can be used to judge whether the adoption of certain changes is feasible for farmers. Nevertheless, there were some problematic issues, mainly regarding the completeness and quality of the data sources, which were often insufficient to include a certain study. To avoid these issues in the future, a more rigorous report of study design details and results and, if possible, the publication of original data and the corresponding statistical analysis is recommended as a new standard for reproducible research. In addition, it needs to be stated that there is a huge body of practical knowledge and experience regarding the effects of crops in rotation on NUE from field experiments that can be utilized to derive detailed and locally adapted management strategies.

6 Conclusion

The data set presented in this chapter contains a broad spectrum of crop rotation, tillage and fertilization systems in organic farming. The N input varied between <50 kg ha^{-1} and more than 350 kg ha^{-1}. Very different intensities (from low-input to high-input systems) were used to analyze and highlight the influence on N balance and NUE. With higher N input, the N output increases sharply. High yields and N output cannot be achieved with very low N input (extensive systems). High N inputs from fertilizers and biological N_2 fixation enable very high N removals (>200 kg ha^{-1}), which reach the level of conventional systems, even under organic farming conditions. A strategy for increasing the productivity of organic systems is therefore the optimal supply of N to crops, since N is often a yield-limiting factor in organic farming.

With increasing N input, the N balances increased. Nevertheless, negative N balances can occur with low N inputs, but negative N balances were also found in some cases with relatively high N inputs (up to more than 200 kg ha^{-1}). In the long-term, negative N balances should be avoided because they can lead to a decrease in soil N stocks and are thus considered unsustainable. In some cases, very high N balances of well over 100 kg ha^{-1}, which are not typical for organic farming systems, were discovered.

A possible explanation may be the origin of the data. Some data were derived from experiments with a different focus and aims, and hence fertilizer intensities, which are not usual for the studied systems and crop rotations.

The results showed that the N balances depend on numerous influencing factors. One way to reduce N surplus is to increase N output without additional N input, e.g. by avoiding over-fertilization, using crop rotation effects and optimizing soil tillage. Reduced tillage resulted in lower N balances than conventional tillage. However, this must always be seen in the context of respective site conditions.

Both very high and extremely low levels of NUE were found. With increasing N input, NUE decreased. This suggested that a specific N input should not be exceeded in relation to site conditions. High proportions of cereals decreased and high proportions of row crops in the crop rotation increased N efficiency. This can also be interpreted well from a plant cultivation point of view. Root crops generally have a long growing season and are able to utilize mineralized N (from organic fertilizers) well at a late growing stage.

Overall, there are many factors influencing the NUE, which always depend on site and management conditions. It is therefore difficult to generalize which management practice has the highest N efficiency and optimal solutions need to be adapted to local site and agronomical conditions.

7 Future trends

Further research is required to understand how best to use organic fertilizers in the crop rotation, taking into account phosphorus requirements, soil organic matter balance, product quality, production costs and potential risks to the environment. Therefore, it will be necessary to look at different organic fertilizer types in more detail. Depending on the diverse contents of nutrients and carbon, different effects on soil, environment and biodiversity as well as yield and product quality are possible. For example, there are differences in the humus balance of soils depending on the organic fertilizer used. Manure including straw would yield better humus balances than biogas slurry or N-rich manure from chickens. It is also important to consider the optimal timing of fertilizer application, especially if the rotations include biological N_2 fixing crops such as clover or alfalfa. Optimized timing of biological N_2 fixation and the often long-term effects of organic manure, e.g. compost, within the rotation will be crucial for future cropping systems. Acquiring knowledge on optimal timing and the combination of crops and fertilizers, and using this knowledge to develop easy and reliable decision support systems for farmers is an important task for the future. Other research topics to consider are how to establish a circularity perspective across different neighboring farms or at regional and national levels, considering other beneficial effects of optimized and diversified crop rotations, e.g. on soil health or biodiversity.

8 Where to look for further information

See the following website: http://www.pilotbetriebe.de/.

9 References

Abdalla, M., Hastings, A., Cheng, K., Yue, Q., Chadwick, D., Espenberg, M., Truu, J., Rees, R. M. and Smith, P. (2019). A critical review of the impacts of cover crops on nitrogen leaching, net greenhouse gas balance and crop productivity. *Global Change Biology* 25(8), 2530–2543.

Alaru, M., Talgre, L., Eremeev, V., Tein, B., Luik, A., Nemvalts, A. and Loit, E. (2014). Crop yields and supply of nitrogen compared in conventional and organic farming systems. *Agricultural and Food Science* 23(4), 317–326.

Anderson, R. L. (2010). A rotation design to reduce weed density in organic farming. *Renewable Agriculture and Food Systems* 25(3), 189–195.

Anglade, J., Billen, G., Garnier, J., Makridis, T., Puech, T. and Tittel, C. (2015). Nitrogen soil surface balance of organic vs conventional cash crop farming in the Seine watershed. *Agricultural Systems* 139, 82–92.

Aronsson, H., Torstensson, G. and Bergström, L. (2007). Leaching and crop uptake of N, P and K from organic and conventional cropping systems on a clay soil. *Soil Use and Management* 23(1), 71–81.

Auerswald, K., Kainz, M. and Fiener, P. (2003). Soil erosion potential of organic versus conventional farming evaluated by USLE modelling of cropping statistics for agricultural districts in Bavaria. *Soil Use and Management* 19(4), 305–311.

Babulicová, M. (2014). The influence of fertilization and crop rotation on the winter wheat production. *Plant, Soil and Environment* 60(7), 297–302.

Barbieri, P., Pellerin, S. and Nesme, T. (2017). Comparing crop rotations between organic and conventional farming. *Scientific Reports* 7(1), 13761.

Barbieri, P., Starck, T., Voisin, A.-S. and Nesme, T. (2023). Biological nitrogen fixation of legumes crops under organic farming as driven by cropping management: a review. *Agricultural Systems* 205, 103579.

Basche, A. D., Miguez, F. E., Kaspar, T. C. and Castellano, M. J. (2014). Do cover crops increase or decrease nitrous oxide emissions? A meta-analysis. *Journal of Soil and Water Conservation* 69(6), 471–482.

Batool, M., Sarrazin, F. J., Attinger, S., Basu, N. B., Van Meter, K. and Kumar, R. (2022). Long-term annual soil nitrogen surplus across Europe (1850–2019). *Scientific Data* 9(1), 612.

Bedoussac, L., Journet, E.-P., Hauggaard-Nielsen, H., Naudin, C., Corre-Hellou, G., Jensen, E. S., Prieur, L. and Justes, E. (2015). Ecological principles underlying the increase of productivity achieved by cereal-grain legume intercrops in organic farming. A review. *Agronomy for Sustainable Development* 35(3), 911–935.

Behnke, G. D., Zuber, S. M., Pittelkow, C. M., Nafziger, E. D. and Villamil, M. B. (2018). Long-term crop rotation and tillage effects on soil greenhouse gas emissions and crop production in Illinois, USA. *Agriculture, Ecosystems and Environment* 261, 62–70.

Benincasa, P., Farneselli, M., Tosti, G., Bonciarelli, U., Lorenzetti, M. C. and Guiducci, M. (2016). Eleven-year results on soft and durum wheat crops grown in an organic and in a conventional low input cropping system. *Italian Journal of Agronomy* 11(2), 77–84.

Benincasa, P., Tosti, G., Tei, F. and Guiducci, M. (2010). Actual N availability from winter catch crops used for green manuring in maize cultivation. *Journal of Sustainable Agriculture* 34(7), 705–723.

Benoit, M., Garnier, J., Anglade, J. and Billen, G. (2014). Nitrate leaching from organic and conventional arable crop farms in the Seine Basin (France). *Nutrient Cycling in Agroecosystems* 100(3), 285–299.

Berntsen, J., Grant, R., Olesen, J. E., Kristensen, I. S., Vinther, F. P., Mølgaard, J. P. and Petersen, B. M. (2006). Nitrogen cycling in organic farming systems with rotational grass-clover and arable crops. *Soil Use and Management* 22(2), 197–208.

Carlsson, G. and Huss-Danell, K. (2003). Nitrogen fixation in perennial forage legumes in the field. *Plant and Soil* 253(2), 353–372.

Chmelíková, L., Schmid, H., Anke, S. and Hülsbergen, K. (2021). Nitrogen-use efficiency of organic and conventional arable and dairy farming systems in Germany. *Nutrient Cycling in Agroecosystems* 119(3), 337–354.

Dal Ferro, N., Zanin, G. and Borin, M. (2017). Crop yield and energy use in organic and conventional farming: A case study in north-east Italy. *European Journal of Agronomy* 86, 37–47.

David M.B., McIsaac, G. F. and Darmody, R. G (2010). Synthetic nitrogen fertilizers deplete soil nitrogen: a global dilemma for sustainable cereal production. In: Mulvaney, R. L., Khan, S. A. and Ellsworth, T. R. in the *Journal of Environmental Quality* 38, 2295–2314.

Doltra, J. and Olesen, J. E. (2013). The role of catch crops in the ecological intensification of spring cereals in organic farming under Nordic climate. *European Journal of Agronomy* 44, 98–108.

Drinkwater, L. E., Wagoner, P. and Sarrantonio, M. (1998). Legume-based cropping systems have reduced carbon and nitrogen losses. *Nature* 396(6708), 262–265.

Fageria, N. K. (2014). *Nitrogen Management in Crop Production*. Boca Raton: CRC Press.

Fageria, N. K. and Baligar, V. C. (2005). Enhancing nitrogen use efficiency in crop plants. *Advances in Agronomy* 88, 97–185.

Fiorini, A., Maris, S. C., Abalos, D., Amaducci, S. and Tabaglio, V. (2020). Combining no-till with rye (Secale cereale L.) cover crop mitigates nitrous oxide emissions without decreasing yield. *Soil and Tillage Research* 196(1), 104442.

Galloway, J. N., Townsend, A. R., Erisman, J. W., Bekunda, M., Cai, Z., Freney, J. R., Martinelli, L. A., Seitzinger, S. P. and Sutton, M. A. (2008). Transformation of the nitrogen cycle: recent trends, questions, and potential solutions. *Science* 320(5878), 889-892.

Garnett, T., Conn, V. and Kaiser, B. N. (2009). Root based approaches to improving nitrogen use efficiency in plants. *Plant, Cell and Environment* 32(9), 1272–1283.

Gaudino, S., Goia, I., Borreani, G., Tabacco, E. and Sacco, D. (2014). Cropping system intensification grading using an agro-environmental indicator set in northern Italy. *Ecological Indicators* 40, 76-89.

Halberg, N., Steen Kristensen, E. and Sillebak Kristensen, I. (1995). Nitrogen turnover on organic and conventional mixed farms. *Journal of Agricultural and Environmental Ethics* 8(1), 30–51.

Halberg, N., Verschuur, G. and Goodlass, G. (2005). Farm level environmental indicators; are they useful? *Agriculture, Ecosystems and Environment* 105(1-2), 195-212.

Hartl, W. and Erhart, E. (2005). Crop nitrogen recovery and soil nitrogen dynamics in a 10-year field experiment with biowaste compost. *Journal of Plant Nutrition and Soil Science* 168(6), 781-788.

Haynes, R. J. (1997). Fate and recovery of 15 N derived from grass/clover residues when incorporated into a soil and cropped with spring or winter wheat for two succeeding seasons. *Biology and Fertility of Soils* 25(2), 130-135.

Holka, M., Kowalska, J. and Jakubowska, M. (2022). Reducing carbon footprint of agriculture–can organic farming help to mitigate climate change? *Agriculture* 12(9), 1383.

Jensen, E. S., Peoples, M. B., Boddey, R. M., Gresshoff, P. M., Hauggaard-Nielsen, H., J.R. Alves, B. and Morrison, M. J. (2012). Legumes for mitigation of climate change and the provision of feedstock for biofuels and biorefineries. A review. *Agronomy for Sustainable Development* 32(2), 329-364.

Jensen, E. S., Peoples, M. B. and Hauggaard-Nielsen, H. (2010). Faba bean in cropping systems. *Field Crops Research* 115(3), 203-216.

Jost, L. (2006). Entropy and diversity. *Oikos* 113(2), 363-375.

Kocira, A., Staniak, M., Tomaszewska, M., Kornas, R., Cymerman, J., Panasiewicz, K. and Lipińska, H. (2020). Legume cover crops as one of the elements of strategic weed management and soil quality improvement. A review. *Agriculture* 10(9), 394.

Kubota, H., Iqbal, M., Quideau, S., Dyck, M. and Spaner, D. (2018). Agronomic and physiological aspects of nitrogen use efficiency in conventional and organic cereal-based production systems. *Renewable Agriculture and Food Systems* 33(5), 443-466.

Ladha, J. K., Jat, M. L., Stirling, C. M., Chakraborty, D., Pradhan, P., Krupnik, T. J., Sapkota, T. B., Pathak, H., Rana, D. S., Tesfaye, K. and Gerard, B. (2020). Achieving the sustainable development goals in agriculture: the crucial role of nitrogen in cereal-based systems. *Advances in Agronomy* 163, 39-116.

Ladha, J. K., Peoples, M. B., Reddy, P. M., Biswas, J. C., Bennett, A., Jat, M. L. and Krupnik, T. J. (2022). Biological nitrogen fixation and prospects for ecological intensification in cereal-based cropping systems. *Field Crops Research* 283, 108541.

Liebman, M., Graef, R. L., Nettleton, D. and Cambardella, C. A. (2012). Use of legume green manures as nitrogen sources for corn production. *Renewable Agriculture and Food Systems* 27(3), 180-191.

Liu, C., Plaza-Bonilla, D., Coulter, J. A., Kutcher, H. R., Beckie, H. J., Wang, L., Floc'h, J., Hamel, C., Siddique, K. H. M., Li, L. and Gan, Y. (2022). Diversifying crop rotations enhances agroecosystem services and resilience. *Advances in Agronomy* 173, 299-335.

Macholdt, J., Piepho, H.-P. and Honermeier, B. (2019). Mineral NPK and manure fertilisation affecting the yield stability of winter wheat: results from a long-term field experiment. *European Journal of Agronomy* 102, 14-22.

Mäder, P., Fliessbach, A., Dubois, D., Gunst, L., Fried, P. and Niggli, U. (2002). Soil fertility and biodiversity in organic farming. *Science (New York, N.Y.)* 296(5573), 1694-1697.

Mayer, J., Gunst, L., Mäder, P., Samson, M., Carcea, M., Narducci, V., Thomsen, I. K. and Dubois, D. (2015). Productivity, quality and sustainability of winter wheat under long-term conventional and organic management in Switzerland. *European Journal of Agronomy* 65, 27-39.

Mayer, S., van Herwijnen, A., Techel, F. and Schweizer, J. (2022). A random forest model to assess snow instability from simulated snow stratigraphy. *The Cryosphere* 16(11), 4593-4615.

Meyer, S., Wesche, K., Krause, B. and Leuschner, C. (2013). Dramatic losses of specialist arable plants in Central Germany since the 1950s/60s - a cross-regional analysis. *Diversity and Distributions* 19(9), 1175-1187.

Morari, F., Lugato, E., Polese, R., Berti, A. and Giardini, L. (2012). Nitrate concentrations in groundwater under contrasting agricultural management practices in the low plains of Italy. *Agriculture, Ecosystems and Environment* 147(1), 47-56.

Mulvaney, R. L., Khan, S. A. and Ellsworth, T. R. (2009). Synthetic nitrogen fertilizers deplete soil nitrogen: a global dilemma for sustainable cereal production. *Journal of Environmental Quality* 38(6), 2295-2314.

Mulvaney, R. L., Khan, S. A. and Ellsworth, T. R. (2010a). Synthetic Nitrogen Fertilizers Deplete Soil Nitrogen: A Global Dilemma for Sustainable Cereal Production, by R. L. Mulvaney, S. A. Khan, and T. R. Ellsworth in the Journal of Environmental Quality 2009 38: 2295-2314. *Journal of Environmental Quality* 39(4), 1530-1532.

Peoples, M. B., Herridge, D. F. and Ladha, J. K. (1995). Biological nitrogen fixation: an efficient source of nitrogen for sustainable agricultural production? *Plant and Soil* 174(1-2), 3-28.

Posner, J. L., Baldock, J. O. and Hedtcke, J. L. (2008). Organic and conventional production systems in the Wisconsin integrated cropping systems trials: I. Productivity 1990-2002. *Agronomy Journal* 100(2), 253-260.

Powlson, D. S., Gregory, P. J., Whalley, W. R., Quinton, J. N., Hopkins, D. W., Whitmore, A. P., Hirsch, P. R. and Goulding, K. W. T. (2011). Soil management in relation to sustainable agriculture and ecosystem services. *Food Policy* 36(156), S72-S87.

Powlson, D. S., Jenkinson, D. S., Johnston A. E., Poulton, P. R., Glendining, M. J. and Goulding, K. W. T. (2010a). Synthetic Nitrogen Fertilizers Deplete Soil Nitrogen: A Global Dilemma for Sustainable Cereal Production, by R. L. Mulvaney, S. A. Khan, and T. R. Ellsworth in the Journal of Environmental Quality 2009 38: 2295-2314. *Journal of Environmental Quality* 39(2), 749-752; author reply 753-756.

Powlson, D. S., Jenkinson, D. S., Johnston A. E., Poulton, P. R., Glendining, M. J. and Goulding, K. W. T. (2010b). Synthetic Nitrogen Fertilizers Deplete Soil Nitrogen: A Global Dilemma for Sustainable Cereal Production, by R. L. Mulvaney, S. A. Khan, and T. R. Ellsworth in the Journal of Environmental Quality 2009 38 2295-2314. *Journal of Environmental Quality* 39(4): 1528-1529.

R Core Team (2020). *A Language and Environment for Statistical Computing*. R Foundation for Statistical Computing, Vienna.

Rahimizadeh, M., Kashani, A., Zare-Feizabadi, A., Koocheki, A. and Nassiri-Mahallati, M. (2010). Nitrogen use efficiency of wheat as affected by preceding crop, application rate of nitrogen and crop residues. *Australian Journal of Crop Science* 4, 363-368.

Ratnadass, A., Fernandes, P., Avelino, J. and Habib, R. (2012). Plant species diversity for sustainable management of crop pests and diseases in agroecosystems: a review. *Agronomy for Sustainable Development* 32(1), 273-303.

Sieling, K. and Christen, O. (2015). Crop rotation effects on yield of oilseed rape, wheat and barley and residual effects on the subsequent wheat. *Archives of Agronomy and Soil Science* 22, 1-19.

Sindelar, A. J., Schmer, M. R., Jin, V. L., Wienhold, B. J. and Varvel, G. E. (2016). Crop rotation affects corn, grain sorghum, and soybean yields and nitrogen recovery. *Agronomy Journal* 108(4), 1592-1602.

Snapp, S. S., Swinton, S. M., Labarta, R., Mutch, D., Black, J. R., Leep, R., Nyiraneza, J. and O'Neil, K. (2005). Evaluating cover crops for benefits, costs and performance within cropping system niches. *Agronomy Journal* 97(1), 322-332.

Stanger, T. F. and Lauer, J. G. (2008). Corn grain yield response to crop rotation and nitrogen over 35 years. *Agronomy Journal* 100(3), 643-650.

Stark, C., Condron, L. M., Stewart, A., Di, H. J. and O'Callaghan, M. (2006). Effects of past and current management practices on crop yield and nitrogen leaching–a comparison of organic and conventional cropping systems. *New Zealand Journal of Crop and Horticultural Science* 34(3), 207-215.

Stenberg, M., Ulén, B., Söderström, M., Roland, B., Delin, K. and Helander, C. A. (2012). Tile drain losses of nitrogen and phosphorus from fields under integrated and organic crop rotations. A four-year study on a clay soil in southwest Sweden. *Science of the Total Environment* 434, 79-89.

Sun, W. and Shahrajabian, M. H. (2022). The effectiveness of Rhizobium bacteria on soil fertility and sustainable crop production under cover and catch crops management and green manuring. *Notulae Botanicae Horti Agrobotanici Cluj-Napoca* 50(2), 1-36.

Swain, E. Y., Rempelos, L., Orr, C. H., Hall, G., Chapman, R., Almadni, M., Stockdale, E. A., Kidd, J., Leifert, C. and Cooper, J. M. (2014). Optimizing nitrogen use efficiency in wheat and potatoes: interactions between genotypes and agronomic practices. *Euphytica* 199(1-2), 119-136.

Tan, Z. X., Lal, R. and Wiebe, K. D. (2005). Global soil nutrient depletion and yield reduction. *Journal of Sustainable Agriculture* 26(1), 123-146.

Thorup-Kristensen, K., Magid, J. and Jensen, L. S. (2003). Catch crops and green manures as biological tools in nitrogen management in temperate zones. *Advances in Agronomy* 79, 227-302.

Tonitto, C., David, M. B. and Drinkwater, L. E. (2006). Replacing bare fallows with cover crops in fertilizer-intensive cropping systems: A meta-analysis of crop yield and N dynamics. *Agriculture, Ecosystems and Environment* 112(1), 58-72.

Torstensson, G., Aronsson, H. and Bergström, L. (2006). Nutrient use efficiencies and leaching of organic and conventional cropping systems in Sweden. *Agronomy Journal* 98(3), 603-615.

Valenzuela, H. (2023). Ecological management of the nitrogen cycle in organic farms. *Nitrogen* 4(1), 58-84.

van Beek, C. L., Brouwer, L. and Oenema, O. (2003). The use of farmgate balances and soil surface balances as estimator for nitrogen leaching to surface water. *Nutrient Cycling in Agroecosystems* 67(3), 233-244.

van Groenigen, J. W., Velthof, G. L., Oenema, O., Van Groenigen, K. J. and Van Kessel, C. (2010). Towards an agronomic assessment of N_2O emissions: a case study for arable crops. *European Journal of Soil Science* 61(6), 903-913.

Watson, C. A., Atkinson, D., Gosling, P., Jackson, L. R. and Rayns, F. W. (2002). Managing soil fertility in organic farming systems. *Soil Use and Management* 18(3), 239-247.

Wilkins, R. J. (2008). Eco-efficient approaches to land management: a case for increased integration of crop and animal production systems. *Philosophical Transactions of the Royal Society of London. Series B, Biological Sciences* 363(1491), 517-525.

Wittwer, R. A., Dorn, B., Jossi, W. and van der Heijden, M. G. (2017). Cover crops support ecological intensification of arable cropping systems. *Scientific Reports* 7, 41911.

Index